Reproductive Biology of Teleost Fishes

Reproductive Biology of Teleost Fishes

Robert J. Wootton

Institute of Biological,
Environmental and Rural Sciences,
Aberystwyth University, Aberystwyth,
Ceredigion SY23 3DA, UK

Carl Smith

School of Biology, University of St. Andrews,
St. Andrews, Fife KY16 8LB, UK

and

Institute of Vertebrate Biology,
Academy of Sciences of the Czech Republic,
Květná 8, 603 65 Brno, Czech Republic

WILEY Blackwell

Registered Office
John Wiley & Sons, Ltd., The Atrium, Southern Gate, Chichester, West Sussex, PO19 8SQ, UK

Editorial Offices
9600 Garsington Road, Oxford, OX4 2DQ, UK
The Atrium, Southern Gate, Chichester, West Sussex, PO19 8SQ, UK
111 River Street, Hoboken, NJ 07030-5774, USA

For details of our global editorial offices, for customer services and for information about how to apply for permission to reuse the copyright material in this book please see our website at www.wiley.com/wiley-blackwell.

Library of Congress Cataloging-in-Publication Data

Wootton, R. J. (Robert J.)
Reproductive biology of teleost fishes / by Robert J. Wootton, Carl Smith.
pages cm
Includes bibliographical references and index.
ISBN 978-0-632-05426-8 (Cloth)
1. Osteichthyes–Reproduction. I. Smith, Carl, 1966–. II. Title.
QL639.2.W66 2015
597–dc23
2014028490

A catalogue record for this book is available from the British Library.

Wiley also publishes its books in a variety of electronic formats. Some content that appears in print may not be available in electronic books.

Cover image: A pair of European bitterling (*Rhodeus amarus*) preparing to spawn in the gills of a freshwater unionid mussel. Photo by Carl Smith.

Set in 9/11.5pt Myriad Pro by SPi Publisher Services, Pondicherry, India
Printed and bound in Singapore by Markono Print Media Pte Ltd

1 2015

For Ella and Alfie

Grandchildren are the dessert of life

Robert John Wootton (1942–2014)

On 3 March 2014 Robert 'Bob' Wootton passed away at his home in Aberystwyth in west Wales. He was 71. Bob was a highly accomplished fish ecologist and evolutionary biologist and an authority on the three-spined stickleback, a species he successfully championed as a model species for research.

Bob suffered progressively declining health over the last 6 years of his life, although his illness did not halt his academic work. From 2009 until a few weeks before his death, Bob and I worked together on the present book, which Bob saw successfully through to completion. It had been a lifetime's ambition of his to write a book devoted to teleost reproduction. Despite his illness, Bob dedicated himself to what would be his last major piece of academic work, and it is testimony to his extraordinary tenacity, and intellectual and physical endurance, that he was able to do so. This book will be one of his many legacies to the field of fish biology.

Carl Smith
St Andrews

Contents

Preface

Teleost fishes show a remarkable diversity of modes of reproduction. This diversity offers great scope for the application of the comparative method, which has proven such a powerful tool in evolutionary biology. There have been catalogues of fish reproductive biology, most notably *Modes of Reproduction in Fishes* by Breder and Rosen (1966), and many reviews of specific aspects of reproduction. But there is no volume that synthesises the genetic, physiological, behavioural, ecological, evolutionary and applied aspects of reproduction in a comparative framework that emphasises the adaptive basis for the diversity. The purpose of the present volume is to fill this gap at a level suitable for advanced undergraduates, postgraduates and research workers requiring a comprehensive overview of fish reproduction.

Acknowledgements

I wish to thank Dr N.F. Beck, Prof. M. Burton, Dr S. Consuegra and Prof. A.P Scott for their comments on early drafts of the manuscript. All errors and misinterpretations are entirely the author's responsibility.

I would also like to thank all the staff of Padarn GP Surgery, Aberystwyth and Bronglais Hospital, who worked so hard to keep me functioning.

Robert J. Wootton

I am extremely grateful to Rowena Spence and Martin Reichard, both of whom read and commented extensively on the entire manuscript, and cared enough about it to be remorselessly critical. Rowena was also responsible for most of the figures in the book and for compiling the index. Thanks also to Mark Warren and Matej Polačik for valuable discussion and for expertly commenting on sections of the text.

I particularly wish to acknowledge the colossal effort made by Bob, my co-author, in completing this book. I have deep admiration for your fortitude and tenacity. Writing a book is a lonely undertaking and, as my Head of Department generously pointed out to me, of limited value to an academic career. However, throughout the time I spent working on this book I had a loyal and gentle companion, Lucy my pitbull, who patiently kept me company through all the long hours and never once had the discourtesy to make dim-witted criticisms about the endeavour.

Carl Smith

Abbreviations and acronyms

3β-HSD	3β-hydroxysteroid dehydrogenase
5HT	serotonin
11β-HSD	11β-hydroxysteroid dehydrogenase
11-KA	11 ketoandrosetnedione
11-KT	11 ketotestosterone
15-K-PGF$_{2α}$	15-keto-prostaglandin F$_{2α}$
17α-MT	17α-methyltestosterone
17α-OHP	17α-hydroxyprogesterone
17,20β-P	17,20βP hydroxyprogesterone
17α,20β-P	17α,20β dihydroxy-4-pregnen-3-one
17α,20β-P-S	17α,20β-dihydroxy-4-pregnen-3-one 20β-sulphate
17β-HSD	17β-hydroxysteroid dehydrogenase
20β-HSD	20β-hydroxysteroid dehydrogenase
17-P	17-hydroxypregnen-4-ene-3,20-dione
20β-S	17,20β,21-trihydroxy-4-pregnen-3-one
α	age at maturity
A	individual production rate
AA	arachidonic acid
ACTH	adrenocorticotrophic hormone
ACT-R	activin receptor
AD	androstenedione
AGS	accessory gonadal structures
AH	adenohypophysis
AMH	anti-Müllerian hormone
ARs	androgen receptors
AVT	arginine vasotocin
BB	blue-barred phenotype
BBS	bombesin
BG	Bateman gradient
bp	before present
BPG	brain–pituitary–gonad
BPI	brain-pituitary inter-renal
CA	cortical alveoli
cAMP	cyclic-3′,5′-adenosine monophosphate
CCK	cholecystokinin
CNS	central nervous system
CRF	corticotrophin-releasing factor
CRH	corticotrophin-releasing hormone
Ct	corticotropes
D$_1$	dopamine receptor 1
D$_2$	dopamine receptor 2
DA	dopamine
DDC	diethyldithiocarbamate
DES	diethylystilbestrol
DHA	docasahexaenoic acid
dpf	days post-fertilisation
dph	days post-hatching
DW	dry mass
E	average adult lifespan
E$_1$	oestrone
E$_2$	oestradiol-17β
EDC	endocrine-disrupting chemical
EE$_2$	ethinyloestradiol
EGF	epidermal growth factor
EGF-R	epidermal growth factor receptor
EMG	electromyogram
ER	oestrogen receptor
ERα	oestrogen receptor α
ERβ1	oestrogen receptor β1

ERβ2	oestrogen receptor β2
ERE	oestrogen response element
ERs	oestrogen receptors
ESC	evolutionarily stable continuum
ESD	environmental sex determination
ESS	evolutionarily stable strategy
E_x	age-specific reproductive effort
EYPP	egg yolk precursor proteins
F	total fecundity
FAA	free amino acid
F_b	breeding season fecundity
F_{batch}	number of eggs produced in a spawning bout
FISH	fluorescent in-situ hybridisation
FNV	frozen niche variation
F_s	fecundity per spawning
FSH	follicle-stimulating hormone
FSH-β	follicle stimulating hormone β subunit
FSH-R	follicle-stimulating hormone receptor
GABA	γ-aminobutyric acid
GAL	galanin
GAP	GnRH-associated peptide
GH	growth hormone
GH-R	growth hormone receptor
GHRH	growth hormone-releasing hormone
GJ	gap junctions
GnRH	gonadotrophin-releasing hormone
GnRH-1	gonadotrophin-releasing hormone type 1
GnRH-2	gonadotrophin-releasing hormone type 2
GnRHa	gonadotrophin-releasing hormone analogue
GnRH-ir	immunoreactive gonadotrophin-releasing hormone
GnRH-R	gonadotrophin releasing hormone receptor
GnRH-R1	gonadotrophin releasing hormone receptor type 1
GPα	gonadotrophic hormone α subunit
GPCR	G protein-coupled receptor
GPER	G protein-coupled oestrogen receptor
GSD	genetic sex determination
GSDF	gonad-soma derived factor
GSI	gonadosomatic index
Gt	gonadotropes
GTH	gonadotrophic hormone
GTH-I	gonadotrophic hormone I
GTH-II	gonadotrophic hormone II
GTH-R	gonadotrophic hormone receptor
GV	germinal vesicle
GVBD	germinal vesicle breakdown
GVM	germinal vesicle maturation
g_x	age-specific growth rate
H	hypothalamus
HPG	hypothalamus pituitary gonad
HPI	hypothalamus pituitary inter-renal
IDH	isocitrate dehydrogenase
IGA	internal gamete association
IGF	insulin-like growth factor
IGF-I	insulin-like growth factor I
IGF-II	insulin-like growth factor II
IGF-R	insulin like growth factor receptor
IP	initial phase
JS	Japan Sea clade
k	growth coefficient
K	threshold stock size above which density dependence limits recruitment
KISS	kisspeptin
L	length
L_∞	asymptotic length
L_a	length at maturity
LD	lipid droplets
L:D	light:dark
LDFR	low-density lipoprotein receptors
LG	linkage group
LH	luteinising hormone
LH-β	luteinising hormone β subunit
LH-R	luteinising hormone receptors
LRE	lifetime reproductive effort
LRS	lifetime reproductive success
L_t	length at age t
l_x	age-specific survival
M	instantaneous natural mortality rate
M_a	mortality rate at maturity
MBT	mid-blastula transition

mER	oestrogen receptor
MHC	major histocompatibility complex
MI	matrotrophic index
MIH	maturation-inducing hormone
MIH-R	maturation inducing hormone receptor
MPF	maturation-promoting factor
mPR	progestagen receptor
mPRα	progestagen receptor α
mRNA	messenger RNA
MVB	multivesicular body
m_x	age-specific fecundity
mya	million years ago
NE	norepinephrine (noradrenaline)
NH	neurohypophysis
NPY	neuropeptide Y
N_s	number of spawnings in a breeding season
OB	orange-blotch phenotype
OHT	hydroxytestosterone
OMC	oocyte maturation competence
OR	olfactory receptor
OSR	operational sex ratio
P	potential fecundity
$P450_{11\beta}$	11β-hydroxylase
$P450_{arom}$	P450 aromatase
$P450_{c17}$	17α-hydroxylase/$C_{17,20}$lyase
$P450_{c17\text{-}I}$	$C_{17,20}$lyase
$P450_{c17\text{-}II}$	17α-hydroxylase
$P450_{scc}$	side-chain cleavage enzyme
PACAP	pituitary adenylate cyclase-activating peptide
pAGS	precursive accessory gonadal structures
PC	phospotidylcholine
PE	phospotidylethanolamine
PGC	primordial germ cell
PGF	prostaglandin
$PGF_{2\alpha}$	prostaglandin F2α
PgR	progestagen receptor
P_i	inorganic phosphorus
PI	pars intermedia
PKA	protein kinase A
PKC	protein kinase C
PMRN	probabalistic maturation norm
PO	Pacific Ocean clade
POA	pre-optic area
PPD	proximal pars distalis
PRP	polypeptide-related peptide
PRR	potential reproductive rate
PUFA	polyunsaturated fatty acid
p_x	probability of survival to age x
r	intrinsic rate of population increase
R	number of recruits
R_0	net reproductive rate
RA	reproductive allocation
r_d	intrinsic rate of population increase in a constant environment
RE	reproductive effort
RF	releasing factor
RHP	resource holding potential
r_m	intrinsic rate of natural population increase
RPD	rostral pars distalis
r_s	intrinsic rate of population increase in a stochastic environment
S	number (or biomass) of spawners
SE	somatic effort
SPR	spawning potential ratio
SRIH	somatostatin
StAR	steroidogenic acute regulatory protein
t	age
T	testosterone
TBA	trenbolone acetate
TCA	triacylglycerol
TGF	tissue growth factor
TGF-β	tissue growth factor β
TH	thryroid hormone
TP	terminal phase
TSD	temperature-dependent sex determination
TSH	thyroid-stimulating hormone
TV	total egg volume
UV	ultraviolet
V_0	reproductive value of an individual at age 0
VBGM	von Bertallanfy growth model
VNR	vomeronasal receptor
VPG	vocal pattern generator
VTG	vitellogenin
V_x	age-specific reproductive value
W_∞	asymptotic mass

W_g	gonad mass
WGD	whole genome duplication
wpf	weeks post-fertilisation
W_s	somatic mass
W_t	mass at age *t*
W_{total}	total body mass
WWTP	waste water treatment plant
x	age
ZP	zona pellucida protein
ZPB	zona pellucida protein B
ZPC	zona pellucida protein C
ZPX	zona pellucida protein X

Chapter 1
Introduction

Introduction

A story, perhaps apocryphal, relates that the eminent British biologist, J.B.S. Haldane, was asked what his biological studies had told him about the nature of the Creator. He replied that it was an inordinate fondness for beetles, reflecting the numerous beetle species that had been described. Perhaps he should have added, the Creator also had a real affection for teleost fishes (Teleostei). There are approximately 30 000 species of teleosts (FishBase: http://www.fishbase.org) and they account for about 50% of all vertebrate species. Two teleost families, the Cyprinidae and the Gobiidae, are the largest vertebrate families in terms of number of species.

Although essentially excluded from the terrestrial environment, the teleosts have shown an exuberant adaptive radiation in the aquatic environment, occupying oceans, lakes, rivers, streams and even temporary pools (Wootton, 1998; Moyle & Cech, 2004; Helfman *et al.*, 2009). They occur from the Antarctic to the Arctic oceans, with species living in water that is at temperatures below 0 °C. In the tropics, they are found in waters

Reproductive Biology of Teleost Fishes, First Edition. Robert J. Wootton and Carl Smith.

close to 40 °C. The altitudinal range of teleosts is about 11 000 m, from the abyssal depths of the oceans to the high plateaus of Tibet in Asia and the Andes of South America. Teleosts occur in waters with salinity ranging from close to distilled water in some streams in the Amazon region of South America to hypersaline waters, such as those found in desert pools.

In parallel to this wide adaptive radiation in habitats occupied, teleosts also show a wide radiation in modes of feeding (Gerking, 1994; Wootton, 1998). The group includes some species, the detritivores, which can survive on decaying organic matter. There are herbivorous teleosts ranging from species feeding on phytoplankton to species browsing on leaves, or even eating seeds and fruits. Carnivores range from scavengers to zooplantktivores to piscivores and include such strange diets as the scales or gill filaments of other fishes.

Teleosts also display a wide range of body forms (Videler, 1993), from the highly streamlined, fast-swimming tunas of the open oceans to the bizarre shape of the seahorses found in inshore habitats. A sector of the adaptive morphological space that is relatively poorly explored by teleosts is gigantism. Compared with the cartilaginous fishes (Chondrichthyes), especially the sharks and rays and the mammalian cetaceans, the median body size of teleostean species is relatively small. It is not clear why the teleosts have not explored the adaptive opportunities of large body size (Freedman & Noakes, 2002).

However, it is in their reproduction that the teleosts display perhaps their most striking diversity. No other vertebrates – indeed few other taxa – rival the teleosts in their extraordinary multiplicity of reproductive modes.

Reproductive modes of the teleosts

Reproduction is the central biological process by which genetic material is passed from one generation to the next. The reproductive biology of the teleosts shows high diversity, far more so than in other vertebrate groups (Table 1.1).

Gender systems of teleosts

About 88% of known teleosts are gonochoristic, with individuals being either male or female (Patzner, 2008). This is the usual condition in vertebrates. The sex of an individual can be determined genetically, known as genetic sex determination (GSD), as is found, for example, in mammals and birds. Sex may also be determined by the environmental conditions encountered in the early stages of development of the individual. This is environmental sex determination (ESD), which was first studied in reptiles (Bull, 1983). Sex in some teleosts is determined by interactions between genes and the environment. Sex determination in teleosts is described in Chapter 2. The differentiation of the gonads in gonochoristic teleosts is the subject of Chapter 3, while the production of the eggs and sperm by the differentiated gonads, the process of gametogenesis, is described in Chapter 4.

Functional hermaphroditism, which is extremely rare in other vertebrates, has been identified in about 2% of teleost species, representing approximately 30 families. Its taxonomic distribution in the teleosts suggests it has evolved independently in several teleost lineages (Patzner, 2008; Avise & Mank, 2009). Interestingly, in some gonochoristic teleost species, the gonad initially differentiates as an ovary, even in individuals that eventually become male. There seems to be indeterminacy in the teleostean gonad in

Table 1.1 Teleost sexuality and mating systems. (Source: Adapted from Wootton 1998. Reproduced with permission of Elsevier.)

Sexuality/mating system	Example
1. Sex determination	
(a) Genetic	Most teleosts
(b) Environmental	*Menidia menidia*
2. Gender systems	
(a) Gonochorism	Most teleosts
(b) Hermaphroditism	
(i) Simultaneous (synchronous)	
Self-fertilising	*Kryptolebias marmoratus*
Outcrossing	*Serranus fasciatus*
(ii) Sequential	
Protogynous	*Thalassoma bifasciatum*
Protandrous	*Amphiprion melanopus*
Serial (bidirectional)	*Gobiodon rivulatus*
3. Mode of fertilisation	
(a) External	Most teleosts
(b) Internal	*Poecilia reticulata*
(c) Buccal	*Oreochromis niloticus*
4. Mating system	
(a) Promiscuity	*Clupea harengus*
(b) Polygamy	
(i) Polygyny	*Oreochromis niloticus*
(i) Polyandry	*Amphiprion melanopus*
(i) Polygynandry	*Rhodeus amarus*
(b) Monogamy	*Amatitlania nigrofasciata*
5. Secondary sexual characteristics	
(a) Monomorphism	*Clupea harengus*
(b) Sexual dimorphism	
(i) Permanent dimorphism	*Betta splendens*
(i) Temporary dimorphism	*Gasterosteus aculeatus*
6. Spawning site preparation	
(a) No preparation	*Phoxinus phoxinus*
(b) Site prepared, but not defended	*Salmo salar*
(c) Site prepared, and defended	*Gasterosteus aculeatus*
7. Parental care	
(a) No care	Most teleosts
(b) Paternal care	*Gasterosteus aculeatus*
(c) Maternal care	
(i) Oviparity	*Oreochromis niloticus*
(i) Viviparity	*Poecilia reticulata*
(b) Biparental care	*Amatitlania nigrofasciata*
(c) Juvenile helpers	*Neolamprologus pulcher*

relation to development as an ovary or testis (Chapter 2), which predisposes them towards hermaphroditism.

Hermaphroditism in the teleosts can be simultaneous (synchronous) or sequential, the latter mode expressed as protandrous, protogynous or serial (bidirectional) hermaphroditism (Avise & Mank, 2009). The commonest forms of hermaphroditism are protandry and protogyny. In simultaneous hermaphroditism, the gonads contain both male (spermatozoa) and female (eggs) haploid gametes at the same time. The co-occurrence of spermatozoa and eggs opens up the possibility of self-fertilisation by an individual, but this has been described for only two closely related species, the cyprinodont mangrove rivulus (*Kryptolebias marmoratus* and *Kryptolebias hermaphroditus*, Rivulidae) (Tatarenkov *et al.*, 2012). Other species with simultaneous hermaphroditism outcross, mating with other individuals. Protandry describes the situation in which an individual functions as a sexually mature male, producing spermatozoa, but at some point in its life history switches to perform as a female, spawning eggs. Protogyny is the reverse of this. When sexually mature, an individual is initially female, spawning eggs, but then later transforms into a male, producing spermatozoa. In serial species, the switch between male and female roles can take place more than once in a lifetime. Hermaphroditism is one of the subjects of Chapter 10.

Parthenogenesis, in which only the female genome is transmitted from generation to generation, also occurs in the teleosts, although it is rare (Chapter 10). A curious feature of parthenogenesis in most of the species that exhibit this mode of reproduction is that they require the presence of males from a closely related gonochoristic species. The male spermatozoon activates the development of the egg of the parthenogenetic female. Teleosts show two modified forms of parthenogenesis: gynogenesis and hybridogenesis. In hybridogenesis, the genetic material of the male is functional in the developing individual, but when that individual reproduces, only the female genome is transmitted to the progeny. In gynogenesis, the role of the spermatozoon is solely to activate the egg, and the genes of the male play no role in the development of the zygote. Thus, parthenogenetic females essentially parasitise the spermatozoa of the gonochoristic males.

Spawning dynamics

Two major temporal patterns of reproductive activity can be recognised in the life history of animals and plants, including teleosts (Roff, 1992, 2002; Stearns, 1992; Wootton, 1998; see Chapter 11). Semelparity describes the condition in which, after attaining sexual maturity, the individual breeds once and then dies. There is no significant post-reproductive survival. In semelparous organisms, the physiological changes associated with reproduction result in consequences that inevitably end in death. Iteroparity is the condition in which, after reproduction, there is some probability that the individual will survive to breed again. Frequently, reproduction takes place at yearly intervals.

Within a breeding season, two temporal patterns of spawning occur in female teleosts (Wootton, 1998; Patzner, 2008). In total spawners, the female spawns all her eggs over a short period of time and no further eggs are shed in that breeding season. If the female is semelparous, she then dies. If iteroparous, her ovaries regress and become quiescent until the physiological (Chapter 5) and environmental conditions (Chapter 6) induce the recrudescence of the ovaries (Chapter 3), leading to the production of eggs for the next breeding season. In batch spawners, the female spawns eggs in batches (or clutches) at intervals during the breeding season, which is typically of an extended duration. The interval between spawnings varies. In some species, the female may spawn at daily intervals, but in others, spawning takes place at intervals of several days. There are two types of batch spawner. Some species have determinate fecundity. The females have all the eggs that are going to be spawned during that breeding season present in the ovaries at the start of the breeding season and there is no addition to the pool of eggs available to be spawned in that season. Other species have indeterminate fecundity. Here, the pool of eggs that can be spawned can be replenished during the breeding season.

In short-lived fishes, the females may be batch spawners, but may not survive to breed in the next breeding season. They thus share the characteristics of semelparous and iteroparous species. However, in such species, there is usually the physiological capacity to survive to the next breeding season if environmental conditions are benign. Such species can be described as showing abbreviated iteroparity.

Modes of fertilisation

The vast majority of teleost species have external fertilisation (Patzner, 2008). The eggs and sperm are released into the water, where fertilisation takes place. In about 500–600 species, fertilisation takes place internally within the female (Chapter 10). The male introduces the spermatozoa into the gonoduct of the female, using an intromittent organ. This is analogous to the mode of fertilisation seen in the cartilaginous fishes, the Chondrichthyes, and in mammals and a few other vertebrates. In most teleosts with internal fertilisation, some development of the fertilised egg takes place within the ovary of the female, but in a few species the eggs are deposited soon after fertilisation (Chapter 10). In some cichlid species in which the female broods the fertilised eggs in her buccal cavity (Chapter 8), the female lays the eggs and then sucks them into her buccal cavity and also sucks in the male's spermatozoa, so the eggs are fertilised within the buccal cavity.

Mating systems

Teleosts also show diversity in the social contexts in which mating occurs (Chapter 8; Taylor & Knight, 2008; Patzner, 2008). These contexts differ in the extent to which they allow some selection of partners by the mating fish. In promiscuity, males potentially fertilise the eggs of several females, and the eggs of a female may be fertilised by several males, with minimal selection of partners. Polygamy refers to the social situation in which an individual mates with several partners, but with the possibility of some selection of the partner at each spawning. Males in polygynous species mate with several females, but each female spawns with only one male. Females in polyandrous species mate with several males, but each male spawns with only a single female. In this system, the male may assume the parental role (Chapter 9). A form of promiscuity, but with the possibility of mate selection, is polygynandry. A male will spawn with several females and a female with several males over the course of a breeding season, but with mate choice operating. In monogamy, a single male and female form a mating pair and show some degree of bonding with the mate. This type of mating system is particularly associated with species in which parental care follows the fertilisation of eggs. The pair bond may be long-lasting or more temporary, only persisting for a single breeding attempt. Even in socially monogamous systems, both males and females may take part in extra-pair spawnings, so socially monogamous matings systems may not be genetically monogamous.

Where a pair bond forms, if only for as long as it takes for successful spawning, the bond may be subverted by other males, who take advantage of the pairing to steal or 'sneak' fertilisations by depositing sperm close to, or at, the site of spawning. The sneaking males are demonstrating an alternative mating tactic (see Chapters 8 and 11).

Secondary sexual characteristics

Sexually mature males and females may differ in appearance, either permanently or only in the breeding season. In sexually monomorphic species, there are no obvious differences between sexually mature males and females, although one or both sexes may produce pheromones (Chapters 5 and 8) to attract mates. The sexual dimorphism may only be present during the breeding season, and outside of breeding, males and females are indistinguishable. However, in some species, the sexual dimorphism is permanent, so sexually mature males and females can be reliably identified.

Parental care

The majority of teleost species show no post-fertilisation care of their progeny. However, in a number of teleost lineages, parental care has evolved, taking different forms in different lineages (Chapter 9). Perhaps the commonest form is paternal care, in which the male cares for one of more clutches of fertilised eggs. An extreme form of this is seen in the pipefishes and seahorses (Syngnathidae). The female transfers eggs to a brood pouch on the abdomen of her male partner and the eggs develop in the male's pouch in a form of male pregnancy (Chapter 10).

In maternal care, the female takes care of the developing eggs. The nature of the maternal care depends on the mode of fertilisation. If the eggs were fertilised externally, the female must protect the eggs from adverse environmental conditions. Incubation in the buccal cavity is one solution to this problem. The female scoops up the eggs in her mouth and the eggs complete their development in their mother's buccal cavity. A second form of maternal care is seen in viviparous species in which the eggs are fertilised and develop in the ovaries of the female (Chapter 10). In some species, the female provides only protection, but in other species, the female provides the developing eggs with nutrients in a way that is analogous to pregnancy in mammals.

In a few species with external fertilisation, there is biparental care, with the female and male cooperating in care of the young. In this situation, the female and male may show some division of labour, often with the male protecting the area around the eggs and the female tending to the eggs and young stages. An even rarer form of parental care has been described for a few cichlid species. The parents are helped in the care of a current brood by juveniles from a previous spawning. This phenomenon of juvenile helpers has been well studied in birds, but less so in teleosts.

Reproductive guilds

The diversity of modes of reproduction has also been described by reproductive guilds on the basis of the nature of the spawning site (Balon, 1975). This classification into reproductive guilds overlaps, to some extent, with the reproductive modes already defined. A simplified version of the original classification is shown in Table 1.2.

Reproductive diversity in teleosts: an explanatory framework

What explains the extraordinary diversity of reproductive modes shown by teleosts? Why has such diversity evolved? It is not yet possible to provide a comprehensive explanation for the extraordinary variability in teleost mating systems and modes of reproduction (Chapter 13). However, two, inter-related explanatory frameworks can be

Table 1.2 Breeding sites of teleosts, based on reproductive guilds defined by Balon (1975). (Source: Balon 1975. Reproduced with permission of Canadian Science Publishing.)

Reproductive guild	Example
1. Non-guarders of eggs and young	
(a) Open substrate spawners	
(i) Pelagic spawners	
Pair spawners	*Zebrasoma scopas*
Group spawners	*Mycteroperca rosacea*
Mass spawners	*Theragra chalcogramma*
(b) Benthic spawners	
(i) Spawners on coarse substrates	
Pelagic free embryo and larvae	*Morone saxatilis*
Benthic free embryo and larvae	*Phoxinus phoxinus*
(ii) Spawners on plants	
Non-obligatory	*Rutilus rutilus*
Obligatory	*Esox lucius*
(iii) Spawners on sandy substrates	
(c) Brood hiders	
(i) Benthic spawners	*Oncorhynchus nerka*
(ii) Cave spawners	*Astyanax jordani*
(iii) Spawners on invertebrates	*Rhodeus amarus*
(iv) Beach spawners	*Leuresthes tenuis*
(v) Annual fishes	*Nothobranchius kadleci*
2. Guarders	
(a) Substrate spawners	
(i) Rock spawners	*Stegastes leucostictus*
(ii) Plant spawners	*Pomoxis annularis*
(iii) Terrestrial spawners	*Copella arnoldi*
(iv) Pelagic spawners	*Channa punctata*
(b) Nest spawners	
(i) Rock and gravel nesters	*Ambloplites rupestris*
(ii) Sand nesters	*Hypsophrys nicaraguensis*
(iii) Plant material nesters	*Hoplosternum littorale*
Gluemakers	*Gasterosteus aculeatus*
Non-gluemakers	*Micropterus dolomieu*
(iv) Bubble nesters	*Betta splendens*
(v) Hole nesters	*Cottus gobio*
(vi) Miscellaneous materials nesters	*Lepomis macrochirus*
(vii) Anemone nesters	*Amphiprion melanopus*
3. Bearers	
(a) External bearers	
(i) Transfer brooders	*Oryzias latipes*
(ii) Forehead brooders	*Kurtus gulliveri*
(iii) Mouthbrooders	*Oreochromis niloticus*
(iv) Gill-chamber brooders	*Amblyopsis rosae* (?)
(v) Skin brooders	*Antennarius commerson*
(vi) Pouch brooders	*Hippocampus zosterae*
(b) Internal bearers	
(i) Viviparous	*Poecilia reticulata*

?, indicates uncertainty with this mode of bearing.

identified: phylogenetic relationships and adaption by natural selection. Phylogenetic lineages map the evolutionary relationships between species. Essentially, they map the changes in the gene pool represented by the lineage. This means that the adaptive responses by a population to its physical and social environment are likely to be constrained to a lesser or greater extent by its genetic inheritance (Taylor & Knight, 2008).

Phylogenetic relationships of the teleosts

There is still uncertainty about the details of the phylogeny of the teleost fishes, but some of the major trends are recognised (Nelson, 2006; Helfman *et al.*, 2009; Orti & Li, 2009). The teleosts belong to the actinopterygian lineage of jawed vertebrates. Early in the evolutionary history of the jawed vertebrates, the Gnathostomata, two lineages separated approximately 450 million years ago (mya) during the Silurian Period, the cartilaginous fishes (Chondrichthyes) and the bony fishes (Osteichthyes). The Chondrichthyes are restricted to the aquatic and largely to the marine environment. Interestingly, they all have internal fertilisation, which is more commonly found in fully terrestrial vertebrates, which cannot use the ambient water to transfer the spermatozoa to the eggs. Chondrichthyian fishes show a range of care for their offspring, ranging from spawning eggs protected by an egg capsule to viviparity with placenta-like structures, by which the female provides nourishment for the developing embryos. There are no reports of chondrichthyian fishes showing any parental care once the eggs have been laid or the female has given birth. In the osteichthyan lineage, there was a further splitting of the lineages during the Silurian Period. One lineage, the lobe-finned fishes, the Sarcopterygii, now has few aquatic representatives, just the coelacanths (Actinistia) and lungfishes (Dipnoi), but it is the lineage that probably gave rise to all the terrestrial vertebrates, the amphibians, reptiles, birds and mammals. The second lineage evolved as the ray-finned fishes, the Actinopterygii. The sturgeons and paddlefish (Acipenseriformes) are living representatives of the evolutionarily more primitive representatives of the Actinopterygii (Fig. 1.1).

Of the contemporary teleost groups, the freshwater Osteoglossomorpha are the earliest to appear in the fossil record and probably appeared 250–290 mya during the Triassic Period (Fig. 1.2). The marine Elapomorpha, which includes the eels (Anguilliformes) and tarpons (Elopiformes), originated 230–260 mya. The Clupeomorpha, which includes the herrings (Clupeiformes), is a largely marine group. However, the related Ostariophysii, which includes the characins (Characiformes), the carps (Cypriniformes) and catfishes (Siluriformes), is the dominant freshwater teleost group on all the continents. The Ostario-Clupeamorpha have their origins about 200–240 mya. Some 80–110 mya ago, the Euteleostei evolved. The more primitive representatives of the Euteleostei include the salmon and trout (Salmoniformes) and the pikes and pickerels (Esociformes). The major radiation of the teleosts in terms of number of species occurred 55–125 mya with the appearance of the Acanthomorpha. There are around 16000 living species of the Acanthomorpha, representing approximately 86% of all non-ostariophysan teleosts. The radiation of the Acanthomorpha in the early Eocene, about 55 mya, has been described as 'the most dramatic evolutionary radiation ever seen in vertebrate history' (Maisey, 1996). Some of the better-known orders in the Acanthomorpha include the cods (Gadiformes), the atherinids (Atheriniformes), the toothcarps (Cyprinodontiformes), the perches (Perciformes) and the flatfishes (Pleuronectiformes). Curiously, although the evolutionary history of the teleosts coincides in time, to a large extent, with that of mammals, the teleosts are still often referred to as lower vertebrates and a common misconception is that they are an evolutionarily primitive group (Fig. 1.2).

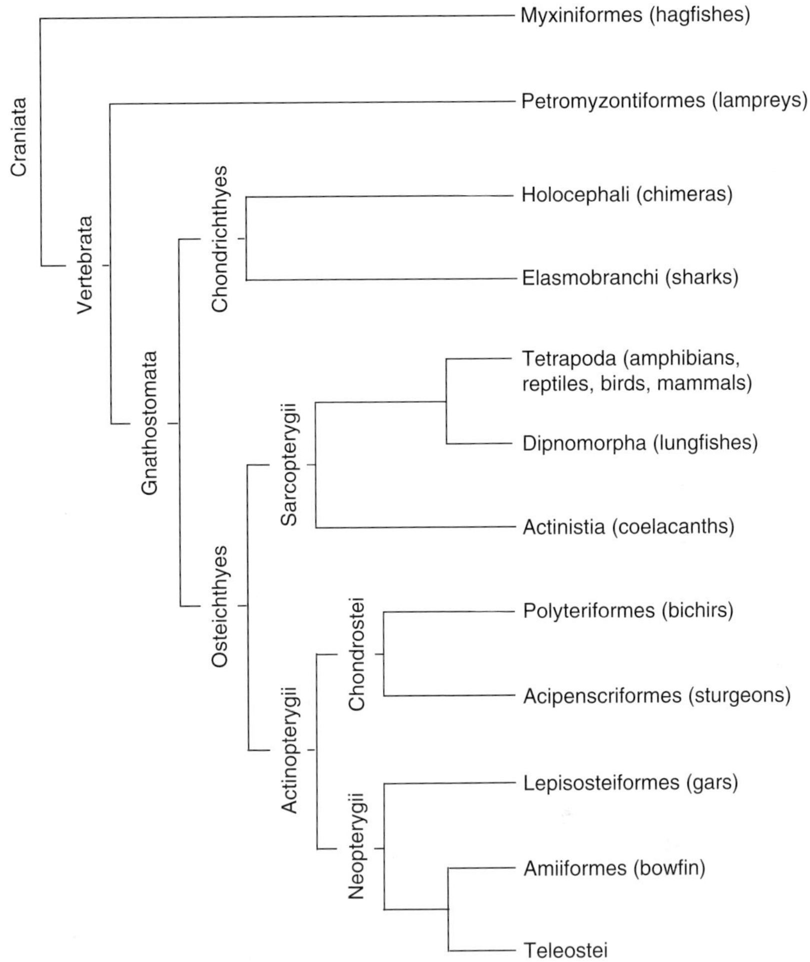

Figure 1.1 Phylogenetic relationships of the vertebrates showing the relationship of the teleosts to other vertebrates. (Source: Adapted from Nelson 2006. Reproduced with permission of John Wiley & Sons.)

A key feature of the evolution of teleosts is that at some point in the early evolutionary history of the teleost lineage, there was a whole genome duplication (WGD; Finn & Kristoffersen, 2007), which results in each gene being duplicated. This duplication raises the possibility that one of the genes will acquire a new function, neo-functionalism. This WGD may help to account for the extraordinary diversity of teleosts and some of their unusual traits compared with other lineages of jawed vertebrates.

Life-history theory

Life-history theory attempts to account for the adaptive relationship between life-history traits and the physical and social environments encountered. The life-history traits typically considered include age and size at sexual maturity, size of eggs, fecundity and parental care (Chapter 11). There have been several approaches to this problem, each of which has provided valuable insights into the evolution of life-history traits.

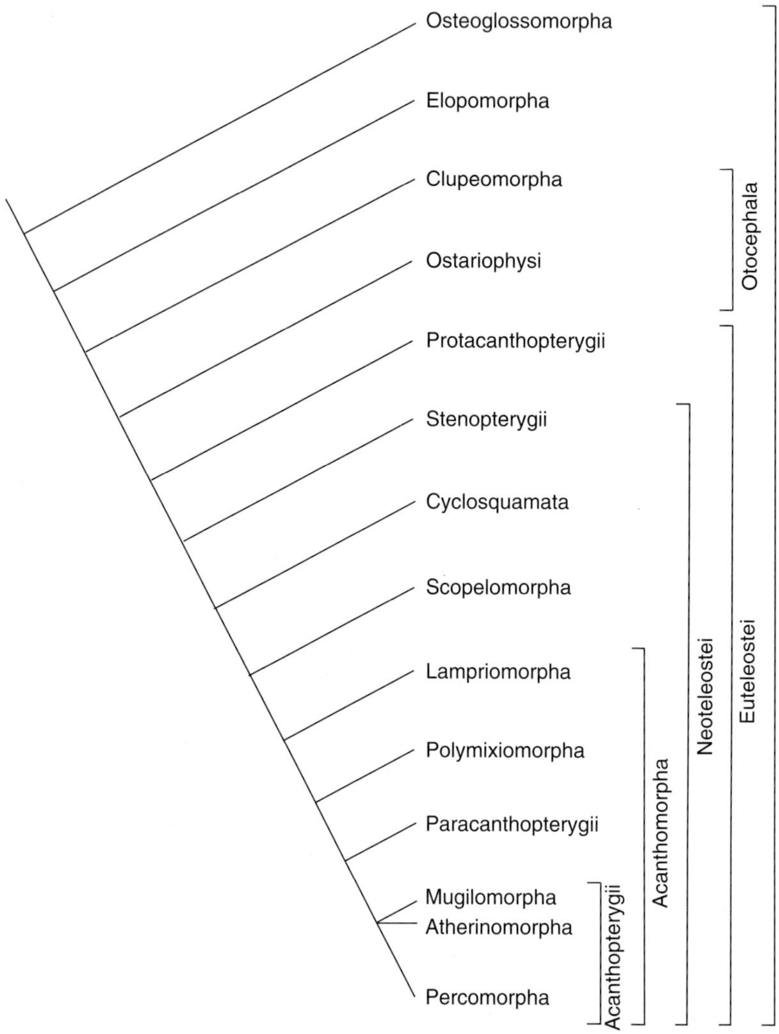

Figure 1.2 Phylogenetic relationships within contemporary teleosts. (Source: Adapted from Helfman *et al.* 2009. Reproduced with permission of John Wiley & Sons.)

Demographic life-history theory addresses the effects of age-specific mortality rates on reproductive traits (Roff, 1992, 2002; Stearns, 1992). It seeks to define the combination of life-history traits that will maximise fitness in a given environment, taking into account trade-offs between traits such as that between age and size at first maturation, or between egg size and fecundity. The approach has been used with particular effect to analyse life-history traits of salmonids (Schaffer, 1979, 2004) and cyprinodonts, especially the guppy (*Poecilia reticulata*) (Reznick, 1996; Reznick *et al.*, 2002a).

A second approach has been to deduce what combination of life-history traits would be adaptive given a set of environmental conditions. This is called the templet model of life histories. The first and most influential example of this model of life histories is the concept of r- and K-selection. This model relates the intensity of density dependence in populations to life-history traits. This approach has been developed by considering other dimensions such as habitat productivity when accounting for the evolution of life-history traits (Southwood, 1988). A scheme that has been applied to terrestrial plants defines three dimensions: a 'disturbance' axis, an 'adversity' axis and a 'biotic interactions' axis. Attempts to apply templet models to teleosts have had limited success (Persat *et al.*, 1994).

A third approach, related to templet models, which has been applied to describing life-history traits in teleosts, is the application of multivariate statistics to define clusters of species that share similar combinations of life-history traits. These combinations can then be related to the environments in which the various clusters tend to occur. An influential example of this approach identified three life-history strategies (Winemiller & Rose, 1992). These represent end points in a continuum, but do capture important combinations of life-history variables. The initial analysis used data on marine and freshwater fishes from North America, but it ranged across many teleost lineages, so the strategies do not reflect phylogenetic constraints. The 'periodic strategy' is characterised by species with delayed maturation, large size at sexual maturation and high fecundities. Some of these species have total spawning, ensuring that eggs are produced in a period most suitable for the development of the early life stages. By contrast, the 'opportunistic strategy' is characterised by early maturation, small adult size, batch spawning over an extended breeding season and rapid early growth. Species with an 'equilibrium strategy' tend to have a small to medium body size, low fecundities, large eggs and some form of parental care.

Finally, an additional approach to understanding life-history evolution comes from identifying regularities in the relationships between life-history traits, termed 'invariants' (Charnov, 1993). The adaptive significance of these different models of life-history evolution is discussed in Chapter 11.

Aims of the volume

The lack of information on the reproductive biology of teleosts and the remaining uncertainty about the phylogenetic relationships within the teleost lineage preclude a comprehensive synthesis of phylogenetic and adaptive influences on their diverse reproductive biology. This book attempts to establish a framework in which such a synthesis could be achieved. It describes the reproductive biology, using, wherever possible, phylogenetic information and life-history theory as means to interpret the information. The pages also celebrate the diversity of modes of reproduction of the teleosts.

Chapter 2
Sex determination

Introduction

In gonochoristic species, a normal, reproductively competent individual is unambiguously either male or female. The inheritance of sex resembles a classic Mendelian trait controlled by a single genetic locus (Mittwoch, 2006). This is reflected in the sex chromosomes of the endothermic vertebrates, mammals and birds. In mammals, with a few minor exceptions, males are the heterogametic sex, XY, and females the homogametic sex, XX. In mammals, a Y-chromosome gene, *Sry*, acts as a master switch, which activates the cascade of events that leads to the differentiation of functional testes (Graves, 2002, 2008). In birds, the female is the heterogametic sex, ZW, and the male the homogametic sex, ZZ. The overwhelming importance for an individual of unambiguously functioning as a male or female suggests that there will be strong stabilising selection on the sex-determining mechanism. The uniformity of the sex-determining mechanism within each of the endothermic vertebrate clades is evidence that this strong selection exists.

Given that the sex-determining mechanisms of birds and mammals are relatively unambiguous, it is disconcerting to discover that in the teleost fishes there is a diversity in the way sex is determined, with evidence of rapid evolution of the mechanisms. The case of hermaphroditic and parthenogenetic teleosts will be considered in Chapter 10, and only gonochoristic species are addressed here.

Reproductive Biology of Teleost Fishes, First Edition. Robert J. Wootton and Carl Smith.

Sex determination is the mechanism by which the sex of an individual is defined as either male or female (Bull, 1983; Purdom, 1993; Penman & Piferrer, 2008). Sex differentiation is the ontogenetic process, which acts on that dichotomous decision to generate functioning male or female gonads (Chapter 3). Across the animal kingdom, a range of sex-determining mechanisms occurs (Bull, 1983). As a group, the teleosts illustrate many, if not all, of these mechanisms (Price, 1984; Devlin & Nagahama, 2002). Sex determination mechanisms can be divided into two major groups. In genotypic sex determination (GSD), sex is determined at fertilisation and genetic differences occur between males and females. In environmental sex determination (ESD), sex is determined after fertilisation by the environmental conditions, typically the temperature regime, experienced by the developing zygote. It is unclear whether these represent two mutually exclusive mechanisms or the end points of a continuum, with differing influences of genetic and environmental factors (Valenzuela *et al.*, 2003; Penman & Piferrer, 2008; Valenzuela, 2008; Baroiller *et al.*, 2009).

Genotypic sex determination

Teleosts illustrate three modes of GSD: monofactorial, multifactorial and polyfactorial (Devlin & Nagahama, 2002). Monofactorial GSD is analogous to the situation in birds and mammals. Sex is inherited as though it is a single-locus, Mendelian character associated with sex chromosomes. The chromosomes act as a transport device for the sex-determining factor. In contrast to birds and mammals, teleosts provide examples of both male heterogamety (XX–XY) and female heterogamety (ZW–ZZ). In some teleosts (e.g. the southern platyfish, *Xiphophorus maculatus*, Poeciliidae), both male and female heterogamety can coexist within a species (Sandra & Norma, 2010). In such a species, a male can be either XY or ZZ depending on the population from which it comes. In multifactorial GSD, there are several but countable sex chromosomes; for example, if the male is the heterogametic sex, females can be defined by $X_1X_2X_1X_2$ and males by X_1X_2Y, with the Y chromosome becoming attached to an autosomal chromosome. In polyfactorial or autosomal GSD, several genetic loci are implicated in sex determination and there are no discrete sex chromosomes.

Monofactorial genotypic sex determination

The karyotype is known for only a small proportion of the approximately 30 000 species of teleost (FishBase: http://www.fishbase.org). Surveys of about 1700 species suggested that 10–11% had morphologically distinct sex chromosomes that were identifiable by cytological methods (Sola *et al.*, 1981; Purdom, 1993; Devlin & Nagahara, 2002). This percentage probably represents an underestimate. When the specific aim of the study was to detect the presence of sex chromosomes cytologically, the percentage of species increased to about 50%. The XX–XY pattern was commoner than the ZW–ZZ pattern.

A problem arises in identifying the heterogametic sex when the sex chromosomes cannot be identified with cytological methods including staining for chromosome bands. Although a variety of methods have been used, developments in genomic techniques, based on the ability to sequence DNA, are providing a powerful methodology.

The demand in aquaculture for rearing cohorts of single-sex fish, which can have the effect of limiting undesirable behaviours such as territoriality and aggression, has stimulated the identification of the heterogametic sex by manipulations of the sex chromosome complement (see Chapter 12). One example is gynogenesis, in which a haploid egg is induced to become diploid without being fertilised by a sperm (Purdom, 1993; Sarder *et al.*, 1999; Chapter 12). If the female is of the homogametic sex, the haploid egg will carry an X chromosome and all eggs induced to become diploid will develop into XX females. If the female is of the heterogametic sex, ZW, the haploid egg will carry either the Z or W sex chromosome. Eggs induced to be diploid will be either WW or ZZ, the latter developing as males (the WW genotype may not be viable). The lability of sex determination in teleosts is such that in some species the appropriate hormonal treatment during early development can reverse the genetically determined sex and produce functional ovaries in genetic males and functional testes in genetic females (Purdom, 1993). The technique has been applied particularly in salmonids, for which monosex female cohorts are favoured, and in mouthbrooding cichlids, for which monosex male cohorts are favoured (Chapter 12). In a species that, for example, has an XX–XY sex determination mechanism, treatment of genetic females with testosterone early in development can cause the development of functional testes with viable spermatozoa, all of which will carry the X chromosome. When this spermatozoa is used to fertilise eggs from normal XX females, all the progeny will be XX females (see also Chapter 12).

The behaviour of the sex chromosomes during meiosis has, in a few examples, been used to infer the heterogametic sex. The heteromorphic sex chromosomes, e.g. X and Y, show delayed pairing in the early pachytene stage in meiosis. At this stage, a synaptonemal complex is formed between paired chromosomes, and unpaired regions of the chromosomes may only occur between heteromorphic sex chromosomes (Carrasco *et al.*, 1999).

Visible sex-linked morphological traits such as colour patterns were the first traits to be used to infer the heterogametic sex, particularly in cyprinodonts such as the guppy (*Poecilia reticulata*, Poeciliidae) and the southern platyfish (*Xiphophorus maculatus*, Poeciliidae) (Yamamoto, 1969; Kallman, 1984; Purdom, 1993). Later, with the development of gel electrophoresis to detect protein and hence genetic variation, biochemical sex-linked markers were detected. Each of the methods outlined is liable to error, so where possible, the use of a combination of these techniques was preferred.

Advances in the use of DNA technology to define the genome of individuals revolutionised the study of sex determination in teleosts. Numerous microsatellite markers were mapped onto linkage groups (LGs), which are linear arrays of genes and other DNA elements that show limited recombination and so are physically linked to each other. Eventually, with full mapping, it is possible to identify LGs with visible chromosomes. Sex-linked markers allow the identification of the LG that carries the sex-determining region and raises the possibility of identifying the genetic locus that is responsible for sex determination (Peichel *et al.*, 2001, 2004). In mammals, the male sex-determining gene has been identified as *Sry*, but in teleosts, five strong candidates for the male sex-determining gene have been identified (Kikuchi & Hamaguchi, 2013). This again illustrates the diversity of sex-determining mechanisms in the teleosts. Where the identity of the sex-determining genes is not known, the locus will be described as *SEX*.

The diversity of GSD mechanisms in teleosts is illustrated by five families that are not closely related phylogenetically. The families are considered approximately along the phylogenetic dimension of basal to advanced teleosts (Mank *et al.*, 2006b).

The Salmonidae is a basal teleost family, with a common ancestor in the Cretaceous. This common ancestor underwent a gene duplication event, doubling the number of chromosomes. Today, the family is distributed in temperate and boreal regions of the Northern Hemisphere. It includes forms that are resident in freshwaters, but also forms that are anadromous, spawning in freshwater but having their major growth phase in the sea (Chapter 7). Sex determination in the salmonids has attracted attention because some species are important in aquaculture (Chapter 12), which prefers to rear all-female cohorts.

The salmonid species studied have XX–XY male heterogamety, although with some evidence of modulation by temperature and other factors (Woram *et al.*, 2003; Davidson *et al.*, 2009). Morphologically distinct sex chromosomes are rare. They have been identified in sockeye salmon (*Oncorhynchus nerka*, Salmonidae), rainbow trout (*Oncorhynchus mykiss*, Salmonidae), lake trout (*Salvelinus namaycush*, Salmonidae) and brook trout (*Salvelinus fontinalis*, Salmonidae). However, not all populations of *O. mykiss* show visible heteromorphic sex chromosomes. In *O. nerka*, the diploid number of chromosomes in males is 57, whereas in females it is 58. In males, the Y chromosome has fused with an autosome. There is evidence that species can differ in the LG that carries the sex-determining locus, but in some species, including the Atlantic salmon (*Salmo salar*, Salmonidae), brown trout (*Salmo trutta*, Salmonidae) and Arctic charr (*Salvelinus alpinus*, Salmonidae), *SEX* has been located at the end of a LG. However, the identity of *SEX* is not known. A candidate for *SEX* in *O. mykiss* is *sdY*, a gene that is highly linked to the *SEX* locus (Yano *et al.*, 2012). It is male-specific and expressed only in the testis.

The sticklebacks (Gasterosteidae) are a small family of fishes, distributed in the temperate and boreal Northern Hemisphere. One species, the brook stickleback (*Culaea inconstans*, Gasterosteidae), occurs in North America where it is resident in freshwater. Another species, the 15-spined stickleback (*Spinachia spinachia*, Gasterosteidae), is a resident of coastal seas of western Europe, while the other species have both freshwater resident and anadromous forms (Wootton, 1976). The three-spined stickleback (*Gasterosteus aculeatus*, Gasterosteidae) has become a model species for the study of the genetic basis of evolution in a vertebrate (Kingsley *et al.*, 2004).

A combination of karyological (Chen & Reisman, 1970; Ocalewicz *et al.*, 2008) and genomic analysis using protein and genetic markers and fluorescent in-situ hybridisation (FISH) staining methods have clarified the sex-determining mechanisms in the Gasterosteidae (Griffiths *et al.*, 2000; Peichel *et al.*, 2004; Ross & Peichel, 2008; Kitano *et al.*, 2009; Ross *et al.*, 2009). These studies of gasterosteids present a vivid picture of the rapid evolution of the sex-determining mechanism in a teleost family.

The four-spined stickleback (*Apletes quadracus*, Gasterosteidae) has female heterogamety (ZZ–ZW), whereas the nine-spined stickleback (*Pungitius pungitius*, Gasterosteidae) and members of the genus *Gasterosteus* have some form of male heterogamety. *G. aculeatus* lacks morphologically distinct sex chromosomes. However, a male-specific polymorphism in the protein isocitrate dehydrogenase (IDH) strongly indicates male heterogamety. The use of genetic markers and *in situ* staining using FISH identified the male sex-determining locus, *SEX*, as located on LG 19 (Peichel *et al.*, 2004). A feature of the sex chromosomes is a lack of recombination between X and Y in the region of *SEX*, with evidence of inversions, deletions and the accumulation of repetitive DNA on the Y chromosome (Ross & Peichel, 2008). *SEX* in *G. aculeatus* has not yet been identified genetically, partly because of the presence of this repetitive DNA.

This model of male heterogamety in *G. aculeatus* was made more complex by a study of interclade variation (Kitano *et al.*, 2009). In the Pacific Ocean (PO) region, molecular phylogenies identified two clades of *G. aculeatus*, a PO clade and a Japan Sea (JS) clade. The JS clade diverged from the PO clade about 1–2 million years ago (mya), when there was a period of geographical isolation. In the JS clade, the Y chromosome has fused with the autosomal LG 9, forming a neo-sex chromosome X_1X_2Y, where X_1 is the ancestral X chromosome as found in the PO clade and X_2 is the neo-X chromosome LG 9.

The North American black-spotted stickleback (*Gasterosteus wheatlandi*, Gasterosteidae) diverged from *G. aculeatus* about 10 mya. In *G. wheatlandi*, a neo-sex chromosome has also evolved, but in this species this occurred by the fusion of the Y chromosome to the autosomal LG 12 (Ross *et al.*, 2009). This fusion may have occurred in the past 10 million years, after the two *Gasterosteus* species diverged. In male *G. wheatlandi*, the diploid number of chromosomes is 41, but in the female it is 42 (cf. *O. nerka*, described earlier). In this species, the sex chromosomes are cytologically distinct.

In *P. pungitius*, *SEX* maps to LG 12, rather than LG 19 as in *G. aculeatus* (Ross *et al.*, 2009; Shapiro *et al.*, 2009). This suggests that the XX–XY heterogamety of *P. pungitius* is not homologous to that of *Gasterosteus*. The situation will be clarified when *SEX* in the two genera is identified. At least in some populations of *P. pungitius*, the sex chromosomes can be identified cytologically (Ocalewicz *et al.*, 2008).

There is no evidence that the ZZ–ZW female heterogamety in *A. quadracus* is related to the situations in *Pungitius* or *Gasterosteus* (Table 2.1). The sex-determining mechanisms in *C. inconstans* and *S. spinachia* have still to be identified (Ross *et al.*, 2009). *S. spinachia* is regarded as phylogenetically the most basal of the Gasterosteidae, so its sex-determination mechanism is of particular interest.

The Cichlidae (Order: Perciformes), in contrast to the species-poor Gasterosteidae, contains more than 700 species. It is a family of subtropical and tropical freshwater teleosts distributed in Central and South America, Africa and parts of the Middle East and in southern India and Sri Lanka. In the Great Lakes of east Africa – Tanganyika, Malawi and Victoria – clades of cichlids have undergone an explosive speciation, with numerous endemic species having evolved from a common ancestor in each lake, in some cases within the last 2 or 3 million years. Mechanisms of sex determination in the Cichlidae have attracted attention because of the importance of a few African species, the tilapias, for tropical aquaculture. Yields in aquaculture systems are higher if all-male cohorts can be raised. Such all-male cohorts can be generated by endocrine manipulation (Chapter 12). For most cichlid species there is no evidence of morphologically

Table 2.1 Sex determination and phylogeny of North American sticklebacks (Gasterosteidae). The scientific name of each species is given, along with the diploid number of chromosomes, type of sex determination system, linkage group to which *SEX* maps, and presence of a heteromorphic sex chromosome pair. In *Gasterosteus wheatlandi*, females have the diploid chromosome number originally reported ($2n = 42$); the number of chromosomes in males (M) is 41. (Source: Adapted from Ross *et al.* 2009. Creative Commons Attribution License 4.0. http://creativecommons.org/licenses/by/4.0/.)

		1970			Present		
		Sex chromosome			Sex chromosome		
Species	2n	System	LG	Hetero?	System	LG	Hetero?
G. aculeatus	42	–	–	No	XY	19	Yes
G. wheatlandi	42	XY	–	Yes	X_1X_2Y	12, 19	Yes
P. pungitius	42	–	–	No	XY	12	Yes
C. inconstans	46	–	–	No	–	–	No
A. quadracus	46	ZW	–	Yes	ZW	–	Yes

Gasterosteidae

–, indicates that no data are available.

distinct sex chromosomes. Evidence for GSD mechanisms has come from gynogenesis, endocrine manipulation, intra- and interspecific hybridisation, and analysis of the synaptonemal complex during meiosis (Carrasco *et al.*, 1999; Sarder *et al.*, 1999; Oldfield, 2005; Dunham, 2011; see Chapter 12). Molecular methods have greatly clarified the situation by examining the association between DNA microsatellite markers and the phenotypic sex.

In a study of six African tilapias, two LGs, LG 1 and LG 3, were identified as important for sex determination (Cnaaci *et al.*, 2008). Two species, the Nile tilapia (*Oreochromis niloticus*, Cichlidae) and the redbelly tilapia (*Tilapia zilli*, Cichlidae), had markers on LG 1 that were linked with the phenotypic sex and indicated XX–XY male heterogamety. *Oreochromis karongae* (Cichlidae) and the spotted tilapia (*Tilapia mariae*, Cichlidae) had markers on LG 3 linked to phenotypic sex in ZZ–ZW female heterogamety. In two species, the situation was more ambiguous. In the blue tilapia (*Oreochromis aureus*, Cichlidae), an Israeli strain had ZZ–ZW hetrerogamety associated with LG 3. However, for an Egyptian strain, markers on LG 1 (XX–XY) explained the sex of most individuals, but markers on LG 3 (ZZ–ZW) were weakly associated with phenotypic sex. In the Mozambique tilapia (*Oreochromis mossambicus*, Cichlidae), sex-linked markers were found on both LG 1 and LG 3 and it was not possible to assign heterogamety.

Fluorescent in-situ hybridisation analysis identified LG 3 with the largest chromosome in the karyotype of the tilapias, while LG 1 corresponded to a smaller, different chromosome. Although the evidence is of strongly genetic mechanisms of sex determination in the tilapias, some anomalous sex ratios in some broods suggest that other genetic and environmental factors may also play a role (see the section on 'Environmental sex determination in teleosts'). The genera that comprise the tilapias shared a common ancestor as recently as 5 mya.

Mechanisms of sex determination in a species flock of haplochromine cichlids from Lake Malawi show a comparable diversity to that of the tilapias, although the two clades probably diverged 10–20 mya (Ser *et al.*, 2010). A study of 19 Lake Malawi cichlid species, focusing particularly on the genus *Maylandia*, found evidence of sex-linked, microsatellite markers on LG 5 and LG 7. Some species had markers on LG 7 linked to phenotypic sex, with XX–XY heterogamety. Four of 19 species studied had markers on LG 5 linked to phenotypic sex with ZZ–ZW heterogamety. This mode of sex determination was associated with a colour morph of females, orange blotch (OB), which was inherited as a dominant allele on the W chromosome. One species, *Maylandia pyrsonotos* (Cichlidae), had both XX–XY and ZZ–ZW systems, with strong epistatic interactions. Fish with the W chromosome (LG 5) were always female, with genotype ZW–XX or ZW–XY, and had the OB phenotype. Individuals lacking the W chromosome, but with the Y chromosome (LG 7) were male, with the genotype ZZ–XY, while fish with the genotype ZZ–XX were mostly female. Both sexes had the blue-barred (BB) colour phenotype. In some experimental families there was no association between phenotypic sex and loci on LG 5 or LG 7, suggesting additional sex-determination mechanisms.

Both the tilapias and the Lake Malawi cichlids show strong evidence for GSD, but both clades had both XX–XY and ZZ–ZW systems and the two clades have species with both mechanisms. However, the sex-determining mechanisms in the tilapias associated with LG 1 and LG 3 are different from those identified for the Lake Malawi cichlids associated with LG 5 and LG 7. In no case has *SEX* been identified, so the evolutionary relationships remain obscure.

Unusually amongst the teleost fishes, the Poeciliidae are livebearers with internal fertilisation (Chapter 10). The male uses a modified anal fin ray, the gonopodium, to insert bundles of spermatozoa, termed spermatozeugmata, into the gonopore of the female. The family is distributed in freshwaters in Central and South America. Sex-linked colour traits were identified early in studies of sex determination in poeciliids (Magurran, 2005). Typically, the males have obvious secondary sexual traits. *P. reticulata* is a popular aquarist species, in which the males are brightly coloured, but the females drab. A variety of evidence has suggested XX–XY male heterogamety, although with some evidence of autosomal effects on the major XX–XY sex chromosome system. The evidence for the XX–XY system includes sex-linked colour patterns in the male, FISH staining and analysis of the synaptonemal complex during meiosis (Schultheis *et al.*, 2009).

In the swordtails and platyfish, *Xiphophorus* spp. (Poeciliidae), also a popular aquarist species, there are a range of sex determination mechanisms (Yamamoto, 1969; Kallman, 1984; Schultheis *et al.*, 2009). In *X. maculatus*, there are three different sex chromosomes: X, Y and W. The female genotype can be XX, XW or YW, while possible male genotypes are XY or YY. By contrast, in the green swordtail (*Xiphophorus helleri*, Poeciliidae), there is evidence that sex determination depends on several alleles, perhaps with a major locus affected by several autosomal modifiers. For the Chiapas swordtail (*Xiphophorus alvarezi*, Poeciliidae), the females are the heterogametic sex with a ZZ–ZW system. As yet, *SEX* has not been identified for any poeciliid.

SEX has been identified in the medaka (*Oryzias latipes*, Adrianichthyidae) and this was the first case of a sex-determining gene to be identified in the teleosts. The medakas are in the family Adrianichthyidae. They occur in fresh and brackish waters in Southeast Asia. *O. latipes* has become one of the model species for genetic and genomic studies of teleost fishes, because of its high rate of reproduction (Kasahara *et al.*, 2007). *O. latipes* has XX–XY male heterogamety, initially deduced using a sex-linked colour. An elegant genomics study identified *dmrt1bY* as the sex-determining gene on the Y chromosome (analogous to the male-determining gene *Sry* in mammals – Matsuda *et al.*, 2002; Schartl, 2004; Matsuda & Sakaizumi, 2009). The same gene was also identified in the closely related Hainan ricefish (*Oryzias curvinotus*, Adrianichthyidae) (Matsuda *et al.*, 2003). *Dmrt1bY* is a duplicate of an autosomal gene *dmrt1a* found on LG 9. This duplication is estimated to have occurred 10–18 mya. Both genes belong to the *DMRT* gene family that has been implicated in sex determination and differentiation in other animals (Ferguson-Smith, 2006).

There are no obvious differences between the X and Y chromosomes in *O. latipes.* Around the sex-determining region of the Y chromosome, there is suppression of crossing-over between sex chromosomes, with the regions around the sex-determining locus consisting mainly of repetitive sequences of DNA. The occurrence of XX males in some natural strains suggests that there may also be autosomal modifiers affecting the sex determination mechanism.

Disappointingly, *dmrt1bY* is not even the sex-determining gene in other species of *Oryzias* (Kondo *et al.*, 2009). Both XX–XY and ZZ–ZW systems occur in the adrianichthyid family, with sex-determining loci located on different LGs compared to *O. latipes.* In *Oryzias luzonensis* (Adrianichthyidae), a strong candidate for *SEX* is *Gsdf*, a variant of the gene for gonadal soma growth factor. In no other teleosts has *dmrt1bY*, or a related gene, been identified as the male-determining locus.

The evidence that emerges from this review of sex-determining mechanisms over a phylogenetically wide range of teleosts is of rapidly evolving systems, in which different LGs can carry sex-determining factors. Both XX–XY and ZZ–ZW systems can be found in closely related species or even within some species, such as *X. helleri.* Table 2.2 lists the strong candidates for *SEX*, the male-determining factor. There are no obvious links between the candidates; although four are variants of genes involved in the differentiation of gonads, the fifth is related to a gene involved in the

Table 2.2 Strong candidates for *SEX*, the male determining gene in teleost species.

Species	Family	Gene	Gene family	Reference
Oncorhynchus mykiss	Salmonidae	*sdY*	Interferon regulatory factor 9	Yano *et al.* (2012)
Oryzias curvinotus	Adrianichthyidae	*dmrt1bY*	*Dmrt*	Matsuda *et al.* (2002)
Oryzias latipes	Adrianichthyidae	*dmrt1bY*	*Dmrt*	Matsuda *et al.* (2002)
Oryzias luzonensis	Adrianichthyidae	*Gsdf*^{Y}	Gonadal soma derived growth factor	Myosho *et al.* (2012)
Odontesthes hatcheri	Atherinopsidae	*amhy*	Anti-Müllerian hormone	Hattori *et al.* (2012)
Takifugu rubripes	Tetraodontidae	*Amhr2*	Anti-Müllerian hormone receptor type II	Kamiya *et al.* (2012)

immunological response. Perhaps as more candidates for *SEX* are identified, a pattern will emerge. This evolutionary lability of the sex-determining mechanisms is peculiar given the importance that an individual in a gonochoristic species is unambiguously male or female.

Multifactorial genotypic sex determination

As the example of the sex-determining mechanisms in *G. aculeatus* illustrated, a multifactorial GSD system can result from rearrangements of the sex chromosomes and autosomes. Several examples of such multifactorial systems occur in species of neotropical freshwater species involving both XX–XY and ZW–ZZ heterogamety (Almeida-Toledo & Foresti, 2001). The darter characin (*Apareiodon affinis*, Parodontidae) has a multifactorial systems based on the ZW–ZZ system, with female heterogamety (ZW_1W_2) and male homogamety (ZZ). A range of multifactorial XX–XY systems occurs. The trahira (*Hoplias malabaricus*, Erythrinidae) has female homogamety ($X_1X_1X_2X_2$) and male heterogamety (X_1X_2Y), whereas in a second, unnamed *Hoplias* sp. (Erythrinidae), the female is XX and the male is X_1X_2Y. It is not clear whether such multifactorial systems of sex determination have adaptive significance or whether the dynamics of the evolution of mono- and multifactorial systems are decoupled from any adaptive consequences.

Polyfactorial (polygenic) genotypic sex determination

In some teleost species, families with different parents but reared under similar conditions have highly variable sex ratios, which contrasts with the 1:1 ratio expected under the XX–XY and ZW–ZZ regimes. These variable sex ratios in families suggest that there is polygenic sex determination, with several genes contributing to sex determination. Even some well-studied species have poorly understood mechanisms of sex determination. The zebrafish (*Danio rerio*, Cyprinidae) is an important model for the study of vertebrate development (Spence *et al.*, 2008). Surprisingly, this species lacks any clearly defined sex chromosomes or evidence for a sex-determining genetic locus. The European seabass (*Dicentrarchus labrax*, Moronidae) is a marine species important for aquaculture in the Mediterranean region, which is characterised by highly variable sex ratios in captivity. The aquaculture industry would prefer to rear all-female cohorts because of the better growth and larger size of the females – hence the interest in the mechanism of sex determination in this species.

For *D. rerio*, breeding and molecular biology studies indicate that sex determination depends on a number of genetic loci distributed across the genome; the determination is polygenic (Liew *et al.*, 2012). Some evidence does suggest that one of the LGs, labelled *chr-4*, has some of the characteristics of a sex chromosome, but does not carry a single sex-determining gene (Anderson *et al.*, 2012). Environmental effects, such as temperature and rearing density, have only minor effects on sex ratios (Liew *et al.*, 2012). Breeding experiments using *D. labrax* reared under aquaculture conditions also generated families with highly variable sex ratios (Vandeputte *et al.*, 2007). The data were compatible with a polygenic model of sex determination, but with some of the variability in sex ratios reflecting environmental effects (see 'Environmental sex determination in teleosts').

An interpretation of the results for *D. rerio* and *D. labrax* is that sex can be regarded as a threshold trait. The effects of several genes, together with any environmental effects, must exceed some threshold to direct differentiation of the gonad in a specific direction that is unambiguously either male or female. However, the circumstances that lead to the evolution of polygenic sex determination remain obscure. It may simply represent a transitional stage in the evolution of monofactorial sex determination. Alternatively, polygenic sex determination may have an adaptive advantage in some circumstances. The situation in teleosts is made even more complicated by the possible effect of environmental factors on the process of sex determination.

Environmental sex determination

The identification of ESD is complicated by two potentially confounding factors. The first is that biased sex ratios may be caused by the differential mortality of genetic males and females before the sex ratio can be assessed. The primary sex ratio is a product of GSD, but the observed ratio is a product of differential mortality. Sex biases in predation have been recorded for males (Matity *et al.*, 1994) and females (Britton & Moser, 1982).

A second complication, relevant to experimental studies of ESD, is that the range of the environmental factor whose effect is being assessed should lie within the range to which the population is adapted and can respond in a physiologically appropriate way (Ospina-Álvarez & Piferrer, 2008). Outside this range, any effects on sex ratios may represent pathological responses during sexual differentiation. If there is evidence of the presence of sex chromosomes in a species, then it should be assumed that GSD rather than ESD is the primary mechanism of sex determination (Ospina-Álvarez & Piferrer, 2008).

Although ESD had been described in a few species of invertebrates and reptiles (Bull, 1983), in teleosts the phenomenon was first analysed in an atherinid species, the Atlantic silverside (*Menidia menidia*, Atherinopsidae) (Conover & Kynard, 1981). Subsequently, compilations identified about 70 species of teleosts for which evidence of ESD has emerged (Baroiller & D'Cotta, 2001; Devlin & Nagahama, 2002; Baroiller *et al.*, 2009). The majority of these examples have identified temperature as the relevant environmental factor. However, pH, growth rate, density and the social environment have also been implicated. A critical review of the data suggested that the number of species identified as showing ESD had probably been overestimated (Ospina-Álvarez & Piferrer, 2008). The number of species with temperature-dependent sex determination (TSD) indicated that TSD is probably the exception in sex determination in teleosts.

A feature of *M. menidia* is its wide geographical range along the eastern seaboard of North America, from Florida to Quebec. This distribution covers a wide range of sea temperatures. *M. menidia* is essentially an annual species, spawning high up the shore in summer, moving to deeper cooler waters to overwinter, before returning to the spawning grounds to reproduce. For a population in Massachusetts, the sex ratio of juveniles recruiting to the population over the summer shifted from a female-biased ratio early in the breeding season, when sea temperatures were cool, to a male-biased ratio later in the season, when sea temperatures were warmer (Conover & Kynard, 1981).

An experimental study of development at a range of rearing temperatures showed that for the Massachusetts population, colder rearing temperatures biased the sex ratio towards females, while warmer temperatures biased the ratio towards males. These effects could not be explained by differential, sex-dependent mortality and the temperatures used were in the range to which the population was adapted, making a pathological effect unlikely (Conover & Kynard, 1981). The sensitive period, during which the temperature regime determined sex, was in the larval phase after the eggs had hatched (Conover & Fleischer, 1986), and the timing of the effect seemed to depend on body size rather than on the age of the larvae. A feature of these experiments was that broods from different parents had different sex ratios, even when reared under the same temperature regimes. This variation suggested that both temperature and genetic factors were interacting to determine sex. An interaction between temperature and genotype was also suggested by the way in which sex ratio changed with temperature. There was not an abrupt change from a female-biased to a male-biased ratio, but as rearing temperature increased, the ratio of females declined (Fig. 2.1).

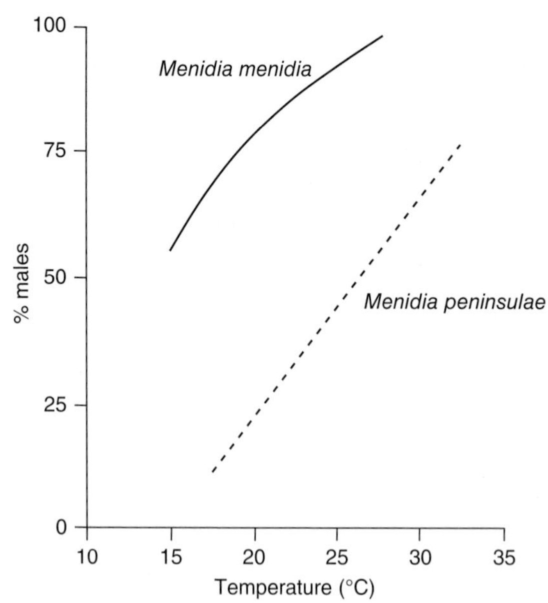

Figure 2.1 Effect of temperature of incubation on the sex ratio of two species of fish with temperature-dependent sex determination: Atlantic silverside (*Menidia menidia*, Atherinopsidae) and tidewater silverside (*Menidia peninsulae*, Atherinopsidae). In both cases, higher temperatures entailed a higher proportion of males. (Source: Adapted from Ospina-Álvarez & Piferrer 2008. Creative Commons Attribution License 4.0. http://creativecommons.org/licenses/by/4.0/.)

Menidia menidia also illustrates the flexibility of sex determination mechanisms in teleosts (Conover & Heins, 1987a,b). When the effects of temperature during the larval period on sex ratios were studied in populations drawn from the geographical range of the species, southern populations (from South Carolina, North Carolina and Virginia) showed strongly female-biased ratios if reared at a low temperature (15 °C), but strongly male-biased ratios if reared at a high temperature (20 °C). The temperature effect was less pronounced in populations from New York State and Prince Edward Island, for which the results were similar to the Massachusetts population. The shape of the relationship between sex ratio and rearing temperature differed between these populations, suggesting a temperature–genotype interaction. However, for a population from Nova Scotia (eastern Canada), which experienced the lowest sea temperatures, there was no evidence of ESD. There was no significant effect of rearing temperature on the sex ratio, and sex ratios did not differ between broods reared at a particular temperature. The sex ratio did not differ significantly from the 1:1 ratio expected with GSD.

Rearing studies confirmed that the Nova Scotia fish showed little evidence of temperature ESD. The results suggest a major sex-determining locus, with no environmental sensitivity. By contrast, for a South Carolina population, temperature affected the sex ratio in all the families studied. The results for this population were compatible with a polygenic ESD system. Results for a New York population suggested an interaction between temperature and major sex-determining loci (Conover & Heins, 1987a,b; Lagomarsino & Conover, 1993). Thus, within this species, the sex-determining mechanism ranges from strong ESD in southern populations to GSD in a northern population.

These studies of *M. menidia* led to a satisfying adaptive explanation for the inter-population differences in the sex-determining mechanism (Conover, 1984). In teleost fishes, female fecundity is strongly related to body size, with bigger females being more fecund (see Chapters 8 and 11). In terms of reproductive success, body size is likely to be of more importance to female *M. menidia* than males. The more southerly populations, experiencing warmer waters, have longer spawning seasons. Young born early in the breeding season have a longer growth period before the onset of winter than those born late in the season. Consequently, it is advantageous to produce females early in the season, when waters are relatively cold, because they will benefit from having a larger body size if they survive to breed in the following summer. To maintain an adaptive 1:1 sex ratio, more males will be produced later in the breeding season, when waters are warmer, although these males are likely to reach a smaller body size than females. In Nova Scotian waters, the growing season in summer is too short for either sex to gain a growth advantage.

Breeding experiments in which *M. menidia* from populations in South Carolina, New York and Nova Scotia were reared over several generations at either constant low or high temperatures showed that the sex ratio converged on the 1:1 ratio predicted by theory (Fisher, 1930). These experiments demonstrated that in *M. menidia*, ESD could evolve along a trajectory that led to a balanced sex ratio, although the way in which ESD changed was not consistent (Conover & van Voorhees, 1990; Conover *et al.*, 1992).

The inter-population differences in the strength of temperature-dependent ESD in *M. menidia* are mirrored in three South American atherinids (Strüssmann *et al.*, 1996a,b,c, 1997). These species differ from *M. menidia* in having a larger adult body size and a longer lifespan. In the freshwater Argentinian silverside (*Odontesthes bonariensis*, Atherinopsidae), a strong temperature-dependent ESD was demonstrated. At a low rearing temperature (17 °C), all female cohorts could be produced, while at a high temperature (29 °C) all male cohorts could be produced. The marine baboso

(*Odontesthes argentinensis*, Atherinopsidae) also showed temperature-dependent ESD, with a higher proportion of females produced at a lower rearing temperature. In the pejerrey (*Odontesthes hatcheri*, Atherinopsidae), the sex ratio was 1:1 for broods reared between 17 and 23 °C, but female-biased at low (13–15 °C) and male-biased at high (25 °C) temperatures. These results were obtained experimentally on stocks that had been transported to Japan, but were consistent with the adaptive explanation of female advantage at low temperatures.

Dicentrarchus labrax belongs to a family of teleosts (Moronidae) that are related to the Serranidae (groupers), which includes many hermaphroditic species (Chapter 10). However, *D. labrax* is gonochoristic, but lacks obvious morphological secondary sexual traits or distinguishable sex chromosomes (Piferrer *et al.*, 2005; Vandeputte *et al.*, 2007). *D. labrax* occurs along the Atlantic coast of North Africa, the coasts of Europe as far north as southern Norway, in the Mediterranean and in the Black Sea. Females tend to grow faster than males and the aquaculture industry has an interest in raising female monosex cohorts. The characteristic temperature at spawning is about 14 °C. Larval and juvenile *D. labrax* reared at 19–22 °C typically show a sex bias towards males, with the ratio about 75%, but potentially reaching 100%, indicating temperature-dependent ESD. In this species, sex differentiation is delayed, not occurring until 128–250 days post-fertilisation (see Chapter 4). There were differences in the sex ratio between families reared at the same temperature, indicating a parental genetic effect that modified the response to temperature. Modelling studies (Vandeputte *et al.*, 2007) suggested that sex is determined by a combination of autosomal polygenic and temperature influences. There is also evidence of a correlation between growth and sex determination (Piferrer *et al.*, 2005; Vandeputte *et al.*, 2007), with a tendency for males to predominate amongst slower-growing fish, and for females to predominate amongst the faster-growing fish. Some data suggested that the same genes were involved in sex determination and growth, or that they were at least tightly linked (Vandeputte *et al.*, 2007). A link between growth rate and sex determination has also been suggested in the anguillid eels (*Anguilla* spp., Anguillidae) and the cyprinid *D. rerio*.

The interest of the aquaculture industry in raising monosex cohorts has also led to analyses of the effect of temperature on sex ratios in cultivated tilapia, most notably in *Oreochromis* spp. Although GSD has been identified in tilapia, both XX–XY and ZZ–ZW systems (see 'Monofactorial genotypic sex determination in teleosts'), in both *O. niloticus* and *O. aureus* high rearing temperatures maintained for 28 days from the initiation of exogenous feeding led to male-biased ratios in the progeny (Baras *et al.*, 2000; Baras *et al.*, 2001). In *O. niloticus* (XX–XY), no significant deviations from a balanced sex ratio were obtained over the temperature range 20–33 °C, but at 35–37 °C, there was a significant skew to male-biased ratios. A strength of these studies was that the analysis eliminated the confounding effect of sex-selective mortality. Again, families raised at the same temperature showed different ratios, indicating a parental genetic effect in addition to the effect of temperature. In *O. aureus* ZZ–ZW, a masculinising effect was obtained at a rearing temperature of 35 °C, and also, but to a lesser extent, with a daily temperature cycle of between 27 and 35 °C.

In the natural habitats of many *Oreochromis* species, high water temperature can be reached in shallow waters. Some genetic females that experience these high temperatures in the thermosensitive period for sex determination will develop as phenotypic males. Later in life, such fish should benefit from the faster growth shown by male tilapia (Baras *et al.*, 2002). However, there is the possibility that the masculinisation is a pathological rather than an adaptive effect of high temperatures in early ontogeny. A further complicating factor is that these studies employed stocks that had been domesticated because of their use in aquaculture, and many cultivated tilapias are derived by hybridising fish from different source populations (Chapter 12).

Temperature and pH have both been implicated in ESD in *Apistogramma* spp. (Cichlidae), a genus of South American cichlids (Römer & Beisenherz, 1996). In this genus, no sex chromosomes have been described. For some species of *Apistogramma*, temperature was the relevant environmental factor, while in a few species, pH was the major factor, and in a third group of species, sex was determined by an interaction between temperature and pH. The proportion of males in a brood tended to increase over the temperature range 23–29 °C. Where it was a factor, an increase in pH from 4.5 to 6.5 tended to decrease the proportion of males in a brood. These results were obtained from laboratory studies. The streams in which *Apistogramma* spp. live naturally have gradients in temperature and pH, but the adaptive significance of ESD in this genus is not understood (Römer & Beisenherz, 1996).

Although *O. latipes* was the first species in which a sex-determining gene, *dmrt1bY*, was identified (see 'Monofactorial genotypic sex determination in teleosts'), a masculinising effect of high temperatures has been revealed in experiments on the highly inbred Hd-rR strain of this species (Hattori *et al.*, 2007). Eggs collected about 3 hours after fertilisation and then reared at constant temperatures that ranged from 15 to 34 °C were kept until the progeny reached sexual maturity. Rearing temperatures of 30, 32 and 34 °C induced some genetic XX females to become phenotypic males. The proportion that changed sex increased with higher temperature, and at 34 °C all genetic females that survived to adulthood were phenotypic males (13 out of 13; see Fig. 2.2). However, at 34 °C, there was a high frequency of embryonic abnormalities. The use of a highly inbred line and the incidence of deformities at the higher temperatures suggest that the masculinisation effect may be a pathological rather than an adaptive response to high

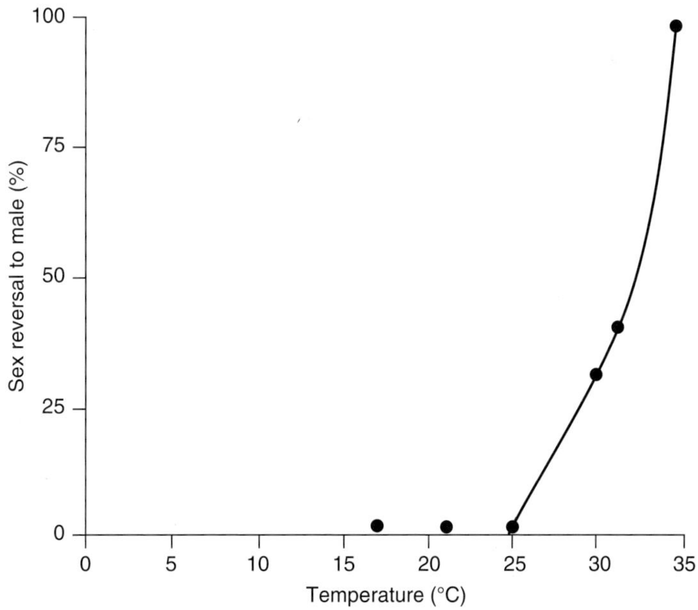

Figure 2.2 Effect of temperature on proportion of genotypic females sex reversing to males in the highly inbred Hd-rR strain of the medaka (*Oryzias latipes*, Adrianichthydae). (Source: Data from Hattori *et al.* 2007.)

temperatures. The phenotypic males with a female genotype produced at 34 °C that survived to adulthood performed courtship, but did not produce viable progeny.

Two further results from the study of *O. latipes* are important (Hattori *et al.*, 2007). The first was that the effect of high temperature on sexual differentiation occurred at a much earlier stage in embryonic development, between stages 5–6 (eight to 16 cells) and stage 36 (heart development). Other studies, including histological examination and exogenous hormone treatment, had suggested that gonadal sex was determined in the latter half of embryogenesis (between stages 36 and 39; see Chapter 3). Secondly, the XX females with male phenotypes did not express the sex-determining gene *dmrt1bY* gene. They did express the autosomal *dmrt1* gene, which is crucial for testis formation (Chapter 3).

Evolution of sex-determining mechanisms

The diversity of sex-determining mechanisms in the teleosts has puzzled even those who are at the forefront of research in the area. The situation has been described as 'chaotic' (Graves & Peichel, 2010) or a consequence of the 'poor state of development of the mechanisms of sex determination' (Hattori *et al.*, 2007).

There is a striking comparison between the teleosts, with their diverse mechanisms of sex determination, and the endothermic birds and mammals with their apparently conservative GSD, with mammals showing male heterogamety (XX–XY) and birds female heterogamety (ZZ–ZW). These differences are paralleled by other differences that may be relevant. Because of their endothermy and the protection provided to the developing young either by an incubated, shelled egg or the womb, birds and mammals are not likely to exhibit ESD. The reproductive traits of birds and mammals are conservative in comparison to the exuberance that is characteristic of the teleosts. Another relevant feature is that early in its evolutionary history, the teleost clade underwent a whole genome duplication (WGD) event (Santini *et al.*, 2009). Subsequently, gene and genome duplications have been unusually common within the teleost clade (Mank & Avise, 2009). These events have released the potential of the duplicated genes to acquire different functional roles, as seems to have occurred in *O. latipes* (see 'Monofactorial genotypic sex determination in teleosts'). Teleosts also differ from tetrapods in key details of the early differentiation of the gonads (Chapter 3).

There is an intimate relationship between population sex ratio and mechanisms of sex determination (Charnov, 1982; Bull, 1983). The sex ratio of a population is relevant to the effective population size, the structure of mating systems and the evolution of sexually selected traits (Shuster & Wade, 2003; Mank & Avise, 2006). The usual evolutionarily stable state in a gonochoristic species is a balanced sex ratio of 1:1 (Fisher, 1930). This is a consequence of frequency-dependent selection, whereby the fitness of a genotype is a function of the frequency of that genotype in the population. In the case of sex ratio, the sex that is in the minority has a fitness advantage over the sex that is in the majority; consequently, the former will tend to increase in frequency relative to the latter. The outcome is that the population converges on a balanced sex ratio (Charnov, 1982; Bull, 1983). Experimental studies on *M. menidia* have demonstrated this process (see 'Environmental sex determination in teleosts').

Heterogametic systems XX–XY or ZW–ZZ achieve this balanced sex ratio through the usual Mendelian process. Heterogamety reduces the impact of environmental influences on the sex ratio. Even if there are wide, unpredictable variations in the environment, the genetic system will still generate a balanced primary sex ratio.

Sex chromosomes will tend to evolve from a state of autosomal control of sex determination, because it is advantageous to prevent crossing-over between the female-determining and male-determining genetic factors. This lack of crossing-over ensures that the X and Y (or Z and W) chromosomes remain genetically distinct, at least over the portion of the chromosome that carries *SEX*, and an unambiguous message initiates sexual differentiation as either a male or a female (Charlesworth, 1991, 2002). A second process may then contribute to the lack of crossing-over and increase the proportion of the sex chromosomes that do not recombine. Sexually antagonistic genes have sex-dependent effects on fitness, with the genes advantageous to one sex, but disadvantageous to the other. Selection favours a close linkage between such genes and the sex-determining genes, and an extension of the non-recombining region of the sex chromosome Y (or W) (Charlesworth, 1991, 2002; Charlesworth *et al.*, 2005). An example of sexually antagonistic genes is provided by *P. reticulata*, in which females are drab but the males are brightly coloured. The colours of the males are important in the competition for sexual partners. However, they are disadvantageous for the females, because they make them visible to predators that hunt by sight (Lindholm & Breden, 2002; Magurran, 2005). A review of sexually selected traits in *P. reticulata* showed that only two were autosomal, with 16 Y-linked, 24 recombining between the X and Y chromosomes, and two X-linked (Lindholm & Breden, 2002).

A disadvantage of the reduced crossing-over between sex chromosomes is that in the heterogametic chromosome (Y or W), the genetic material not directly related to sex traits, but in the region over which crossing-over is suppressed, undergoes deterioration as a consequence of several processes, including Müller's ratchet, hitchhiking genes, accumulation of transposable elements and repetitive DNA. In *G. aculeatus*, there is evidence of suppression of crossing-over along a region of the Y chromosome near *SEX* (Peichel *et al.*, 2004; Charlesworth *et al.*, 2005). The identification of *SEX* in this species has been hampered by the repetitive DNA that has accumulated in the region of *SEX*.

Thus, factors involved in the evolution of heterogamety include the need to stabilise the population at a balanced sex ratio, the insulation of the sex-determining mechanism from environmental disturbances and selection for sexually antagonistic genes.

There are conditions that favour the evolution of ESD, rather than GSD (Charnov, 1982; Bull, 1983; Uller *et al.*, 2007). The first is that the fitness of a sex is a function of the environment during sexual differentiation. Some environmental conditions during differentiation favour males, while other environmental conditions favour females. A second condition is that mating is random with respect to these environmental conditions. A third condition is that parents and offspring have limited control over the environmental conditions the offspring will experience during sexual differentiation. The best-described example of ESD in teleosts is *M. menidia* (see 'Environmental sex determination in teleosts'). A disadvantage of ESD is that if the environmental conditions vary too much, the system generates large variations in sex ratio, either spatially or temporally, while GSD is resistant to such environmental fluctuations.

In a pure ESD system, it is assumed that there is some threshold operating in the sex-differentiation pathway. Below the threshold, the individual develops unambiguously as one sex, above the threshold as the other sex (Bull, 1983). If there is no genetic variation in the threshold, there should be a sharp division, with one set of environmental conditions producing one sex and another set producing the other sex. However, if there is an interaction between environmental conditions and genotype, the position of the threshold will vary depending on the genotype of the individual. In *M. menidia*, there is good evidence of a genotype × temperature interaction in the sex determination of some populations (see 'Environmental sex determination in teleosts').

In a survey of several species that showed temperature-dependent ESD, there was a positive, linear relationship between the percentage of males and temperature (Ospina-Álvarez & Piferrer, 2008), rather than a clear threshold effect, which suggests that the mechanism of ESD in teleosts is not yet fully understood.

For gonochoristic species, there should be strong selection for stable sex-determining mechanisms to ensure the stability of the male–female dichotomy. Interestingly, both theoretical and empirical studies suggest that the transition between GSD and ESD can be rapid, as can transitions between male and female heterogamety (van Dooren & Leimar, 2003; Uller *et al.*, 2007; Baroiller *et al.*, 2009).

When interpreting the patterns of sex determination in teleosts, several dimensions have to be considered. The first is that the patterns are constrained by the dynamics of Mendelian genetic systems, as is common to other animals. A second dimension is the phylogenetic inheritance of sex-determining mechanisms. In birds and mammals, the uniformity of the GSD reflects an event early in the evolutionary history of the two clades, generating the XX–XY system in mammals and the ZW–ZZ system in birds. A third factor is the environmental conditions during which sexual differentiation takes place. A fourth factor is the intensity of sexual selection and the prevalence of sexually antagonistic genes. However, the analysis of sex determination in teleosts has yet to develop a coherent theoretical framework.

In contrast to endothermic vertebrates, the genomic architecture of the teleosts has had a dynamic evolutionary history (Mank & Avise, 2006). Although, the frequency distribution of chromosome numbers in teleost species shows a sharp peak at a diploid number of chromosomes of 48 or 50, there is evidence that in addition to the gene duplication at the phylogenetic base of the Teleostei, duplications have also occurred at the base of several clades within the teleosts, e.g. the family Salmonidae (Mank & Avise, 2006). The high rates of genomic change are correlated with the diversity of sex determination mechanisms. One possibility is that the high rates may permit a rapid adaptive change in sex determination mechanisms as the ecological or the social environment changes. This hypothesis suggests that sex-determining mechanisms in the teleosts are much more responsive to selection than those of birds and mammals. One possible approach to investigating the adaptive significance of sex-determining mechanisms would be to map, as data become available, sex-determining mechanisms onto the three-dimensional model of teleostean life-history traits (Winemiller & Rose, 1992) described in Chapter 11.

The phylogenetic analysis of sex-determining mechanisms is hampered by a lack of information on the mechanisms in many teleost species and the provisional nature of models of the phylogeny of the teleosts. However, broad outlines can be sketched (Mank *et al.*, 2006a; see Fig. 2.3). Even at this broad level, the survey revealed a striking diversity of mechanisms. Fifteen orders of teleosts contained gonochoristic species with sex chromosomes and no known hermaphrodites. Nine orders had both gonochorists with sex chromosomes and hermaphroditic species. Gonochoristic species with XX–XY male heterogamety occurred in 22 orders, but only two orders had exclusively ZW–ZZ female heterogamety. However, eight orders included both XX–XY and ZW–ZZ heterogamety. Seven orders had species with ESD, but six of these also included species with chromosomal sex determination and the number of species with ESD may have been over-estimated (Ospina-Álvarez & Piferrer, 2008). The mechanisms of sex determination of the ancestral teleosts are not known. Heterogametic sex chromosomes probably evolved from autosomal sex determination (Ohno, 1967; Charlesworth, 1991), but ESD or hermaphroditism also provide potential starting conditions for the evolution of GSD and heterogamety.

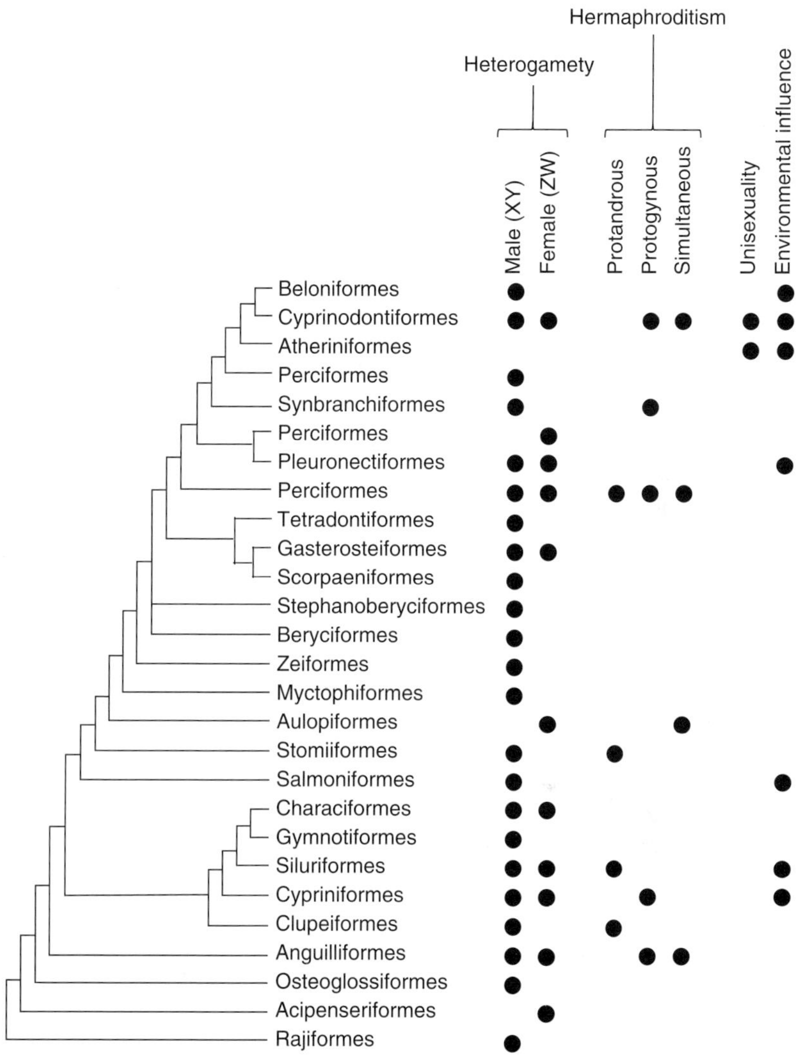

Figure 2.3 Phylogenetic distribution of modes of reproduction in the teleosts. (Source: Mank *et al.* 2006a. Reproduced with permission of John Wiley & Sons.)

Estimates of the proportion of teleostean species with sex chromosomes have ranges from 10% to 50%. As methods of identifying sex chromosomes improve, it is likely that the higher percentage will be approached. There seem to be no equivalent estimates for the percentage of species with autosomal or ESD mechanisms. Even within teleostean clades, the same picture of diversity emerges. The clades that have been studied so far are similar in that they all show a diversity of mechanisms. Several clades have a mixture of XX–XY and ZW–ZZ heterogamety and some also have ESD (see examples earlier in the chapter), suggesting rapid selection of adaptive mechanisms of sex determination. These phylogenetic analyses indicated that heterogamety has evolved independently in several teleostean clades. One possibility is that there are a small number of ancestral sex-determining genes that re-emerge in different clades. The genome duplications that characterise the teleosts mean that there will be several genes derived from these ancestral genes available for selection. Another possibility is that there are only a restricted number of genes that can function as initiators of sexual differentiation and these repeatedly emerge in clades (Graves & Peichel, 2010). If and when the identity of *SEX* is established in several teleostean species, a better understanding of the evolutionary history of sex determination should become apparent.

A factor in such selection is likely to be the mating system and intensity of selection for secondary sexual traits (Chapter 8). The outcome might be selection for sexually antagonistic genes leading to reductions in crossing-over and the evolution of heterogamety, with eventually morphologically distinct sex chromosomes. Thus the genetic architecture of sex determination has implications for sexual conflict and the evolution of sexually selected traits (Arnqvist & Rowe, 2005). A preliminary analysis found no correlation between sex chromosome systems XX–XY or ZW–ZZ and well-developed male secondary sexual traits (Mank *et al.*, 2006b). However, there is a need for a study that covers a wide range of secondary sexual traits. It is probably not coincidental that species with weak or nonexistent secondary sexual traits, such as *D. rerio* and *D. labrax*, also have sex determined by an interaction between autosomes and environmental conditions.

The diversity of sex-determining mechanisms of the teleosts will provide an important testing ground for theories of the evolution of sex-determining mechanisms in animals in general (Kikuchi & Hamaguchi, 2013). In the teleosts, the diversity is coupled with a developmental flexibility displayed in the differentiation of the gonads (Chapter 3), which the aquaculture industry uses to its advantage with the production of monosex cohorts by the treatment of the developing zygote by masculinising or feminising hormones (Chapter 12).

Chapter 3

Sex differentiation

Introduction

In gonochoristic species, whatever the mechanism of sex determination (Chapter 2), sex differentiation is the ontogenetic process by which potentially functioning males and females are generated by a cascade of gene regulations that result in either ovaries or testes (Piferrer & Guiguen, 2008). In the gonads, the process of gametogenesis generates the haploid gametes, eggs in the ovaries and spermatozoa in the testes, from diploid precursors. These gametes are the only vehicles carrying parental genes into the offspring generation, with a haploid spermatozoa from the male fertilising a haploid egg from the female, to form the diploid zygote of the next generation. This is the crucial biological event and all the biochemical, physiological and behavioural mechanisms associated with reproduction have evolved to achieve this successfully, although in some groups, additional parental resources are devoted to protecting that zygote through its early life-history stages (Chapter 9). In vertebrates, the process of gametogenesis proceeds according to a relatively uniform plan (Blüm, 1986). Examples of the major patterns are described here, but there are still substantial gaps in our knowledge of these important processes.

Reproductive Biology of Teleost Fishes, First Edition. Robert J. Wootton and Carl Smith.

Embryology of the gonads

In comparison with other jawed vertebrates (Teleostomi), the teleosts display some unique developmental characteristics. The functional and adaptive significance of these characteristics remains obscure. For example, in teleosts and some more basal ray-finned fishes, the forebrain develops by a process of eversion, rather than the inversion that occurs in the developing brain of other jawed vertebrates (Romer & Parsons, 1986). In the development of the gonads, the teleosts also differ in significant ways from other jawed vertebrates.

The earliest appearance of the gonads in the developing vertebrate embryo is as two longitudinal epithelial ridges in the dorsal wall of the coelomic cavity, one ridge on either side of the embryonic gut (Blüm, 1986). The pair of genital ridges form the gonadal anlagen or gonadal primordia, the initial stage of what will eventually differentiate into functional male or female gonads. As the paired gonadal anlagen develop, they come to hang in the coelom from its dorsal wall, suspended by mesenteries. The mesentery supporting a testis is its mesorchium, while that supporting an ovary is its mesovarium (Blüm, 1986).

The gonads of vertebrates have two cellular components, the somatic cells and the germ cells. It is the latter that give rise to the gametes. In most lineages of jawed vertebrates, the somatic cells have two embryological origins. One population originates from the cells of the walls of the peritoneum, lining the coelomic cavity. This population forms the cortex of the developing gonad. The second population, which forms the medulla of the gonad, is derived from mesonephric blastema, a feature of the developing kidney. In females, the medulla component of the gonad degenerates, while the cortical component expands and gives rise to the ovaries. In males, the development of the testes involves the medulla, while the cortex degenerates (Atz, 1964; Merchant-Larios, 1978; Nakamura *et al.*, 1998). In terms of the embryological tissues involved, the cortex has a mesodermal origin, while the medulla has an endodermal origin (Francis, 1992). It has been argued that the position of cells, cortical or medullary, determines whether they develop as cells characteristic of males or females. However, recent evidence from studies on the mouse (*Mus musculus*) suggests that the outcome is determined by each cell rather than by the position of the cells (Maatouk & Capel, 2008).

In teleosts, the gonadal somatic cells are derived only from the peritoneal cells; both ovaries and testes develop from the gonadal cortex and hence have a mesodermal origin. The somatic tissues of ovaries and testes of teleosts have a common embryological origin (Nakamura *et al.*, 1998). There is a correlation between this distinctive mode of development of the gonads in teleosts and the relative sensitivity of the developing gonad to factors that cause changes in sex of the differentiating gonad, e.g. the presence of exogenous hormones (Francis, 1992; see Chapter 5). The evidence suggests that teleosts tend to show a protogynous pattern of sexual development, even in gonochoristic species (Francis, 1992). In some teleostean lineages, even males pass through a stage in which the gonad develops as a proto-ovary, before switching to differentiate as testis. The position of the cells in the gonadal anlagen is not a factor in their pattern of differentiation.

The second cellular component of the gonad is the germ cells. As in all jawed vertebrates, the germ cells are initially identified as primordial germ cells (PGCs), which can be first identified at sites far from the site of the primordial gonad (Yamaha *et al.*, 2010). The PGCs migrate to the gonadal anlagen, where they encounter the somatic cells. At this stage, both the PGCs and the somatic cells are bipotential and capable of developing as male-characteristic or female-characteristic cells. The process of differentiation determines which of these fates the cells experience, with the process depending on cross-talk between the somatic cells and PGCs, which communicate by chemical signals.

Origin and migration of primordial germ cells

Primordial germ cells are unique, as they are the only embryonic cells with the potential to transmit genetic information from the parental to the offspring generation (Yamaha *et al.*, 2010). Histologically, PGCs differ from gonadal somatic cells, with a rounder shape, distinct outline and a larger nucleus (Devlin & Nagahama, 2002; Kondo *et al.*, 2009; see Fig. 3.1). The PGCs also contain distinctive granular cytoplasmic inclusions called nuage. These features were used in attempts to describe the source of the PGCs and the route by which they reached the gonadal anlagen. Such attempts have been greatly enhanced by the subsequent use of gene markers that are characteristic of the PGCs. A key gene has been *vasa* (*vas*), but other genes, including *daz-like* (*dazl*), *nanos1* (*nos1*) and *dead end* (*dnd*) and their mRNA transcripts, have also been used as marker genes for PGCs (Yamaha *et al.*, 2010). Much of the research on this subject has used the medaka (*Oryzias latipes*) or zebrafish (*Danio rerio*), because of their effectiveness as model species for studying the genetic basis of development.

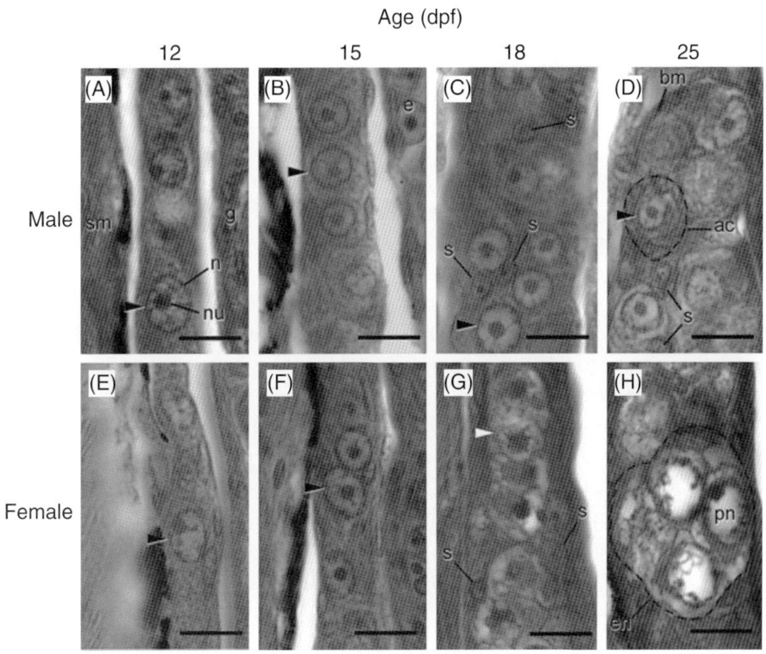

Figure 3.1 Initial gonadal differentiation illustrated by the gonads of the three-spined stickleback (*Gasterosteus aculeatus*). Longitudinal sections taken between 12 and 25 days post-fertilisation (dpf). (A–D) Differentiation of testis; (E–H) differentiation of ovary. ac, spermatogenetic cysts; bm, basement membrane; e, erythrocyte; en, egg nest; g, gut; n, nucleus; nu, nucleolus; pn, perinucleolar oocyte; s, somatic cell; sm, skeletal muscle. (Source: Lewis *et al.* 2008. Reproduced with permission of John Wiley & Sons.)

The use of genetic markers has suggested that cells destined to become PGCs are predetermined by maternal factors in the egg (Kondo *et al.*, 2009; Lubzens *et al.*, 2010; Yamaha *et al.*, 2010). The transcripts of genes such as *vasa* that characterise the PGCs are not distributed at random between the cells, as the egg and its daughter cells divide to produce the multicellular zygote. In *D. rerio*, the *vasa* transcripts, originally in the egg, aggregate at the terminal regions of the early cleavage furrows, formed where the cells are dividing (Yamaha *et al.*, 2010). As the dividing cells form the blastodisc, initially *vasa, nanos1* and *dead end* transcripts aggregate around the animal pole of the egg. These aggregates then move from the animal pole to a peripheral section of the blastoderm. At the 32-cell stage, the aggregates are localised in four cells (blastomeres). At subsequent divisions of these four cells, the aggregates are passed to only one of the two daughter cells. This asymmetric division of the aggregate establishes the germ line lineage. In *D. rerio*, the nuage structure contains the *vasa* transcript (Yamaha *et al.*, 2010). In other species, the pattern is different. The mRNA transcripts of *vasa* and *nanos1* in *O. latipes* are evenly distributed until the embryonic shield stage. At the four-cell stage in the development of *O. latipes*, nuage material is in the area of cytoplasm where the four cells are in contact, again suggesting that PGCs are determined by cytoplasmic factors in the egg (Kondo *et al.*, 2009). The adaptive and functional consequences of the interspecific differences in the distribution of the transcripts are not known. Too few species have been studied to reveal any phylogenetic patterns.

In *D. rerio*, the PGCs are specified about 3 hours post-fertilisation (Blaser *et al.*, 2005). The number of PGCs containing *vasa* transcripts increases only slowly (Lubzens *et al.*, 2010). At the 32-cell stage of the embryo, four cells contain the transcripts. When the total number of cells in the embryo reaches about 4000, four to 12 cells contain the transcript, this number increasing to 16–25 cells at the embryonic shield stage. These cells are in four clusters close to the margin of the blastodisc. At this stage, the PGCs have a simple morphology and show no evidence of migration. But shortly afterwards, the PGCs show morphological changes, with both small and elongated protrusions, but they are still not migratory. Then, starting about 4.5 hours post-fertilisation, a few cells become polarised in shape and start their migration. Over the next 2 hours, the proportion of PGCs showing migratory behaviour increases, until over 90% are actively migrating (Blaser *et al.*, 2005). About 25–30 PGCs reach the gonadal anlagen, representing two to three mitotic divisions of the PGCs before their migration and arrival at the gonadal anlagen.

Studies on *D. rerio* provide evidence as to the role of genes in these events (Schulz *et al.*, 2010). If the *dead end* gene is not functional, the PGCs fail to migrate to the gonadal anlagen. Loss-of-function mutations in the genes *ziwi* and *zili* result in the complete loss of germ cells later in the differentiation of the gonads. The genes code for proteins in the *Argonaut* family of proteins. These proteins are involved in the metabolism of small non-coding RNAs specific to the germ line.

Based on studies of *O. latipes* using transcripts of *vasa* and *nanos1*, the migration of the PGCs can be divided into three phases (Kondo *et al.*, 2009). Phase I is characterised by the migration of PGCs from the animal pole of the developing egg to the marginal zone of the embryonic shield. In phase II, PGCs move from the periphery towards the embryonic body at the time when the central nervous system and the somites are developing (neurulation and somitogenesis). During phase III, the PGCs migrate to the lateral plate mesoderm and start to colonise the site of the gonadal anlagen. At the same time, somatic cells from the lateral plate mesoderm on both sides of the embryonic axis are also moving into the gonadal anlagen, resulting in the paired gonadal primordia, with the PGCs surrounded by somatic cells. The migration of PGCs in *D. rerio* follows essentially the same pattern, although six phases have been identified (Yamaha *et al.*, 2010).

During their migration, the PGCs are guided by molecular factors. The chemokines SDF1a and SDF1b from somatic cells exert their guiding influence in the early and late phases of migration respectively, in *O. latipes* (Kondo *et al.*, 2009). The PGCs have two receptors for the chemokines, CXCR4b and CXCR7. The role of CXCR4b is to guide the PGC, whereas the role of CXCR7 is to remove the chemokine and so enhance the directionality of the movement (Lubzens *et al.*, 2010). During the migration, the PGCs'fate as the precursors of the germ cells is confirmed. However, at this stage, they are still bipotential, having not yet encountered the factors that will determine whether they go on to produce eggs or spermatozoa. In addition to active movement of the PGCs, their movement to the gonadal anlagen is also influenced by the morphogenetic movements that characterise the embryonic development (Devlin & Nagahama, 2002).

All vertebrates have the characteristic that the PGCs originate at locations distant from the site where the gonads develop. Consequently, the PGCs must move over relatively large distances to reach the site where they will undergo gametogenesis. It is not clear why in vertebrates the PGCs cannot be determined close to, or at, the site of gonadal anlagen. It is possible that the initial peripheral location of the PGCs protects them from cell movements and chemical signalling in the early stages of embryogenesis.

Sex differentiation in the gonadal anlagen

Differentiation of the gonadal anlagen into definitive ovaries or testes provides an example of how the totipotential cells of the early embryo give rise to a variety of functional organs. When the PGCs and somatic cells first meet in the gonadal anlagen, both populations have the potential to develop into either female or male characteristic cells. However, in the process of oogenesis, the PGCs in the female develop into haploid oocytes that become yolk-filled eggs, while the somatic cells produce the granulosa and thecal layers that surround a developing oocyte, with the three cellular components comprising an ovarian follicle complex. The granulosa and thecal cells are important in the endocrine control of the development of the oocyte (see Chapter 5). In males, the PGCs develop into small, motile spermatozoa. During most of the process of spermatogenesis, the germ cells are contained in cysts, enveloped by Sertoli cells, derived from the somatic population. In males, the somatic population also gives rise to the Leydig cells, which produce steroid hormones characteristic of the male (Chapter 5).

The cell movements bring the PGCs and somatic cells to the site of the developing gonads and so set the stage for the process of sexual differentiation. Two processes are involved in the differentiation of the gonadal cells. Autonomous changes occur in cells, independent of the environment in which they occur, whereas other changes require information provided by the environment external to the cell, usually information being sent by other cells.

The first histological sign of sex differentiation is that the PGCs in gonads destined to become ovaries proliferate at a faster rate than PGCs in the gonads destined to be testes. Although similar numbers of PGCs arrive in the gonadal anlagen, the rate of mitosis of PGCs in gonads destined to become ovaries is faster than the mitotic rate in gonads destined to become testes. In the three-spined stickleback (*Gasterosteus aculeatus*, Gasterosteidae), sex is genetically determined, with male heterogamety (see Chapter 2). A sex-linked marker was used to identify the sex of larval and juvenile *G. aculeatus* and the number of PGCs present in the gonadal anlagen recorded at ages ranging from 11 days post-fertilisation (dpf) to 26 dpf (Lewis *et al.*, 2008). Even at 11 and 12 dpf, females had more than double the number of PGCs than males, with mean numbers of approximately 160 PGCs in females and 60 in males. In the females, by 15 dpf, the rapid proliferation of the PGCs was accompanied by high rates of cell death (apoptosis). By contrast, the rate of apoptosis was low in male *G. aculeatus* (Fig. 3.1). This differential rate of proliferation of PGCs in males and females has been inferred for other teleosts. In *O. latipes*, also with genetic sex determination and male heterogamety, at 1 day post hatching (dph), the female gonads contain more PGCs than do male gonads (Kondo *et al.*, 2009). In males, the PGCs cease to divide after hatching, with the gonads containing about 100 PGCs and no evidence of meiotic cells. By contrast, in female gonads, the PGCs continue to divide and some become meiotic cells in the first prophase of the reduction division that will eventually produce the haploid gametes.

In *D. rerio* and *O. latipes*, individuals in which the proliferation of PGCs is blocked develop as phenotypic males in terms of the morphology of the gonads, general morphology and behaviour (Slanchev *et al.*, 2005; Siegfried & Nüsslein-Volhard, 2008). The results suggest that in such species, there is a threshold number of PGCs required for the gonads to develop as ovaries. However, this effect of PGCs may not be general for teleosts. In the pond loach (*Misgurnus anguillicaudatus*, Cobitidae), genetically female individuals deficient in PGCs still developed as phenotypic females, and developed ovary-like gonads, but lacked the typical ovarian follicles (Fujimoto *et al.*, 2010). *M. anguillicaudatus* belongs to the same order as *D. rerio*, the Cypriniformes, but is a member of the family Cobitidae rather than the Cyprinidae.

Initial differentiation of the gonads

Two modes of differentiation of the gonads have been defined for teleosts (Devlin & Nagahama, 2002). In differentiated, gonochoristic species, the initial differentiation is from the gonadal anlagen directly to ovaries or testes. Species that follow this pattern come from a variety of teleostean clades, including sockeye salmon (*Oncorhynchus nerka*, Salmonidae), common bream (*Abramis brama*) and common carp (*Cyprinus carpio*), both cyprinids, and the gasterosteid, *G. aculeatus* (Fig. 3.1). There are no obvious phylogenetic or ecological characteristics that link these species. In undifferentiated gonochorists, the gonadal anlagen develop initially as ovaries, but in approximately half the population, the ovarian tissue subsequently degenerates and masculinisation of the gonad occurs, resulting in functional testes. This pattern occurs in the cyprinids *D. rerio* and Sumatra barb (*Puntigrus tetrazona*, Cyprinidae). *D. rerio* lacks sex chromosomes and sex seems to be determined by autosomal genes and environmental factors (Chapter 2). The postponement of the definitive sex determination in undifferentiated gonochorists may allow time for the sex ratio to be adjusted to the prevailing environmental or social conditions.

The pattern in *D. rerio* illustrates the undifferentiated mode of gonadal differentiation (Maack & Segner, 2003). At 26 °C and 4 weeks post-fertilisation (wpf), indifferent gonads contain PGCs whose morphological features resemble oogonia (type 1), the initial stage in the process of oogenesis, and cells whose morphological features correspond to the early meiotic stage of primary oocytes (type 2). Ovaries contain both type 1 and type 2 cells, but also oocytes in arrested prophase (Type 3). Over the next 2 weeks, ovarian differentiation continues, with a population of densely packed oocytes. After about 6 wpf, in a proportion of the fish (the factors that determine the proportion are not clear), the size and number of oocytes decrease and oocytes become isolated from each other and show signs of degeneration. The number of somatic cells also increases. Over the following weeks the oocytes disappear and groups of PGCs appear arranged

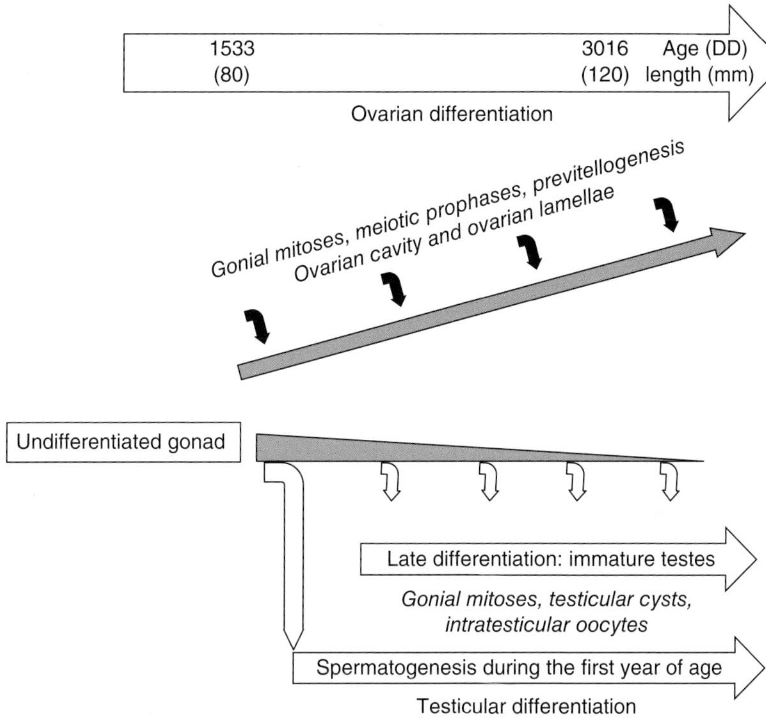

Figure 3.2 Schematic of the gonadal differentiation in the European seabass (*Dicentrarchus labrax*) illustrating the delayed differentiation. Age of fish is indicated by length in mm and by degree-days (DD; growth is assumed to be zero at 10 °C). (Source: Saillant *et al.* 2003. Reproduced with permission of John Wiley & Sons.)

in cyst-like clusters. After 8 wpf, in addition to the germ cells arranged in cysts, there is evidence of active spermatogenesis, with spermatocytes and spermatids present (see later). By 10 wpf, individual fish have differentiated either ovaries or testes.

A variant of this pattern occurs in the European seabass (*Dicentrarchus labrax*, Moronidae) (Saillant *et al.*, 2003; Piferrer *et al.*, 2005; see also Fig. 3.2). In this species, the gonad remains in an undifferentiated state for over 100 dpf, by which time individuals have reached a standard length (distance from the tip of the snout to the origin of the tail fin) of about 80 mm. One study detected the first signs of ovarian differentiation at about 170 dpf. In a small proportion of individuals, the gonads showed signs of differentiating as testes at a similar time. However, the timing of the initiation of testes varied widely among individuals, with some individuals not showing signs of testis differentiation as late as 250 dpf. A proportion of fish differentiating as males had testes that contained identifiable oocytes, these males showing a phase of juvenile intersexuality. The proportion of males with testes containing oocytes remained relatively constant until the end of testicular differentiation, but then declined.

A third pattern of differentiation in teleosts occurs in secondary gonochorists in which the gonads are initially in an intersexual condition prior to differentiating as either ovaries or testes (Devlin & Nagahama, 2002). Such species seem to be developing as functional hermaphrodites before they undergo a definitive differentiation as males or females. Santer seabream (*Cheimerius nufar*, Serranidae) and Nassau grouper (*Epinephelus striatus*, Serranidae) are secondary gonochorists.

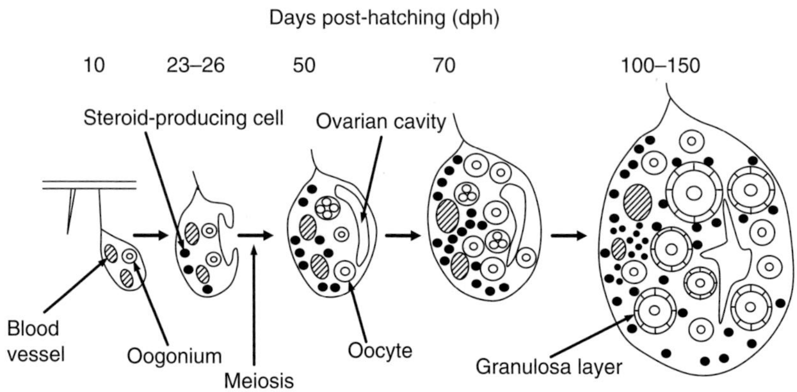

Figure 3.3 Schematic of early ovarian differentiation in the Nile tilapia (*Oreochromis niloticus*). (Source: Nakamura *et al.* 1998. Reproduced with permission of John Wiley & Sons.)

Initial differentiation of ovaries

The initial stage of the differentiation of the ovaries is the transformation of PGCs into oogonia, the diploid cells that initiate the meiotic divisions that will generate the haploid oocytes (Nakamura *et al.*, 1998; Devlin & Nagahama, 2002; Lubzens *et al.*, 2010). Structural changes in the PGCs lead to the formation of oogonia. These then proliferate by mitotic divisions and form oogonial (egg) nests. Within a nest, the oogonia are connected by cytoplasmic links and each nest is surrounded by a thin layer of somatic cells, which are the precursors of the granulosa cells. There are also undifferentiated PGCs, distinguishable from the cells in the nests and isolated from each other by somatic cells (Fig. 3.3). These undifferentiated PGCs provide a reservoir of germ cells, which give rise to successive populations of oogonia and oocytes when the female becomes sexually mature. The early stages of ovarian differentiation take place faster than the differentiation of the testes. In the short-lived *O. latipes*, the formation of oogonial nests starts about 5–6 dpf, but the differentiation of the somatic cells takes place later at about 1–2 wpf. The further progress of oogenesis is described in detail in Chapter 4.

Initial differentiation of testes

In teleosts, the differentiation of the testes tends to lag behind that of the ovaries. For example in *O. latipes*, in genetic females, the PGCs continue to proliferate and differentiate into oogonia from hatching, whereas in genetic males, PGCs cease to divide at hatching and only after about 15 dph does the mitotic rate of the PGCs increase so the number reaches about 200 by 20 dph (Kondo *et al.*, 2009). In the developing testes of *O. latipes*, differentiation of the somatic cells is detectable three to four wpf and spermatogenesis begins with the resumption of mitosis of spermatogonia at five to six wpf (Saito & Tanaka, 2009).

The first phase of spermatogenesis is defined by proliferation and differentiation of the diploid spermatogonia (Schulz *et al.*, 2010; Fig. 3.4). Undifferentiated spermatogonia (type A_{und}) produce differentiated type A spermatogonia (A_{diff}), which have a reduced capacity for increase by mitotic division. There then follows an irreversible commitment

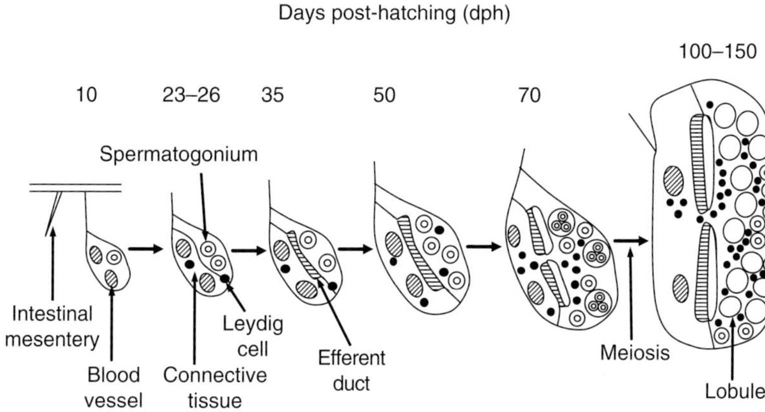

Figure 3.4 Schematic of early testicular differentiation in the Nile tilapia (*Oreochromis niloticus*). (Source: Nakamura *et al.* 1998. Reproduced with permission of John Wiley & Sons.)

to differentiation as male cells when the A_{diff} spermatogonia give rise to the B-type spermatogonia, which divide more rapidly. In the transitions from A_{und} to B-type spermatogonia, there are morphological changes in the cells (Schulz *et al.*, 2010). In the tilapias (*Oreochromis* spp.), there is a decrease in cell volume, but an increase in nuclear size. The A_{und} cells still contain the nuage material characteristic of the PGCs, but A_{diff} spermatogonia contain little or no nuage material.

The number of generations of B-type spermatogonia varies between species and may be genetically determined. In the guppy (*Poecilia reticulata*, Poeciliidae), there are 14 generations but in *D. rerio* there are nine. When a spermatogonium divides, the daughter cells maintain cytoplasmic bridges. Consequently, a clone of spermatogonia is produced, which forms a spermatogenic cyst. After the appropriate number of mitotic divisions, the spermatogonia in a clone initiate meiosis and differentiate as primary spermatocytes, in the pre-leptotene phase of meiosis. Spematocytes do not proliferate, and the spermatocytes of the clone in a spermatogenic cyst differentiate in synchrony, eventually forming the motile spermatozoa (Chapter 4).

Sertoli cells are essential for the formation and development of the spermatogonial cysts (Schulz *et al.*, 2010). The Sertoli cells derive from the same gonadal, somatic cells that, in ovaries, give rise to the granulosa, and the way in which Sertoli cells envelop a spermatogonium at the start of the clonal divisions of the spermatogonia resembles the way in which granulosa cells envelop an oogonium. Sertoli cells determine the capacity of the testes to produce spermatozoa, act as a target for chemical signalling systems that regulate spermatogenesis and act as nurse cells for the differentiating clone. In this way, the Sertoli cells and the germ cells constitute the unit, the spermatogonial cyst, that is responsible for the production of the haploid spermatozoa.

An early morphological indication of testicular differentiation is the formation of the efferent ducts, which will eventually carry the spermatozoa to the gonoducts. The ducts appear as narrow spaces in the stomal tissue of gonadal anlagen (Nakamura *et al.*, 1998). The further process of spermatogenesis is described in Chapter 4.

A marked contrast between the patterns of differentiation in the ovaries and testes is the timing of germ cell proliferation. In the presumptive ovary, the rate of proliferation of PGCs exceeds that in the presumptive testis. However, once an oogonium starts to be surrounded by granulosa and thecal cells, no further proliferation takes place and the oogonium differentiates as an oocyte. By contrast, the spermatogonium enveloped by Sertoli cells undergoes several cycles of mitotic division, generating a clone of spermatogonia in the spermatogonial cyst. In this way, the number of spermatozoa produced in the testes far exceeds the number of eggs produced by the ovaries.

Genetic control of early gonadal differentiation

The cascades of gene activation that result in the unambiguous differentiation of either ovaries or testes in gonochoristic teleosts have not been fully mapped (Siegfried, 2010). Because there is no teleostean equivalent of the male-determining *Sry* gene of mammals, the initial stages of the process of sex differentiation are obscure (see Chapter 2). Consequently, any description of the genetic control in teleosts is provisional and will be subject to much revision. Although the factors that initiate sex determination in teleosts are diverse (see Chapter 2), the subsequent gene cascade controlling the gonadal differentiation may have similarities to the control in other vertebrates (Herpin & Schartl, 2011). The gene cascades are probably comparable to those that control the differentiation of gonads in the better studied mammals (Wilhelm *et al.*, 2007).

A key element in the differentiation of ovaries is the role of the enzyme P450 aromatase ($P450_{arom}$; van Hofsten & Olsson, 2005; Christiansen *et al.*, 2008; Guiguen *et al.*, 2010). Aromatase is a microsomal enzyme found in the smooth endoplasmic reticulum of steroidogenic cells. The enzyme catalyses the conversion of androgen to oestrogens, particularly oestradiol-β17 (E_2; Guiguen *et al.*, 2010). The balance between the levels of androgens and oestrogens in the differentiating gonad is probably crucial in determining whether the gonad develops as testes (androgens dominant) or ovaries (oestrogens dominant). Aromatase shifts the balance towards oestrogens. There are two genes that code for aromatase, *cyp19a1* and *cyp19b1*, which probably reflect the gene duplication that is characteristic of the teleost clade. The gene *cyp19a1* is characteristic of the gonads, whereas *cyp19b1* is characteristic of the brain. In the differentiating gonad, *cyp19a1* is expressed initially in the stromal cells, but later it is expressed in the granulosa cells of the follicle (Guiguen *et al.*, 2010). Thereafter, the oestrogens generated by aromatase help to control the expression of the genes whose products are required for the development of mature ovaries.

This conformity begs the question of how the expression of the *cyp19a1* gene is activated. A candidate is the protein product Foxl2 of the gene *foxl2* (Siegfried, 2010). The protein is a transcription factor, which belongs to the family of forkhead transcription factors that can bind to DNA. Foxl2 is important in the differentiation and development of the ovaries in mammals. The binding of Foxl2 to the appropriate stretch of DNA may activate expression of *cyp19a1* and hence the production of aromatase in stromal and granulosa cells.

In the Nile tilapia (*Oreochromis niloticus*, Cichlidae), expression of *foxl2* and *cyp19a1* was detected exclusively in the gonads of XX females at 5 dph (Kobayashi & Nagahama, 2009). In *O. latipes*, the timing of the expression of *foxl2* and *cyp19a1* is different, because *cyp19a1* expression was not detected until after oogenesis had started (Saito & Tanaka, 2009). It is likely that, at least in some species, the earliest stages of oogenesis are not dependent on the presence of oestrogens (Saito & Tanaka, 2009). Other factors may also be important in the expression of *cyp19a1* in the developing gonad. An example is the protein steroidogenic factor 1 (Sf1), which is coded by the gene *nr5a1*. This protein plays an important role in the expression of genes involved in the synthesis of steroids.

A candidate gene for directing the differentiation of the testis is *dmrt1*. This gene codes for the Doublesex- and Mab-3-related (DM) domain in the resultant protein, which acts as a transcription factor for other genes. Genes containing the DM domain have been implicated in the process of sex differentiation in both invertebrates and vertebrates (van Hofsten & Olsson, 2005; Ferguson-Smith, 2006; Herpin & Schartl, 2011). In *O. niloticus*, *dmrt1* expression was detected at 5 dph in XY males, before any histological indication of testis differentiation (Kobayashi & Nagahama, 2009). Expression occurred in gonadal somatic cells, both somatic cells around the PGCs and those in the interstitial tissue of the gonad. The expression of *dmrt1* is also restricted to males in other teleost species, including the North African catfish (*Clarias gariepinus*, Clariidae) and the rare gudgeon (*Gobiocypris rarus*, Cyprinidae) (Herpin & Schartl, 2011). In *D. rerio*, *dmrt1* is expressed in the testes during the differentiation from the juvenile's ovary condition. In some species, including *D. rerio* and Atlantic cod (*Gadus morhua*, Gadidae), *dmrt1* expression does occur in the ovaries, but only in the germ cells and not in gonadal somatic cells, as occurs in testes. However, in *O. latipes*, expression of *dmrt1* does not occur until late in testis differentiation (Saito & Tanaka, 2009). In this species, the closely related gene, *dmrt1bY*, is the male-sex determining gene (see Chapter 2). Expression of *dmrt1bY* was detected from the embryonic stages through to sex differentiation, i.e. well before the expression of *dmrt1*. This temporal pattern suggests that *dmrt1bY* has taken over some of the functions of *dmrt1* in the earlier phases of testis differentiation. The product of *dmrt1bY* may also be responsible for the inhibition of mitotic increase of PGCs that characterises the early development of the testes.

A second gene that is probably important in the differentiation of testis is *amh*, which codes for the anti-Müllerian hormone (AMH; Siegfried, 2010). AMH is a glycoprotein and a member of the tissue growth factor family (TGF). In most of the jawed vertebrate clades, one of the functions of AMH is to suppress the development of the Müllerian ducts, which eventually differentiate as oviducts in the female. This function is not relevant in the teleost clade, because the oviducts do not differentiate from the Müllerian ducts (Chapter 4). However, AMH is implicated in testis differentiation in teleosts (van Hofsten & Olsson, 2005; Siegfried, 2010). As with *dmrt1*, the temporal pattern of expression of *amh* differs between teleostean species. AMH may have a role in inhibiting the mitotic division of PGCs, thus contributing to the differential rates of proliferation of PGCs in differentiating ovaries and testes.

In *O. niloticus*, from about 5–10 dph, both XX females and XY males show a low expression of *amh* (Kobayashi & Nagahama, 2009). After 15 dph, levels of expression of *amh* increase in the XY males but remain low in the XX females. This sexual dimorphism in the expression of *amh* occurs after the sexual dimorphism in the rate of proliferation of PGCs, but before morphological differentiation in the somatic cells of the gonad. The gene is expressed in all individuals during the juvenile ovary stage seen in *D. rerio*, but as sex differentiation occurs, the expression of *amh* becomes restricted to the developing testes. However, in *O. latipes*, *amh* does not show a sexually dimorphic pattern of expression in association with sex differentiation (Siegfried, 2010).

The *sox9* gene encodes a transcription factor (van Hofsten & Olsson, 2005; Siegfried, 2010). This gene is of interest because it is related to the male-determining gene, *Sry*, in mammals. *Sox9* is also associated with the differentiation of testis in some invertebrates. The temporal pattern of *sox9* expression in *O. niloticus* and *O. latipes* suggests that it is not directly involved in the initiation of sex differentiation, but is important in the morphogenesis of the testes (Chapter 4).

A comparison with mammals, which have been better studied, makes it probable that other genes are implicated in the control of differentiation of the gonads of teleosts.

Synthesis

Preliminary models of the dichotomous gonadal differentiation in gonochoristic teleosts focus on possible inhibitory relationships between male and female sex-determining factors (Piferrer & Guiguen, 2008; Guiguen *et al.*, 2010; Herpin & Schartl, 2011). The crux of these models is the role of aromatase. If aromatase activity is low, androgens are not converted to oestrogens and the balance between androgens and oestrogens favours testis differentiation. With high activity of aromatase, the balance is shifted in favour of oestrogens, particularly E_2, and ovarian differentiation proceeds. Interestingly, in studies in which high temperatures have induced sex ratios strongly biased towards males, even in species with a genetic basis for sex determination, the high temperatures have led to an inhibition of aromatase activity (Fernando *et al.*, 2008). One possibility is that the expression of *cyp19a1* is inhibited in the presence of the products of a gene such as *dmrt1* or *amh*. A second possibility is that the expression of genes such as *dmrt1* and *amh* is inhibited by the expression of genes such as *foxl2*. A third possibility is that there is a mutually inhibitory relationship between male and female determining factors. The effectiveness of the system is indicated by the low frequency of intersexes that are detected in natural populations of gonochoristic teleosts.

A methodological problem that constrains studies of gonad differentiation in teleosts is the small size of the embryos and larvae. This makes it technically difficult to measure levels of components such as growth factors and hormones. It also makes it difficult to localise where factors are having their effects. An important question is at what stage in gonadal differentiation and in which cells are androgens first produced so that the action of aromatase in converting androgen to oestrogen can begin. The small number of teleost species studied in detail makes it difficult to identify any phylogenetic and adaptive patterns in the temporal patterns of gonad differentiation.

A comparison of the sex differentiation of *O. latipes* (differentiated gonochorist) and *D. rerio* (undifferentiated gonochorist) identified possible commonalities in the pattern of gonadal differentiation, even though the detailed temporal pattern differs (Saito & Tanaka, 2009). These commonalities include the similar sequence in the major developments in the germ and somatic cells. In both species, oogenesis is the earliest indication of gonadal differentiation (as it is in *D. labrax*). This oogenesis consists of consecutive mitotic divisions of the germ cells, which form nests of oogonia, with the oogonia then entering early meiosis as they transform into oocytes. These events occur before the expression of aromatase is detected and before there is evidence of the differentiation of gonadal somatic cells, both of which occur a few weeks later. In both species, differentiation of the testes involves suppression of oogenesis. In *O. latipes*, this suppression is total, while in *D. rerio* it is incomplete. In *D. rerio*, all individuals pass through an indifferent gonad stage and a juvenile ovary stage before testis differentiation starts. Also, in both species, the presence or the relative abundance of oocytes at about 4 wpf seems to be critical for the subsequent differentiation of the ovaries. An early expression of aromatase in somatic cells begins in both species at about 2–3 wpf. In *O. latipes*, this only occurs in females, whereas in *D. rerio* it occurs in all individuals. The expression is then down-regulated in those individuals that start to differentiate testes. At 4–5 wpf, both species show a female-specific pattern of aromatase expression, when developing oocytes become numerous. In *D. rerio*, this phase of aromatase expression signals the definitive differentiation of the ovaries. In both species at about 4–5 wpf, the gonadal somatic cells show expression of male-specific genes, signalling the definitive differentiation of the testes. Oogenesis starts before spermatogenesis in both species.

The consequence is that in both species, gonadal differentiation is routed towards either female or male pathways, a process that has been called canalisation of sex differentiation (Saito & Tanaka, 2009). However, a characteristic of teleosts is that the pathway of sex differentiation can be switched between male and female routes by environmental factors such as temperature (see Chapter 2) and exogenous hormones (see Chapter 5). This suggests that over a critical period of time, the divergence of pathways is not sufficient to prevent developmental trajectories from switching between them. Such switching is seen naturally in hermaphroditic species (see Chapter 10).

With the envelopment of the oogonium by the granulosa and thecal cells in the differentiating ovary and the envelopment of the spermatogonium by Sertoli cells to form a spermatogenic cyst, the stage is set for the processes that end with the spawning of haploid eggs or haploid spermatozoa (Chapter 4).

Chapter 4
Gametogenesis

Reproductive Biology of Teleost Fishes, First Edition. Robert J. Wootton and Carl Smith.

Introduction

A consequence of the initial differentiation of the gonads in teleosts with separate sexes (i.e. gonochoristic species) is the canalisation of the development of the gonads into either ovaries or testes. The cell lineages derived from the primordial germ cells (PGCs) and gonadal somatic cells, which initially have the potential of developing either as female- or male-typical cells (i.e. they are bipotential), undergo a change, which determines their subsequent development (Chapter 3). The developmental outcome is the production of functional ovaries or testes characterised by major morphological differences. There is also the dramatic difference in the characteristics of the germ cells. The female cells, the oocytes, increase massively in size but retain the approximately spherical shape of the original PGC. The male cells undergo an extraordinary change that results in the formation of flagellated, motile spermatozoa, produced in vast numbers by the testes. As an aside, the differentiation of the gonads of teleosts provides a good model for studying the development of complex organs from the totipotential embryonic cells in vertebrates.

The structures of the testes and ovaries have some similarities, but there are also major differences. In comparison to the ovaries, the testes generate vast numbers of haploid, motile gametes. The size of the testes and the abundance of spermatozoa that they produce are related to the mode of mating of the male (Chapter 8). The ovaries and testes produce gametes, but also are endocrine organs, producing a range of steroid hormones (Chapter 5) (Grier & Aranzábal, 2009).

The development of the gonads during gametogenesis can be described at a macroscopic level using two basic indices, maturation staging and the gonadosomatic index (GSI). Both of these indices have to be supported by histological studies of the developing gonads. The basic indices are most valuable when large samples of fish are taken; for example, in surveys to estimate the reproductive potential of a population subjected to fishing (see Chapter 12). The sample sizes in such surveys are too large for each individual to be the subject of a histological study.

Maturational staging uses macroscopic visual examination of the gonads to identify the stage of development of the ovaries and testes. In Table 4.1, an example is shown based on studies of Neotropical species (Núñez & Duponchelle, 2009). Several different staging schemes have been developed for different species, so comparisons need to be made with care. For a given species, if a consistent staging scheme is used, then it provides useful information on the reproductive status of a population. A disadvantage of the method is that it depends of a subjective assessment of the visual appearance of the gonads.

The GSI is a more objective index of gonadal development. The index has two definitions, one of which should be preferred to the other. A common definition is

Table 4.1 Example of maturation staging of teleost gonads based on a study of Neotropical species. Species are classified as total or batch spawners. (Source: Adapted from Núñez & Duponchelle 2009. Reproduced with permission of Springer Science + Business Media.)

Females

Total spawners	
Stage 1	*Immature* – ovaries have a circular section, small, thin, partly translucent and opaque or pinkish; oocytes are invisible to the naked eye
Stage 2	*Maturing* – ovaries are much lower, occupying a significant volume of the abdominal cavity; they are filled with white or yellowish oocytes of different sizes
Stage 3	*Advanced maturation* – ovaries are fuller and larger; oocytes are larger and more homogeneous in size
Stage 4	*Ripe* – similar to stage 3, but oocytes are partially ovulated and can be stripped by gentle pressure on the fish flanks; ovary reaches its maximal development
Stage 5	*Spent* – the ovaries are large, but are flaccid and often bloody; some remaining mature oocytes still visible along with atretic follicles
Stage 6	*Resting* – ovaries are similar to stage 1, but they are usually larger, wider and pink to dark red in colour; ovarian wall is also thicker
Batch spawners	
Stage 1	*Immature* – ovaries have a circular section, small, thin, partly translucent and opaque or pinkish; oocytes are invisible to the naked eye
Stage 2	*Maturing* – ovaries much lower, occupying a significant volume of the abdominal cavity; they are filled with white or yellowish oocytes of different sizes
Stage 3	*Advanced maturation* – ovaries are fuller and larger; oocytes are larger, and vitellogenic oocytes of different sizes occur between the larger oocytes to be released at the next spawning
Stage 4	*Ripe* – similar to stage 3, but oocytes are partially ovulated and can be stripped by gentle pressure on the fish flanks; vitellogenic oocytes of different sizes occur in between oocytes undergoing maturation and ovulated oocytes are partially filling the ovocoel
Stage 5	*Spawned and recovering* – ovaries relatively large and flaccid with remaining empty spaces and sometimes atretic follicles; developing vitellogenic oocytes of different sizes occur. This stage characterises the beginning of one or more new spawning cycles until the end of the breeding season
Stage 6	*Resting*: ovaries are similar to stage 1, but are usually larger, wider and pink to dark red in colour; the ovarian wall is also thicker
Males	
Stage 1	*Immature* – testes are two silvery or translucent threads, thinner and longer than stage 1 ovaries.
Stage 2	*Maturing* – testes are longer, wider, often of triangular or circular section and whitish to pinkish in colour; resting adults usually at this stage
Stage 3	*Ripe* – testes are larger, fuller and completely white; sperm is emitted with gentle pressure on the abdomen
Stage 4	*Spent* – testes are still large, as at stage 3, but they are flaccid and empty

the mass of the gonad (M_G) divided by total body mass of the fish (M_T), and is usually expressed as a percentage [$(M_G/M_T) \times 100$]. A better definition is obtained by first defining the somatic mass of the fish, M_S as the total mass of the fish less the mass of the gonads, i.e. $M_S = (M_T - M_G)$. The GSI is defined as $100 \times (M_G/M_S)$. The advantage of this second definition is that it separates out the gonadal and somatic contributions to the total mass of the fish (Wootton, 1998). A disadvantage of the use of GSI is that for species in which an individual may spawn several times in a breeding season, GSI may not distinguish different stages of gonadal development such as 'just spawned' and 'resting' (Table 4.1). A second disadvantage of GSI as an index is that ratios such as GSI present serious difficulties for statistical analyses (Wootton, 1998).

Structure of ovaries

As the oocytes develop, the ovaries assume their adult form, although until an individual passes through puberty and becomes sexually mature, the ovaries form only a small proportion of the total mass of the body and can be seen as thin threads of tissue on the dorsal wall of the peritoneal cavity. During the juvenile phase, the growth and development of somatic tissues take precedence over the growth of the gonads (Wootton, 1985).

Typically there are paired ovaries, one on either side of the midline of the fish each suspended by a mesentery, the mesovarium, from the dorsal wall of the peritoneum. However, in some species, the paired ovaries fuse to form a single ovary as in the medaka (*Oryzias latipes*, Adrianichthyidae) and in some viviparous clades, such as the livebearers (Poeciliidae) and goodeids (Goodeidae) (Grier *et al.*, 2009).

Gross morphology of the ovaries

In some species, an early indication of the differentiation of the gonad as an ovary is the formation of an internal ovarian cavity or ovocoel (Nakamura *et al.*, 1998). However, not all teleosts develop an ovocoel and two forms of ovary are identifiable (Blüm, 1986; Grier *et al.*, 2009; see Fig. 4.1). In the gymnovarian condition, an ovocoel is not formed, and at ovulation the eggs are released directly into the body cavity, as occurs in the mammals. The gymnovarian condition is seen in phylogenetically more basal teleosts such as the salmonids (Salmonidae) and freshwater eels (Anguillidae). In the cystovarian condition, an ovocoel does form into which the oocytes are released at ovulation.

The details of the formation of the ovocoel differ among species (Nakamura *et al.*, 1998). In the goldfish (*Carassius auratus*, Cyprinidae) and the pond loach (*Misgurnus anguillicaudatus*, Cobitidae), an aggregation of somatic cells develops on the peritoneal wall and this aggregation eventually fuses with the extended edge of the ovary to form a narrow ovocoel between the ovary and the peritoneal wall. In the mottled sculpin (*Cottus bairdii*, Cottidae) and the tilapias (*Oreochromis* spp., Cichlidae), somatic tissue elongates towards the lateral coelomic wall to form a dorsal and ventral fold. The ventral fold then develops dorsally while the dorsal fold develops ventrally so they meet and fuse creating an ovocoel along the lateral side of the ovary. The functional significance, if any, of the different methods of forming an ovocoel in cystovarian ovaries is not known. A general explanation for the cystovarian condition is that the high number of small eggs that are produced at ovulation by teleosts would block the opening to the oviduct that allows eggs to pass from the coelomic cavity into the oviduct in other jawed vertebrates.

As an ovary develops, longitudinal folds, the ovarian lamellae, also develop (Grier *et al.*, 2009) (Figs 4.1 and 4.2). In cystovarian species, the lamellae protrude into the ovocoel, while in gymnovarian species, the lamellae form only on the germinal side.

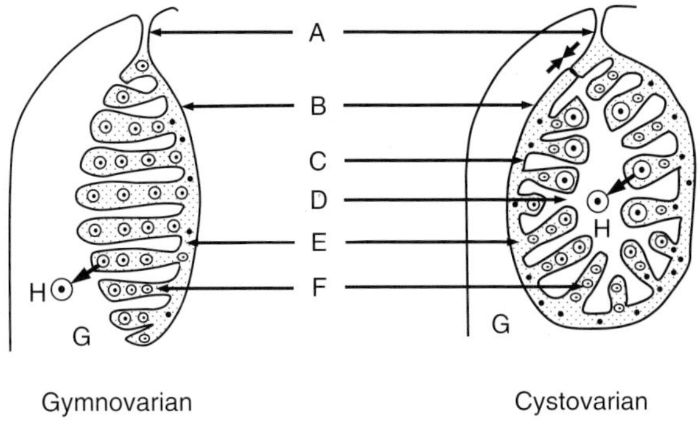

Figure 4.1 Diagram illustrating the gymnovarian and cystovarian conditions of the ovary of teleosts: A, mesovarium; B, peritoneal epithelium; C, internal (germinal) epithelium; D, ovocoel; E, ovarian stroma; F, follicle containing an oocyte (follicular complex); G, coelomic cavity; H, oocyte. Arrows show the direction of ovulation. Opposing arrows indicate where the lateral fold meets the body of the ovary. (Source: Blüm 1986. Reproduced with permission of Springer Science + Business Media.)

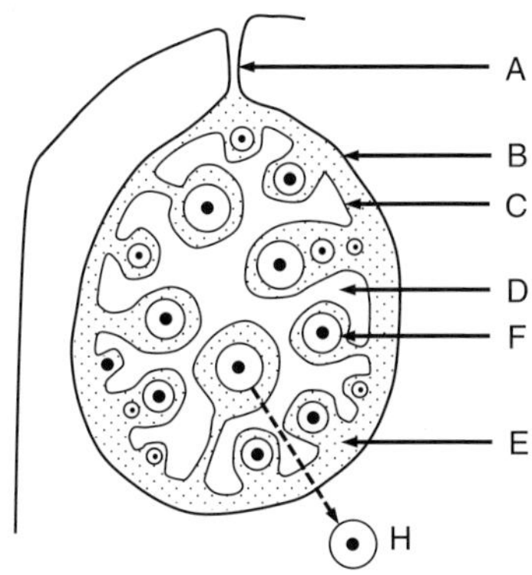

Figure 4.2 Diagram of transverse section through a cystovarian ovary of a teleost, illustrating the ovarian lamellae: A, mesovarium; B, peritoneal epithelium; C, internal (germinal) epithelium; D, ovocoel; E, ovarian stroma; F, follicle containing oocyte (follicular complex); H, oocyte. Dashed arrow shows the direction of ovulation. (Source: Blüm 1986. Reproduced with permission of Springer Science + Business Media.)

The effect of the formation of ovarian lamellae is to increase the surface area exposed to the ovocoel or coelom. Each lamella has a germinal epithelium and, internal to this, the stroma of the ovary. Between the germinal epithelium and the stroma, a basement lamina (basement membrane) develops, which divides the ovary into two compartments, epithelial and stromal.

The germinal epithelium consists of somatic epithelial cells, but is invaded by oogonia, which then develop into early oocytes (Grier *et al.*, 2009; Lubzens *et al.*, 2010). Although the germinal epithelium is usually a monolayer of cells, the structure becomes multi-layered where oogonial nests develop (see Chapter 3). The germinal epithelium is the site where the process of oogenesis takes place. The stromal compartment

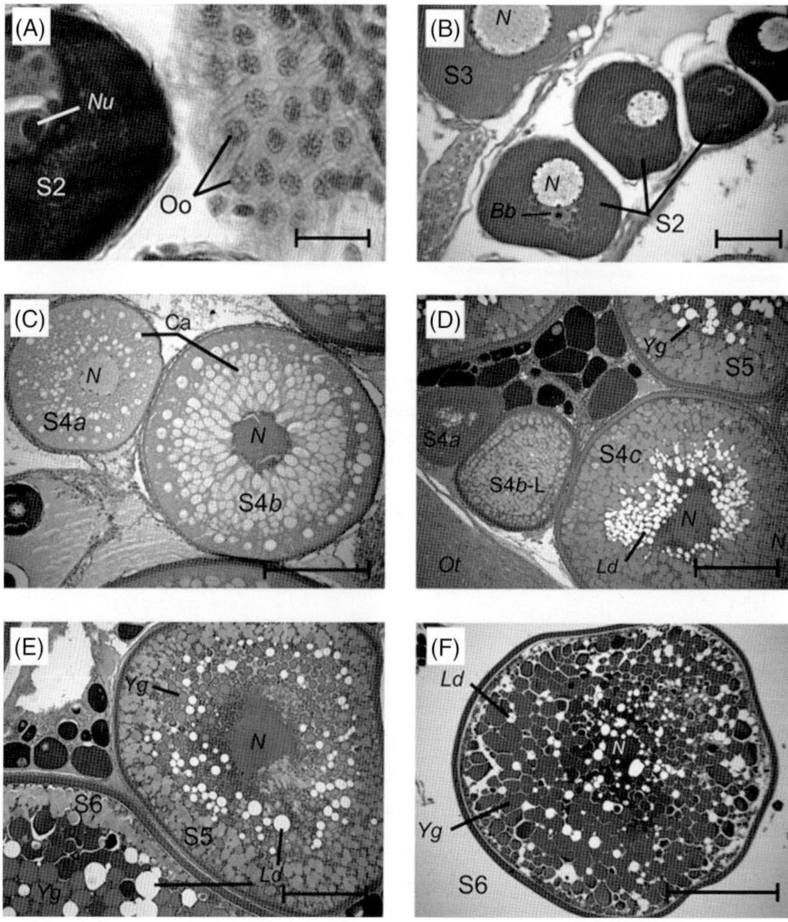

Figure 4.3 Transverse sections of Murray cod (*Maccullochella peelii*) ovaries showing oogonia and oocyte development stages. (A) Oogonia (Oo) and early perinuclear oocyte (S2); (B) early (S2) and late (S3) perinuclear oocytes; (C) early (S4a) and mid-stage (S4b) cortical alveoli oocytes; (D) early (S4a) and late (S4b-L) cortical alveoli oocytes, lipid droplet stage oocyte (S4c), and early vitellogenic oocytes (S5); (E) early (S5) and late (S6) vitellogenic oocytes; (F) late vitellogenic oocyte (S6). Nu, nucleoli; N, nucleus; Bb, Balbiani bodies; Ld, lipid droplets; Yg, yolk granules; Ot, ovarian tunica. Scale bars: A, 10 μm; B, 50 μm; C, 200 μm; D, 300 μm; E, 300 μm; F, 1000 μm. Chromatin nucleolar stage oocytes (S1) not shown. (Source: Newman *et al.* 2007. Reproduced with permission of Elsevier.)

contains gonadal somatic cells, fibroblasts, granulocytes, collagen fibres and blood vessels. These two compartments are not isolated from each other, so materials including hormones, paracrines and the precursors of the yolk that accumulates in the oocyte can pass between them (Fig. 4.3).

A variant of ovarian structure is seen in the pipefishes and seahorses (Syngnathidae), fishes in which the males incubate the eggs in brood pouches or other structures (Wallace & Selman, 1981; Sogabe & Ahnesjö, 2011; see Chapter 10). The ovary is a rolled structure, formed by a tubular sheet, called the follicular lamina, which is sandwiched between the epithelium of the lumen and the outer ovarian wall. In some species, including the northern pipefish (*Syngnathus fuscus*, Syngnathidae), there is a single germinal ridge along the edge of the follicular lamina, which contains all the oogonia. From this edge, oocytes of increasing size occur, with the most advanced oocytes at the other edge of the follicular lamina. In other species, exemplified by the straight-nose pipefish (*Nerophis ophidion*, Syngnathidae), there are two germinal ridges along the opposite edges of the follicular lamina and follicles are generated from each so the most advanced follicles are near the centre of the lamina.

Development of female gonoducts

The formation of the ovocoel presents the problem of how the eggs are to be carried to the exterior to be fertilised by spermatozoa at spawning. Again, the teleosts demonstrate their diversity in reproductive biology and their divergence from other jawed vertebrates. In the latter, the ovulated eggs from each ovary are released into the coelomic cavity and collected through a funnel into the oviduct. In this case, the paired oviducts are derived embryologically from the Müllerian ducts (Romer & Parson, 1986). Teleosts have evolved female gonoducts that are new structures not derived from Müllerian ducts and so are not termed oviducts (Romer & Parsons, 1986; Cole, 2010a). In most, but not all, teleostean species, the ovocoel opens directly into the gonoduct, so the eggs never pass into the coelom. Species with gymnovarian ovaries, such as the salmonids, do release eggs into the coelom, but the gonoducts are still new structures and not derivatives of Müllerian ducts.

In cystovarian species with paired ovaries, there are small differences in the arrangement of the gonoducts. In some species, the paired ovaries fuse posteriorly and the ovocoels unite and open into a single gonoduct. For other species, the ovaries remain separate and each opens into a separate gonoduct, which then fuse to form a single gonoduct. The single gonoduct opens into the external environment at the tip of a genital papilla, just posterior to the gonopore opening.

A variety of tissues may contribute to the formation of the gonoducts (Cole, 2010b). In the gymnovarian salmonid, the sockeye salmon (*Oncorhynchus nerka*, Salmonidae), which is a total spawner (Chapter 1), the anterior portion of each gonoduct is derived from a portion of the peritoneum lining the ovary and the mesovarium. These paired gonoducts merge into a single gonoduct. More posteriorly, the gonoduct is formed by a folding of the dorsal mesentery. However, the gonoduct remains open dorsally, and so forms a trough. The three-spined stickleback (*Gasterosteus aculeatus*, Gasterosteidae) is a batch spawner (see Chapter 1), with paired ovaries. In this species, paired gonadal ridges that do not contain PGCs or their derivatives develop posteriorly along the dorsal wall of the coelom. From each ridge, extensions develop, which merge to form a structure with a lumen. Anteriorly, this tubular structure becomes continuous with the ovary, completing the gonoduct. Even within a clade, the origin of the gonoducts can vary, as illustrated by the viviparous Poeciliidae, which have a single ovary, formed by the fusion of the paired gonadal anlagen. The gonoduct of the green swordtail (*Xiphophorus helleri*, Poeciliidae) derives from two components. Anteriorly, cells proliferate at the caudal end of the ovary. The resulting cell mass widens and develops a groove, which is covered by a portion of the mesovarium to enclose a lumen. Posterior to this development, a second proliferation of peritoneal cells forms a cord of cells along the dorsal midline. This cord fuses with the anterior portion of the gonoduct. Within the cord, a lumen develops, which merges with the lumen of the anterior component. In the guppy (*Poecilia reticulata*, Poeciliidae), the gonoduct develops from the dorsal mesentery and dorsal peritoneum. Lateral folds extend away from the midline and then upwards. With the fusion of the two folds in the midline, a paired lumen is formed, and with the disappearance of the internal partition between the two, a single lumen is formed, which is continuous with the ovocoel.

A feature of the gonoducts of the teleosts is that they do not develop specialist sections that secrete egg capsules and shells or act as a uterus, features that have evolved in jawed vertebrates that use the Müllerian ducts as the basis for the oviducts. Consequently, in viviparous teleosts, the development of the embryos takes place within the ovary and not within a uterus (Chapter 10).

Oogenesis

Oogenesis is the developmental process by which oogonia are transformed into mature oocytes ready to be ovulated (Wallace & Selman, 1981; Tyler & Sumpter, 1996; Le Menn *et al.*, 2007; Grier *et al.*, 2009; Lubzens *et al.*, 2010; see Fig. 4.4). Although the eggs produced differ in size from species to species and reflect life history (Chapter 11), the stages through which the oocytes go are relatively constant across species, but the time spent in each phase varies between species (Tyler & Sumpter, 1996). In the zebrafish (*Danio rerio*, Cyprinidae), oogenesis takes a few days, but in salmonids it takes a few months (Knoll-Gellida & Babin, 2007). Examples of detailed descriptions of oogenesis for an identified species include the mummichog (*Fundulus heteroclitus*, Fundulidae) (Selman & Wallace, 1986) and rainbow trout (*Oncorhynchus mykiss*, Salmonidae) (Bromage & Cumaranatunga, 1988). On the basis of such studies, a generic description of oogenesis in teleosts can be given.

After an egg is fertilised by a spermatozoon, there is a period of time during the development of the zygote when the zygote's genes are not activated (Lyman-Gingerich *et al.*, 2007). The start of the zygotic gene programme typically occurs during the blastula stage of embryonic development at the mid-blastula transition (MBT). Before this transition, the developmental programme of the embryo depends solely on products deposited in the developing oocyte by the female. Many of these products

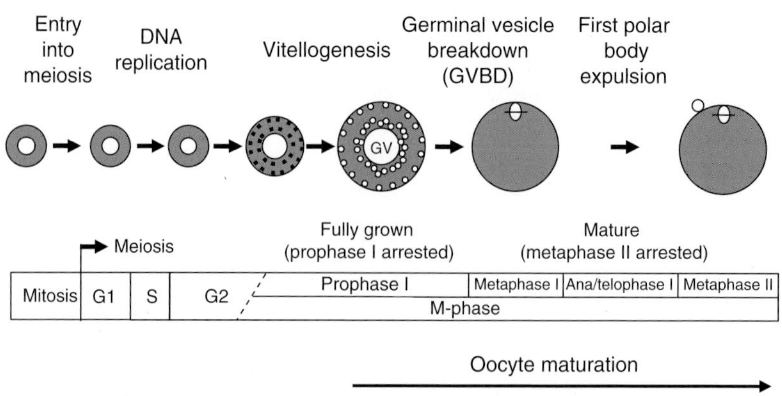

Figure 4.4 Developmental sequence of oocytes from primary oocyte to the post-vitellogenic phase showing the sequence of chromosomal changes. GV, germinal vesicle. (Source: Lubzens *et al.* 2010. Reproduced with permission of Elsevier.)

are synthesised endogenously in the oocyte, including mRNAs, growth factors and other biomolecules. Some crucial products are synthesised exogenously in organs of the female other than the ovaries. The most important of these products is vitellogenin (VTG), which is synthesised in the liver and transported to the ovary in the blood supply. The VTG is the main source of the proteins and lipids needed by the developing embryo. Another product, which in many species is synthesised in the liver of the female, is the group of proteins, the zona pellucida proteins (ZPs), that will form the protective envelope, the zona pellucida, around the oocyte.

The timing of events during oogenesis is largely controlled by gonadotrophins produced by the pituitary (Chapter 5). Of the two gonadotrophins, follicle-stimulating hormone (FSH) and luteinising hormone (LH), FSH is important in the earlier stages of oogenesis, while LH is more important in the later stages that lead to the maturation and ovulation of the oocyte.

The quality of the egg in terms of its capacity to be fertilised and develop as a zygote until it becomes a successful, actively feeding young fish will be partly determined by the quality of the maternal products with which it is endowed (Brooks *et al.*, 1997). These maternal effects can have significant impacts on the performance of the early life stages (Agbali *et al.*, 2010).

Oogonial proliferation and oogonial nest formation

The oogonia proliferate by mitotic division and daughter cells aggregate in oogonial nests, in which the oogonia are connected by cytoplasmic links (Chapter 3). A single layer of somatic pre-granulosa cells surrounds the nest. Each oogonium then becomes isolated from the other oogonia in a nest and enters the prophase of the first meiotic division.

At the transition between an oogonium and an oocyte, each chromosome divides to produce two daughter chromatids linked by a centromere. This step is common to both mitosis and meiosis, but now the cell enters the first prophase of meiosis, becoming a primary oocyte. Individual oocytes then start to become enveloped by a single layer of somatic cells, which form the granulosa layer. These somatic cells secrete a basement lamina that separates the oocyte and its granulosa from the stroma (Le Menn *et al.*, 2007; Grier *et al.*, 2009; Lubzens *et al.*, 2010).

It is a feature of teleosts that oogonial proliferation continues in sexually mature adults, so there is a pool of oogonia from which oocytes can be recruited in reproductively active females. This is in contrast to mammals, in which a pool of oocytes is defined early in development. This is the entire pool from which mature oocytes can be recruited during the period of female reproductive activity and the pool is not replenished (Blüm, 1986). The high fecundities seen in most teleosts (Chapter 8) would not be possible if this oogonial replenishment did not occur. In iteroparous teleosts (Chapter 11), oogonial numbers typically peak at, or soon after, spawning (Grier *et al.*, 2009).

Chromatin nucleolus stage

As the granulosa layer completely surrounds the oocyte, the formation of an ovarian follicle consisting of oocyte, granulosa and basement lamina is completed. Each follicle may remain attached to germinal epithelium by a section of shared basement lamina. Outside a follicle, a single layer of somatic cells forms the theca, typically closely associated with blood vessels. The follicle and its associated theca form the ovarian follicular complex (Le Menn *et al.*, 2007; Grier *et al.*, 2009). This is the unit that now defines the process of oogenesis. Both the thecal and granulosa cells are steroidogenic and so the unit also has an endocrine function. This process of folliculogenesis may not be fully completed at the end of the chromatin nucleolus stage. Later in the development of the ovarian follicle complex, the theca may become multi-layered.

The nucleus of the oocyte now shows changes in morphology, characteristic of the prophase of meiosis (Le Menn *et al.*, 2007; Lubzens *et al.*, 2010). In the leptotene stage of meiosis, the chromosomes condense and become visible. The zygotene stage is defined by a shortening of the chromosomes, and homologous chromosomes form pairs and become linked at the synaptonemal complexes. In the pachytene stage, the chromosomes condense and crossing-over between homologous chromosomes occurs at the synaptonemal complexes. These complexes break down in early diplotene and the chromosomes become extended in the lampbrush structure characteristic of oocytes (Callan, 1986). The oocyte now becomes arrested in the diplotene of meiotic prophase. At this stage, the oocyte is approximately spherical, the cytoplasm (ooplasm) is clear and the nucleus has a single prominent nucleolus (Figs 4.3 and 4.4).

Primary growth

Early in the primary growth phase of the oocyte, the formation of the follicle complex is completed. The main feature of this stage is the synthesis of components that have important roles in the further development of the oocyte. Histologically, the ooplasm becomes basophilic, because of the accumulation of RNA, especially messenger RNAs (mRNA), required for protein synthesis and the receptors required for the uptake of VTG, the precursor of yolk (Babin *et al.*, 2007). The nucleus, now called the germinal vesicle, increases in size and multiple nucleoli become arranged around its periphery. Another histological feature at this stage is a Balbiani body (Fig. 4.3; Le Menn *et al.*, 2007; Grier *et al.*, 2009). This is a complex of cell organelles and associated RNA. As the oocyte continues to develop, several Balbiani bodies may form, but then disperse. The organelles involved can include mitochondria, Golgi bodies and smooth endoplasmic reticulum. The functional significance of the Balbiani body is unclear (in the older literature, the Balbiani body was sometimes called the yolk nucleus).

The main feature of the primary growth phase is the appearance of cortical alveoli. These alveoli are vesicles bounded by a membrane. Their contents are synthesised in the oocyte with the Golgi bodies playing a part in the synthesis. These contents will eventually play a role in the events that follow the fertilisation of an oocyte by a spermatozoon (see the section 'Fertilisation'). In some species, the cortical alveoli come to fill most of the cytoplasm, whereas in other species they tend to be found in the periphery of the ooplasm.

A second type of cell inclusion that appears in the primary growth phase of some species is the oil droplet that will contribute to the lipid content of the oocyte (Grier *et al.*, 2009). Oil droplets are probably derived from very low-density lipoproteins (VLDL) transported to the ovaries in the bloodstream and taken up by the follicular complex. In some species, oil droplets may also contribute to the buoyancy of the egg, because of the low specific gravity of oils. The timing of the appearance of cortical alveoli and oil droplets varies between species and may even be delayed until after vitellogenesis has started. The size, colour and distribution of oil droplets are characteristic of some species and can be an aid in identifying the species that spawned the egg (M'Intosh & Masterman 1897; Russell, 1976).

During this phase, the oocyte and the granulosa cells begin to separate, with the intervening space becoming filled with extracellular matrix. The oocyte extends microvilli into the space and around these microvilli, a proteinaceous structure, the zona pellucida, is laid down (Le Menn *et al.*, 2007; Modig *et al.*, 2007). With the completion of the zona pellucida, the oocyte is protected by this acellular envelope, but the microvilli maintain contact with the granulosa layer outside the zona pellucida. The granulosa cells also extend microvilli towards the oocyte, although these are fewer in number. Species vary in the origin of the ZPs. The site of synthesis can be the oocyte itself, the liver or a combination of both (in the literature, the nomenclature of the zona pellucida is confused – it also called the zona radiata, vitelline envelope or the chorion).

Oocytes increase in volume during the primary growth phase, but the ovaries continue to form only an insignificant proportion of total body mass. In *O. mykiss*, oocytes increase in diameter from 10–20 μm to 100–200 μm, but the ovaries form less than 0.1% of the total body mass (Tyler & Sumpter, 1996).

Secondary growth - vitellogenesis

Once the oocyte reaches a critical size, the secondary growth phase can begin (Tyler & Sumpter, 1996). In *G. aculeatus*, the critical diameter is about 0.6 mm, with a final oocyte size of about 1.2 mm. In the salmonid *O. mykiss*, the critical diameter is also about 0.6 mm, but the final diameter is about 5.0 mm. The secondary growth phase is the major growth period of the oocyte, when growth is measured in terms of the dry mass of the oocyte. This growth primarily reflects the accumulation of yolk by the oocyte. The proportion of the total oocyte volume at spawning that is contributed by the yolk varies between species (Tyler & Sumpter, 1996), with the proportion tending to be higher in demersal than in pelagic eggs. *O. mykiss* is a demersal spawner in freshwater (Chapter 7) in which the yolk accounts for about 95% of the egg volume, whereas in the Atlantic cod (*Gadus morhua*, Gadidae), a pelagic spawner in seawater, the yolk accounts for about 12% of the egg volume. However, in almost all teleost species, the yolk is the major, maternal nutritional contribution to the zygote, providing the resources required for the differentiation and growth of the embryo within the egg and for the larva after hatching (phase of endogenous feeding), until the larva can support itself by its own feeding activity (phase of exogenous feeding).

In the secondary growth phase, the follicular complex is complete, but the oocyte remains arrested in the diplotene stage of prophase. The zona pellucida continues to thicken and the accumulation of oil droplets continues. However, the dominant process is vitellogenesis. In a broad sense, this is the synthesis, transport, uptake and processing of VTG by the oocyte, processing that transforms the VTG into the lipoprotein and protein stored in yolk globules.

Vitellogenin is the main precursor of yolk (Babin *et al.*, 2007; Mommsen & Korsgard, 2008). It is a high-molecular-weight (300–640 kDa) glycophospholipoprotein. It is synthesised outside the ovaries, in the liver. This synthesis is stimulated by an oestrogen, oestradiol-17β (E_2), produced by the granulosa cells of the follicular complexes of the ovaries (see Chapter 5). VTG is carried from the liver to the ovaries in the bloodstream and arrives at the thecal blood vessels. It leaves the bloodstream, passes between the thecal cells and through the basement lamina, then between the granulosa cells and along the oocyte microvilli. The VTG is sequestered by specific VTG receptor molecules located in clathrin-coated pits in the oolemma. Clathrin is a fibrous protein found on the surface of coated pits and vesicles. By the process of endocytosis, the pits move into the ooplasm, thus forming vesicles. The VTG-containing vesicles fuse with lysosomes, which are multivesicular bodies (MVBs) originating from the Golgi bodies of the oocyte. In the lysosomes, the VTG is broken down into the yolk components by proteolytic cleavage. The lysosomes increase in size and become small yolk globules and subsequently large yolk globules. The production of VTG receptor mRNAs starts before the onset of the secondary growth phase (Mommsen & Korsgard, 2008), so VTG receptor molecules are synthesised during the primary growth phase and may be recycled during vitellogenesis. The density of VTG receptors per unit area of the oolemma increases during vitellogenesis, supporting the high rate of transfer of VTG into the oocyte, but the density then declines shortly before ovulation (Tyler & Sumpter, 1996). The factors that bring the process of vitellogenesis to a conclusion, including how the follicular complex signals that the oocyte should cease accumulation of VTG, are not well understood.

In a few teleost species, such as the striped bass (*Morone saxatilis*, Moronidae), the yolk is stored as relatively insoluble crystalline yolk platelets. In the majority of species, the yolk is stored in fluid-filled globules. These globules may remain distinct. In other species, such as *G. aculeatus*, the globules fuse to form a continuous mass of fluid yolk. The fusion takes place from the central region of the oocyte and spreads outwards to the periphery. This results in the ooplasm occupying a narrow region in the outermost region of the oocyte. In the Atherinimorpha, a group that includes viviparous species (Chapter 10), the fusion of the yolk globules during vitellogenesis is one of the characteristic traits of the group. The adaptive significance, if any, of these inter-species differences in the pattern of yolk deposition is not understood.

The capacity of the fertilised egg to subsequently develop as a viable zygote depends to a large extent on the quality of resources provided by the female to her eggs, which will manifest themselves at maternal effects (Brooks *et al.*, 1997). VTG is not the only material that the female transfers to her eggs. The microvilli of the oocyte form heterologous gap junctions with the microvilli of the granulosa cells. At gap junctions, gap junction proteins form a continuous aqueous channel, linking the cytoplasm of the two cells (Grier *et al.*, 2009). These channels allow ions and small molecules to pass between the cells permitting the transfer of factors that regulate vitellogenesis and maturation.

The female must also provision the oocyte with vitamins and immunologically active molecules that are required by the embryo and larva. Lipid-soluble vitamins can be transferred to the oocyte in the lipoprotein fraction originating in the liver. There may also be pathways that are specific to the transport of vitamins as demonstrated for retinol, a component of vitamin A, in *O. mykiss* (Wiegand, 1996).

This secondary growth phase is the period over which the mass of the ovaries relative to the total mass of the fish greatly increases, so that at the end of the phase, the ovaries in some species account for over 20% of the total body mass. This represents a massive allocation of resources away from the somatic body to the gonads (Chapter 11).

Vitellogenin and the zona pellucida proteins

Vitellogenin is synthesised in the liver in response to oestrogen (Chapter 5). There is a family of vertebrate VTGs and these are the main egg yolk precursor protein (EYPP) in vertebrates. The vertebrate VTGs are probably derived from a common ancestor (Babin *et al.*, 2007). In the genome, there are multiple versions of the *vtg* gene. Twenty copies have been identified in *O. mykiss* and at least seven in *D. rerio*. Some evidence suggests that ancestral teleosts had at least two VTG isoforms (Ding, 2005). Now, three or even four lineages of VTG have been detected in teleosts, although it is likely that no species has the capacity to produce all three (or four) (Mommsen & Korsgard, 2008). Within the teleosts there is interspecific variation in the distribution of these genes. A survey of

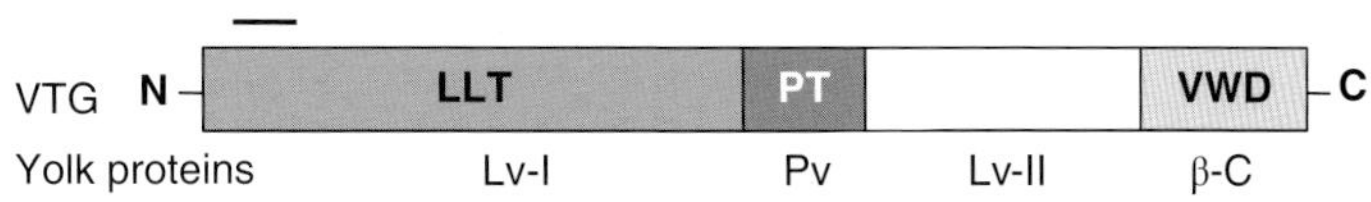

Figure 4.5 Subunit structure of vitellogenin (VTG), showing domains indicated by capital letters, and yolk proteins generated by enzymatic cleavage of VTG in the oocyte indicated by lower-case letters. LLT, large lipid transfer module; PT, polyserine track domain; VWD, von Willebrand factor type D domain indicated from the N-terminal to the C-terminal ends of the plasma precursor protein; Lv-I, lipovitellin I; Lv-II, lipovitellin II; Pv, phosvitin; β-C, β-component. Horizontal black bar indicates the receptor-binding region. (Source: Babin *et al.* 2007. Reproduced with permission of Springer Science + Business Media.)

11 perciform species suggested that interspecific variation reflected variation in the size and number of the subunits of the VTG molecule (Ding, 2005).

The basic structure of VTG has several subunits (Fig. 4.5), which are destined to be separated in the oocyte by enzymatic cleavage to generate the yolk proteins. After the synthesis of the protein backbone of the VTG molecule, there is post-translational modification of the molecule, as it is phosphorylated, lipidated and glycosylated before release into the bloodstream from the liver (Babin *et al.*, 2007). The VTG also carries minerals such as calcium, iron, magnesium and zinc, which are required by the developing zygote. Even lipophilic hormones, such as the thyroid hormones, can be transported with VTG.

Vitellogenin is transported in the blood as a dimer. At the surface of the oolemma, it binds to the VTG receptor molecule. This receptor belongs to a family of proteins, the low-density lipoprotein receptors (LDFR), which are coded by a supergene family. Once the VTG is in the multivesicular body (MVB) within the oocyte, it is enzymatically cleaved by a protease, cathepsin D, and possibly other enzymes to generate the yolk proteins (Fig. 4.5) (Ding, 2005; Babin *et al.*, 2007). These yolk proteins include lipovitelline I and II, phosvitin and phosvettes, which are incorporated in the yolk globules or, in a minority of species, yolk platelets. The β-component deriving from the VTG (Fig. 4.5) is retained in the cytoplasm soluble fraction.

The lipids of the yolk include the polar phospholipids and neutral lipids, especially triacylglycerol (TCA; Wiegand, 1996). The polar lipids include phosphatidylcholine (PC) and phosphatidylethanolamine (PE), which are generally rich in (n-3) polyunsaturated fatty acids (PUFAs) and arachidonic acid (AA). In the developing zygote, the phospholipids are used as both energy reserves and as a source of structural lipids, such as those involved in membranes. The neutral lipids act primarily but not exclusively as an energy source. The quality of the eggs, as assessed by their capacity to give rise to zygotes that develop into independently feeding juveniles, is related to the effectiveness with which the oocytes have been supplied with essential fatty acids such as docasahexaenoic acid (DHA; Brooks *et al.*, 1997). There is some evidence that there is significant quality control of the materials transferred to the oocyte during vitellogenesis (Brooks *et al.*, 1997). However, there is a wide interspecific range in the lipid content of teleost eggs, ranging in terms of dry mass from 50% in a mouthbrooding cichlid (*Labeotrophus* sp., Cichlidae) to 0.1% in the plaice (*Pleuronectes platessa*, Pleuronectidae), a pelagic, marine spawner.

A second class of proteins is that used in the formation of the zona pellucida, the ZPs (Modig *et al.*, 2007). ZPs (sometimes called choriogenins) are glycoproteins, and are found throughout the vertebrates. Of the four groups that occur in the vertebrates, teleosts have three: ZPB, ZPC and ZPX. In many teleosts, the ZPs are synthesised in the liver in response to oestrogen. In some species, such as *D. rerio*, the ZPs are synthesised in the ovary while in other species, including *O. latipes*, they are synthesised in both the liver and ovary (Lubzens *et al.*, 2010).

In the oocytes, the zona pellucida glycoproteins are incorporated into secretory vesicles that fuse with the oolemma. The proteins are then deposited at the base of the microvilli of the oolemma at the inner layer of the zona pellucida. Microscopically, the zona pellucida has visible layers; for example, in gilthead seabream (*Sparus aurata*, Sparidae), four layers are visible during early vitellogenesis, but later the outermost layer is reduced and eventually lost.

Polarity of the oocyte

A further feature of the secondary growth phase (vitellogenesis) is the increasing evidence of the development of a polarity in the oocyte (Lyman-Gingerich *et al.*, 2007). The oocyte polarises along an animal–vegetal axis. The animal pole is indicated by the developing micropyle, down which a spermatozoon will enter to fertilise the oocyte. The opposite, vegetal pole will eventually contain the yolk mass as development of the embryo proceeds. A further indication of the development of polarity is that the pattern of mRNA localisation in the oocyte becomes defined as the oocyte develops. Some mRNAs are uniformly distributed throughout the oocyte. Other mRNAs are localised in a wedge-shaped region of the animal pole. A third pattern of localisation is shown by mRNAs, which are distributed in the cortical region of the oocyte. A small number of mRNAs are localised in the cortical region of the vegetal pole. These spatial distributions help to determine the early development of the embryo. In this way, a spatial maternal influence is exerted over the embryo in addition to the material resources provided maternally, typically resulting in significant additive genetic maternal effects on early development with consequences for subsequent offspring performance.

Oocyte maturation

Whatever mechanisms signal that an oocyte has reached its final dry mass, the outcome is the termination of the uptake of VTG by that oocyte and the transfer of material resources from the female essentially ends. The oocyte now enters the phase of maturation, which ends with the ovulation of the oocyte into the ovarian lumen (or body cavity in gymnovarian taxa such as the salmonids) ready to be spawned. Oocyte maturation is a complex of processes not yet fully understood and is defined by two major events, which occur concomitantly (Bobe *et al.*, 2008).

The first of these events is similar to that seen in the oocyte maturation of other vertebrates. The arrest of the oocyte in diplotene I of meiosis I ends and meiosis proceeds until a second arrest in metaphase II (Fig. 4.4). The second event, which is characteristic of teleosts, is hydration of the oocyte by the uptake of water. This hydration results in an increase in the volume of the oocyte, but no increase in its dry mass.

A sign that meiosis is to resume is that the nucleus of the oocyte, the germinal vesicle, takes up an eccentric position and then migrates towards the animal pole, taking up a position in the periphery of the ooplasm. This is close to the micropyle through which a spermatozoon must enter to fertilise the egg. Shortly after reaching the animal pole, the germinal vesicle breaks down with the dissolution of the nuclear membrane. This dissolution is required for meiosis to proceed. Meiosis I is now completed, with the formation of two, unequally sized cells, the large oocyte and a small first polar body, which degenerates. The oocytes then enter meiosis II, until arrest at metaphase II. The final steps of meiosis II do not occur until the oocyte is fertilised (or activated by some alternative stimulus) and becomes an egg. As these meiotic events are occurring, the microvilli of the oocyte withdraw and lose contact with the granulosa cells, so the heterologous gap junctions are lost. A result is that the oocyte becomes isolated from the other components of the follicular complex.

Although rapid oocyte hydration is a characteristic of teleosts, its extent varies with the life history of the species (Cerda *et al.*, 2007; Le Menn *et al.*, 2007; Lubzens *et al.*, 2010). In marine species with pelagic eggs (Chapters 1 and 7), a high proportion of the final volume of an egg, about 70%, is acquired by the process of water uptake during oocyte maturation, so water accounts for 90–95% of the fresh mass of the egg. In marine species with demersal eggs, the increase in volume through hydration is less, with increases of one- to threefold, with water accounting for approximately 70–80% of the mature egg mass. In freshwater species, the hydration is much less. For some freshwater species, there is only a 5–6% increase in water content. In the salmonid, *O. mykiss*, a demersal spawner, the wet mass of the oocyte increases by about 25% during maturation (Bobe *et al.*, 2008).

Hydration of the oocyte depends on molecular water channels, or aquaporins. Aquaporins are proteins, widely distributed in organisms, whose function is water transport down an osmotic gradient. An osmotic gradient between an oocyte and the maternal tissues is generated by two processes. The first is the breakdown of some yolk proteins to free amino acids (FAAs) and small peptides, increasing the osmolality of the ooplasm. In marine species with pelagic eggs, FAAs account for about half of the osmolality. The other half is caused by inorganic ions, including potassium (K^+), calcium (Ca^{2+}), magnesium (Mg^{2+}), ammonium (NH_4^+) together with inorganic phosphorus (P_i). In *F. heteroclitus*, a marine, demersal spawner with a capacity to withstand wide variation in salinity, hydration depends more on the accumulation of potassium ions than on FAAs derived from yolk proteins (Lubzens *et al.*, 2010).

Freshwater has a lower density than sea water, so the massive hydration of pelagic marine eggs renders them buoyant in the sea (Cerdà *et al.*, 2007). Although the eggs are not motile, the buoyant eggs are transported by water currents. This results in dispersion of the eggs and also ensures that the water layer depleted of oxygen by the respiration of the egg is replenished. However, even marine and freshwater demersal spawners show some hydration, which seems to be an essential component of oocyte maturation in teleosts. The evolution of an effective means of hydrating the oocyte may have been essential for the teleosts to invade the open ocean, because there are no surfaces on which eggs can be deposited, so the eggs have to be pelagic. Duplication of a VTG gene may have allowed the specialisation of a VTG whose function is to be proteolysed to FAA (Finn & Kristoffersen, 2007).

Lipids also have a lower density than water and so could contribute to egg buoyancy. In marine pelagic species whose oocytes contain oil droplets, these droplets coalesce during maturation to form larger oil droplets, and in some species just a single oil droplet. The coalescence of oil droplets is associated with a change in the visible properties of the yolk, which becomes more transparent. However, some marine pelagic species lack oil droplets, while some species with demersal eggs have lipid contents equivalent to those in pelagic species. These observations suggest that any effect on buoyancy of lipids is secondary to their role as reserve materials. For freshwater species with pelagic eggs, hydration cannot contribute to egg buoyancy, so it must depend mainly on their lipid content.

For oocyte maturation to occur, the oocyte must achieve a state called oocyte maturation competence (Bobe *et al.*, 2008). Only if that state has been achieved can the follicular complex respond to gonadotrophic hormones (GTHs) produced by the pituitary (Chapter 5). The primary gonadotrophin-stimulating oocyte maturation is LH (Suwa & Yamashita, 2007; Bobe *et al.*, 2008). In response to LH, the follicular cells synthesise and release a progestin steroid, the maturation-inducing hormone (MIH). In turn, this progestin stimulates the oocyte to produce a maturation-promoting factor (MPF). An MPF is present in eukaryotic cells, promoting the mitotic phase (M-phase) of the cell cycle. In teleosts, MPF is a complex of two subunits, Cdc2 and cyclin B. Neither of these units on its own can stimulate maturation of the oocyte. The integration of the events during oocyte maturation probably involves other molecules such as insulin-like growth factors (IGFs) and their receptors, but the details remain to be defined.

Ovulation

At ovulation, the mature oocyte is released into the ovarian lumen (or body cavity in gymnovarian taxa). For the release to take place, the basement lamina must break down, opening the follicle to the ovarian lumen. The breakdown of the basement lamina involves proteases, including serine proteases, and other factors are also implicated in ovulation. In *O. latipes*, gelatinase and metalloproteins 1 and 2 are the hydrolytic enzymes responsible for the rupture of the follicle (Clelland & Peng, 2009). In *C. auratus*, a prostaglandin (PGF2α) has been implicated as a regulator of ovulation. Ovulation has been likened to an inflammatory reaction. Although both maturation and ovulation depend on MIH, different pathways are involved (Suwa & Yamashita, 2007). Even during ovulation, some limited oocyte hydration continues, but the increase in oocyte volume is small. The process of oocyte development and maturation are summarised in Figure 4.6.

Atresia

Atresia is the breakdown of ovarian follicular complexes, allowing recovery and recycling of the materials (Tyler & Sumpter, 1996; Habibi & Andreu-Vieyra, 2007; Lubzens *et al.*, 2010). Atresia occurs in two contexts: post-ovulatory and pre-ovulatory atresia.

At ovulation, only the oocyte is released into the ovarian lumen, leaving behind the granulosa and thecal components of the follicular complex. The follicular cells become a continuum with the epithelial cells from which they were derived, but the follicle loses its shape as a result of its collapse as the oocyte is released. At this point, the lumen of the follicle is continuous with the ovarian lumen. The follicle complex is still a source of steroids, but it is not clear whether this is just a remnant of its capacity when the oocyte was present or has a functional significance. In mammals, the endocrine activity of the post-ovulatory complex is important in maintaining pregnancy, but this is not relevant to oviparous teleosts. Degeneration of the post-ovulatory follicle is due to cell death (termed apoptosis). Species differ in the length of time for which post-ovulatory follicles can persist. In the mouthbrooding Mozambique tilapia (*Oreochromis mossambicus*, Cichlidae), they can persist for up to 3 weeks after spawning, while in the pelagic spawning California anchovy (*Engraulis mordax*, Engraulidae), the follicles persist for only 1 or 2 days. Such information is useful in a fishery context when estimating how recently an individual female has spawned (Chapter 12).

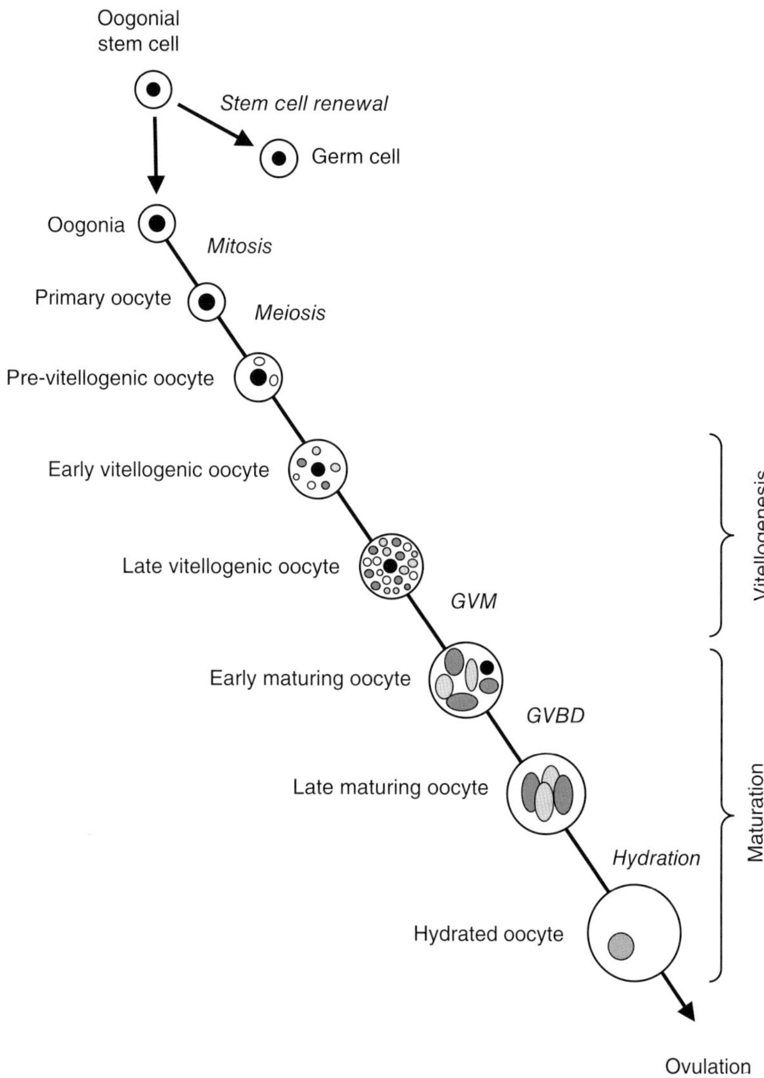

Figure 4.6 A summary of the process of oocyte development and maturation in female fish. GVM, germinal vesicle migration; GVBD, germinal vesicle break down. (Source: Mañanós *et al.* 2009. Reproduced with permission of Taylor and Francis Group.)

The second form of atresia involves the breakdown of a follicular complex that includes an oocyte. This form of atresia can occur at all stages of oogenesis, but predominantly in the vitellogenic phase. Also, follicle complexes in which the oocyte has failed to ovulate prior to spawning become atretic. In pre-ovulatory atresia, the oocyte nucleus and cellular organelles break down and the zona pellucida starts to disintegrate. The follicle cells proliferate, hypertrophy and phagocytose the oocyte. Once the digestion of the oocyte is complete, the follicle cells are, in turn, phagocytosed by stromal cells of the ovary.

Pre-ovulatory atresia can be a response to adverse conditions, including poor nutrition of the female, unfavourable abiotic conditions and pollution, including endocrine disrupters (Chapter 5). Pre-ovulatory atresia may also provide a mechanism by which the fecundity of the female is adjusted in the light of prevailing conditions, but the importance of atresia as a mechanism for adjusting fecundity varies among species (Chapter 6).

Fertilisation

Although not part of oogenesis, the story of the oocyte/egg can be completed by describing the process of fertilisation (Kinsey *et al.*, 2007; Murata, 2009). After ovulation into the ovarian lumen, or in some species the coelomic cavity, the oocytes are carried along the gonoducts to the genital papilla for release into the water. The exception is viviparous species, with internal fertilisation (Chapters 1 and 10).

A spermatozoon can only gain access to the oolemma by passing through the micropyle, the channel that penetrates the zona pellucida at the animal pole of the oocyte (see the section 'Oocyte maturation'). The form of the micropyle varies between species, although the adaptive significance, if any, of this variation is unclear (Murata, 2009). The basic structure of the micropyle is funnel-shaped, with an exterior vestibule and the canal (Fig. 4.7). A type I micropyle has a deep vestibule and a short canal. Type I micropyles are seen in the eggs of some cyprinids, such as the European minnow (*Phoxinus phoxinus*, Cyprinidae) and tench (*Tinca tinca*, Cyprinidae). The Atlantic salmon (*Salmo salar*, Salmonidae), has a type II micropyle, with a flat vestibule and a longer canal. The closely related brown trout (*Salmo trutta*) has a type III micropyle, which lacks a vestibule and the canal tends to widen at the internal opening. A type IV micropyle,

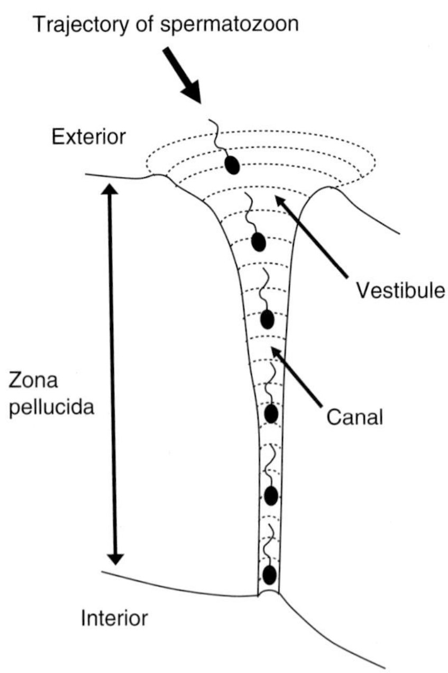

Figure 4.7 Diagram of micropyle of oocyte showing vestibule and canal and indicating the trajectory of spermatozoon. (Source: Adapted from Mellinger 2002. Reproduced with permission of CNRS Editions.)

such as that of the painted notie (*Notothenia larseni*, Nototheniidae), has both an inner and outer vestibule connected by a short canal. The size of the canal is usually such that it is only sufficiently wide for one spermatozoon to pass through it to reach the oolemma, with the spermatozoon travelling preferentially along grooves in the micropyle. This physical characteristic helps to prevent invasion of the oocyte by more than one sperm, a condition termed polyspermy.

While spermatozoa are unable to locate the micropyle at longer distances, in close proximity spermatozoa appear to be 'attracted' to the micropyle. The attractant is a glycoprotein bound to the chorion surface in the region of the micropyle. Protease treatment to remove this glycoprotein results in reduced fertilisation success (Yanagimachi *et al.*, 2013), implicating a role in successful fertilisation, and potentially serving as a mechanism for cryptic female choice (Eberhard, 1996) (Chapter 8).

The spermatozoon binds to the oolemma, probably because there is a sperm-binding molecule on the surface of the plasma membrane. In the presence of Ca^{2+} ions, the plasma membrane of the spermatozoon fuses with the oolemma. This results in the activation of the oocyte. In addition to insemination, an oocyte can also be activated by a physiological or mechanical stimulus. In *O. mykiss*, activation is a response to the environmental change the oocyte experiences at spawning, but in *Oryzias latipes* the oocyte can be activated by pricking with a needle. In the activated oocyte, the cortical alveoli release their contents into the gap between the egg and the zona pellucida, and a perivitelline space is formed as the zona pellucida separates from the oolemma.

The plasma membrane of the spermatozoon that has navigated down the micropyle to the surface of the oocyte fuses with microvilli of the oolemma. This fusion is completed within 15–20 seconds. Where the two membranes fuse, a fertilisation cone forms, blocking the exit of the micropyle and thereby helping to prevent more spermatozoa from gaining access to the oolemma. This also has the effect of drawing the fertilising spermatozoon deeper into the ooplasm. This first fertilisation cone collapses, but a second cone forms that persists for 3–4 minutes, disappearing when the entire spermatozoon is in the ooplasm.

The fusion of the spermatozoon plasma membrane with the oolemma evokes changes in the electrical properties of the oocyte. The resting potential of the oocyte, e.g. in *Oryzias latipes*, is –47 mV. But, with the fusion, there is a depolarisation to +4 mV for about 20 seconds, followed by a hyperpolarisation of +31 mV in amplitude, lasting about 2 minutes. Following depolarisation of the oocyte, there is a wave of release of intracellular Ca^{2+} from the endoplasmic reticulum in the cortical region of the oocyte. The release of Ca^{2+} starts near the animal pole and spreads across the oocyte. This wave initiates the release of the cortical alveoli by exocytosis, and the perivitelline space begins to form. The initial transient wave of Ca^{2+} release can be followed by further waves of release. In *D. rerio*, three of four transient waves follow the initial wave. Zones of increased Ca^{2+} concentration form cytoplasmic channels through which cytoplasm moves towards the animal pole and, as the cytoplasm accumulates at the animal pole, it forms the protrusion of the blastodisc. At the vegetal pole, the yolk accumulates.

When the redistributions of cytoplasm and yolk are complete, the blastodisc is ready to initiate the rounds of mitotic cell division that build the embryo. The stage is set for the zygote to initiate the process of sex determination and differentiation described in Chapters 2 and 3.

Meanwhile, the oocyte has emerged from arrest in the metaphase II stage of meiosis II and completed meiosis II. This results in the formation of two haploid cells. One of these is the oocyte, the other is a small second polar body, which, like the first polar body, degenerates. Thus, the meiotic division of the oocyte produces only a single functional haploid gamete. By contrast, in the male, the meiotic division of the spermatocyte produces four haploid gametes, the motile spermatozoa (see the section 'Spermatogenesis'). The nuclear membrane of the fertilising spermatozoon breaks down, releasing the chromosomes into the ooplasm. These chromosomes show decondensation and form the male pronucleus. The male and female pronuclei move towards the animal pole and remain close as the blastodisc enlarges until the first mitotic division of the now diploid zygote begins. The cytoplasmic components of the sperm atozoa, including the flagellum, break down within the ooplasm. The transition from oocyte to egg is defined by the transition from the haploid to the diploid chromosomal state.

The egg is protected by the chorion, which is a hardened derivative of the zona pellucida of the oocyte. The chorion may also have hair-like structures or other ornamentation on the surface. Alveolin and transglutaminase released by the cortical alveoli into the perivitelline space catalyse the reactions of subunits of the ZPs, resulting in the hardening. This modification of the zona pellucida also prevents polyspermy by blocking the access of other spermatozoa to the egg surface. The hardened chorion also has antimicrobial properties, which inhibit the colonisation of the egg by microorganisms.

Polyspermy is also prevented by other mechanisms. The structure of the micropyle limits the access of more than one sperm atozoa at a time. The formation of a fertilisation cone plugs the internal opening of the micropyle. Lectins released by the cortical alveoli interact with the excess spermatozoa, causing them to agglutinate and become immobilised. Perivitelline fluid pushes excess spermatozoa out of the micropyle canal and can form a 'fertilisation plug' in the vestibule of the micropyle. These mechanisms combine to ensure that only a single spermatozoon enters the egg.

In some species, including salmonids, there is evidence that when they are ovulated, the oocytes may not have full developmental competence. In *O. mykiss*, the success of fertilisation is highest about 4–5 days after ovulation (Springate *et al.*, 1984; Lubzens *et al.*, 2010).

A major variation in the site of fertilisation is seen in the small number of species, which show some form of viviparity (Chapter 10).

Genetic control of oogenesis

The analysis of the genetic control of oogenesis is in its infancy, with the techniques of molecular biology now being applied to the question. The so-called '-omics' – genomics, transcriptomics and proteomics – will be valuable in untangling the network of gene activations (up-regulation) and repressions (down-regulation) that control the process of oogenesis and the temporal patterns of gene regulation. Examples provide an early insight into such studies.

The liver is the source of VTG, and in many species, the ZPs. A study of the transcriptome (mRNAs) of the liver of *D. rerio* females in the early and later vitellogenic phases identified some of the genes regulated by E_2 (Levi *et al.*, 2009). These included the *vtg3* VTG gene and *err1*, the gene for an oestrogen receptor. The transcript of the VTG gene (*vtg3*) was the most up-regulated transcript in vitellogenic females. Females in the earlier and later phases of vitellogenesis had similar expression patterns, but the expression pattern of early-phase females was intermediate between pre-vitellogenic and later-phase females.

The transcriptome of *D. rerio* ovarian follicles identified the expression of several groups of genes (Knoll-Gellida & Babin, 2007). Not surprisingly, a group of 'housekeeping' genes was identified. These are required for the routine metabolism of the oocyte. There was a group of ovarian-related genes, including ZPs and groups related to metabolism involving metals: metallothionein (*mt2*) and ferritins (*fth1*). The analysis of the proteome of vitellogenic follicles is hampered because the protein content of the oocyte is vastly dominated by VTG and its derivatives (Chapovetsky *et al.*, 2007).

Dynamics of oocyte development

The story of oocyte development has been described sequentially and in some species this sequence is essentially followed by a cohort of oocytes, which thus develop in synchrony. The rate at which the oocytes grow may vary, but essentially at a given point in time they are all at the same stage. There is no replenishment of this cohort of developing oocytes from the earlier stages. This synchrony means that all the oocytes mature at the same time and are spawned over a short period of time. The species showing this synchronous pattern of ovarian development are total spawners (Wallace & Selman, 1981; see Chapter 1). In this pattern of ovarian development, the physiological control of oocyte development can also follow a sequential pattern, with the appropriate control operating at each successive stage.

Some species with synchronous oocyte development are semelparous (Chapters 1 and 11), dying after their first spawning. However, many species with synchronous oocyte development are iteroparous, so the oocytes destined to be spawned at the next spawning, usually in the following year, are recruited either from the pool of oogonia or from early-stage oocytes in the ovary.

In other species, more than one cohort of developing oocytes may be present in the ovaries simultaneously, which complicates the pattern of physiological control. In species with group synchronous development, at least two populations of oocytes are present. A cohort of larger oocytes develops synchronously so the oocytes mature and ovulate together. The second population consists of a more heterogeneous population of smaller oocytes from which another cohort of maturing oocytes is recruited in the same breeding season. The result is a series of spawnings, with successive clutches being produced. This mode of development is termed group synchronous (Wallace & Selman, 1981). In a variation of this pattern, more than one cohort of developing oocytes can be identified by their size distribution. Typically, the cohort of larger oocytes is maturing, whereas the second cohort of smaller oocytes is in the vitellogenic stage. Such a pattern is seen in *G. aculeatus*. In this species, clutches are spawned at intervals of a few days (Fletcher & Wootton, 1995). Such species are called multiple or batch spawners (Chapters 1 and 11).

A third pattern is that of asynchronous oocyte development (Wallace & Selman, 1981). Here there is no obvious cohort of developing oocytes. Oocytes at all stages of development are present in the ovary. This pattern is shown by *F. heteroclitus*. In these species, small batches of eggs are spawned at short intervals, even daily.

The Syngnathidae (pipefishes and seahorses) have unusual ovarian structures (see the section 'Structure of the ovaries'). Some species have an asynchronous pattern of oocyte development. In the broadnosed pipefish (*Syngnathus typhle*, Syngnathidae) females, there are mature oocytes present in the ovary irrespective of the timing of spawning and all stages of oocyte development are present. By contrast, in the straightnosed pipefish (*Nerophis ophidion*, Syngnathidae), development is group synchronous with a cohort of oocytes reaching maturity about 4 days after spawning (Sogabe & Ahnesjö, 2011).

In group synchronous and asynchronous ovaries, the physiological control of oocyte development is more complicated than in synchronous ovaries. There is the problem of simultaneously regulating oocytes at different stages of their development, so the relevant physiological controls must also be operating simultaneously (Chapter 5).

Spawning dynamics and fecundity

The dynamics of spawning have a major consequence for the fecundity of the female. In fisheries management, estimates of the total egg production by a population in a given year are important for the development of predictive models of the dynamics of the population (Chapter 12). For species that have synchronous development of the oocytes, with the eggs all spawned over a restricted period of time, the estimation of fecundity is uncomplicated. The number of oocytes in the developing cohort can be counted close to the time of spawning.

Where there is batch (multiple) spawning, the situation is more complicated. In such species, the breeding season fecundity (F_b) is defined by two variables: the fecundity per spawning (F_s) and number of spawnings performed in a breeding season (N_s), so:

$$F_b = F_s \times N_s$$

Two types of batch spawners are identified (Murua *et al.*, 2003; Korta *et al.*, 2010). In determinate spawners, the total number of oocytes that can be spawned in a breeding season is determined before the start of the breeding season. In indeterminate spawners, there can be recruitment to the vitellogenic cohorts during the breeding season. This potential recruitment during the breeding season complicates the estimation of fecundity (Chapter 12).

Oogenesis is only successful if at least some of the oocytes ovulated and spawned are successfully fertilised. This requires the presence of sexually mature males producing viable spermatozoa. The process of spermatogenesis is the process by which haploid spermatozoa are produced from the primordial germ cells.

Structure of the testes

A pair of testes develops from the gonadal anlagen, with one testis on either side of the midline. Each testis is supported from the dorsal wall of the peritoneum by a mesentery, the mesorchium. In immature males, the testes are inconspicuous, longitudinal strands of tissue lying dorsally in the body cavity. In mature males, there is considerable variation in the proportion of the total body mass that is contributed by the testes and this variation can be related to life history (Chapter 11) and the reproductive behaviour and mating system typical of the species (Chapters 7 and 8). In mature male salmonids and cyprinids, the testes account for 5–10% of body mass, in northern pike (*Esox lucius*, Esocidae) 2–3% and in tilapias (*Oreochromis* spp., Cichlidae), 0.2–2% (Billard, 1986).

Gross morphology of a testis

The testis is contained in a capsule that consists of the epithelium and an underlying layer of connective tissue, which together form the tunica albuginea (Blüm, 1986; Schulz *et al.*, 2010). The body of the testis has two compartments separated by a basal lamina (or basement membrane). As in the ovary, one compartment, the interstitium, consists solely of somatic cells, whereas the second compartment, the tubular component, contains the germinal epithelium. The tubular component consists of a complex of tubules or quasi-tubules separated by connective tissue (Parenti & Grier, 2004; Grier & Aranzábal, 2009; Schulz *et al.*, 2010). These open into the testicular ducts, which transfer the spermatozoa to male gonoducts.

The interstitial tissue contains macrophages and mast cells, neural and connective tissue, but it also contains Leydig cells. The Leydig cells are steroidogenic, with the characteristic well-developed smooth endoplasmic reticulum of steroidogenic cells (Chung *et al.*, 2010). Leydig cells are responsible for the synthesis of androgens (Chapter 5). They are comparable to the thecal cells of the ovary (see the section 'Gross morphology of the ovaries'). In some gobies (Gobiidae), there is an aggregation of Leydig cells in each testis, forming the mesorchial gland (Cole, 2010a). The cells secrete steroids, some of which act as pheromones (Chapter 5), which may play a role in mate choice (Chapter 8).

Delineated by the basal lamina and peritubular myoid cells, the tubular component contains the germinal epithelium. This has two cellular types, Sertoli cells and the germ cells. Sertoli cells are somatic gonadal cells comparable to the granulosa cells of the ovary. Germ cells form the lineage that generates the haploid spermatozoa. The Sertoli cells and germ cells have an intimate relationship. Sertoli cells surround a spermatogonium, forming a spermatocyst, and ensure that, thereafter, the spermatogonium is isolated from the basement lamina and from other cells.

Three distinct arrangements of the germinal epithelium have been recognised in the testes of teleost fishes: anastomising, tubular; unrestricted lobular; and restricted lobular testes (Parenti & Grier, 2004). The distribution of these three arrangements across teleostean shows a clear phylogenetic pattern (Fig. 4.8).

Taxa from the base of the teleostean phylogeny, including the Elopomorpha [e.g. the tarpon (*Megalops atlanticus*, Megalopidae)], Cypriniformes [e.g. the common carp (*Cyprinus carpio*, Cyprinidae)] and Esociformes (e.g. *E. lucius*), have anastomising tubular testes. The germinal compartments form tubules that loop around close to the tunica albuginea, doubling back and forming a branched, anastomising network, which open into the testicular ducts of the testis. In this arrangement, spermatogonia and Sertoli cells are distributed along the length of the tubules. As the spermatogenic spermatocysts develop, they remain in approximately the same position along the tubule (Fig. 4.9A). Eventually, the spermatocyst releases its spermatozoa into the lumen of the tubule, from where they are transported to the sperm ducts. In the Salmonidae and Esocidae, the tubules open into efferent testicular ducts, and hence into the testicular main ducts. These efferent ducts form a transitional phase between the tubules and the testicular main ducts. In other teleosts, the tubules open directly into the ventrally located testicular main ducts (Lahnsteiner & Patzner, 2009).

Phylogenetically more advanced teleosts such as the Perciformes have unrestricted lobular testes. The germinal compartment extends to close to the tunica albuginea, ending blindly close to the periphery of the testis forming a lobule. Each lobule opens into the testicular duct. As in the anastomising tubule testis, the spermatocysts remain in approximately the same position along the lumen as they develop. Spermatocysts at different stages of development of their germ cells are distributed throughout the tubules, with no obvious spatial pattern in the distribution (Fig. 4.9B).

Gadus morhua has a variant of the unrestricted tubule pattern of testis development (Almeida *et al.*, 2008). Each testis is composed of several lobes. Spermatozoa are drained from the tubules by a central testicular duct. For spermatocysts within a lobe, there is a gradient of development. Spermatocysts with germ cells in their early stages of spermatogenesis develop at the blind end of a tubule close to the periphery of the lobe. As more spermatocysts are recruited from the germinal epithelium, earlier

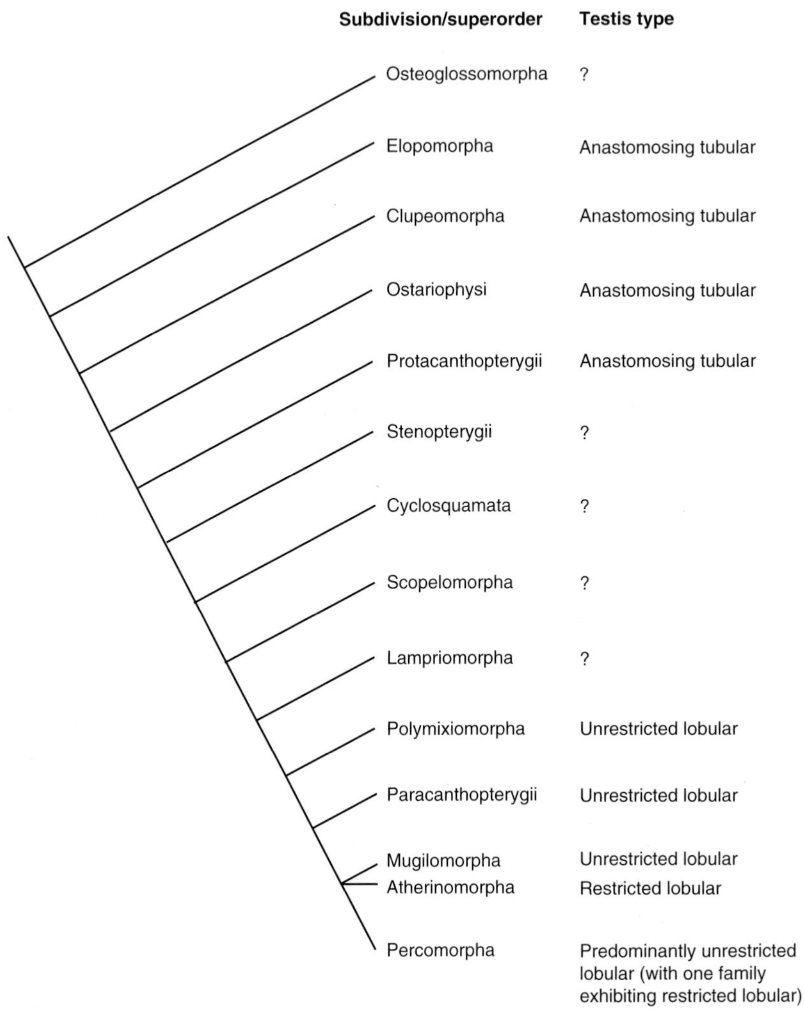

Figure 4.8 Phylogenetic relationships within contemporary teleosts showing testis type. In the Percomorpha, a restricted locular testis is exhibited in the infantfishes (Schindleriidae), a highly paedomorphic group. (Source: Adapted from Helfman *et al.* 2009. Reproduced with permission of John Wiley & Sons. Data from Parenti & Grier 2004.)

spermatocysts are forced down the tubule by appositional growth. As a consequence, spermatocysts with germ cells in the later stages of spermatogenesis accumulate near the testicular duct. As this process takes place, the testis grows rapidly. In August in Norwegian coastal populations of *G. morhua*, the GSI, the gonad mass as a proportion of total mass, is about 0.2%, but the following March it is over 8%. At the end of this growth phase, recruitment of spermatocysts ceases and tubules come to contain only spermatocysts in the later stages of spermatogenesis, so the spatial gradient of development is lost.

The Atherinomorpha, which includes the Atheriniformes; e.g. Atlantic silverside (*Menidia menidia*, Atherinopsidae), Cyprinodontiformes; e.g. *P. reticulata*, and

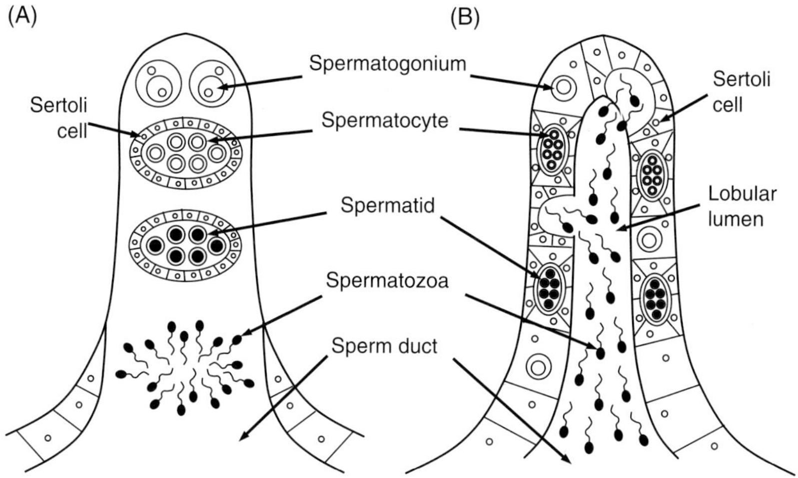

Figure 4.9 Generalised diagram of testicular structure illustrating: (A) the tubular type; (B) the lobular type. (Source: Adapted from Nagahama 1983. Reproduced with permission of Elsevier.)

Beloniformes; e.g. *O. latipes*, have restricted lobular testes. This type of testis is characteristic of this phylogenetic grouping. The defining characteristic of the restricted lobule testis is that spermatogonia only occur around the blind ends of the lobules close to the tunica albuginea. As spermatogenesis progresses, the spermatocysts migrate towards the testicular ducts and release the spermatozoa into the ducts. The lobules lack a lumen because processes from the Sertoli cells bridge the width of the lobules.

The three different arrangements of testes show a clear phylogenetic pattern, but it is not clear whether the differences have an adaptive significance in relation to sperm production (Fig. 4.8; Parenti & Grier, 2004). It may be coincidental that the Atherinomorpha includes, at least for the teleosts, a relatively high proportion of viviparous species (Chapter 10).

A feature that distinguishes the teleosts from mammals is that in teleosts the population of Sertoli cells is replenished. This, and the mitotic replenishment of spermatogonia, means that the germinal epithelium is active throughout life, or at least until senescence of the individual sets in.

Development of male gonoducts

In most jawed vertebrates, the sperm duct that transfers sperm from the testis to the exterior or an intromittent organ, is derived from the Wolffian duct, an embryonic component of the developing urinary system (Blüm, 1986). However, in the teleosts, the genital system is completely separate from the renal system (Romer & Parsons, 1986; Lahnsteiner & Patzner, 2009). The gonoducts that transfer the sperm from the testis are not derived from the kidney–urinary complex. As in female teleosts, the male gonoducts are new structures and their mode of origin varies from species to species (Cole, 2010a). Typically, the gonoduct has three cellular layers, external to its lumen: an outer peritoneum, a middle layer of connective tissue and an internal epithelium. This epithelium and that of the testicular ducts secrete into the lumen material that provides the spermatozoa with nutrition and defines the environment for the spermatozoa prior to spawning. Behind the testis, the epithelium of the gonoduct becomes folded and the folds may fuse, forming a complicated system of tubules (Lahnsteiner & Patzner, 2009). This folding increases the surface area of the secretory epithelium. More posteriorly, the gonoducts fuse to form a single duct with a simple epithelium.

In the cyprinodont *X. helleri*, the male gonoduct develops in a pattern similar to the female (see the section 'Development of female gonoducts'). Anteriorly, the proliferation of peritoneal cells at the caudal end of the testis becomes an extension of the efferent testicular duct and becomes continuous with the extragonadal gonoduct. *P. reticulata* is also a cyprinodont, but in this species the male gonoducts develop from masses of somatic cells that arise from the caudal end of the testes. Initially, these masses are extended posteriorly as solid, bilateral ridges, which then fuse to form a single ridge. A lumen develops, which becomes continuous with the efferent testicular duct system. As with the female gonoducts, it is not clear why the teleosts have evolved male gonoducts *de novo* rather than following the pattern found in other jawed vertebrates.

Accessory structures associated with the testes and gonoducts

In some teleosts, there are modifications of the testicular duct or the gonoduct. In catfish (Siluriformes), such as the African sharptooth catfish (*Clarias gariepinus*, Clariidae), tubular structures develop from the caudal end of the testis, where spermatogenic tissue merges into the testicular duct. These tubular structures, called seminal vesicles, have an epithelium that has the same origin as the Sertoli cells (Schulz *et al.*, 2010).

In some blenny species (Blenniidae), each testis has a testicular gland (Lahnsteiner & Patzner, 2009). This gland has tubules, with a monolayer epithelium and an interstitium, which consists of fibrocytes, blood capillaries and Leydig cells. The testicular glands show much interspecific variation. In the peacock blenny (*Salaria pavo*, Blenniidae), the gland represents 15–30% of the spermatogenic portion of the testis, whereas in *Microlipophrys* spp. (Blenniidae) it represents 220–250%. Some blenniid species have testicular blind pouches. These occur as a pair of evaginations of the gonoduct. A notable modification of the male gonoduct is found in many gonochoristic gobies (Gobiidae), as well as toadfish (Batrachoididae), cottids (Cottidae), blennies (Blenniidae), bitterling (Cyprinidae) and several families of catfish (Siluriformes). A pair of transparent, secretory structures, called 'sperm duct' glands (Miller, 1984) open into the gonoduct (Fig. 4.10). Each gland consists of large lobules with a simple columnar epithelium, which produce a sialoglycoprotein-rich fluid (mucin). In some species, the secretions of the sperm-duct glands mix with spermatozoa and seminal fluids and the male lays down sperm trails on the spawning surface. Spermatozoa are slowly released from this sperm trail over an extended period after ejaculation (Miller, 1984; Marconato *et al.*, 1996; Scaggiante *et al.*, 1999; Petersen *et al.*, 2005), prolonging the period over which fertilisation can occur. Male gobies typically nest in holes or small caves. Female egg deposition can take several hours, and sperm trails ensure efficient fertilisation of eggs deposited in the nest. Males are also liberated for mate and nest guarding during oviposition (Marconato *et al.* 1996) (Chapter 8). In some species, the sperm-duct gland is augmented with a discrete, glandular mass immediately in front of where the gland unites with the gonoduct. Other species lack sperm-duct glands or they are reduced (Miller, 1984; Cole, 2010b). There is a need to examine how these interspecific variations relate to variation in reproductive behaviour and mating system.

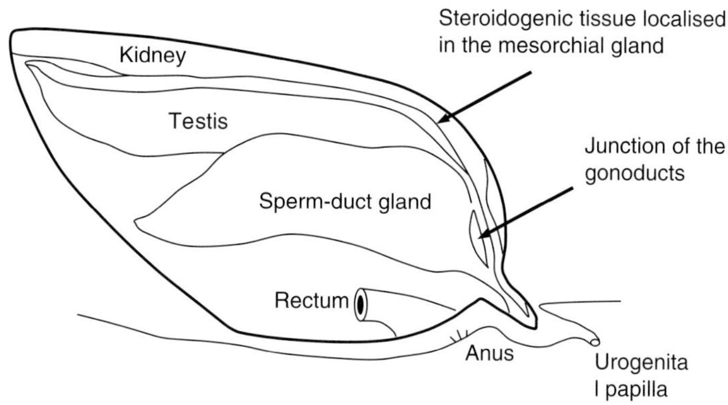

Figure 4.10 Diagrammatic longitudinal section of the giant goby (*Gobius cobitis*, Gobiidae) showing testis and 'sperm duct gland'. (Source: Mellinger 2002. Reproduced with permission of CNRS Editions.)

Spermatogenesis

As for oogenesis, the temporal sequence of spermatogenesis is regulated by the gonadotrophins secreted by the pituitary gland, FSH and LH (Chapter 5). There is also a parallel with oogenesis in that the effect of FSH tends to dominate during the earlier stages of spermatogenesis and the effects of LH play a role later in the process.

Spermatocysts

The unit of development in the testis that is equivalent to the follicular complex in the ovary (see the section 'Chromatin nucleolus stage') is the spermatogenic spermatocyst (Billard, 1986; Grier & Aranzábal, 2009; Schulz *et al.*, 2010). Each spermatocyst has two cellular components: Sertoli cells and germ cells. Sertoli cells are derived from somatic cell precursors in the germinal epithelium and so are comparable to the granulosa cells of the ovarian follicle complex. The germ cells are ultimately derived from the PGCs (Chapter 3).

A spermatocyst is formed when a type A spermatogonium in the germinal epithelium is enveloped by extensions of a Sertoli cell (Chapter 3), converting it into a type B spermatogonium. This then undergoes a species-typical number of mitotic divisions, generating an isogenic clone of spermatogonia. These spermatogonia are linked by cytoplasmic bridges and the divisions of the spermatogonia are synchronised (Fig. 4.11).

Sertoli cells

These cells are essential for the successful development of a spermatogenic spermatocyst. As the clone of germ cells of each spermatocyst increases in abundance, the number of Sertoli cells also increases. This increase is, at least partially, a result of the proliferation of Sertoli cells. The number of Sertoli cells associated with a given stage of germ cell development is species-specific. The number of Sertoli cells in a testis quantitatively limits sperm production. The proliferation of Sertoli cells in the teleost testis contrasts with the situation in the mammalian testis, in which the Sertoli cells are recycled during spermatogenesis.

The Sertoli cells of *D. rerio* illustrate their structure (Rupik *et al.*, 2011). The nucleus, with one or two nucleoli, is located close to the basement lamina. The cytoplasm

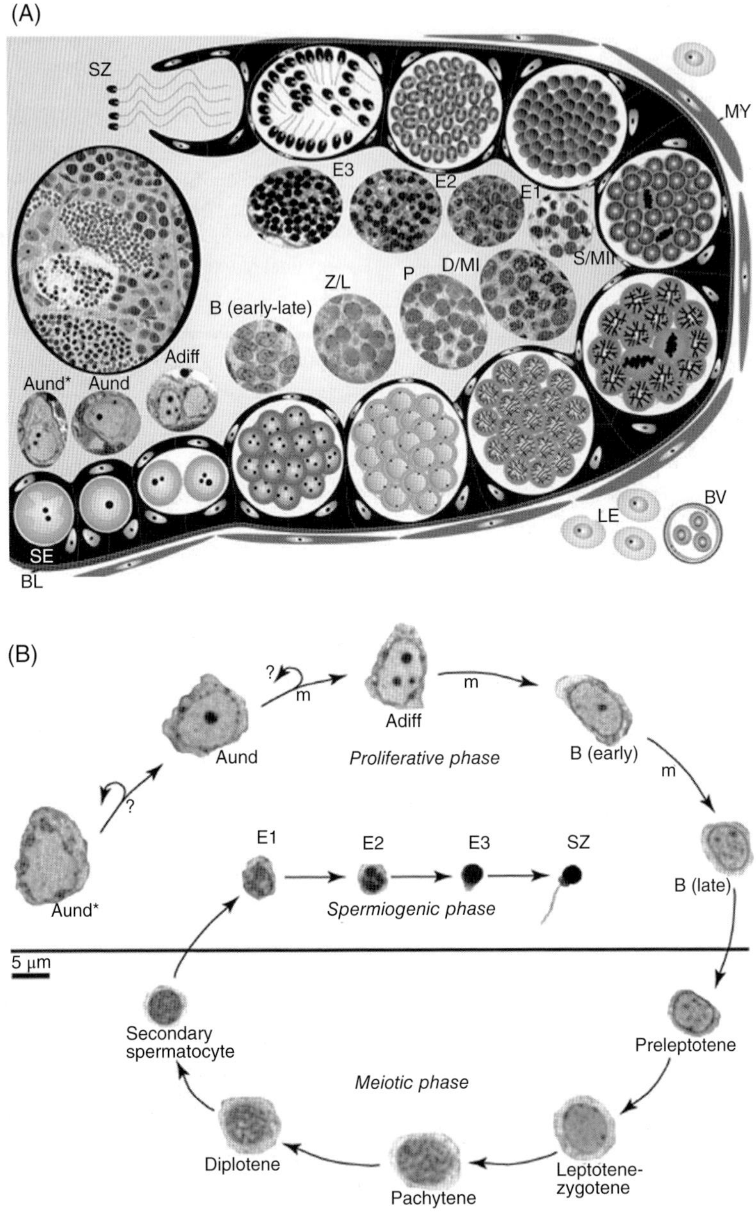

Figure 4.11 Spermatogenesis. (A) Developmental sequence during spermatogenesis from spermatogonia to spermatozoa in the zebrafish (*Danio rerio*), illustrating spermatogenic cysts in a tubule. (B) Developmental sequence of spermatogonia to spermatozoa. The first question mark (?) indicates ambiguity about whether A_{und}* and A_{und} are separated by a mitosis or represent different stages of the same cell cycle (see Figure 11.4A). The second question mark indicates ambiguity as regards the extent to which A_{und} has stem cell-like competence. Abbreviations: A_{und}*, undifferentiated spermatogonia (stem cell?); A_{und}, undifferentiated spermatogonia; A_{diff}, differentiated spermatogonia; B, spermatogonia type B; Z/L, leptotenic/zygotenic primary spermatocytes; P, pachytenic primary spermatocytes; D/Mi, diplotenic spermatocytes/metaphase I; S/MII, secondary spermatocytes/metaphase II; E1, E2, E3, early, intermediate and final spermatids; SZ, spermatozoa; SE, Sertoli cells; BL, basal lamina; MY, peritubular myoid cells; LE, Leydig cells; BV, blood vessels. (Source: Schulz *et al*. 2010. Reproduced with permission of Elsevier.)

contains both rough and smooth endoplasmic reticulum, other vesicular structures and numerous mitochondria. Tight gap junctions are eventually established between the Sertoli cells of a spermatocyst, which ensure that the clone of germ cells is completely isolated from cells exterior to the spermatocyst (Fig. 4.11A). This isolation is established when the germ cells have become haploid. Thus, all communication between the germ cells and the rest of the fish's body is mediated by the Sertoli cells.

Sertoli cells have numerous roles during spermatogenesis. They are essential for the survival, development and functioning of germ cells. They secrete the fluid that fills out the lumen of the tubules. They also act as phagocytes, efficiently engulfing any spermatozoa left over after spawning. They also engulf residual bodies that are ejected by the spermatids as they transform into spermatozoa. Cells similar to Sertoli cells also form the epithelium of the efferent testicular ducts.

Stages of spermatogenesis within a spermatogenic spermatocyst

During the development of germ cells within a spermogenetic spermatocyst, the cells pass through successive stages: differentiated spermatogonia, primary spermatocytes, secondary spermatocytes, spermatids and spermatozoa (Fig. 4.11). The initiation of meiosis marks the transition between spermatogonia and spermatocytes (Roosen-Runge, 1977; Blüm, 1986; Schulz *et al.*, 2010).

Differentiated spermatogonia

When a type A spermatogonium becomes associated with a Sertoli cell, it is one of the largest cells in the testis. In *Oreochromis* spp., it has a volume of about 2300 μm^3. Type A spermatogonia also retain the cytoplasmic, electron-dense material, nuage, that is characteristic of the PGCs (Chapter 3). When a type A spermatogonium converts to the type B, a series of mitotic divisions is initiated. Type B spermatogonia are smaller and have a smaller nucleus, but still retain nuage material. A type B spermatogonium in *Oreochromis* spp. has a volume of about 160 μm^3, but after the series of mitotic divisions, there are about 120 per spermatocyst, so the total volume is around 19 200 μm^3. This growth explains why, in many species, there is a large increase in GSI as the testes mature. Note that, in the female, the increase in GSI is largely the result of the accumulation of yolk in the oocytes, a process for which there is no equivalent in the testes. Type B spermatogonia are still diploid cells.

Primary spermatocysts

At the end of the cycle of mitotic divisions, each spermatogonium in the clone enters meiosis I, and so becomes a primary spermatocyte. Primary spermatocytes lack the nuage material. Features of meiosis I include the doubling of each chromosome to form chromatids and crossing-over between homologous chromosomes, which are linked by synaptonemal complexes. At the end of meiosis I, two secondary spermatocytes have been produced, each with a haploid number of chromosomes, but with two chromatids per chromosome.

Secondary spermatocysts

These secondary spermatocytes rapidly pass through meiosis II, which generates four haploid spermatids. Thus for each spermatogonium entering meiosis I, four haploid spermatids emerge. The rapidity with which meiosis II is completed means that secondary spermatocytes are relatively rare in the testis at any one time. This pattern in the male is in marked contrast to the events in the female, in which meiosis I and II generate only a single haploid oocyte, with the first and second polar bodies degenerating and making no contribution to the pool of haploid gametes. There is probably some cell death as the cells of a clone of a spermatocyst increase in abundance and these are phagocytosed, but the process of spermatogenesis is efficient. In *G. morhua*, there are about 1900 spermatocytes per spermatocyst at the pachytene stage of meiosis I.

Spermatids

The haploid spermatids go through spermiogenesis, a major morphological rearrangement that produces the spermatozoa. In *D. rerio*, spermiogenesis can be divided into four arbitrary stages (Rupik *et al.*, 2011), and these are probably typical of other teleosts. The early spermatids remain linked by cytoplasmic bridges. Each spermatid has a spherical nucleus, but the nuclear membrane has a shallow groove, or fossa, on one side of the nucleus, where the centrioles are eventually located. In the second stage, the chromatin in the nucleus of the spermatid is more compact and the spermatid is smaller. Two centrioles develop and are located in the fossa. A flagellum starts to form, with the distal centriole forming the basal body of the flagellum. The chromatin in the nucleus in the third stage is even more compact. The flagellum becomes linked to the distal centriole and is surrounded by mitochondria. In the fourth stage, the nucleus is a regular sphere, with condensed chromatin. The flagellum is attached at about 110° to the spermatozoon head. Excess cytoplasm, called the residual body, is ejected from the spermatid and is phagocytosed by Sertoli cells. During the third and fourth stages, the spermatids lose their intercellular links and their association with the Sertoli cells. At the completion of spermiogenesis, the links between Sertoli cells weaken. The spermatocyst ruptures, releasing spermatozoa into the tubule lumen (Fig. 4.11a).

In some species, germ cells are released into the tubule lumen at an earlier stage of development and spermiogenesis is completed in the tubule lumen. This is known as semi-cystic development and is associated with adaptations for high rates of sperm production (Chapter 8). In blenniids with a testicular gland, the final stages of spermiogenesis are completed in the testicular gland, which regulates the final stages of spermatid differentiation (Lahnsteiner & Patzner, 2009). The secretions of the testicular gland provide nutrition for the spermatids and spermatozoa and add sialomucins to the seminal fluid (Lahnsteiner *et al.*, 1993). A number of teleost families produce mucin-rich seminal fluids (Chapter 8). Like the Sertoli cells, the epithelial cells of the testicular gland phagocytise spermatids remaining after spawning.

In teleosts, three types of spermiogenesis have been defined based on the position of the flagellum in relation to the nucleus (Quagio-Grassiotto & Oliveira, 2008; Schulz *et al.*, 2010; see Fig. 4.12). In type I spermiogenesis, the early spermatid has a central nucleus, with two centrioles and one or more mitochondria. The centrioles migrate in the direction of the nucleus and draw with them the flagellum and the cell membrane, which creates a cytoplasmic canal. The nucleus develops a fossa and rotates through 90° with respect to the axoneme of the flagellum, which is perpendicular to the nucleus. The centrioles move, either fully or partially, into the fossa. The centriole closest to the nucleus is the proximal centriole and the other is the distal centriole. *D. rerio* exemplifies type I spermiogenesis (Rupik *et al.*, 2011). There is no nuclear rotation in type II spermiogenesis. The centrioles migrate in the direction of the nucleus but do not enter the fossa, and the flagellum is tangential to the nucleus. This type of spermiogenesis is typical of perciform fishes. In type III spermiogenesis, the flagellum is

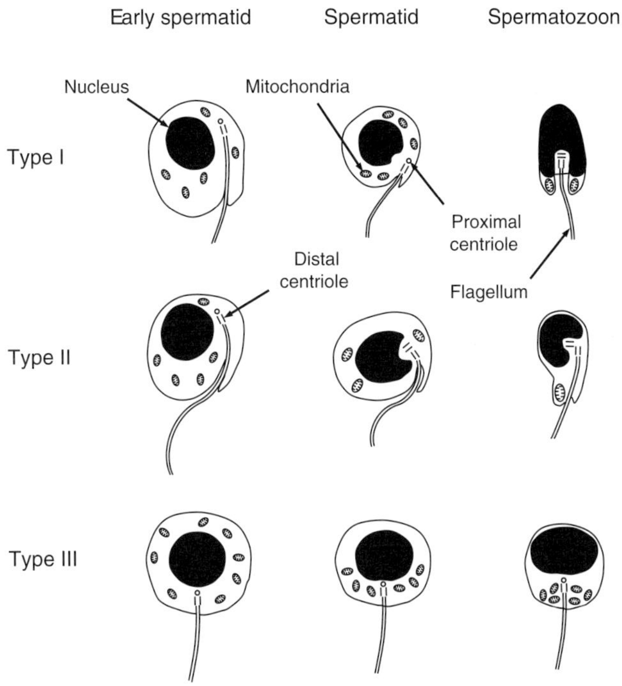

Figure 4.12 Stylised examples of types I, II and III spermiogenesis. (Source: Adapted from Billard 1986 and Quagio-Grassiotto & Oliveira 2008.)

central, but there is no nuclear rotation. Type III spermiogenesis is described in the barred sorubim (*Pseudoplatystoma fasciatum*, Pimelodidae) and long-whiskered catfish (*Pimelodus maculatus*, Pimelodidae) (Quagio-Grassiotto & Oliveira, 2008), and appears common among siluriform catfishes, as well as occurring in characiform fishes. Different patterns of spermiogenesis reflect variation in spermatozoa structure and are conserved within taxa, permitting their use in phylogenetic analyses (Jamieson, 1991). The possible adaptive significance of this variation for fertilisation efficiency, sperm competition and mating system evolution (Chapter 8) is unclear (see the section 'Structure and phylogenetic distribution').

Spermiation and capacitation

The breakdown of the connections between Sertoli cells and the release of spermatozoa into the tubule lumen, and hence into the efferent sperm ducts, is the process of spermiation (Schulz *et al.*, 2010). At this point, in many species, sperm (termed 'milt' in aquaculture) can be expressed by gently squeezing the abdomen of the male. This technique of stripping is used, e.g. in the aquaculture of salmonids, to fertilise eggs stripped from gravid females (Chapter 12).

The gonoducts and testicular ducts fill with spermatozoa. At this stage, the spermatozoa lack motility and this state is maintained by the ionic composition of the seminal fluid. The final stage in the development of a spermatozoon capable of fertilising an egg is the process of capacitation, including the acquisition of motility (Longo, 1987). These changes are physiological and do not require a morphological change (Schulz *et al.*, 2010). This final maturation of the spermatozoa can take place

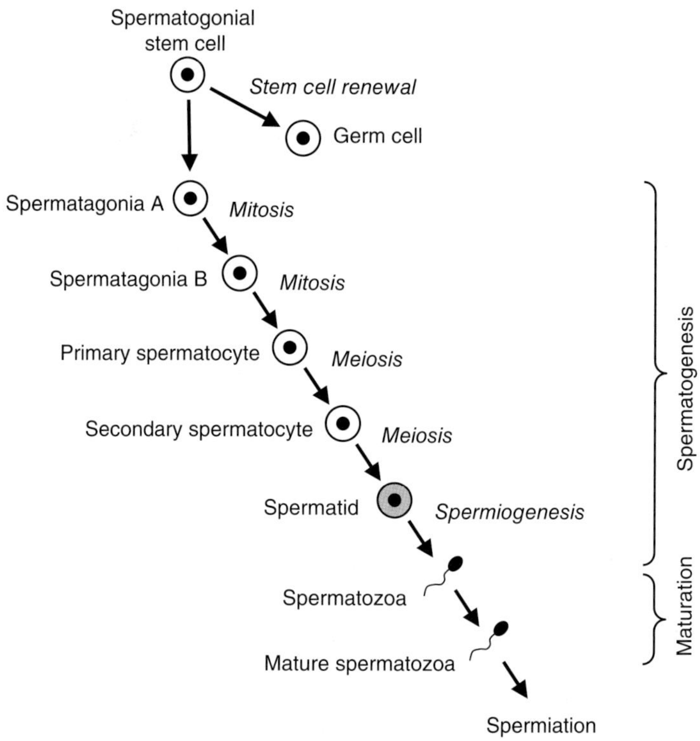

Figure 4.13 A summary of the process of spermatozoa development and maturation in male fish. (Source: Mañanós et al. 2009. Reproduced with permission of Taylor and Francis Group.)

in the testicular ducts, the gonoduct and the external environment. The final acquisition of motility can be a response to a change in the osmotic environment, the pH or the concentrations of inorganic ions such as K^+ or Ca^{2+} (Kinsey *et al.*, 2007). Chemical cues from the egg may also play a role in promoting the motility of spermatozoa (see also Chapter 8). The process of spermatozoa development and maturation in males is summarised in Figure 4.13.

Types of spermatozoa

Structure and phylogenetic distribution

As in other vertebrates, a teleost spermatozoon has three components: a head containing the nucleus; a middle piece containing mitochondria and a central core; and a tail, which contains the flagellum. The flagellum consists of the axoneme, with the typical pattern of 9 + 2 microtubules and the surrounding plasma membrane. The flagellum may have fins running down its length. Teleosts differ from other vertebrates in lacking an acrosome at the anterior end of the spermatozoon. This lack reflects the mode of fertilisation in teleosts, in which the spermatozoon enters an egg through the micropyle (see the section 'Fertilisation'), rather than penetrating the zona pellucida of the oocyte.

The diversity of the teleosts is partly reflected in the structure of their spermatozoa (Jamieson, 2009). The pattern of spermiogenesis (type I, type II and type III) is mirrored in the structure of the spermatozoa in terms of the positions of nucleus, centrioles and flagellum, so even without information on the pattern of development of the spermatozoa, the type of spermiogenesis may be inferred from the structure of the spermatozoa (Fig. 4.12).

The structure of spermatozoa reflects the effects of natural and sexual selection within the constraints of phylogeny. Although the phylogenetic distribution of spermatozoa types is well documented (Jamieson, 1991), the correlation between the form of the spermatozoa and the events surrounding fertilisation is still poorly understood. The Elapomorpha, one of the basal teleost groups in terms of phylogeny (Chapter 1), is unusual in that the flagellum has a 9 + 0 formula for the microtubules. Extensions of the proximal centriole run as bundles of four and five microtubules towards the tip of an elongate nucleus. In the Ostariophysi, types I, II and III all occur. Amongst the Acanthopterygii, the species so far described have types I and II, but groups such as the Blennioidea and Gobioidea show a wide range of modifications of these basic patterns. In many species with internal fertilisation (Chapter 10), there is a pronounced elongation of the spermatozoa head, which presumably confers some advantage in the internal environment of the female reproductive tract.

A number of species across a range of teleostean orders have spermatozoa with two flagella rather than the usual one. Among them are catfish in the genus *Ictalurus*. In the catfish family Ariidae, there are two axonemes, but these are contained within a single flattened flagellum. A rare variation is found in the dwarf pencilfish (*Lepidogalaxias salamandroides*, Lepidogalaxiidae), whose taxonomic status is uncertain. The spermatozoa of this species have an acrosome-like structure at the anterior end of the spermatozoon, which correlates with the lack of a micropyle in the egg. Presumably, fertilisation involves an interaction between the 'acrosome' and plasma membrane of the egg (Leung, 1988). The most unusual spermatozoa occur in two families of the Osteoglossomorpha, again a basal group of the teleosts (Chapter 1). In the Mormyridae and the Gymnarchidae, the spermatozoa are rounded cells that lack flagella, although two centrioles are present in the cytoplasm.

Spermatozoa viability

In externally fertilising species, once released into the external environment containing the eggs, the spermatozoa only have a short period of activity during which they must make contact with an egg. They quickly lose their capacity for active locomotion by the sinusoidal beating of the flagella. When released into the ambient environment, the spermatozoa experience a major change in the osmotic environment compared with that of the seminal fluid. For freshwater species, the spermatozoa move into a hypo-osmotic environment, while in marine species the move is into a hyperosmotic environment. This change triggers the onset of motility of the spermatozoa. There is a tendency for the period of motility to be longer in marine than in freshwater species. In the former, the period is about 2–10 minutes, but in the latter, it is 0.5–2 minutes (Alavi & Cosson, 2006; Cosson *et al.*, 2008; Cosson, 2010; Lahnsteiner, 2011). The decline in motility reflects the damage done to the structure of the spermatozoa by the osmotic shock and the exhaustion of the energy reserves of the spermatozoa.

The spermatozoa swim at species-typical velocities – e.g. in *G. morhua*, 65–100 μm/s; in the Atlantic halibut (*Hippoglossus hippoglossus*, Pleuronectidae) 150–180 μm/s;

in the turbot (*Scophthalmus maximus*, Scophthalmidae) 220 µm/s. These velocities and the duration of motility translate into the distances over which the spermatozoa can travel in search of an egg: in *G. morhua*, about 14 mm; *H. hippoglossus*, 9 mm; in *S. maximus*, 12 mm. These distances require that the spermatozoa are released by the male in close proximity to where the eggs are spawned and so have implications for the courtship and spawning behaviour of the male and female (Chapter 8).

Movement of the spermatozoa is driven by waves that propagate down the flagellum from the head of the spermatozoon to, at least initially, the tip of the flagellum (Cosson, 2010). The waves are typically arcsine in form, which means that straight segments alternate with curved segments. Several waves occupy the flagellum at any one time. Initially, once the spermatozoon is activated, the waves run the complete length of the flagellum, but as the spermatozoon ages, the waves become restricted to the section close to the head as the spermatozoon loses its motility. The beat frequency of the flagellar waves is related to the rate of ATP hydrolysis, which controls the dynein-ATPase activity that generates the waves. The ATP content of a spermatozoon rapidly depletes during its period of motility.

Beat frequency is sensitive to the water temperature, with the frequency higher at higher temperatures, but this leads to a shorter period of motility (Cosson, 2010). In *O. mykiss*, the beat frequency is 55 Hz, with a period of motility of 14–20 seconds at 26 °C. At 5 °C, the beat frequency is 32 Hz with a motility period of 140–180 seconds. The spawning site preferences of *O. mykiss* mean that the values at the lower temperature are more ecologically relevant.

In species whose eggs are laid in a confined volume, the period of motility of the spermatozoa may be lengthened by the viscous, protein-rich ovarian fluid that surrounds the egg mass at spawning. Female *G. aculeatus* spawn their eggs in the confined volume of the male's nest. The ovarian fluid surrounding the eggs protects, to some extent, the change in the osmotic pressure experienced by the spermatozoa and may extend the period of motility when the eggs are spawned in freshwater, enabling a high proportion of the eggs to be fertilised (Elofsson *et al.*, 2003; Le Comber *et al.*, 2004a). This property of the ovarian fluid may have been a factor in allowing *G. aculeatus* to invade freshwater environments, although other populations breed in the sea or in brackish water (Elofsson *et al.*, 2003). Ovarian fluid may also play a role in biasing fertilisation success towards certain males (Yeates *et al.*, 2013), a form of cryptic mate choice (Chapter 8).

When a spermatozoon reaches an egg, it tends to move close to the surface of the egg, essentially moving in a two-dimensional plane rather than in the three dimensions of the open water (Cosson *et al.*, 2008). This pattern makes it more likely that a spermatozoon will reach the micropyle. Some eggs have grooves in the chorion that tend to guide a spermatozoon towards the micropyle (Fig. 4.7). An unusual situation is seen in the Pacific herring (*Clupea pallasi*, Clupeidae), which spawns in huge aggregations close to the coast. In this species, the spermatozoa only become motile if they reach the zona pellucida of an oocyte (Cosson *et al.*, 2008). The zona pellucida contains a herring spermatozoa-activating peptide and a spermatozoa motility initiation factor.

Once at the micropyle, a successful spermatozoon moves down the micropylar canal to the surface of the egg and initiates the events described earlier (see the section 'Fertilisation'). The haploid gametes are combined and can form a diploid zygote of the offspring generation. The ontogenetic processes that lead to the differentiation of the gonads described in Chapter 3 are thereby initiated.

Genetic control of spermatogenesis

Studies on the transcriptome of the testis at successive stages of spermatogenesis are starting to provide a picture of the genes involved in the process of spermatogenesis (Schulz *et al.*, 2010). Clusters of genes that are involved in the control of mitosis and meiosis show temporal patterns of expression that correlate with the phases of mitotic proliferation of the differentiated spermatogonia and the meiotic divisions of the primary and secondary oocytes. Another cluster of genes code for growth factors that belong to the tissue growth factor β (TGF-β) superfamily of growth factors, including anti-Müllerian hormone (AMH) and gonad-soma derived factor (GSDF). These genes are expressed in the Sertoli cells and provide a system by which the Sertoli cells control the growth of the germ cells in the spermatogenic cyst. Some evidence suggests that AMH inhibits the proliferation of spermatogonia, but this inhibition is released by androgens produced by the Leydig cells stimulated by FSH from the pituitary. The proliferation is then promoted by the growth factor activin. Other factors implicated in the proliferation of spermatogonia and spermatocytes are insulin-like growth factor (IGF-I and IGF-II) and growth hormone (GH). The participation of factors such as IGF-I, IGF-II and GH in spermatogenesis may reflect the need for the organism to integrate the activities of the reproductive and growth functions, but the details are poorly understood.

Dynamics of spermatogenesis

Generally, for males, reproductive success depends on the number of eggs the male can successfully fertilise (see Chapter 8). Even in teleost species that are total spawners (Chapter 1), in which females deposit all their eggs over a short period in the spawning season, males will tend to stay on the spawning grounds longer than the females, in order to encounter as many gravid females as possible. Consequently, a male must have a sufficient store of spermatozoa in the testes and gonoducts to give it a chance of spawning with several females. In species in which the female is a batch spawner (Chapters 1 and 4), the male must have a sufficiently large store of spermatozoa to ensure it can fertilise several batches of eggs during the breeding season. In other species, the breeding season is not strongly seasonal but is prolonged over several months. In such species, the females may be able to spawn virtually every day.

Consequently, two main patterns of spermatogenesis can be recognised. In seasonal breeders, whether total or batch spawners, spermatogenesis occurs before the start of the spawning season, so at the start of the season a male has a large volume of spermatozoa available (Wootton, 1998; Knapp & Carlisle, 2011). This pattern can be compared with the synchronous pattern of oocyte development (see the section 'Dynamics of oocyte development'). In autumn spawners such as many salmonids, spermatogenesis occurs over the summer and early autumn. In early spring spawners, such as *E. lucius* and European perch (*Perca fluviatilis*, Percidae), spermatogenesis starts in autumn and continues over winter. In many species that spawn in summer, such as the cyprinid *T. tinca*, spermatogenesis occurs in spring and early summer. In *G. morhua*, a marine, batch-spawning species that reproduces in spring, spermatogenesis occurs over autumn and winter (Almeida *et al.*, 2008). Progressively, the testicular tubules and gonoducts fill with spermatozoa, while the volume occupied by the earlier stages of spermatogenesis, the spermatogonia and spermatocytes, decreases. After the spawning season, spermatogonia are the only germ cells present in the testes. Sertoli cells complete the phagocytosis of any residual spermatocytes and spermatozoa.

By contrast, *O. latipes* has a prolonged breeding season, with females spawning virtually every day. The testes of sexually mature males contain, simultaneously, all stages of spermatogenesis from spermatogonia to spermatozoa, so there is continual production of spermatozoa (Iwasaki *et al.*, 2009). Even during the non-breeding season, the testes contain all stages of spermatogenesis, although the numbers of spermatogonia and spermatocytes decrease in comparison to the situation in the breeding season. The ratio of primary spermatocytes to type B spermatogonia is also lower in the non-breeding season. This temporal pattern is close to the asynchronous pattern of oocyte development (see the section 'Dynamics of oocyte development').

These different dynamics of spermatogenesis raise the question of how the temporal patterns of spermatogenesis are controlled by the neuroendocrine system (Chapter 5).

Quantitative analysis of sperm production

Individual eggs are comparatively easy to count, given their size, so the fecundity of a female when it spawns can be quantified (Chapter 11). In comparison, the quantitative analysis of sperm production is technically more difficult. If a mature male fish can be stripped of sperm by pressure on the abdomen, it is not clear whether the volume produced is the same as the volume that would be produced by the male spawning naturally. Stripping sperm from males has been used routinely in studies with teleosts (e.g. Gage *et al.*, 1995; Janhunen *et al.*, 2009), on the assumption that it provides an index of the total spermatozoa available to a male for mating. Nevertheless, there are some estimates of sperm production both in terms of number of spermatozoa produced per year per gram of body mass (/g per year) and the number of spermatozoa produced per egg available to be fertilised (Billard, 1986). The numbers are astronomical. In the total spawner *O. mykiss*, the annual production was estimated at about 7×10^9/g per year and 10^9 per egg. *P. reticulata* is a viviparous cyprinodont with internal fertilisation (Chapter 11). The annual production is about 3×10^9/g per year with approximately 9×10^7 spermatozoa per egg. These figures indicate the large difference in gamete production by males and females (Chapter 8) and indicate the difficulty of a spermatozoon finding an unfertilised egg. The courtship and mating behaviour of a male can be seen as mechanisms for maximising the chance of one of its spermatozoa successfully fertilising one or more eggs (Chapter 8).

There are relatively few studies on fertilisation success. For pelagic spawners, in which gametes are dispersed by water currents, quantitative estimates of fertilisation success are technically difficult to make. However, studies have suggested fertilisation success in excess of 90% (Petersen *et al.*, 2005). Higher rates of success may be achieved by demersal spawners, especially when the gametes are released in a confined space such as a nest (Chapters 7 and 9). The participation of multiple males at a spawning may also ensure higher rates of fertilisation success (Shapiro *et al.*, 1994; Smith & Reichard, 2005).

Chapter 5
Endocrinology of reproduction

Introduction

Regulation of the complex processes of oogenesis and spermatogenesis (Chapter 4) is achieved by the neuroendocrine and endocrine systems of the fish (Kah & Dufour, 2011; Yaron & Levavi-Sivan, 2011). Neuroendocrine refers to hormones that are released by neurons of the nervous system, while endocrine refers to the tissues and their hormonal products that do not form part of the nervous system, although this distinction is not absolute (Adkins-Regan, 2005). Together, they form, in the striking phrase of Stephen Stearns, components of the 'wet machinery' by which physiological control is exerted over biological processes. The neuroendocrine component of the system is critical, because it allows both internal and external stimuli to be integrated by the central nervous system (CNS) in such a way that the timing and intensity of reproduction are adaptive given the conditions the individual fish is experiencing. The role of external environmental factors is described in Chapter 6.

Hormones are chemical messengers that are transported in the bloodstream from the endocrine tissues producing the hormones to the target tissues of those hormones. Their production and their effects are modulated by more local chemical messengers. These local factors may have paracrine effects, influencing neighbouring cells, or autocrine effects, in which the producing cell is also the target of its product. The role of paracrine and autocrine factors in the regulation of teleost reproduction is still poorly understood compared with the effects of hormones.

In early development, hormones can act as organising factors directing the differentiation of tissues (Chapter 3). Once the tissues have differentiated, hormones can then take the role of activating factors, operating as causal agents in directing the functioning of the tissues. In teleosts, as in other vertebrates, once the gonads have differentiated, hormonal stimulation is required for the differentiated gonads to produce functional eggs and sperm. Hormones may also act directly on non-gonadal somatic tissues, influencing characteristics such as morphology and colour. The hormones that regulate gametogenesis also regulate the formation of secondary sexual characters. A classic example is provided by the male three-spined stickleback (*Gasterosteus aculeatus*, Gasterosteidae) in which androgens, the typically male hormones, stimulate the appearance of the red throat and belly of the sexually mature male and also induce the hypertrophy of cells in the kidney, which synthesise the glue (termed spiggin) that the male uses to glue his nest together (Wootton, 1976; Borg, 2007).

Natural selection acts through its effects on the survival and reproduction of organisms (Chapter 11). The control of the reproductive process by the neuroendocrine–endocrine system defines the timing and intensity of reproduction and hence important components of the life-history characteristics on which selection acts. Selection must act on components of the 'wet machinery' of reproduction to achieve the shifts in reproductive traits required by the selective regime operating. An example is provided by the timing of puberty. Life-history theory (Chapter 11) predicts that if the mortality rates experienced by adults increase, the age of first reproduction will be selected to decrease, i.e. puberty occurs earlier. This shift in selection can only be achieved by alterations in the control of timing by the endocrine system.

Brain–pituitary–gonad reproductive axis in vertebrates

In vertebrates, the brain does not communicate directly with the gonads to control reproduction. Instead, neurohormones produced in the brain, especially the hypothalamus, affect the pituitary, the major endocrine organ in vertebrates. Under the influence of these neurohormones, some pituitary cells produce hormones, which are secreted into the bloodstream to be transported to the target organs. In the case of the reproductive system, the gonads are the target organ. This arrangement is called the brain–pituitary–gonad reproductive axis (BPG axis), or, to emphasise the role of the hypothalamus of the brain, the hypothalamus–pituitary–gonad (HPG) axis (Fig. 5.1). In the ovary, the thecal and granulosa cells of the ovarian follicle complex are the target cells. In the testes, the Leydig cells are the main target cells. Modulation and modification of the activity of the BPG reproductive axis can take place at a variety of levels in the axis, including the brain, the pituitary and the gonads (Van der Kraak *et al.*, 1998).

In comparison with the other jawed vertebrates, the teleosts have an idiosyncratic organisation of the brain–pituitary relationship (Kah & Dafour, 2011). In teleosts, hypophysiotropic neurons from the brain pass into the pituitary and release their neurohormones close to the pituitary cells (gonadotrophs) that synthesise the hormones (Fig. 5.2). In other jawed vertebrates, by contrast, the neurohormones are released into the bloodstream and pass into the pituitary in a hypophysial portal system. The adaptive significance of the difference between teleosts and other vertebrates is unclear, but the direct neural connection in teleosts may allow a faster and more precise control over the secretion of pituitary hormones.

In the vertebrates, the release of reproductive hormones by the pituitary is stimulated by the gonadotrophin-releasing hormone (GnRH), the neurohormone

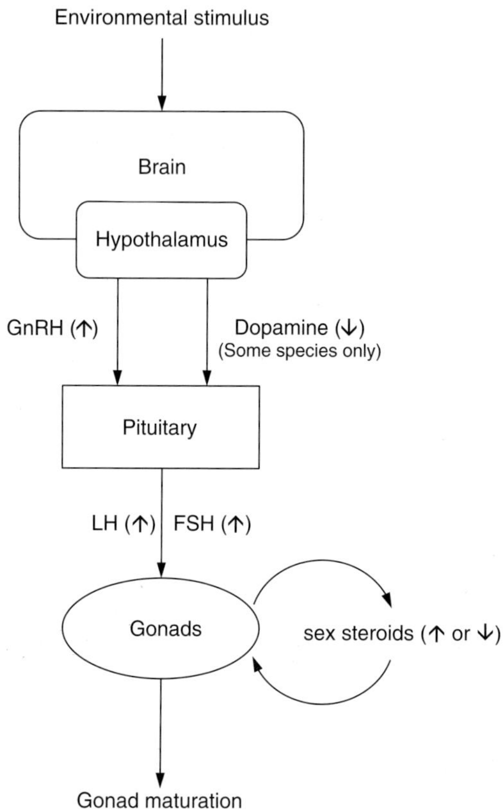

Figure 5.1 Simplified summary of the hypothalamus–pituitary–gonad (HPG) axis regulating reproduction in teleosts. GnRH, gonadotrophin-releasing hormone; FSH, follicle-stimulating hormone; LH, luteinising hormone. ↑ indicates a positive effect and ↓ a negative effect. In some species, the neurotransmitter dopamine can exert an inhibitory effect on GnRH release. (Source: Adapted from Mylonas & Zohar. Reproduced with permission of Springer Science + Business Media.)

produced by the brain (Sherwood & Adams, 2005). Two reproductive hormones, the gonadotrophins (GTHs), are produced in the pituitary by cells called gonadotrophs. The two hormones are the follicle-stimulating hormone (FSH) and the luteinising hormone (LH). This duality of control of the reproductive axis is characteristic of all vertebrates (Chauvigné *et al.*, 2010). These hormones are transported in the bloodstream to the target cells in the gonads. In the gonads, FSH and LH stimulate cells to produce steroid hormones, including androgens, oestrogens and progestagens, which then regulate the processes of oogenesis and spermatogenesis (Planas & Swanson, 2008). Another important function of these steroid hormones produced in the gonads is to provide information to the brain on the present state of the gonads. The steroids have both positive and negative feedback effects on the BPG reproductive axis.

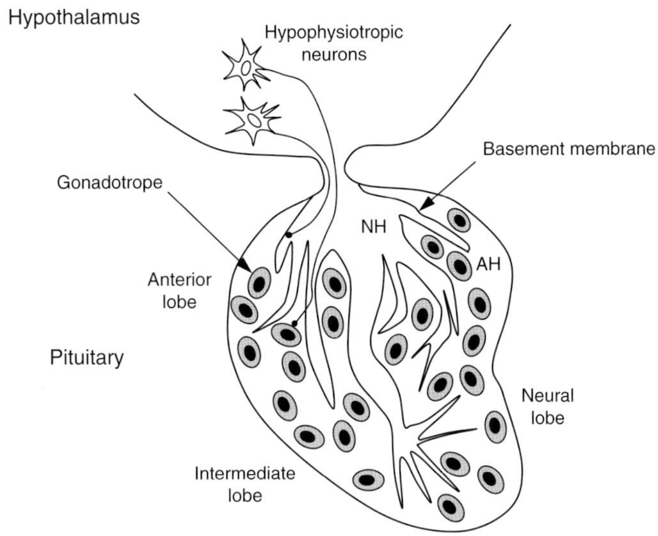

Figure 5.2 Relationship between hypothalamus and the pituitary gland in teleost fishes, showing the direct neuroendocrine link. Hypophysiotropic neurons end either at the basement membrane separating the neurohypophysis (NH) from the adenohypophysis (AH) or directly at secretory cells in the anterior lobe. (Source: Adapted from Kah & Dufour 2011 and Zohar *et al.* 2010. Reproduced with permission of Elsevier.)

Levels of analysis

In analyses of the hormonal control of reproduction, decisions have to be made about the levels within the BPG axis that are to form the focus of the study. One essential of such studies is that they are accompanied by histological studies on the status of the gonads. The highest focus of analysis is the level of GnRH in the brain and the rate of transmission to the pituitary. The effect of the GnRH on the pituitary cells will depend on the density of the receptors for GnRH in the gonadotrophic cells. The rate of synthesis of GTHs can be monitored by the level of expression of the mRNAs for GTHs and the levels of GTHs in the pituitary. Plasma levels of GTHs will indicate their availability to the target organs, the gonads. However, the effect of the GTHs at the level of the gonads will depend on the density of the receptors for FSH and LH in the target cells. These target cells include the steroidogenic cells, the Leydig cells in the testis, and the theca and granulosa cells of the ovary. The levels of activity of the enzymes in the steroidogenic pathway will determine the rate of production of sex steroid hormones. This rate of synthesis will then determine the levels of sex steroids in the blood plasma, but also the levels available for paracrine functions within the gonad. The effect of the sex steroids will depend on the concentration of suitable receptors in the target cells, whether within the gonad or in extragonadal sites such as the liver, pituitary and brain. The effect of sex steroids may also be mediated by growth factors such as insulin-like growth factors (IGFs) and activin. The ultimate goal of the studies is to develop quantitative models that predict the sequence of gonadal maturation under natural conditions or under the conditions imposed in aquaculture facilities. The interpretation of the studies on the endocrinology of reproduction in teleosts, which now use the highly developed techniques of molecular biology, has been hampered by the unsophisticated analyses of the quantitative results and the lack of predictive, quantitative models based on the experimental results.

Gonadal steroids

In females, the cells of theca and granulosa of the ovarian follicle complex (Chapter 4) are steroidogenic, synthesising the sex steroids. The thecal cells are in the interstitium of the ovary, whereas the granulosa cells derive from somatic cells in the germinal epithelium of the ovary. In males, the Leydig cells of the interstitium are steroidogenic. However, the somatic cells of the germinal epithelium, the Sertoli cells, lack a clear steroidogenic function.

Three classes of steroids can be recognised (Pankhurst, 2008). The first class contains those steroids that have clear biological effects such as oestradiol-17β (E_2) and testosterone (T). A second class contains steroids that are produced as intermediate products in the synthesis of functional steroids. A third class is formed by steroids that are modified for the purposes of inactivation and excretion. These classes are not mutually exclusive. For example, T has both biological effects and can act as an intermediate product in the synthesis of E_2.

Traditionally, steroid sex hormones are regarded as typically male, the androgens, or typically female, the oestrogens and progestagens. However, in teleosts, some androgens, particularly T, have a role in the female reproductive cycle, and both oestrogens and progestagens have roles in the male reproductive cycle. Androgens contain 19 carbon atoms and so are classified as C19 steroids. In teleosts, T and 11-ketotestosterone (11-KT) are the main androgens. T is an aromatisable androgen, which means it can be converted to E_2 by the aromatase enzyme. This converts a C19 steroid to a C18 steroid. 11-KT is not aromatisable. E_2 is the main oestrogen in teleosts. The other class of 'female' hormones, the progestagens, are C21 steroids. The progestagen, 17α,20β dihydroxy-4-pregnen-3-one (17α,20β-P), has a prominent role as a sex steroid, especially in the context of the maturation of the gametes (Chapter 4). In some species, the role of 17α,20β-P is taken by another progestagen, 17,20β,21-trihydroxy-4-pregnen-3-one (20β-S).

There is also the possibility that other steroids act as functional sex steroids (Kime, 1993), but problems of detection can cause problems in identifying such steroids, and T, 11-KT, E_2, 17,20β-P and 20β-S continue to be identified as the major functional sex steroids in teleosts.

Sex steroid synthesis

In steroidogenic cells (Young *et al.*, 2005; Pankhurst, 2008), the first step in the synthesis of functional sex steroids requires the transfer of cholesterol from the cytoplasm to the inner membrane of a mitochondrion. This transfer requires the presence of the steroidogenic acute regulatory protein (StAR). After the transfer, the cholesterol is converted to pregnenolone by the enzyme $P450_{scc}$ (side-chain cleavage enzyme). Pregnenolone acts as the starting point for a number of steroidogenic pathways. A key enzyme in these pathways is $P450_{c17}$, which exists in two forms, one with C_{17-20}lyase activity ($P450_{c17-I}$) and the other with 17α-hydroxylase activity ($P450_{c17-II}$). When the activity of $P450_{c17-I}$ dominates, the pathway leads to androgens, whereas when $P450_{c17-II}$

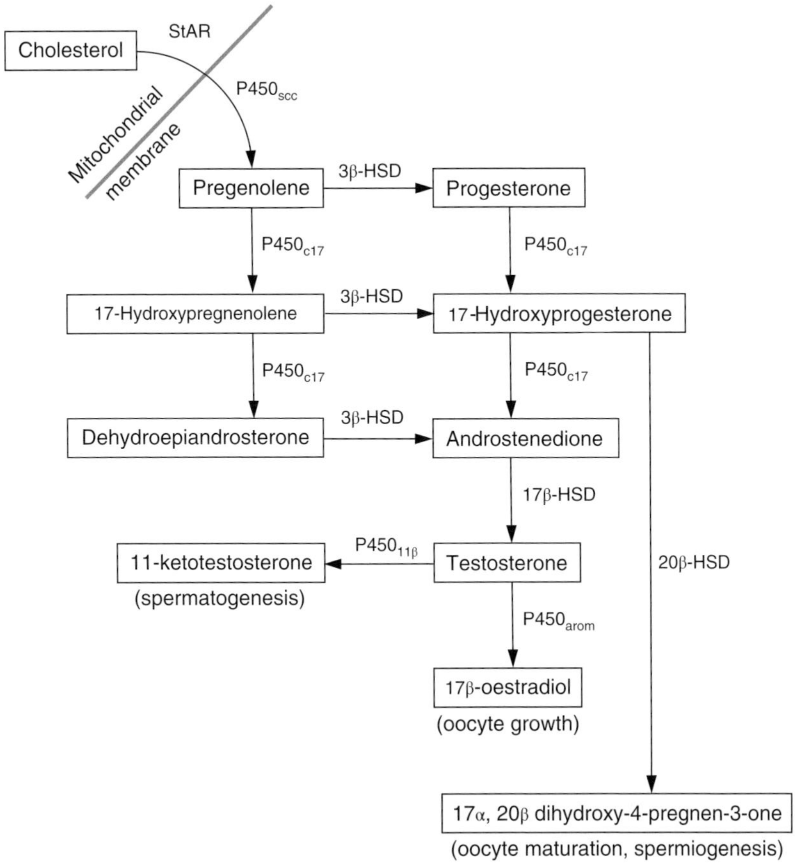

Figure 5.3 Pathways of steroid synthesis in teleost fishes. Steroids are in boxes, enzymes are in normal type, and the key function of the steroid is in parentheses. StAR, steroidogenic acute regulatory protein; $P450_{scc}$, side-chain cleavage enzyme; 3β-, 17β- and 20β-HSD, 3-, 17- and 20β-hydroxysteroid dehydrogenase, respectively; $P450_{c17}$, 17α-hydroxylase/$C_{17,20}$lyase; $P450_{11}$β, 11β-hydroxylase; $P450_{arom}$, aromatase. (Source: Adapted from Pankhurst 2008. Reproduced with permission of Taylor and Francis Group.)

dominates, the pathway leads to progestagens. Because of the ability of aromatase to convert T to E_2, the androgen pathway can lead to oestrogens (Fig. 5.3).

There are interspecific variations in the pathways; for example, in red seabream (*Pagrus major*, Sparidae), the main pathway leading to E_2 is through oestrone, E_1, rather than T (Matsuyama, 2008).

Sex steroid receptors

For sex steroids to have a biological effect, they have to access the target cells. The 'classical' or genomic pathway by which a steroid produces an effect involves the activation of steroid-sensitive genes, such as the vitellogenin genes of the liver in females (Chapter 4) (Menuet, *et al.*, 2005; Pankhurst, 2008). The hydrophobic steroid enters the cell through the plasma membrane and within the cell binds to a receptor, forming a steroid-receptor complex. The binding of the steroid, the ligand, to the receptor cause changes in the shape of the receptor and may result in dimerisation, in which two steroid–receptor complexes become associated. The steroid–receptor complex then interacts with the promoter region of the steroid-sensitive gene, causing a change in the activity of the gene, typically activating the gene. This activation usually involves other coactivators and transcription factors. This pathway takes hours or days to become effective.

The three main classes of sex steroid receptors are oestrogen receptors (ERs), androgen receptors (ARs), and progestagen receptors (PgRs). The number of steroid receptors partly reflects the whole-genome duplication that occurred early in the evolution of teleosts (Chapter 1). In teleosts, there are three ERs: ERα, ERβ1 and ERβ2. The ER binds to an oestrogen response element (ERE), in the promoter region of the steroid-sensitive gene. ERs have a modular structure with six domains, labelled A to F, and it is the C-domain that interacts with the ERE element, activating transcription of the gene. Teleosts have up to two ARs, reflecting gene duplication (Olsson *et al.*, 2008). However, in some clades, one of the ARs has been lost. The zebrafish (*Danio rerio*, Cyprinidae) has only a single AR. In most teleosts studied, including *D. rerio* and the Atlantic salmon (*Salmo salar*, Salmonidae), only a single progestagen receptor has been found, although in the Japanese eel (*Anguilla japonica*, Anguillidae), two PgRs occur.

For some responses to sex steroids, the time frame is too short for the 'classical' genomic pathway to be operating (Pankhurst, 2008). This has led to the identification of sex steroid receptors in the plasma membrane. When these bind to the relevant steroid, they stimulate a cascade of events in the cell that do not require activation of the steroid-sensitive gene. This non-genomic pathway is particularly important in the maturation and ovulation of the oocyte, which involves the lifting of the arrest of meiosis I and the completion of meiosis I (Chapter 4).

Endocrine control of oogenesis

The endocrine control of oogenesis can be followed most clearly in semelparous species that have a synchronous development of the oocytes, so that essentially there is only a single cohort of oocytes developing, all at approximately the same stage of development (Chapter 4). In semelparous species, as spawning is followed by death, there does not need to be a mechanism maintaining a reservoir of early oogonia that can provide the oocyte cohorts for future years. This mechanism has to be present in iteroparous species with synchronous development of oocytes. More complicated is the endocrine control of oogenesis in species with group synchronous or asynchronous development. In such species, there are cohorts at different stages in oogenesis present simultaneously, so the endocrine controls of the stages are confounded. The initial account is largely based on semelparous salmonids, such as Coho salmon (*Oncorhynchus kisutch*, Salmonidae) (Campbell, *et al.*, 2006).

Oogonial proliferation and primary growth

The physiological control of the earliest stages of oogenesis is not well defined. The initial stage of oogenesis is the proliferation of diploid oogonia by mitotic division. This proliferation may be stimulated by E_2 (Miura *et al.*, 2007), but the origin of any E_2 at this stage of oogenesis is not yet clear. The key step is that the oogonium becomes surrounded by the granulosa and thecal layers forming the ovarian follicle complex (Chapter 4), and meiosis is initiated, defining the transition from an oogonium to an oocyte. Experimental studies on the Japanese huchen (*Hucho perryi*, Salmonidae) and common carp (*Cyprinus carpio*, Cyprinidae) suggest that the initiation of meiosis is triggered by the progestagen 17α,20β-P (Miura *et al.*, 2007), synthesised by the follicular cells that are stimulated by a brief surge of the gonadotrophin LH produced in the pituitary. Subsequently, the progress of meiosis I in the oocyte is arrested in the prophase stage (Chapter 4). At this stage, the gene for anti-Müllerian hormone, *amh*, is expressed in the granulosa, but expression declines as vitellogenesis proceeds. Anti-Müllerian hormone may suppress the activity of aromatase and hence the production of E_2, which stimulates vitellogenesis (Clelland & Peng, 2009).

Oncorhynchus kisutch typically spawn in autumn at the age of 3 years (Campbell *et al.*, 2006). In fish aged 2 years, the ovary in August contained oocytes at the perinucleolar stage and some at the cortical alveoli stage, indicating that the onset of meiosis had already occurred, but by December all oocytes had reached the cortical alveoli stage. Between August and December, the levels of pituitary FSH, plasma FSH and plasma E_2 had all increased. An increase in the rate of sex steroid, particularly E_2 synthesis, was indicated by an increase in transcript levels of ovarian *star*, the gene for StAR.

By May, in the spring prior to spawning, some females had accumulated lipid droplets in the oocytes in addition to the cortical alveoli (LD group). A second group of females had a smaller body size and had only reached the cortical alveoli stage (CA group). The LD group had significantly higher levels of pituitary and plasma FSH, plasma E_2 and plasma insulin-like growth factor I (IGF-I) than the CA group.

The results for *O. kisutch* suggest that FSH and the E_2, whose synthesis is stimulated by the FSH, play a role in promoting the production of cortical alveoli and lipid droplets in the oocytes. The growth of the oocytes at this stage may also be mediated by IGF-I. The combined actions of E_2 and IGF-I could be an example of the cross-talk between the reproductive and growth axes, which would allow for the adjustment of the rate of oocyte growth in response to the prevailing nutritional conditions experienced by the fish. However, at the end of this primary growth phase, the ovaries only represent a small proportion of the total mass of the female. In May, the ovaries of the LD females represented about 1% of the body mass. The major growth of the ovaries between May and the autumn is the consequence of the accumulation of yolk in the oocytes.

Secondary growth: vitellogenesis

The key feature of this stage of oocyte development is the endocrine influence that the ovaries have on the liver (Ding, 2005; Pankhurst, 2008; Lubzens *et al.*, 2010). Under the influence of pituitary FSH, the thecal cells of the follicle complex synthesise T. This passes to the granulosa cells of the follicle, where the aromatase enzyme converts T to E_2. The E_2 is transported to the liver in the bloodstream. In the liver cells, E_2 has three main effects. At low levels, it stimulates the synthesis of ERs, so the abundance of ERs in the liver cell increases, allowing more E_2–ER complexes to form. This leads to the stimulation of the *vtg* genes that direct the synthesis of vitellogenin, the precursor of the yolk (Chapter 4), which is transported to the ovaries in the bloodstream. A third effect of E_2 is to stimulate the proteins that are destined to form the zona pellucida, which separates the granulosa layer from the oocyte. The circulating E_2 also has feedback effects on E_2-sensitive cells in the brain and pituitary, and plays a role in regulating the BPG reproductive axis.

In autumn-spawning *O. kisutch*, the plasma levels of FSH increase from March up to a plateau, but then decline rapidly as the spawning season approaches. The plasma levels of E_2 lag behind the levels of FSH, and rise rapidly to a peak before declining equally rapidly (Swanson, 1991).

In the iteroparous rainbow trout (*Oncorhynchus mykiss*, Salmonidae), which spawns in spring rather than autumn, the plasma levels of FSH start to rise during the initial phase of vitellogenesis about 8 months before ovulation. Levels of FSH decline as vitellogenesis proceeds, but then there is a second peak at the onset of maturation and ovulation (Prat *et al.*, 1996). The levels of vitellogenin in the plasma, which reflect levels of E_2, increase from late spring to reach a peak in late winter (Tyler & Sumpter, 1996). The levels of plasma E_2 increase from basal levels in the late summer to reach a peak in late autumn, but then decline rapidly, falling to low levels around the time of ovulation in late winter (Scott *et al.*, 1980a). Belying its reputation as a 'male' androgen, the levels of plasma T exceed those of E_2, but reach a peak level at the same time as E_2. The levels then decline more slowly so high levels are still present when ovulation occurs in winter and baseline levels are reached in late spring. The second peak in FSH levels may be associated with the iteroparity of *O. mykiss*, with FSH initiating the growth of the next cohort of oocytes that will be spawned in the following breeding season. This provides a contrast with the pattern seen in the semelparous *O. kisutch*.

There is increasing evidence that the effects of FSH and E_2 are modulated by non-steroid growth factors, including IGFs. IGF-I can influence steroid synthesis in the thecal and follicular layers (Lubzens *et al.*, 2010). Activin βA is a member of the tissue growth factor β superfamily (TGF-β). In *D. rerio*, the expression of activin βA increased from a low level in the primary growth phase to reach a peak in mid-vitellogenesis. The role of these growth factors and others in the process of vitellogenesis has still to be clarified.

Maturation and ovulation

After the completion of vitellogenesis, there may be a shorter or longer pause before the process of oocyte maturation is initiated (Lubzens *et al.*, 2011; Urbatzka *et al.*, 2011). A key feature of oocyte maturation is the resumption of meiosis I, which has been arrested in diplotene I since the transition from oogonium to oocyte. This initiation depends on the individual female achieving the appropriate physiological state and an external environment providing the suitable conditions for spawning. A change in the gonadotrophin and sex steroid environment characterises the onset of oocyte maturation. It is not clear how an oocyte signals that is has completed vitellogenesis or how its completion by the entire cohort of oocytes is advertised. Maturation is initiated by a relatively short-term surge of LH secretion by the pituitary while the level of FSH falls. In goldfish (*Carassius auratus*, Cyprinidae), the LH surge lasts about 15 h (Munakata & Kobayashi, 2010). At the level of sex steroids, the production of E_2 is replaced by the production of maturation-inducing hormone (MIH), a progestagen. However, to respond to the presence of MIH, the oocytes must have achieved oocyte maturation competence (OMC). Oocytes earlier in their development have not achieved this competence and so are refractory to the action of MIH.

The action of MIH depends on the presence of PgRs. Maturation is rapid, which suggests that the process of maturation is driven by a non-genomic rather than the genomic pathway of steroid regulation. The non-genomic pathway depends on the presence of membrane steroid receptors on the oolemma, rather than the entry of steroids into the oocyte. For oocyte maturation, two classes of membrane steroid receptors are relevant, oestrogen receptors (mERs) and progestagen receptors (mPRs). The identification of these two classes of membrane receptors has led to the development of a dual model of the control of maturation (Pang & Thomas, 2010; Urbatzka *et al.*, 2011).

Evidence suggests that E_2 binds to a membrane ER called G protein-coupled ER (GPER). This binding triggers a cascade of events, which maintains the arrest of meiosis I. As the levels of E_2 decline and the levels of MIH increase, the inhibitory effect of E_2 declines. The binding of MIH with the membrane receptor mPRα now triggers a cascade, which leads to a resumption of meiosis I (see Chapter 4). One of the early effects of the LH surge may be to up-regulate the mPRs in the follicle cells, thus contributing to the development of OMC.

The shift from the synthesis of E_2 to MIH depends on a shift in the activity of the two forms of $P450_{c17}$, $P450_{c17\text{-}I}$ and $P450_{c17\text{-}II}$ in the steroid biosynthetic pathways in the follicle cells (Fig. 5.4; Clelland & Peng, 2009; Lubzens *et al.*, 2010; Hanna & Zhu, 2011). With $P450_{c17\text{-}I}$ dominant, the pathway leads to T and hence to E_2, through the activity of aromatase. With $P450_{c17\text{-}II}$ dominant, the pathway leads to 17α-hydroxprogesterone and hence to 17α,20β-P, which is the MIH in species such as salmonids. However, in other species, a modification of the pathway leads to 20β-S as the MIH. The synthetic steps leading to 17α-hydroxprogesterone takes place in the theca cells, while the final step leading to the MIH occurs in the granulosa cells. In the medaka (*Oryzias latipes*, Adrianichthyidae), the relative activity of $P450_{c17\text{-}I}$ and $P450_{c17\text{-}II}$ is mediated by transcription factors, including Ad4BP/SF-1 and Foxl2. During vitellogenesis, Ad4BP/SF-1 acts primarily on *cyp17a1*, the gene for $P450_{c17\text{-}I}$, and hence favours the pathway that generates E_2. Before maturation, a spike in the Foxl2 level up-regulates *cyp17a2*, coding for $P450_{c17\text{-}II}$, thus favouring the pathway leading to MIH.

Paracrine growth factors, including epidermal growth factor (EGF), activin and follistatin, are also implicated in the onset of OMC and the fine control of maturation by cross-talk between the oocyte and the follicular layers (Fig. 5.4; Yaron & Levavi-Sivan, 2011). In *D. rerio*, activin βB increases the rate of oocyte maturation, but follistatin from the oocyte can inhibit this effect, forming a feedback system, which may be mediated by EGF. The details of the paracrine roles of these growth factors are still not fully understood, but they are likely to be important in synchronising the behaviour of oocytes in the process of maturation and ovulation. Ovulation in *D. rerio* is induced by LH and MIH. But MIH probably acts by the genomic pathway of steroid control by binding to a nuclear progestagen receptor.

Endocrine control of oogenesis in species with batch spawning

For species such as *O. kisutch* and *O. mykiss* with synchronous development of the oocytes and a single spawning in a breeding season, the endocrine control of oogenesis is clear, with two major events, vitellogenesis stimulated by FSH and maturation by LH. For species with synchronised group and asynchronous ovaries (Chapter 4), the endocrine control is more complicated, because several phases of oogenesis occur at the same time, especially vitellogenesis and maturation, which are controlled by different sex steroids (Pankhurst, 2008).

In the study of such species, protocols of sampling have to be carefully designed. For changes over an annual reproductive cycle, sampling at approximately monthly intervals may be appropriate, as it would be for a species with synchronous oocyte development. However, a second protocol with a finer timescale of sampling is required to monitor changes between successive spawnings, which may be separated by a few hours or a few days. There is a dearth of studies that have applied both long- and short-term sampling protocols to the same species.

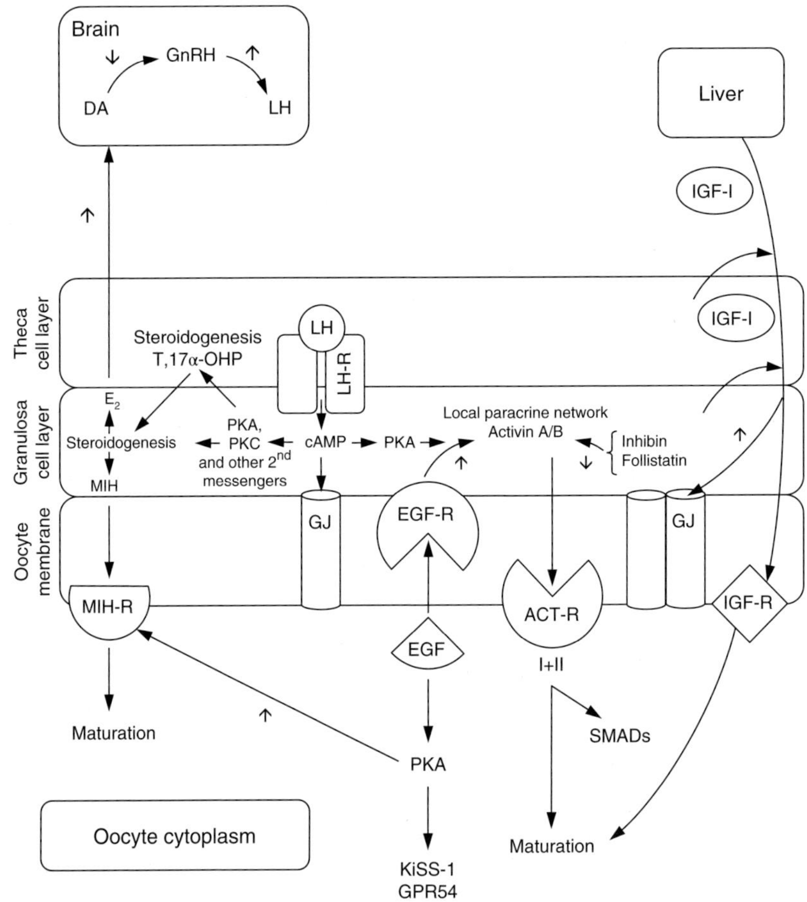

Figure 5.4 Preliminary model of the regulation of oocyte maturation in teleosts. Neuroendocrines, hormones and receptors: DA, dopamine (not in all species); GnRH, gonadotrophin-releasing hormone; LH, luteinising hormone; LH-R, luteinising hormone receptor; T, testosterone; 17α-OHP, 17α-hydroxprogesterone; E_2, 17β-oestradiol; MIH, maturation-inducing hormone; MIH-R, maturation-inducing hormone receptor. Cellular signalling molecules and receptors: PKA, protein kinase A; PKC, protein kinase C; cAMP, cyclic-3′,5′-adenosine monophosphate; EGF, epidermal growth factor; EGF-R, epidermal growth factor receptor; ACT-R, activin receptor. Growth factors and receptors: IGF-I; insulin-like growth factor 1; IGF-R, insulin-like growth factor receptor. GJ, gap junctions. ↑ indicates a positive effect and ↓ a negative effect. (Source: Urbatzka *et al.* 2011. Reproduced with permission of Elsevier.)

A study using long-term sampling protocols comparing closely related cyprinid species that had different patterns of oocyte development (Rinchard *et al.*, 1997) provided information on the changes in E_2 over the annual reproductive cycle for three European cyprinids. Roach (*Rutilus rutilus*, Cyprinidae) has synchronised oocyte development, with spawning over a few days in late spring. Bleak (*Alburnus alburnus*, Cyprinidae) is a batch spawner with indeterminate fecundity. Recruitment of vitellogenic oocytes from the cortical alveoli continues over the breeding season, which occurs during the early summer. White bream (*Blicca bjoerkna*, Cyprinidae) is also a batch spawner, but has determinate fecundity. All the vitellogenic oocytes are recruited before the start of the breeding season in early summer.

Plasma E_2 levels in *R. rutilus* were low from mid-May to August, when the gonads are quiescent. In September, plasma E_2 increased, but the level stabilised over the winter, when low temperature inhibited ovarian growth. When ovarian growth resumed in spring, plasma E_2 levels increased quickly, but then decreased rapidly during the brief spawning period in May. Plasma levels of a potential MIH, 17α,20β-P, reached a maximum in the oocyte maturation phase, and levels were high but variable in the initial post-spawning phase. By contrast, *A. alburnus* and *B. bjoerkna* maintained high levels of plasma E_2 over the extended breeding season. After completion of spawning, levels of E_2 declined and only started to recover in autumn; then, after a pause over winter, they increased again in spring. There was no clear pattern in levels of 17α,20β-P. This meant that, at the timescale of sampling used, the patterns of plasma E_2 and MIH levels could not be disentangled.

Gasterosteus aculeatus is a batch spawner with indeterminate fecundity, with the physiological capacity of producing 10 or more clutches in an extended breeding season in late spring and summer. Levels of the messenger RNAs for the subunits of the GTHs (see later), FSH-β mRNA and LH-β mRNA, in the pituitary of female *G. aculeatus* had distinct seasonal patterns over a year, when monitored at monthly intervals (Hellqvist *et al.*, 2006). FSH-β mRNA peaked in January and February and then declined steadily, reaching a minimum in July, after the breeding season. The pattern for LH-β mRNA closely followed that of the gonadosomatic index (GSI), the gonad mass as a proportion of total mass, peaking in the breeding season and then declining sharply. The female spawns at intervals of a few days over the breeding season, so the monthly sampling interval was too long to detect the changes between successive spawnings. The same study also monitored plasma levels of 11-KT and T. 11-KT levels were always low, but T levels peaked in the breeding season. The high T levels may reflect the role of T in the synthesis of E_2 during the vitellogenic phase of oogenesis.

Changes over shorter periods were illustrated by the mummichog (*Fundulus heteroclitus*, Fundulidae), which has a semi-lunar cycle of spawning, relating to the tidal cycle experienced by this coastal species. For about 6 days after spawning, E_2 levels were low, but then increased over the next 6 days as vitellogenesis and oocyte maturation occurred. A finer timescale of sampling showed that the related Gulf killifish (*Fundulus grandis*, Fundulidae) displayed diurnal changes in plasma steroid levels, with a peak in plasma T levels at midnight on the day prior to spawning (Mylonas & Zohar, 2007; Pankhurst, 2008; Mylonas *et al.*, 2010).

Species that spawn daily also show diurnal changes in levels of plasma sex steroids. *Pagrus major* ovulates in the early afternoon, spawning in the early evening. An early morning peak in 17α,20β-P was associated with the onset of oocyte maturation and a peak in E_2 with the recruitment of vitellogenic oocytes that formed the next batch of eggs.

The data show that changes in the levels of plasma steroids can occur over short time frames, with correspondingly shorter periods when levels are high. However, the temporal patterns suggest an association of E_2 levels with vitellogenesis and MIH levels with maturation, as is seen in species of synchronous development of oocytes.

Endocrine control of spermatogenesis

Spermatogenesis, like oogenesis, is under the control of the pituitary GTHs, FSH and LH (Fig. 5.5), and there are some parallels between the control of spermatogenesis and the control of oogenesis. This is perhaps not surprising given that initially both the PGC and

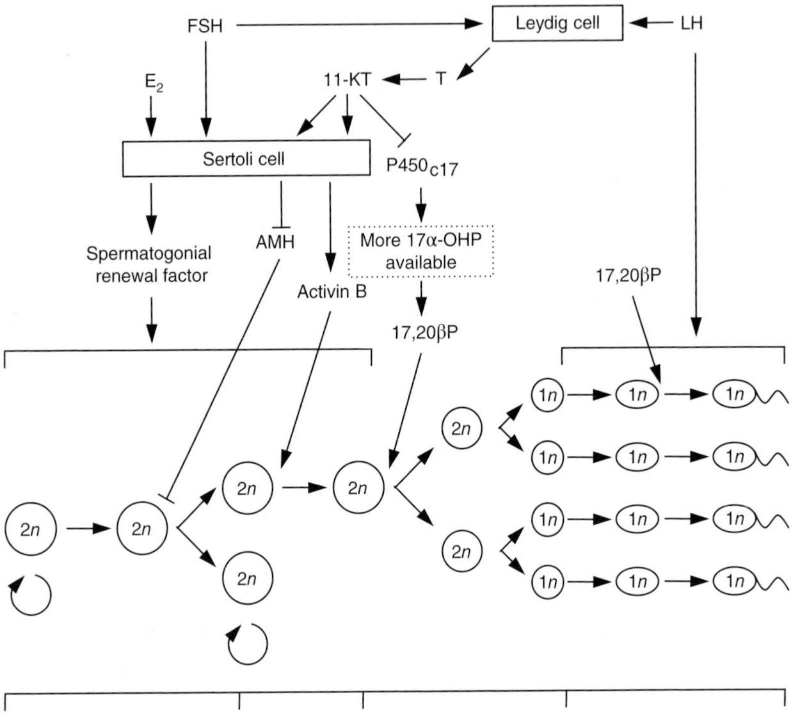

Figure 5.5 Preliminary model of the regulation of spermatogenesis in teleosts. Different stages of spermatogenesis are shown at the bottom of the figure, corresponding to different germ cell stages with ploidy levels indicated as 2*n* (diploid) and 1*n* (haploid). Circles with arrows signify points of cell proliferation. Hormones: FSH, follicle-stimulating hormone; LH, luteinising hormone; E_2, 17β-oestradiol; 11-KT, 11-ketotestosterone; T, testosterone; 17,20βP; 17,20β hydroxyprogesterone; 17α-OHP, 17α-hydroprogesterone. Enzymes: $P450_{c17}$, 17α-hydroxylase/$C_{17,20}$lyase. Cellular signals: AMH, anti-Müllerian hormone. (Source: Knapp & Carlisle 2011. Reproduced with permission of Elsevier.)

the gonadal somatic cells are bipotential, having the capacity to differentiate as either female or male characteristic cells (Chapter 3). There is evidence that FSH receptors on the Leydig and Sertoli cells may also bind with LH, when the latter is present at high, but physiological, levels. By contrast, the LH receptors are more specific to LH, except at unnaturally high levels of FSH. The steroidogenic Leydig cells of the testis interstitium are regulated by both FSH and LH, whereas the Sertoli cells are predominantly regulated by FSH (Schulz *et al.*, 2010).

As for oogenesis, the endocrine control of spermatogenesis is easiest to describe for a semelparous species such as *O. kisutch*. Spermatogenesis is completed just prior to spawning and there is no subsequent episode of spermatogenesis because the adults die during or soon after spawning. In *O. kisutch*, levels of plasma FSH showed a transient peak during spermatogonial proliferation. Levels then increased again at spermiation, but decreased before the spawning season started. Plasma LH did not increase until close to the spawning season. These patterns are comparable to those shown during oogenesis (Campbell *et al.*, 2003).

Two 'female' sex steroids are implicated in spermatogenesis. Plasma levels of E_2 increase for a short period of time at the beginning of the reproductive cycle. At this stage, E_2 promotes the slow proliferation of undifferentiated spermatogonia, which in iteroparous species provide the source of the next cohort of differentiating germ cells. The progestagen, 17α,20β-P, is involved at two stages of spermatogenesis. Its first role is to stimulate the onset of meiosis, which marks the transition between spermatogonia and spermatocytes (Chapter 4). The levels of the progestagen rise again during spermiation and the maturation of spermatozoa. Presumably, both E_2 and 17α,20β-P are synthesised in the Leydig cells under the influence of the GTHs. However, some evidence suggests a two-cell model for the production of 17α,20β-P, in which the Leydig cells synthesise 17-hydroxypregnen-4-ene-3,20-dione (17-P) in response to LH, and the 17-P is converted to 17α,20β-P in the spermatogonia, or later in the spermatozoa by the enzyme 20β-hydroxysteroid dehydrogenase (Scott *et al.*, 2010). Leydig cells synthesise androgens, particularly 11-KT and T, in response to FSH. Leydig cells, as well as Sertoli cells, also have ARs (Schulz *et al.*, 2010; Knapp & Carlisle, 2011).

Spermatogonial proliferation

Spermatogonial proliferation by mitotic division takes two forms. In iteroparous species, there is a slow proliferation of undifferentiated spermatogonia. This proliferation provides a reservoir of undifferentiated spermatogonia that can provide recruits for the bouts of spermatogenesis generating spermatozoa for successive breeding seasons. There is evidence that the oestrogen, E_2, promotes this slow proliferation. The source of E_2 is presumably the steroidogenic Leydig cells or by aromatisation of T in the undifferentiated spermatogonia (Schulz & Miura, 2002; Schulz *et al.*, 2010). In semelparous species, this slow renewal of a population of undifferentiated spermatogonia is not required, because all the adults die during or soon after their first and only spawning season.

The second form of proliferation of spermatogonia takes place within the spermatogenic cysts. A single spermatogonium is surrounded by Sertoli cells and goes through a species-typical number of mitotic divisions (Chapter 4). This proliferation is dependent on androgen, typically 11-KT stimulation. In the semelparous *O. kisutch*, levels of plasma FSH show a peak during this proliferation, with the FSH stimulating androgen synthesis by Leydig cells (Campbell *et al.*, 2003).

The androgen essentially acts as a paracrine factor influencing the Sertoli cells, which have ARs that the spermatogonia lack. The action of the androgen is mediated by growth factors produced by the Sertoli cells, including IGF-I and activin, which stimulate the mitosis of the spermatogonia. Another effect of the androgen is a down-regulation of AMH, which acts as an inhibitor of mitotic proliferation. Thus, this down-regulation of AMH marks the start of spermatogenesis (Schulz *et al.*, 2010; Knapp & Carlisle, 2011).

Initiation of meiosis and formation of spermatocysts

The transition from mitotic to meiotic divisions of the germ cells in a spermatogenic cyst is triggered by a brief surge in LH, which leads to the production of the progestagen 17α,20β-P. This steroid is an essential factor in the initiation of meiosis and hence the formation of primary spermatocytes (Scott *et al.*, 2010). It plays a similar role in inducing meiosis in oogenesis. However, once meiosis has started, further development of the spermatocytes is dependent on the presence of androgens, usually 11-KT, again controlled by FSH acting on the Leydig cells. In contrast to oogenesis, there is no arrest of meiosis I in the primary spermatocytes. The transition from meiosis I to meiosis II marks the transition from primary to the short-lived phase of secondary spermatocytes, which is completed with the transformation into spermatids. Spermiogenesis, the process of the transformation of spermatids into the spermatozoa, continues to be androgen-dependent.

It is not clear whether the effect of androgens is continuous or that their presence at certain critical times during spermatogenesis is sufficient, with spermatogenesis continuing between these critical times by processes autonomous to the spermatogenic cyst.

Spermiation

Once a spermatogenic cyst is full of spermatozoa, the next phase is spermiation, with the breakdown of the wall of the spermatogenic cyst. Plasma levels of LH increase and result in the synthesis of 17α,20β-P, while 11-KT levels continue to be high. Under the influence of these steroids, the cyst walls rupture, releasing the spermatozoa into the testis tubules (Schulz & Miura, 2002; Schulz *et al.*, 2010).

In *O. kisutch*, these events are accompanied by changes in levels of plasma GTHs. Plasma FSH increases transiently during spermatogonial proliferation, but then increases again at spermiation. Levels of FSH decrease before the spawning season starts. Plasma LH levels do not increase until close to the start of the breeding season. These changes in the levels of FSH and LH are comparable to those seen in oogenesis. There is a clear parallel between the role of 17α,20β-P in the ovulation of oocytes and its role in spermiation (Campbell *et al.*, 2003, 2006).

In iteroparous *O. mykiss*, with spermiation at a peak in winter, the levels of plasma androgens, T and 11-KT, increased from late summer (Scott *et al.*, 1980b). The level of T reached a peak in autumn and declined back to basal levels by spring. By contrast, the levels of 11-KT continued to increase, not reaching a peak until winter, and declining thereafter. Peak levels of 11-KT were higher than those of T.

In some species, such as *A. japonica*, 17α,20β-P affects the composition of the seminal fluid, leading to a decrease in the density of spermatozoa per unit of seminal fluid and an increase in the pH of the fluid. This increase is associated with an increase in the percentage of motile spermatozoa in the fluid (Schulz & Miura, 2002).

Patterns of spawning in relation to spermatogenesis

The pattern of spermatogenesis seen in the salmonids, with spermatogenesis completed before the start of the breeding season, is a common pattern in teleosts and is seen in species with both total and batch spawning (Chapter 1). Males enter the spawning season with a full complement of spermatozoa and spermatogenesis does not resume until after the completion of the spawning season. The contrast between total and batch spawners is less clear in males than in females. Even in total spawners, the males often spend much longer periods on the spawning grounds than females, as the males seek to maximise the number of females with which they successfully spawn (Chapter 8).

The Atlantic cod (*Gadus morhua*, Gadidae) is a batch spawner, with the female releasing eggs at intervals over an extended breeding season in spring. A study of male *G. morhua* held in captivity found that at the start of the breeding season, spermatozoa filled the lobular space and efferent ducts in the testis (Dahle *et al.*, 2003). After the spawning season, only spermatogonia and spermatocytes occurred in the distal (spermatogenic) component of the testis. Plasma levels of 11-KT and T increased in maturing fish, reaching a peak in spermiating males, with 11-KT levels higher than levels of T. Androgen levels were low in the first stages of spermatogenesis after the spawning season.

Senegalese sole (*Solea senegalensis*, Soleidae) in southern Europe show a peak in spawning in spring with a secondary peak in autumn (García-López *et al.*, 2006). However, ripe males occur throughout the year. The testes show little change in size over the reproductive cycle and values of GSI remain low. Monthly samples showed that plasma levels of 11-KT and T had similar seasonal profiles, but with 11-KT levels much higher than levels of T. The levels of the androgens increased from August, peaking in January, but then declined, reaching the lowest levels in late spring. The highest levels were recorded in males in late spermiogenesis, when the seminiferous tubules were almost filled with spermatids. *S. senegalensis* has a semi-cystic mode of spermatogenesis (Chapter 4), in which spermatids rather than spermatozoa are released in the testis tubules. The evidence suggested that the population of spermatids differentiated into spermatozoa in successive batches, analogous to the pattern seen in group synchronous development in oogenesis (Chapter 4).

Spermatogenesis in teleosts shows some variation in timing and the patterns seen in plasma FSH, LH, 11-KT and T, but it is usually associated with androgens, particularly 11-KT and T. Male *G. aculeatus* seem to illustrate an anomalous pattern (Borg, 2007). The species is a batch spawner, with an extended breeding season in spring and summer. In the male, spermatogenesis is completed in the autumn prior to spawning, although the spermatozoa at this stage are not capable of fertilising eggs, because changes in the nuclear proteins have to occur for the spermatozoa to become functional. Spermatogenesis does not start until after the completion of the breeding season and occurs when the levels of plasma 11-KT and T are low, whereas in the breeding season, when no spermatogenesis is occurring, the androgen levels are extremely high. Monthly sampling of the expression of mRNA for FSH-β and LH-β (see later) in the pituitary showed that there were some seasonal differences in expression (Hellqvist *et al.*, 2006; Borg, 2007). FSH-β expression increased between December and January, and then remained stable, decreasing from May onwards to low summer levels. The changes in LH-β mRNA lagged somewhat behind. An increase started in December and reached a peak in May, at the height of the breeding season, but then declined and remained at low levels until autumn. The temporal patterns of FSH-β and LH-β suggest that FSH-β is important in the initiation of sexual maturation and LH-β in its maintenance. However, there is a general correlation between levels of plasma 11-KT and the expression of FSH-β and LH-β mRNA, but androgen-dependent processes are apparently decoupled from spermatogenesis.

Male *G. aculeatus* have an elaborate reproductive biology (Wootton, 1976; see also Chapters 7– 9). At the start of the breeding season, a male claims a territory and builds a nest within this territory. This nest consists of fragments of vegetation held together by a glue deposited by the male (Chapters 7 and 8). This glue consists of a protein, spiggin, synthesised by kidney epithelial cells, which become hypertrophied. At this time, the male also develops nuptial colouration of a red throat and belly and blue eyes. These reproductive behaviours, the production of spiggin and the nuptial colours are all androgen (particularly 11-KT)-dependent. In *G. aculeatus*, the functions of the GTHs and androgens seem to relate to reproductive behaviour and secondary sexual traits and not to the regulation of spermatogenesis. It is possible that the levels of androgen required to regulate spermatogenesis are far smaller than those to regulate the synthesis of spiggin in the non-gonadal environment of the kidney.

The pituitary and the gonadotrophins

Brain–pituitary relationship

The GTHs that regulate the activities of the steroidogenic gonadal cells are synthesised by cells, the gonadotrophs, in the pituitary gland or hypophysis (Kah & Dufour, 2011). This small gland, unique to the vertebrates, is the main endocrine organ of the body, releasing hormones that regulate the reproductive axis, the growth and energy balance axis, and long-term responses to stress, amongst others (Cerdá-Reverter & Canosa, 2009). These activities of the pituitary are controlled by the brain through neurohormones produced in the pre-optic area (POA) of the brain and the hypothalamus. These neurohormones are synthesised in the cell bodies of the neurons and travel down the axons to be released from the terminus of the axon. These neurohormones can be releasing factors that stimulate the activity of the target pituitary cells, or in some cases the neurohormones exert an inhibitory effect on the pituitary cells. This control of the pituitary by the hypothalamus reflects their close morphological relationship. The pituitary lies in a bony chamber just posterior to the optic chiasma and just below the hypothalamus (Fig. 5.2). A short stalk connects the pituitary to the hypothalamus. In teleosts, this stalk consists of the axons of neurosecretory neurons, whose cell bodies lie in the hypothalamus (Romer & Parsons, 1986).

The pituitary gland consists of two components, which have completely different embryological origins (Romer & Parsons, 1986; Kah & Dufour, 2011). The neurohypophysis develops as a down-growth of the ventral diencephalon of the brain. The adenohypophysis develops from an ectodermic pouch in the roof of the embryonic mouth. It is this glandular component that is the source of the GTHs as well as a suite of other hormones, including growth hormone (GH) and thyroid-stimulating hormone (TSH). The adenohypophysis is divided into three major parts: a rostral pars distalis, a proximal pars distalis (PPD) and a pars intermedia (PI).

In teleosts, uniquely amongst the jawed vertebrates, the neurons that regulate the activities of the cells of the adenohypophysis enter the pituitary, so the neurohormones are released from the terminals of the axons in close proximity to their target cells (Kah & Dufour, 2011). In other jawed vertebrates, the products of the hypothalamic neurons are transported to the adenohypophysis in the bloodstream through the hypophysial portal system. This difference would suggest that the CNS of the teleost can exert a finer control over the activities of the pituitary than is seen in other vertebrates and may be a factor contributing to the striking diversity in modes of reproduction in teleosts (Adkins-Regan, 2005).

The adenohypophysis is separated from the neurohypophysis by a double basement membrane. In teleosts, the neurohypophysis lies dorsal to the adenohypophysis and is tightly associated with the PI. In some species, such as the salmonids and anguillid eels, the neurohypophysis is completely separated from the adenohypophysis by the double basement membrane, with the terminals of the neurosecretory axons releasing the neurohormones against the basement membrane. It is assumed the neurohormones diffuse between the cells to their target cells in the adenohypophysis. In other species, including cyprinids, the basal membrane has gaps through which the axons invade the adenohypophysis, where they can release the neurohormones adjacent to the target cells (Fig. 5.2; Kah & Dufour, 2011).

The relationship between the hypothalamus and the pituitary allows, in a process of amplification, a small number of neurons, synthesising tiny quantities of neuropeptides, to finely control the synthesis and release of larger quantities of hormones that are released into the bloodstream to influence distant targets in the body (Kah & Dufour, 2011).

Gonadotrophs and gonadotrophins

Separate populations of gonadotrophs produce FSH and LH in the PPD, with the cells producing LH tending to be more peripheral than those producing FSH (Levavi-Sivan *et al.*, 2010). There can be variability in the distribution of the gonadotrophs in relation to the reproductive cycle. In some salmonids, gonadotrophs invade the periphery of the PI during sexual maturation (Zohar *et al.*, 2010).

Initial studies suggested that teleosts had only a single gonadotrophin, but as analytical techniques improved, it was recognised that there were two distinct GTHs produced by the teleost pituitary (Swanson, 1991). One, associated with vitellogensis, was labelled GTH-I. The second, associated with oocyte maturation and ovulation, was labelled GTH-II. Comparisons of the structural and functional properties of GTH-I and GTH-II with mammalian FSH and LH identified GTH-I as a teleost FSH and GTH-II as a teleost LH (Suzuki *et al.*, 1988; Kawauchi *et al.*, 1989; Yaron *et al.*, 2003).

Follicle-stimulating hormone and LH are both heterodimeric glycoproteins with a molecular weight in the range 30–50 kDa (Levavi-Sivan *et al.*, 2010). The two GTHs share a common α subunit (GPα), which is non-covalently but specifically linked to a β subunit, the subunits arranged head to tail. The β subunit is unique to each GTH, determining the biological activity and specificity of the hormone. However, the functional hormones require both the α and β subunits. The three subunits are coded by three separate genes.

The GTHs belong to the superfamily of cystine knot growth factors, as does the TSH, which also contains the α subunit (Weltzien *et al.*, 2004; Levavi-Sivan *et al.*, 2010; Zohar *et al.*, 2010). Phylogenetic analysis suggests that, although GPα, FSH-β, and LH-β have a remote common ancestor, in vertebrates they form distinct monophyletic groups, reflecting their distinct functional roles.

The synthesis of GTHs is largely controlled by a GnRH released by the neurohormone axons from the hypothalamus, but the influence of GnRH may be modulated by other growth factors, including IGFs and activin. The GnRH released by the neurons binds to the receptors for GnRH (GnRH-R). This binding between hormone and receptor initiates a series of intracellular events within the gonadotroph involving one or more signalling pathways (Chong *et al.*, 2005). Such pathways lead to transcription factors binding to the promoter region of the GTH gene and stimulating the gene to transcribe the mRNA for the appropriate subunit of either FSH or LH, depending on the gonadotroph (Chong *et al.*, 2005; Wang *et al.*, 2009). The details of the promoter activation differ for the three genes coding the GPα, FSH-β and LH-β. In Chinook salmon (*Oncorhynchus tshawytscha*, Salmonidae), the synthesis of LH-β is also stimulated by E_2, whose plasma levels increase during vitellogenesis. Control of the synthesis of LH-β may involve cross-talk between GnRH and E_2 pathways at the level of the promoter region of the LH-β gene (Wang *et al.*, 2009).

The synthesis of FSH-β and LH-β is also modulated by growth factors including activin and follistatin (Ge, 2005; Aroua *et al.*, 2012). In the European eel (*Anguilla anguilla*, Anguillidae), activin A and activin B stimulated synthesis of FSH-β, but had little or even an inhibitory effect on the synthesis of LH-β. Follistatin inhibited the effect of activin on FSH-β synthesis. Activin acts by an intracellular signalling pathway that is separate from the pathways activated by GnRH, and the effects of the two pathways are integrated at the promoter sequence of the FSH-β gene (Chong *et al.*, 2005).

The cross-talk between the intracellular pathways stimulated by GnRH, E_2 and growth factors ensures that release of FSH and LH occurs at appropriate times and appropriate quantities to regulate the activities of the gonads during the reproductive cycle. It thus plays an important role in defining the reproductive traits of a given species.

Gonadotrophin receptors

In the blood, the GTHs are transported to their remote targets in the gonads, the theca and granulosa cells in the ovary, and the Leydig and Sertoli cells in the testes. For these GTHs to have an effect on the target cells, they must bind with the appropriate receptors in the plasma membrane of the cells. Thus, the effect of GTHs is a function not only of the levels in the plasma, but also of the availability of binding sites provided by the polypeptide gonadotrophin receptors, GTH-Rs.

In teleosts, two genes, positioned on two separate chromosomes, code for the GTH-Rs, one for the FSH receptor (FSH-R) and one for LH receptor (LH-R; Chauvigné *et al.*, 2010; Levavi-Sivan *et al.*, 2010). Both GTH-Rs belong to the superfamily of G protein-coupled receptors (GPCRs), as do the GnRH-Rs that bind to the GnRH produced in the brain. The general structure of a GTH-R has three components: a long extracellular component that carries the N-terminal amino acid; seven transmembrane helices and an intracellular chain that carries the C-terminus of the polypeptide. The GTH binds to the extracellular component of the GTH-R, in which multiple repeats rich in the amino acid leucine are crucial.

A study that focused on the phylogenetic position of the GTH-Rs of the flatfish *S. senegalensis*, a phylogenetically advanced teleost in the Pleuronectiformes, suggested that FSH-Rs in teleosts were monophyletic, evolving from a common ancestor at the base of the teleost clade. However, there was evidence of two separate LH-Rs, with a more complex evolutionary history, with two mutually exclusive genes, whose functions have converged. Even within the flatfishes, the Atlantic halibut (*Hippoglossus hippoglossus*, Pleuronectidae) and *S. senegalensis* (Soleidae) have different forms of LH-R (Chauvigné *et al.*, 2010).

In teleosts, the GTH-R genes are mostly expressed in the gonads. In the ovary, they are expressed in the steroidogenic cells of the theca and granulosa. In the testis, they are expressed in the steroidogenic Leydig cells, but also in the Sertoli cells of the spermatocyst. There is some evidence that after binding with FSH, the FSH-Rs stimulate the synthesis of T and, as a result of aromatisation, E_2 in the ovarian follicle cells. By contrast, LH-R stimulates the synthesis of progestins, such as 17α,20β-P, an MIH (Chapter 4). In males, the picture is less clear, but the temporal pattern of FSH in the plasma suggests that FSH-R stimulates the Leydig cells to produce 11-KT and T, while LH-R stimulates the same cells to produce 17α,20β-P, to initiate meiosis in the spermatogonia, and later to stimulate spermiation and the maturation of the spermatozoa (Chapter 4). However, some results are not consistent with this model. In *S. senegalensis*, expression of the gene *fshr* was localised to the Sertoli cells, whereas *lhr* was localised to the Leydig cells and not detected in the Sertoli cells (Chauvigné *et al.*, 2010).

The pattern of expression of the mRNAs for the GTH-Rs offers some clues relevant to the control of gametogenesis in species that have group synchronous or asynchronous oocyte development (Chapter 4). In such species, oocytes in the process of vitellogenesis may coexist in the same ovary with oocytes completing their maturation and ovulating. Consequently, the oocytes will be exposed to hormones and growth factors relevant to their stage of oogenesis, but also to hormones and growth factors regulating a different stage.

Solea senegalensis is a batch spawner with an extended breeding season. The males have a semi-cystic mode of spermatogenesis, so spermatids rather than spermatozoa are released into the testis tubules (Chapter 5). In this species, expression of *fshr* in the follicle cells of the ovary was closely associated with vitellogenesis, but declined at the completion of vitellogenesis and during maturation. By contrast, increased expression of *lhr* occurred in late vitellogenesis, just prior to the initiation of oocyte maturation and subsequent ovulation. In the testis, expression of *fshr* and *lhr* peaked in winter, when spermatogenesis was stimulated. During the spawning season in spring, expression of *fshr* declined to levels characteristic of the resting stage of the testis in summer. The level of *lhr* expression also declined over the breeding season, but was maintained at levels higher than in the resting period (Chauvigné *et al.*, 2010).

A species with asynchronous development of oocytes and daily spawning is the bamboo wrasse (*Pseudolabrus sieboldi*, Labridae), a protogynous hermaphrodite (see Chapters 1 and 10). In the female phase, spawning takes place daily in the early morning. This contrasts with batch spawners with group synchronous oocytes, in which spawning takes place at intervals of some days. In *P. sieboldi*, vitellogenesis and oocyte maturation occur on the same day. Expression of *fshr* in the ovarian follicles was high in the active vitellogenic stages. As vitellogenesis came to end, expression of *fshr* decreased, while that of *lhr* increased just prior to oocyte maturation and ovulation. These observations suggest that the temporal pattern of oocyte development over the day is defined by shifts in the abundance of FSH-R and LH-R in the plasma membrane of the oocytes. There were also diurnal fluctuations in FSH-β and LH-β subunits in the pituitary (Kitano *et al.*, 2011).

Hypothalamic control of the pituitary

Gonadotrophin-releasing hormone

The synthesis and release of the GTHs FSH and LH are stimulated by GnRH. This decapeptide is synthesised in the cell bodies of neurohormone neurons in the hypothalamus and adjacent areas of the brain and transported along the axons of these neurons through the pituitary stalk to the pituitary (Fig. 5.2). Other neurohormones and neurotransmitters synthesised in the brain can modulate the stimulatory effect of GnRH; most notably, the neurotransmitter dopamine (DA) exerts an inhibitory effect on gonadotrophin release in some species (Kah & Dufour, 2011).

This picture is complicated by the presence of at least three forms of GnRH in the teleost brain (Sherwood & Adams, 2005; Zohar *et al.*, 2010; Kah & Dufour, 2011). Some species have all three forms, whereas other species have retained two of the three. Cyprinids and salmonids, for example, have only two forms of GnRH. Separate genes code for each of three GnRHs: *gnrh1*, *gnrh2* and *gnrh3*. The nomenclature of the three GnRHs coded by these genes is confused by the use of two systems, one based on the species in which the form was recognised; cGnRH-II was first identified in the chicken, whereas sGnRH was first described in salmonids and sbGnRH in seabream. An alternative terminology uses a numerical system: GnRH-1 is sbGnRh; GnRH-2 is cGnRH-II; and GnRH-3 is sGnRH. Phylogenetic analysis suggests that the GnRH-2 lineage separated from the lineage of GnRH-1 and GnRH-3 before the origin of the teleosts, whereas GnRH-1 and GnRH-3 reflect the whole-genome duplication at the base of the teleost clade (Chapter 1). Because of variants within each GnRH group, there are about eight variant forms of GnRH in the teleosts. Variants of GnRH-2 occur in teleosts, amphibians, birds and mammals, whereas variants of GnRH-1 and GnRH-3 occur in the teleosts.

Each of the three GnRH genes codes for a large mRNA, whose product is a precursor peptide, prepro-GnRH. This consists of a signal peptide, which is involved in the intracellular transport of the peptide, the mature GnRH decapeptide, a processing tripeptide and a GnRH-associated peptide (GAP). After synthesis in the cell body of the neuron, the mature decapeptide is concentrated in secretory granules together with the GAP and transported down the axon of the neuron to the axon terminal, where the GnRH is released close to its target gonadotrophs. Thus GnRHs form the main link between the CNS and the endocrine system in the BPG reproductive axis.

Functions of gonadotrophin-releasing hormone

There is evidence that the GnRH forms have different functions (Zohar *et al.*, 2010; Kah & Dufour, 2011). Changes in the levels of GnRHs in the pituitary correlated with the reproductive cycle suggest that, in the species that have it, GnRH-1 is the releasing hormone that stimulates the gonadotrophs. GnRH-2 and GnRH-3 possibly have neuroendocrine roles other than the direct control of the gonadotrophs. However, in species with only two forms, if GnRH-1 is the absent GnRH, one of the other two plays the role of GnRH-1. In the Siluriformes, it is GnRH-2 that is found in the pituitary.

In striped bass (*Morone saxatilis*, Moronidae), pituitary levels of GnRH-1 and GnRH-2 increased during recrudescence of the gonad, peaking around the time of spawning in mating males and females. In juvenile individuals, these two GnRHs showed similar seasonal peaks as though in rehearsal for events during sexual maturation. Although present in the pituitary, GnRH-3 showed a seasonal peak only in maturing fish. European seabass (*Dicentrarchus labrax*, Moronidae) had all three forms of GnRH in the pituitary, but only GnRH-1 increased at spawning. In the pituitaries of spawning male and female gilthead seabream (*Sparus aurata*, Sparidae), GnRH-3 was not detectable and GnRH-2 was present only at low levels throughout the reproductive cycle. Levels of GnRH-1 were higher and correlated with levels of LH in the plasma. The pituitaries of *P. major* showed an increase in the level of GnRH-1 just before the spawning season, while GnRH-3 levels were low and GnRH-2 was not detectable. The significance of these interspecific differences is not clear.

During the breeding season, *S. aurata* spawns daily. Reflecting this, the relevant neuroendocrine neurons showed a daily pattern of gene expression closely related to the daily spawning cycle. Eight hours before spawning, the synthesis of mRNA peaked for the three forms of GnRH. These peaks correlated with peaks in plasma LH and the production of the MIH by ovarian follicle cells. It is not clear whether the gene activity was related to the current spawning or to spawning events on the next or subsequent days. The temporal relationships between gene activity and events downstream in the BPG reproductive axis have still to be clarified.

Gonadotrophin-releasing hormone receptors

For the GnRHs to influence the activity of other cells such as the gonadotrophs, they must bind as a ligand to a suitable receptor molecule in the plasma membrane of the target cell (Blomenröhr *et al.*, 2005; Schneider & Poehland, 2009; Levavi-Sivan *et al.*, 2010). The patterns of synthesis and release of FSH and LH will depend on the arrival of GnRH at the gonadotrophs and the dynamics of GnRH-R synthesis and breakdown by the target cells. Although GTHs and GnRHs are very different peptides, the receptors for GnRHs belong to the same super-family of GPCRs as the receptors for FSH and LH. The receptor is synthesised in the target cell, and then occupies a position in the plasma membrane. The GnRH-R molecule has an extracellular component, which carries the N-terminal amino acid, seven transmembrane domains linked by extracellular and intracellular loops and a long intracellular component that terminates with the COOH amino acid. The ligand, GnRH, binds to the extracellular component and this binding activates an intracellular signalling cascade, which in gonadotrophs stimulates the synthesis of the GTH subunits. There are variants of the GnRH-R molecule; five subtypes have been identified in *D. labrax*, but only one variant is strongly expressed in the pituitary.

Gonadotrophin-releasing hormones may play a role in the early ontogeny of the BPG reproductive axis, although their role is not fully understood. In *S. aurata*, there are correlated increases in the levels of mRNAs for all three GnRH forms, GnRH receptors, FSH-β, FSH receptors, LH receptors and the products of the *vasa* gene (Chapter 3) in the first 36 days after fertilisation. This suggests there is an integrated development of the main neuroendocrine and endocrine components of the BPG in early ontogeny (Zohar *et al.*, 2010).

Distribution of gonadotrophin-releasing hormone neurons in the brain

With the cell bodies of the neurons producing GnRH located in the brain, synaptic connections between these neurons and the axons of neurons belonging to other functional systems in the brain can form. The activities of the GnRH neurons can be coordinated with the status of other systems in the body and the state of the external environment.

As in all vertebrates, the brain of a teleost can be divided into three units: the posterior rhombencephalon (the hindbrain), which includes the cerebellum, the middle mesencephalon and, most anterior, the prosencephalon. The prosencephalon can be further subdivided into a posterior diencephalon and an anterior telencephalon (Davis & Northcutt, 1983; Romer & Parsons, 1986). There is a general correlation between the nature of the sensory input to the brain and these three units. Information on the position of the body and the lateral-line system projects to the rhombencephalon, with the cerebellum playing a major role in the control of body orientation. Visual information from the eyes projects to the mesencephalon, particularly the optic tectum. Olfactory information projects to the prosencephalon, especially the telencephalon. The hypothalamus forms the floor of the diencephalon immediately posterior to the optic chiasma.

The neurons synthesising the different forms of GnRH have somewhat different distributions in the teleost brain (Dubois *et al.*, 2002; Zohar *et al.*, 2010; Kah & Dufour, 2011). In a variety of species with the three forms of GnRH, including the perciform *D. labrax* and *S. aurata*, neurons expressing GnRH-1 and GnRH-3 occur in olfactory bulbs, the terminal nerve extending from the olfactory bulbs to the telencephalon, ventral telencephalon and the pre-optic nucleus and hypothalamus of the diencephalon. Cells expressing GnRH-2 occur in the syencephalon, which is a dorsal transitional region between the diencephalon and the mesencephalon. GnRH-1 neurons with cell bodies located in the POA and the hypothalamic area provide the major source of innervation of the pituitary for the reproductive axis. In *D. labrax*, some GnRH-3 axons do reach the pituitary, but compared with the GnRH-1, the innervation is much reduced. Fibres of GnRH-2 and GnRH-3 neurons have wide distributions in the brain and these GnRHs may not play a direct role in the control of gonadotroph activity. Axons may even extend outside the brain; for example, in *D. labrax*, some axons from GnRH-3 neurons in the terminal nerve project into the retina of the eye (Servili *et al.*, 2012). In several species, axons of GnRH-2 neurons in the syncephalon project into the pineal gland, also a photosensitive organ in teleosts, which suggests that GnRH-2 and GnRH-3 may play some role in the sensory control of reproduction.

Danio rerio has only two GnRHs, and GnRH-3 neurons are present in regions occupied by both GnRH-1 and GnRH-3 neurons in those species with three GnRHs. GnRH-3 axons in *D. rerio* project into the pituitary, suggesting that this is the form that controls the activity of the gonadotrophs. In this species, GnRH-3 presumably fills the roles of both GnRH-1 and GnRH-3 in species with three GnRHs. Studies on another cyprinid, the goldfish (*C. auratus*, Cyprinidae), in which small areas of the brain were lesioned, indicated that the nucleus lateralis tuberis of the hypothalamus was involved in GnRH secretion, because of the effect of the lesion on gonadal recrudescence. However, lesions in other parts of the nucleus lateralis tuberis seemed to abolish inhibition of GTH secretion, suggesting that in *C. auratus*, the hypothalamus can exert both stimulatory and inhibitory control over the gonadotrophs (Peter & Fryer, 1983).

The duplication of genes at the base of the teleost clade (Chapter 1) has allowed a diversification of functions of GnRHs. Too few species have been studied to identify a relationship between the number of GnRHs and GnRH-Rs possessed by a species and its reproductive characteristics.

The ontogeny of the neurons that synthesise the GnRHs reflects their evolutionary history and functional diversity (Zohar *et al.*, 2010). In species with three forms of GnRH, the initial development of the GnRH-1 and GnRH-3 neurons takes place in the olfactory placode at the extreme anterior of the brain. During embryonic development, the cells migrate posteriorly to their final positions. The GnRH-2 cells originate in the synencephalon at the boundary between the diencephalon and the mesencephalon. In species such as *D. rerio* with only two GnRHs, GnRH-3 neurons appear initially in the olfactory region, but then migrate to the terminal nerve, the ventral telencephalon, the POA and the hypothalamus.

Role of dopamine as an inhibitor of reproduction

Dopamine is a catecholamine, which acts as a neurotransmitter and is produced by dopaminergic neurons. These neurons have a wide distribution in the vertebrate brain and have a range of functions. Studies on *C. auratus* and other species indicate that, at least in some species, a dopaminergic system is an inhibitor of gonadotrophin release, in contrast to the stimulatory effect of GnRH (Popesku *et al.*, 2008; Dufour *et al.*, 2010; Zohar *et al.*, 2010; Kah & Dufour, 2011). Species with a DA inhibitory network come from a range of teleostean groups, including cyprinids, silurids, salmonids, some perciforms, among them tilapias (*Oreochromis* spp., Cichlidae), and the flathead grey mullet (*Mugil cephalus*, Mugilidae). However, not all species have an inhibitory dopamine system acting on the GTHs. Species that lack a DA inhibitory pathway include the Atlantic croaker (*Micropogonias undulatus*, Sciaenidae) and *S. aurata*. For those species that do have an inhibitory dopaminergic system, the gonadotrophs are under a dual control, with a stimulatory GnRH and an inhibitory DA pathway.

Dopamine inhibition of GTH release takes place at two levels (Fig. 5.6). Within the brain, neurons in the dopaminergic system synapse with GnRH neurons, with an inhibitory effect on the latter. Cell bodies of the dopaminergic system in the POA project axons into the pituitary, so DA released from the terminals can interact with the appropriate receptors on the gonadotrophs. The DA inhibition can have a role at two different phases of the reproductive cycle. In adults of species with DA inhibition, it acts to block the final stages of oogenesis. In some species, the DA inhibitory system blocks entry into puberty (see the section 'Physiological control of puberty').

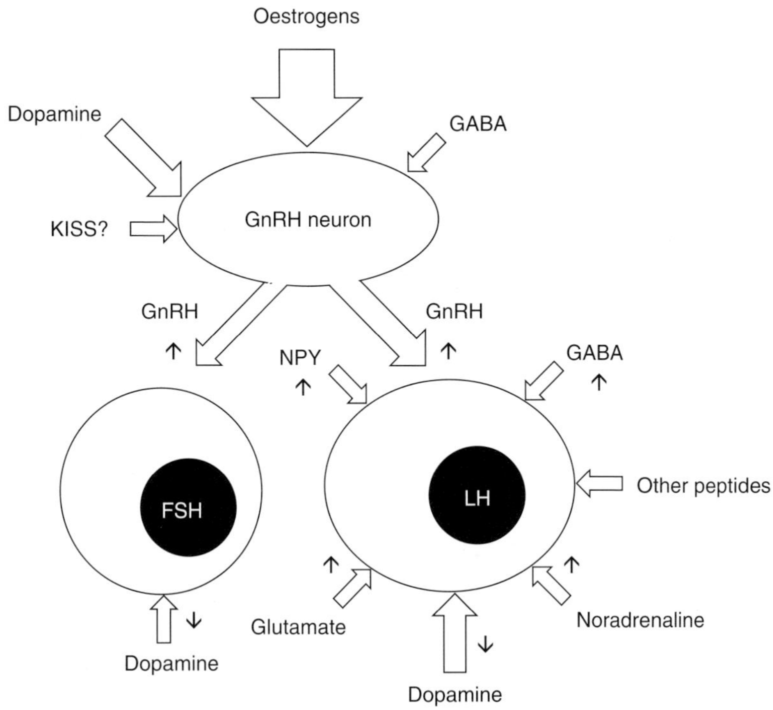

Figure 5.6 Model of the inhibitory effect, present in some teleost species, of dopamine on release of gonadotrophins from the pituitary gonadotropes for follicle-stimulating hormone (FSH) and luteinising hormone (LH). Other possible regulators of gonadotrophin production are also shown. GABA, γ-aminobutyric acid; KISS, kisspeptin; GnRH, gonadotrophin-releasing hormone; NPY, neuropeptide Y. ↑ indicates a positive effect and ↓ a negative effect. (Source: Kah & Dufour 2011. Reproduced with permission of Elsevier.)

In female *C. auratus*, DA inhibits the surge in LH required for ovulation, while in male *C. auratus* it inhibits spermiation before potential female partners ovulate. In salmonids, the premature release of LH during vitellogenesis is inhibited by DA. The role, if any, of DA inhibition on the dynamics of synthesis and the release of FSH is unclear.

For DA to have its effect on GnRH neurons and gonadotrophs, it must bind to the appropriate receptors in the plasma membrane of the target cells. DA receptors belong to the GPCR family, as do the receptors for GTH and GnRH. Two subtypes of DA receptors exist, D_1 and D_2, which seem to have evolved convergently the ability to bind DA independently. Gonadotrophs carry D_2 receptors, whereas D_1 receptors are carried by GnRH neurons.

Role of other brain factors in the control of gonadotrophs

Studies on *C. auratus* and subsequently on other species have also implicated other neurotransmitters and neurohormones in the control of the pituitary gonadotrophs (Popesku *et al.*, 2008).

γ-Aminobutyric acid (GABA) is a major neurotransmitter in the brain of vertebrates. In the POA and the hypothalamic area of the brain, GABA stimulates GnRH neurons, but inhibits DA neurons, thus promoting the synthesis and release of LH. GABA has a stimulatory effect on the secretion of FSH and LH in *O. mykiss*, although the effect varies with sex and the stage in the reproductive cycle. Neuropeptide Y (NPY) is a peptide that plays a role in the control of appetite and feeding in teleosts, and may also play a role in the coordination of the reproductive and growth axes mediated by the pituitary gland. Another candidate for the control of GTH release is a polypeptide, pituitary adenylate cyclase-activating polypeptide-related peptide (PRP) (Tam *et al.*, 2011). In the brain, PRP increases the expression of GnRH mRNA. In the pituitary, it increases the expression of mRNA for both FSH-β and LH-β, whereas in the ovary it increases the expression of mRNA for FSH-β, but decreases the expression of LH-β.

The kisspeptin system in the brain operates upstream of the GnRH neurons. Kisspeptins are a family of peptides encoded by the *kiss1* gene. The receptor for *kiss1*, originally called GPR54, is a product of two genes, *kiss1ra* and *kiss1rb*, which are both expressed in the brain, while *kiss1ra* is also expressed in the gonads and *kiss1rb* in the pituitary and other tissues. The relevance of the kisspeptin system is that, in mammals, it is implicated in the control of puberty and the reproductive cycle by acting on GnRH neurons (Kah & Dufour, 2011).

The situation in teleosts is more complicated because of the whole-genome duplication that occurred at the base of the teleost lineage (Chapter 1). Two kisspeptin genes, *kiss1* and *kiss2*, have been identified in teleost species. Both genes are expressed in *D. rerio* and *D. labrax*, amongst other species. Only the *kiss2* gene has been identified in other species, such as *G. aculeatus* (Kah & Dufour, 2011; Migaud *et al.*, 2012). *Kiss1* and *kiss2* are expressed in different neurons and their distribution suggests that *kiss2* has a role in the regulation of reproduction, with *kiss1* having other functions (Migaud *et al.*, 2012).

An example of the evidence that implicates the kisspeptin system in the control of reproduction in teleosts was provided by the changes in the expression of *kiss1* and *kiss2* during the annual reproductive cycle of male *D. labrax* (Migaud *et al.*, 2012). The highest levels of expression occurred in males in the mid- and late recrudescent stages of the testes (see Chapter 4). Levels of *kiss2* were correlated with expression of the genes responsible for the production of GnRH-2. There was evidence that the *kiss2* gene was implicated in the positive feedback control of reproduction by gonadal steroids. In *O. latipes*, *kiss1* neurons express ERs and may be directly involved in the oestrogenic regulation of reproduction (Oka, 2009; Mitani *et al.*, 2010). It is likely there are interspecific differences in the details of kisspeptin control of reproduction in teleosts, reflecting the diversity of their modes of reproduction.

It is hypothesised that, in teleosts, the kisspeptin system helps to integrate signals related to the nutritional status of the fish and the external environmental conditions in order to regulate the GnRH neurons. A problem with assessing the effects of this variety of potential influences on the synthesis and release of GTHs is that there are no quantitative assessments of their effects in intact animals. There may be redundancy in the fine control of GTH production, the factors may operate independently or interact, or their effects on GTH production may be an artefact of experimental conditions employed.

Feedback control in the brain–pituitary–gonad reproductive axis

Control systems have two basic forms. In an open-loop control system, the output of the system has no effect on the effectors of the system during the output. Once a snake has initiated a strike at a prey, it cannot alter the focus of the attack even if the prey moves. There is no sensory feedback on the position of the prey to the brain of the snake, which is controlling the strike. In closed-loop systems, the output feeds back to the effectors, modifying their output. If the feedback is negative, the activity of the effectors is reduced as the output increases, as is seen in a thermostat controlling the temperature of an oven. Negative feedback tends to stabilise a system. In positive feedback, the output enhances the activity of the effectors, leading to a further increase in the output. Such positive feedback is heard in the unpleasant noise generated by an amplifier if the microphone is placed in the wrong position. Positive feedback tends to destabilise a system. The BPG reproductive axis shows both positive and negative feedbacks during the reproductive cycle.

An important feedback in the BPG axis is provided by the sex steroids synthesised by the gonads, the target organs of the BPG axis (Levavi-Sivan *et al.*, 2010; Zohar *et al.*, 2010). During the reproductive cycle, E_2 is released into the bloodstream by the follicle cells of the ovary. A major function of E_2 is to stimulate the synthesis of vitellogenin and zona pellucida proteins (ZPs) in the liver (Chapter 4). However, E_2 also reaches the brain and the pituitary in the bloodstream. Androgens, including T, are also carried in the blood. These sex steroids, as output of the ovaries, provide the brain and pituitary with information about the current state of the gonads.

Evidence that feedback in the BPG axis is provided by the sex steroids comes from experiments in which females are wholly or partially gonadectomised, thereby changing the level of circulating sex steroids, or by treatment with exogenous steroids. Unilateral gonadectomy is a particularly valuable tool for examining feedback in the BPG axis. It involves the removal of one of the gonads, leaving the other intact, so the consequences of the manipulation can be observed in the remaining organ.

If the BPG system was an open-loop control system, the brain and pituitary would not respond to the change, so the remaining gonad would develop as normal. In female *O. mykiss*, the removal of one of the ovaries (unilateral ovariectomy) was followed by a compensatory growth of the surviving ovary, indicating that the brain and the pituitary had responded to the loss of one of the ovaries by changing their activities, showing that the BPG axis had closed-loop control (Tyler *et al.*, 1994, 1996, 1997). The compensatory growth response in the remaining ovary depended on the stage in oogenesis when the ovariectomy took place. If the ovary was removed in the pre-vitellogenic or early vitellogenic stage, the remaining ovary grew. Thus, close to ovulation there were no differences in total ovary mass or total fecundity between intact females with two ovaries and ovariectiomised females with one ovary. The compensation had been completed and had been achieved by the recruitment of a second cohort of vitellogenic oocytes. If the ovariectomy took place about 4 months before ovulation, in the late vitellogenic stage of oogenesis, the compensation by the remaining ovary was only partial. Close to ovulation, either the surviving ovary either had a normal number of ovarian follicles, but the oocytes were larger than in the intact females, or it had a higher number of oocytes, which fell into two size distributions. Oestradiol levels in the plasma of females ovariectomised early in oogenesis did not differ from those in control females with two ovaries. The follicles in the remaining ovary of operated females were producing E_2 at a much higher level than each individual ovary in the intact females.

The removal of an ovary in the early stage of oogenesis in a species that has synchronous ovarian development with total spawning (Chapter 4) caused the recruitment of a second cohort of vitellogenic oocytes. This outcome suggests that the transition between synchronous and group synchronous ovarian development seen in batch spawners is not necessarily a major physiological change.

Unilateral ovariectomy of female *D. labrax*, a batch spawner with group synchronous ovarian development, in the pre-vitellogenic or early vitellogenic stage of oogenesis before the first spawning led to complete compensation by the surviving ovary (García-López *et al.*, 2011). Levels of gene expression in the pituitary suggested that the compensatory growth was controlled by the FSH and FSH-R system. The surviving ovary had a higher rate of synthesis of sex steroids. There was also up-regulation of the transcription of *gnrh1* (the gene for GnRH-1) and *gnrhr* in the pituitary, implicating the GnRH pathway in the compensatory response. The unilateral ovariectomy also caused the up-regulation of the *star* gene in the ovary, which codes for StAR, a protein involved in a rate-regulating step in steroid synthesis. The expression of *igf1*, *igf2*, *amh* and other growth factors was also affected, but their role in the compensatory growth of the ovary is unclear.

One level at which feedback occurs in teleosts is the brain (Zohar *et al.*, 2010; Kah & Dufour, 2011). Receptors for E_2 and T occur in the brain, although GnRH neurons do not have receptors for E_2 (ERα). In the brain, T may act in two ways, directly by binding to ARs or indirectly through aromatisation to E_2. The brain form of aromatase, aromatase B, is coded by the gene *cyp19ab1* and is distinct from the gonadal aromatase (Chapter 3). Unexpectedly, aromatase B is expressed not in the neurons, but in the radial glial cells. These cells are important in the embryonic development of the brain and may play a role in the plasticity of the brain of juvenile and adult teleosts. Notably, aromatase B mRNA and protein also occur in the pituitary.

Given the lack of receptors for E_2 of the GnRH neurons, the feedback of E_2 at the level of the brain must operate through neurons that synapse on the GnRH neurons, such as the dopaminergic (inhibitory) and kisspeptin (stimulatory) circuits. In *D. rerio*, ERs are strongly distributed in the POA and the mediobasal hypothalamus, although they are also found in other areas of the brain.

In *O. mykiss* and *C. auratus*, E_2 and T can stimulate the GnRH-3 content of the brain, which suggests a positive feedback effect, enhancing the activity of the steroidogenic cells in the gonads. Oestradiol or a combination of E_2 and T caused an increase in brain and pituitary GnRH-1 levels, while in North African catfish (*Clarias gariepinus*, Clariidae), T accelerated the development of GnRH-1 neurons. A pathway for negative feedback in *C. auratus* and *O. mykiss* is provided by the presence of ERα receptors on the neurons of the dopaminergic system, which inhibits the release of GTHs, particularly LH from the pituitary. The GnRH neurons have D_1 receptors for binding DA released by the dopaminergic neurons. This negative feedback pathway cannot operate in teleost species that lack an inhibitory dopaminergic pathway. By contrast, in *A. anguilla*, androgens rather than oestrogens up-regulated DA synthesis. Unusually for a teleost, the brain of *A. anguilla* does not express high levels of aromatase B, suggesting that the effect on DA synthesis must be mediated by ARs. The pituitary of teleosts has receptors for both oestrogens and androgens; for example, androgens can increase the sensitivity of gonadotrophs producing LH.

During sexual maturation and the reproductive cycle, the balance of positive and negative feedback responses to gonadal sex steroids changes. *M. saxatilis* spawn in spring and show clear seasonal changes. During puberty, there is an increasing responsiveness to sex steroids at the level of the brain and pituitary, with E_2 having a positive feedback effect at the level of the pituitary on the expression of the FSH-β subunit and GnRH receptors. This positive feedback, in a sense, destabilises the juvenile stage leading to sexual maturation. In adult female *M. saxatilis*, feedback regulation of FSH-β changes from negative in the autumn to positive in the spring, as the females show the advanced stages of oogenesis. Feedback at the level of the brain is restricted to the spring, when E_2 up-regulates gene expression of GnRH-1, but down-regulates the expression of GnRH-2 and GnRH-3. Low levels of E_2 in juvenile *M. saxatilis* may cause modifications in the GnRH neurons or their environment in the brain that helps to create the neuroendocrine profile of a mature fish.

Other non-steroidal factors, such as growth factors, may also have feedback effects on gonadotrophs in the pituitary of teleosts. However, the lack of adequate quantitative models of the BPG axis hinders the quantitative analysis of the relative importance of the feedback pathways present in the axis.

A provisional model for the control of the developmental dynamics of ovaries and testes

The control of the gonadotrophs in the pituitary involves complex pathways in the brain, which integrate information about the internal and external environments. These pathways include the kisspeptin, in some species the dopaminergic, and in all species the GnRH pathways, and perhaps other pathways yet to be identified. The integrated effects of these pathways are orchestrated into a controlled temporal and quantitative pattern of release of FSH and LH by the pituitary and distribution and abundance of GTH receptors.

Spermatogenesis is essentially regulated by FSH binding with FSH-R on Leydig cells, stimulating the synthesis and release of androgens, particularly 11-KT. The effects of the androgens are further modulated and fine-tuned by the effect of growth factors such as activin produced in the Sertoli cells after stimulation by GTHs. Meiosis and then (later in spermatogenesis) spermiation and maturation of the spermatozoa are regulated by MIH synthesised under the stimulation of LH acting through LH-R.

Oogenesis shows a similar pattern of distinct functional differences between FSH and LH. In total spawners such as the salmonids, with clear phases of pre-vitellogenesis, vitellogenesis and oocyte maturation succeed each other with little or no overlap. FSH is dominant in vitellogenesis, while LH is dominant in maturation and ovulation. The effects of the GTHs are modulated by paracrine and autocrine growth factors.

In group synchronous species such as *D. labrax*, the phases are less clear and with greater overlap, but processes are governed by the levels of FSH and LH and the density of FSH-R and LH-R on the relevant target cells. In asynchronous species such *F. heteroclitus*, the phases almost completely overlap, with vitellogenesis and maturation occurring simultaneously, but there is still a FSH-R/LH-R dichotomy in terms of the density of GTH-R on the target cells.

Different patterns of oocyte dynamics represent shifts in temporal patterns, with a 'crushing together' of phases in species with group synchronous and asynchronous spawning, rather than radically different control mechanisms. A key question is what selective factors favour the different dynamics.

Puberty in teleosts

Puberty is a major life-history event for an individual fish. It is the transitional phase between the non-reproductive, juvenile stage and the sexually mature adult stage (Taranger *et al.*, 2010). After sexual differentiation (Chapter 3), there is a shorter or longer juvenile period during which the priority of the individual is investment in somatic growth prior to maturation (Chapter 11). During this juvenile period, the gonads are often represented by mere threads of tissue suspended from the dorsal wall of the abdominal cavity. After puberty, a significant portion of the available resources is allocated to reproduction. This is signalled by the increase in the size of the gonads during the reproductive cycle. In species such as *S. salar*, the GSI of females can increase from less than 0.3% of total body weight at the start of puberty to 20–25% just before spawning. In male *S. salar*, the increase is from about 0.1% to 5–10% at the beginning of the spawning period. Most of this gonadal mass is then lost as the gametes, eggs or sperm, are released into the external environment. The end of puberty is well defined by the production of fertile gametes, eggs or spermatozoa, for the first time. The start of puberty is less well defined. One possible definition of the start is when the rapid proliferation of spermatogonia is initiated in males, or when the oocytes begin to accumulate cortical alveoli in females (Taranger *et al.*, 2010) (Chapter 4). A more conservative identification of the start of puberty is the first appearance of spermatocytes in the male and the onset of vitellogenesis in females (Manning *et al.*, 2004; Nocillado & Elizur, 2008).

The age and size at which puberty occur shows great interspecific variation in the teleosts. In some species, puberty is attained within a few weeks of fertilisation, whereas in others, an individual may be more than 10 years old before becoming mature (Chapter 11). Even within a species, there may be both inter- and intra-population variation in the age or size at which puberty starts. In poeciliid species in the genus *Xiphophorus*, the age of maturity can vary from 8 to 104 weeks (Adkins-Regan, 2005). Some of the variation within a species is the consequence of genetic differences, whilst some is an effect of the environment, with the trait showing phenotypic plasticity (Taranger *et al.*, 2010).

Physiological control of puberty

From its initiation, puberty is controlled by the BPG axis. The brain integrates information on the internal and external environments and initiates puberty by activating the GnRH neuronal system that controls the gonadotrophs of the pituitary. It is probable that a genetically determined threshold has to be exceeded for the initiation to occur, although the physiological nature of the threshold has still to be fully defined. In some species, juveniles may show similar changes to those seen during puberty but without becoming sexually mature, as though the BPG axis is 'rehearsing' the process. In species such as the white grouper (*Epinephelus aeneus*, Serranidae) and *M. saxatilis*, juveniles undergo an incomplete pubertal cycle before a completed cycle subsequently leads to maturity (Manning *et al.*, 2008).

Some evidence in teleosts suggests that the kisspeptin neuronal network plays a role in the integration of the internal and external factors and stimulates the GnRH neuronal system (Nocillado & Elizur, 2008). In mammals, the kisspeptin has been described as the 'gatekeeper' of puberty. In some teleost species, the dopaminergic system acts to inhibit the initiation of puberty. In *A. anguilla*, DA inhibits the onset of puberty for long periods of time. This DA inhibition has to be brought to an end before LH synthesis and release are stimulated by GnRH. It is likely that specific cues encountered by *A. anguilla* in its downstream spawning migration are required to overcome the inhibition of the dopamine system. A model for the regulation of puberty in *M. cephalus* suggests that the resolution of the conflict between the stimulatory effect of the kisspeptin systems and the inhibitory effect of the dopaminergic system is resolved by the effects of E_2 produced by the ovary or by aromatisation of T in the brain (Nocillado & Elizur, 2008).

A consequence of the activation of the GnRH neurons is the stimulation of gonadotrophs to synthesise and eventually release FSH. In males, this stimulates the Leydig cells to produce androgens, particularly 11-KT, while in females it stimulates the ovarian follicle cells to produce E_2. This activation of the BPG axis requires that the receptors for GnRH and the GTHs have been synthesised by the target cells in sufficient numbers. Thus, spermatogenesis and oogenesis are initiated for the first time. In juveniles, the hormone-producing cells of the gonads, e.g. the Leydig cells in the male, may have the histological characteristics of steroidogenic cells, but then develop their full potential only slowly. The testes of immature *C. carpio* produced mainly 11-ketoandrostenedione (11-KA), but during puberty this switched to 11-KT, the biologically active androgen. However, in late-maturing female *M. saxatilis*, T and E_2 were not detectable in juveniles over the first 2 years of life.

Initially, the sex steroids, particularly T in the males and E_2 and T in females, have a positive feedback effect on the brain and the pituitary, leading to further increases in the synthesis and release of FSH (Taranger *et al.*, 2010). Partially dormant GnRH are present in juvenile *C. gariepinus* neurons. If juveniles were castrated, the GnRH neurons could be activated by treatment with exogenous T. The release of FSH in turn led to increased production of the sex steroids, enhanced by the proliferation of the steroid-producing cells. This positive feedback drives the transition from the stable state of the juvenile phase to the stable state of the sexually mature adult. As the levels of sex steroids increase still further, they start to have a negative feedback effect, tending to dampen the production of sex steroids, through effects both at the level of the brain and pituitary and at the level of the steroidogenic cells in the gonads (Taranger *et al.*, 2010).

Most species of Pacific salmon (*Oncorhynchus* spp.) are semelparous, which means that start and completion of puberty are the only reproductive period experienced by individuals, which die immediately or shortly after spawning (Campbell *et al.*, 2003). The gonads show a synchronous pattern of development, with only a single cohort of germ cells undergoing maturation. Studies on cultivated salmon suggest that there may be two key decision points in the process of puberty. The first is in the autumn, a year before spawning, when puberty is initiated. The second, a permissive point, is in the following spring about 8 months before spawning, when the continuation of the process to full maturation is determined. The outcome at both of these points may be determined by the preceding growth rate or the level of lipid reserve accumulated.

A study of cultivated male spring-run *O. tshawytscha* sought to determine whether there were physiological correlates of these key points (Campbell *et al.*, 2003). Pituitary FSH content in maturing males increased gradually from a year before spawning to the following June and then more rapidly to September, when the males were spermiating. Plasma FSH increased from late autumn to a peak in June, but then decreased back to the initial levels in September. It is not clear whether this decrease represented a decrease in release by the pituitary or an increase in turnover by the target cells. The changes in pituitary and plasma FSH were related to changes in the testis. The autumnal increase, a year before spawning, was associated with an increase in type B spermatogonia and the continuing increase with the continuation of spermatogenesis to the production of spermatids and then spermatozoa. The process of spermiation was marked by a large increase in pituitary FSH, but a decline in plasma FSH. Maturing males had higher levels of plasma FSH than non-maturing males, except at spermiation. The plasma level of LH, the other GTH, did not differ between immature and mature males in the summer before spawning, but was significantly higher in spermiating males. Plasma levels of 11-KT increased from the late summer of the year before spawning, reaching a peak immediately before spawning in autumn. By the spring before spawning, the level of 11-KT in the plasma was significantly lower in the non-maturing cohort.

The first key point, about a year before spawning, was marked by an increasing level of plasma FSH and 11-KT. The second key point in the following spring coincided, loosely, with the entry of spermatogonia into meiosis forming the spermatocytes. The levels of IGF-I, an important growth factor, increased in the early autumn a year before spawning, suggesting that this growth factor was having a stimulatory role in the early stages of spermatogenesis. Subsequently, the levels of plasma IGF-I fell as levels of FSH and 11-KT continued to increase and spermatogenesis proceeded. In the spring before spawning, levels of IGF-I were higher in maturing than in non-maturing males and reached a peak in spermiating males. The changes in IGF-I suggest some link between the growth factor and spermatogenesis, and hence a link between the reproductive and growth axes controlled by the pituitary, but the link is unclear.

A contrast is provided by an iteroparous species, the yellowtail flounder (*Limanda ferruginea*, Pleuronectidae) (Manning *et al.*, 2004, 2008). This iteroparity allows a comparison of the events during puberty with those in subsequent reproductive cycles in the mature adults. Spawning in cultivated *L. ferruginea* took place from June onwards. Immature males had testes with primary spermatogonia, but mitotic proliferation of spermatogonia was not observed and 11-KT was undetectable. In immature females, the oocytes were in the primary growth phase, but E_2 was detectable in the plasma, though not T, indicating that the ovarian follicles at this stage did have some steroidogenic capacity and could respond to GTHs. The start of puberty in males was signalled by mitotic proliferation of the spermatogonia, the appearance of spermatocytes and detectable levels of 11-KT. In females, the start of puberty was indicated by an increase in plasma E_2 above the baseline level seen in the immature females. These pubertal females developed oocytes containing cortical alveoli, and some showed signs of vitellogenesis. Plasma T was also detected in females with cortical alveoli oocytes. Spermatogenesis in pubertal males led to a GSI in December of 4%, whereas in the previous summer, the GSI of immature males was less than 0.3%. Over winter, the GSI decreased to about 2% at pre-spawning in April and spawning, which started in June. This decrease probably reflected the breakdown of the cytoplasm of the spermatids as they transformed into spermatozoa. Spermatozoa were present in the testes from December onwards and the males in June were producing running milt. Levels of plasma androgen tended to increase, reaching a peak in the pre-spawning period. The dominant androgen was 11-KT, but this showed greater variations than T. In females, early vitellogenic oocytes were largely absent by April and completely absent in June as the developing cohort of oocytes completed vitellogensis and started maturation. In June, the ovaries of newly matured females also contained oogonial nests and early pre-vitellogenic oocytes, setting the stage for the next reproductive cycle. Plasma levels of E_2 in developing pubertal females were relatively stable during vitellogenesis, but increased in May and June at the beginning of spawning. The pattern in adults over their second reproductive cycle was similar to that of pubertal fish, although the adults had somewhat larger GSIs and levels of E_2 and some adult males were spermiating as early as April. This similarity indicates that the crucial event in puberty is its initiation and, once initiated, the pattern of gametogenesis and its BPG control is essentially similar to that seen in adult fish. Once initiated, the pubertal maturation in *L. ferruginea* seems to go through to completion, culminating in spawning, suggesting there is not the permissive key point in the process that seems to be present in salmonids.

A feature that *L. ferruginea* shares with some other pleuronectids, such as winter flounder (*Pseudopleuronectes americanus*, Pleuronectidae), is that the testes contain spermatozoa up to 6 months before spawning. In other species, including the European plaice (*Pleuronectes platessa*, Pleuronectidae), the spermatozoa are present only a short time before spawning.

Puberty marks a shift from investment in somatic growth and the accumulation of lipid reserves to a state in which there is also significant investment in the gonads and the production of viable gametes. This means that there must be interactions between the controls being exerted by the brain and pituitary on reproduction, growth and metabolism.

Interactions between growth and reproductive axes

In an individual fish, the resources available for growth generated by feeding behaviour can be allocated to three components: structural growth, stored growth and gonadal growth (Wootton, 1998). Structural or somatic growth builds the body and the functioning systems that sustain the survival of the individual. This structural growth is represented by changes in the length of the fish (axial growth) and changes in the mass of the fish (ponderal growth). Resources, especially lipids, can be allocated to storage as reserves that can be called upon when feeding is not sufficient to meet the resource demands of the individual. In gonadal growth, resources are allocated to gametogenesis (Chapter 4), and eventually a portion of those resources are voided by the individual as gametes and are no longer available to the individual. Life-history theory (Chapter 11) seeks to predict the optimal allocation between structural, storage and gonadal growth in relation to the age (and size) of the individual and the prevailing internal and external environmental conditions. The optimal allocation is that which tends to maximise the Darwinian fitness, often quantified as lifetime production of offspring by the individual (but see Chapter 11).

During the juvenile phase of an individual's life, priority is given to structural growth and, if resources are abundant, to storage growth. Puberty marks the point when significant allocation of resources to gonadal growth is initiated. A complicating factor in teleosts is that most species have an indeterminate growth pattern (Wootton, 1998). In such species, structural growth does not cease around the time of puberty. This is in contrast to endothermic vertebrates, in which determinate growth is the usual pattern, with structural growth essentially ceasing around the time of puberty and only allocations related to storage and reproduction continuing. In teleosts, structural, storage and gonadal growth coexist, which raises the question of how that coexistence is regulated.

Two patterns of resource allocation during the post-juvenile phase of a teleost can be identified, although they probably represent end points in a continuum of allocation (Bonnet *et al.*, 1998; Stephens *et al.*, 2009). In capital breeders, the resources required for gonadal growth are initially stored as reserves and then transferred to gonads during gametogenesis. In many teleost species with capital breeding, feeding ceases or becomes much reduced during the later stages of gametogenesis and during the spawning season. This pattern of feeding is characteristic of many total spawners or species with determinate fecundity and batch spawning (Chapter 4). The salmonids are a notable example of anorexia developing before spawning (Chapter 7).

In income breeders, the resources required for gametogenesis and activities associated with spawning are acquired from the ongoing feeding behaviour. This pattern is characteristic of species with indeterminate fecundity and batch spawning. Species may fall along the continuum, with gametogenesis being supported from feeding but also by transfer from the reserves if the income from feeding is insufficient.

From puberty onwards, the coordination of reproductive, structural and storage growth must be achieved by the neural and neuroendocrine systems. Reproduction, growth and metabolism are regulated by brain–pituitary–target organ axes. The key pituitary hormones for the three axes are the GTHs (FSH and LH) produced by the gonadotrophs, GH produced by the somatotrophs and the TSH produced by the thyrotrophs. The activities of these cells of the anterior pituitary are regulated by neurohormones produced by neurons projecting into the pituitary from the brain, especially the POA and the hypothalamus. This means there are three levels at which cross-talk between the axes can coordinate their activities: the brain, the pituitary and the target organs.

Growth hormone has an essential role in the regulation of the somatotrophic axis, i.e. growth and development (Rousseau & Dufour, 2007). It can act directly on the target cell, or indirectly through IGFs, especially IGF-I. This growth factor can act as an endocrine transported to its target cells by the blood circulatory system. The main source of endocrine IGF-I is the liver, which produces it under the influence of GH. However, tissues can also produce IGF-I locally, so it can act as a paracrine/autocrine factor. The hypothalamic control of GH operates through both a stimulatory factor, growth hormone-releasing hormone (GHRH) and an inhibitory factor, somatostatin (SRIH). This dual control recalls the dual control of GTHs found in some teleosts such as *C. auratus*. Regulation of GH levels is also achieved by the negative feedback effects of endocrine IGF-I.

The target organ of TSH is the thyroid tissue, located in the pharyngeal region of the fish, and which produces thyroid hormones (THs), important regulators of metabolism. It is possible that THs have a key role in the allocation of resources for structural, storage and gonadal growth, but this role has yet to be defined (Raine, 2011).

It is unclear how the reproductive, growth and metabolic axes interact to regulate the initial switch to gonadal maturation at puberty and then, in iteroparous species, the reversible switch between structural and gonadal growth. Gonadal growth shares some characteristics of structural growth, including cell division and the synthesis of new material, especially the synthesis of vitellogenin and ZPs by the female. Consequently, gonadal growth may require collaboration of the growth axis with the reproductive axis, either directly or indirectly through endocrine or local IGF-I. Receptors for GH (GH-R) occur in immature oocytes and in the theca and granulosa cells of the ovarian follicles. Thus, GH might influence the local synthesis of IGF-I in the ovaries (Reinecke, 2010). IGF can stimulate spermatogenesis, particularly in the presence of 11-KT. In ovaries, IGFs have been implicated in oocyte maturation and may also modulate steroidogenesis in theca and granulosa cells.

Maturing male 1+ *O. tshawytscha* had higher growth rates than immature males from the December before spawning to the following summer (Shearer & Swanson, 2000). The maturing males also had higher levels of plasma IGF-I, with levels reaching a peak in March and then staying relatively stable. These high levels of IGF-I suggest the involvement of the growth axis, at least in the early stages of gametogenesis. A possible role for endocrine IGF-I in the coordination of the reproductive and growth axes was indicated in studies that used cultured pituitary cells of *D. rerio* (Lin & Ge, 2009). In these cultured cells, IGF-I suppressed the expression of the *gh* gene, stimulated the expression of *fshb* (*fsh-β*), but had no effect on the expression of *lhb* (*lh-β*). The same study also identified another potential player in the coordination of reproductive and growth axes. The neuropeptide, pituitary adenylate cyclase-activating peptide (PACAP) had a stimulatory effect on the expression of *gh*, but suppressed the expression of *fshb* and *lhb*. These results suggest mutual inhibitory and stimulatory interactions between the reproductive and growth axes, but it is not clear whether these studies on cell cultures have relevance for the interactions in intact fish.

Studies on salmonids suggest that fish must achieve a threshold level of growth rate or lipid content at a critical time if they are to mature. In juvenile *S. salar*, males had to reach a threshold lipid level in spring if they were to mature the following autumn (Chapter 11) (Rowe *et al.*, 1991). However, these maturing males also had higher growth rates (Rowe & Thorpe, 1990a), so it was unclear whether the rate of structural growth or level of lipid reserves was the critical factor. A high rate of structural growth and a high rate of lipid storage by an individual both indicate effective foraging by an individual and so indicate the capacity to support the expenditure of resources on reproduction. An attempt to tease apart the effects of rapid structural growth and high levels of lipid storage used transgenic juvenile *S. salar* (Moreau & Fleming, 2012). Such transgenic fish have higher rates of structural growth than normal fish. In the second year of life, the number of transgenic males that were maturing was half that of normal males, suggesting that the rate of structural growth does not play a role in causing precocial maturation in this species. The high growth rate in the GH transgenic males is associated with high metabolic rates and reduced energy reserves, suggesting that the critical threshold for maturation is a storage threshold, specifically the level of stored lipids. This puts the focus back on the poorly understood brain–pituitary regulation of the allocation to structural and storage growth.

A further complication is that growth depends on food consumption (Wootton, 1998), and thus on the regulation of appetite. The control of appetite in teleosts is poorly understood and complex (Volkoff *et al.*, 2005). Brain peptides such as NPY and orexin promote increases in appetite, as does the gastric hormone, ghrelin (Hoskins *et al.*, 2008). Leptin, a hormone that in vertebrates serves to signal levels of stored lipid, may inhibit appetite. All of these factors have been implicated in the regulation of the synthesis and release of GTHs, but their roles are poorly understood.

Teleosts show two main patterns of appetite change in the reproductive cycle. In some species, the latter stages of gametogenesis are associated with a decline in appetite, and feeding may stop before or during the spawning season. This pattern is pronounced in the anadromous salmonids, including *Oncorhynchus* spp. and *S. salar*. These are total spawners. Individuals cease to feed before they enter freshwater on their upstream spawning migration. All the resources expended on the migration (Chapter 7), the final stages of gametogenesis and reproductive behaviour on the spawning ground (Chapter 8) must be met from the stored reserves. Before the onset of this reproductive anorexia, the maturing fish have higher growth rates and rates of food consumption than immature fish (Kadri *et al.*, 1995, 1996; Shearer & Swanson, 2000). This indicates that the initial stages of maturation in salmonids may involve coordinated activity in both the reproductive and growth axes. In *S. salar*, the maturing fish have to reach thresholds levels of lipid reserves and lean body mass before the reproductive anorexia starts (Kadri *et al.*, 1995, 1996). The physiological mechanisms by which the anorexia is initiated and maintained are not clear. Curiously, although anorexic, the migrating fish will still strike at lures presented by anglers.

At the other extreme are species that are batch spawners, such as *G. aculeatus* with indeterminate fecundity (Chapter 1 and 6), which feed voraciously during the spawning season, thus supporting the production of successive clutches by the food income (Fletcher & Wootton, 1995; Wootton & Fletcher, 2009). During the interval of a few days between successive spawnings, female *G. aculeatus* have high rates of food consumption initially, but then consumption declines on the day or so before spawning, when the ovaries occupy virtually all of the abdominal cavity. The changes that take place in the reproductive and growth axes and their interactions with the regulators of appetite are not known. In *C. auratus*, orexin, a hypothalamic neuropeptide, inhibits GnRH-1, while promoting an increase in appetite. In this case, there is an inhibitory relationship between the physiological control of appetite and the BPG axis (Hoskins *et al.*, 2008).

An interaction between feeding and the reproductive and growth axes also occurs if there is a lack of food during gametogenesis. Food deprivation of immature female *O. kisutch* for 17 weeks in the late spring, which is before the initiation of the reproductive anorexia, evoked both growth and hormonal effects (Yamamoto *et al.*, 2011). The fasting arrested growth in body mass and length, and also reduced ovarian growth compared with normally fed females. Fasted females did show some growth of ovarian follicles, but their growth lagged behind that achieved by fed females. The ovaries of fasted females had a higher proportion of atretic follicles than in fed females. Fasted females had lower levels of plasma IGF-I and E_2 and a lower level of pituitary FSH than fed fish, but the levels of plasma FSH did not differ. Expression of some genes involved in steroidogenesis in the theca and granulosa cells, including *star*, *hsd3b* and *cyp19a1a*, were also lower in fasted females. By contrast, the expression of genes involved in apoptosis (cell death) was higher in the ovaries of fasted females, which may be related to a role for cell death in pre-ovulatory atresia. These results indicate that, at least at some phases of gametogenesis, an inadequate supply of resources inhibits reproduction.

A physiologically poorly understood comparable effect of food availability occurs in some species with batch spawning. In female *G. aculeatus*, the number of spawnings is positively correlated with the availability of food during the spawning season, with low food levels leading to fewer spawnings (Fletcher & Wootton, 1995; Wootton & Fletcher, 2009). An understanding of the relationships between the reproductive, growth and metabolic axes controlled by the brain and the pituitary together with the control of appetite is key to understanding the physiological basis for life-history patterns (Chapter 11), although a synthesis of this sort is currently lacking.

Endocrines, behaviour and secondary sexual characteristics

Gonadal sex steroids carried in the bloodstream to the brain can modulate the activity of the BPG reproductive axis, providing a regulatory feedback . Another effect of at least some sex steroids is to modulate the behaviour of the individual by evoking reproductive behaviour. By this mechanism, appropriate reproductive behaviour can be synchronised with appropriate states of the gonads, particularly ovulation in the ovaries and spermiation in the testes (Chapter 4). The external environment also provides relevant information to the brain through the sense organs and the associated sensory neurons. On the basis of this information, the brain regulates the activity of the BPG axis and hence the production of sex steroids. Abiotic factors such as photoperiod and temperature can play a major role in determining the timing of events in the reproductive cycle (Chapter 6). Food availability can influence the investment made by an individual in reproduction in a given reproductive cycle (Chapter 6). A third factor is the socio-sexual environment provided by male and female conspecifics, with the behaviour of conspecifics influencing the activity of the BPG axis. There is, then, a reciprocal relationship between the BPG axis and the biological and physical environment (Liley & Stacey, 1983). The endocrines of the axis regulate the behaviour required for reproduction, but they are responsive to social, biotic and abiotic stimuli, with behaviour proving a link between the individual and its environment.

Gonadal sex steroids have two effects relevant to the behaviour of individuals. The first is the modulation of reproductive behaviour. The effect of hormones on behaviour can be classified into three categories (Munakata & Kobayashi, 2010). The first is where the hormone acts as a physiological trigger, which activates behaviour within a few minutes. The second category is as a primer, in which the presence of the hormone is necessary, but not sufficient, for the behaviour. The occurrence of the behaviour in the presence of the hormone still requires a relevant stimulus such as the presence of a potential mate. In the third category, the hormone acts a potentiator. The hormone does not trigger the behaviour and may not even be essential, but it does have an enhancing effect on the behaviour.

The second effect of sex steroids is to stimulate the appearance of secondary sexual characteristics, including traits, such as breeding colours and morphological traits, that act as signals in sexual and related encounters (Chapter 8).

In the CNS, sex hormones can potentially act at three levels. The first level is the sensory system, including sense organs and the neuronal circuitry that carries information from the sense organs to the integrative neural systems of the brain. These integrative systems form the second level of hormone effect. They control the patterns of neural output to the motor systems. The third level is these motor systems that carry information to the effector organs, principally the muscle systems that generate the behaviour.

The principal androgens in teleosts are T and 11-KT, but most studies suggest that 11-KT is more potent in regulating behaviour and inducing secondary sexual traits than T (Borg, 1994). In the brain, T can be transformed into E_2 by aromatase, so the effect of T is ambiguous, it may operate directly or indirectly through E_2. The role of oestrogens and progestagens in the control of reproductive behaviour is less well understood. Other potential relevant endocrines or neuroendocrines include prostaglandins and the neurohypophysial arginine vasotocin (AVT) and isotocin.

The analysis of the reciprocal relationships between the sex hormones and the socio-sexual environment in teleosts has depended heavily, if not exclusively, on four model species: *C. auratus*, *G. aculeatus*, plainfin midshipman (*Porichthys notatus*, Batrachoididae) and Burton's mouthbrooder, *Astatotilapia burtoni* (Cichlidae) Brief accounts of their relevant biology are given here, but fuller accounts of the reproductive behaviour of teleosts are given in Chapters 7–9.

Four model species

The goldfish, *C. auratus*, is a cyprinid. Increasing water temperature in the presence of aquatic vegetation triggers the release of a surge of LH in the female in the afternoon (late photophase). Ovulation occurs about 12 h later (late scotophase) and the female becomes sexually active for several hours. She repeatedly enters the aquatic vegetation, spawning batches of eggs. Mature males engage in a scramble competition for access to the spawning female, with intense sperm competition between the males (Stacey & Sorensen, 2008; Stacey, 2011; see Chapter 8).

Gasterosteus aculeatus (Gasterosteidae) has a breeding season in spring and summer. Mature males develop breeding colours of a red throat and abdomen and bright blue eyes. The male starts to defend a territory from which other males are excluded. Within its territory, the male builds a nest from fragments of vegetation, which are glued together with a proteinaceous secretion, spiggin, produced by a hypertrophied kidney. Once the nest is complete, the male courts a gravid female, approaching the female in a characteristic zig-zag dance (Tinbergen, 1951). If the female is responsive, she follows the male back to his nest and spawns there, with the male entering the nest to fertilise the eggs. The male then cares for the eggs, principally by fanning the nest to ventilate them (Chapter 9). A male will initially add more clutches to the nest, but courtship declines as fanning increases. After the eggs have hatched and the young have dispersed, the male may build another nest and resume courtship of gravid females (Wootton, 1976; Mayer & Páll, 2007).

Porichthys notatus is a coastal, marine species. It has three reproductive phenotypes: females, type I males and type II males. In the reproductive season, type I males defend a territory containing a nest site. A male attracts a gravid female to the nest by emitting a characteristic vocalisation, the 'hum'. Type II males are smaller and do not defend a nest site. They rely on stealing fertilisations from type I males by mimicking females to enter the nests of type I males, an example of male mating polymorphism (Chapter 8). Both male morphs and females produce agonistic 'grunts' (Bass & McKibben, 2003; Bass, 2008).

In the cichlid *A. burtoni*, dominant males defend a territory within which they dig a nest. The male courts a gravid female, which enters the nest. The female picks up the spawned eggs in her mouth, where they are brooded until they hatch. The eggs may be partly fertilised by sperm taken into the mouth with the eggs. Non-territorial, subordinate males are unable to mate. However, if space opens up, a subordinate male may rapidly become territorial and assume the status of a dominant male (Barlow, 2000; Forlano & Bass, 2011).

Role of endocrines in regulation of reproductive behaviour

Sensory systems

Some evidence, largely indirect, suggests that sex steroids can affect the sensory system, either at the level of the sense organs or at the level of the sensory neurons serving the sense organs, or, a third possibility, at both.

Both male and female *G. aculeatus* become more sensitive to the colour red during the breeding season (Boulcott & Braithwaite, 2007). Red is a major component of the breeding colouration of territorial males and an important signal in both intra- and inter-sex interactions (Chapter 8). The physiological basis for this change in visual sensitivity is not known, but it is associated with a period of increased levels of sex steroids in the plasma.

In *P. notatus*, there is an improvement, associated with the breeding season, in the frequency encoding by the saccular hair cells in the inner ear of type I males. A similar improvement is also seen in breeding females. The change in sensitivity in type I males is associated with increasing androgen levels in the plasma. It is possible that androgens act directly on the saccular hair cells and associated neurons through ARs, or indirectly through ERs after conversion of an aromatisable androgen to E_2 (Rohmann & Bass, 2011).

The olfactory bulbs have neuronal connections with the telencephalon and other brain areas, with information travelling both from and to the olfactory bulbs. In male *C. auratus*, androgens increase the sensitivity to pheromones released by the mature female (Forlano & Bass, 2011).

Central integrative systems

There is still a poor understanding of the neural circuits in the teleost brain that integrate the information coming from the sensory systems and the physiological state of the individual to generate the neural patterns that stimulate the motor systems coordinating reproductive behaviours. In comparison with studies on mammals, there have been relatively few studies that have used electrical stimulation of localised brain areas to map relevant neuronal circuits. Particularly relevant to the endocrine modulation of behaviour is the identification of a neurochemically rich 'core area' of the brain that includes the POA and the hypothalamus. Neurons in the core area synthesise a wide range of neuropeptides and have receptors for both androgens and oestrogens. Interconnected with the core area is a 'paracore area' of the brain stem, containing cells that are especially rich in monoamines. The associated neural fibre tracts extend caudally and include the midbrain periaqueductal grey nuclei in the caudal medulla and the central grey of the spinal cord. This neuronal system, core and paracore areas, is highly conserved in vertebrates and probably has a key role in modulating the activity of the brain and the regulation of adaptive behaviour, including behaviour associated with reproduction (Forlano & Bass, 2011).

Electrical brain stimulation of sunfish species (Centrarchidae), the bluegill sunfish (*Lepomis macrochirus*) and green sunfish (*L. cyanellus*), have identified some of the relevant neural circuitry (Demski, 1983). Stimulation of the POA inhibited aggressive behaviour, but facilitated courtship behaviour. At the termination of the stimulation, there was a transient increase in aggression. A mutually inhibitory relationship between aggressive and courtship behaviour is also seen in nesting male *G. aculeatus*. A nesting behaviour, sweeping, in territorial male *L. macrochirus* was evoked by stimulation of sites in the dorsal telencephalon and the POA. Stimulation of the POA also evoked courtship. Electrical stimulation identified a pathway that stimulated sperm release extending from the POA, through the dorsal hypothalamus (both core areas) into the mid-brain tegmentum and ending in the rostral spinal cord, which is associated with a presumed primary motor system that controls sperm release. Overall, the evidence suggests that neuronal circuitry in the telencephalon, POA and hypothalamus control the complex of reproductive behaviours.

Sex steroids act as primers for these central neuronal systems, so that, in the presence of suitable stimuli, the appropriate behaviours will occur. Presumably, the neuronal circuits are largely laid down during the development of the brain and so are available to be primed as the steroids, at suitable levels in the blood plasma, become available. Typically, male reproductive behaviours are more complex than those of females, and androgens, particularly 11-KT, are the most effective at priming the neuronal circuits. Male *G. aculeatus* provide a good example (Páll *et al.*, 2002a, 2002b; Mayer & Páll, 2007). In this species, the testes synthesise T and 11-KA, but the 11-KA is converted to 11-KT by blood cells, so 11-KT is the major plasma androgen (Borg, 2007). Outside the breeding season of *G. aculeatus*, levels of plasma 11-KT are low or not detectable. With the onset of maturation, levels of 11-KT rise to extremely high levels. Levels of T also increase, but not to such high levels. The levels of 11-KT are highest during the nest-building and courtship phases. As the male accumulates eggs in its nest, the level of plasma 11-KT declines. Behaviourally, the levels of courtship also decline while parental care, dominated by nest fanning, increases to a peak just before the eggs hatch. As the eggs hatch, fanning declines. When the young disperse from the nest, the level of 11-KT increases, and the male builds another nest and resumes courtship. During these fluctuations of 11-KT, the levels always remain higher than those seen outside the breeding season. If males are castrated at the start of a parental cycle, but implanted with 11-KA so plasma levels of 11-KT remain high, the males still show the decline in courtship and increase in nest fanning. There is some weak evidence that the parental behaviour may be regulated by prolactin, a hormone produced in the anterior pituitary and associated with parental care in mammals (Le Rouzic & Prunet, 2005).

In other species that show parental care, plasma levels of 11-KT are higher in the courtship phase, while tending to decline during the parental phase. Such a change in levels occurs in a reef fish, the garibaldi damselfish (*Hypsypops rubicundus*, Pomacentridae), which may complete about four broods during a breeding season (Sikkel, 1993).

In salmonid males, the progestagen 17α,20β-P, a maturation-inducing stimulus, has been implicated in the regulation of courtship. In castrated male *O. mykiss*, treatment with 11-KT failed to restore courtship behaviour (Gonçalves & Oliveira, 2011), although androgens are required for the development of secondary sexual traits. Peptides produced by the neurohypophysis, AVT and isotocin, may also have roles in the regulation of reproductive behaviour, but the details are unclear. Both AVT and GnRH or their receptors occur in peripheral sensory organs and the central, integrative areas of the brain (Forlano & Bass, 2011).

In contrast to the importance of androgens for male reproductive behaviour, especially reproductive territoriality and courtship, oestrogens such as E_2 do not seem to play a role in the reproductive behaviour of females, at least in oviparous species with external fertilisation. This situation may be different in the relatively few species that have internal fertilisation (see Chapter 10). The reproductive behaviour of females is typically far less complex than that of males (Chapter 8). In female *C. auratus*, the prostaglandin F2α ($PGF_{2α}$) is produced in the ovaries when oocytes are ovulated and appear in the gonoducts. The $PGF_{2α}$ is transported to the brain in the blood and activates sexual receptivity and spawning behaviour. It acts within minutes, indicating it is functioning as an activating factor rather than a priming factor (Stacey & Sorensen, 2008).

Motor systems

Endocrines can also affect the neurons and muscular systems responsible for the motor patterns expressed as reproductive behaviour. The effects on the muscles themselves form a component of the secondary sexual traits regulated by endocrines.

The vocalisations of sexually mature *P. notatus* are produced by the simultaneous contractions of a pair of sonic muscles attached to the walls of the swim bladder (Bass, 2008). The contraction of the sonic muscles is controlled by a vocal pattern generator (VPG) in the hindbrain–spinal cord. The neural output of the VPG matches 1:1 with each muscle contraction. The onset of the advertising vocalisations ('hums') of the type I males is paralleled by a rise in plasma androgens. An injection of 11-KT caused a significant increase in the duration of calls within 5 minutes. The speed of this response suggests that the 11-KT is acting via a non-genomic mechanism. Experiments in which the VPG was isolated from other areas of the brain showed that the hindbrain–spinal region could account for the rapid effect of 11-KT and was not dependent on endocrine effects on other brain areas.

Endocrines and secondary sex characteristics

In addition to their effects on the nervous system, sex endocrines also have effects on other body traits, particularly secondary sexual traits (Chapter 8). These are traits that differ between the sexes, but do not belong to the complex of gonads, gonoducts and other associated structures. Most of these secondary sexual traits have functions in the context of interactions between individuals, either between individuals of the same sex that are competing for access to mates or as signals between the sexes during courtship (Chapter 8). However, some secondary sexual traits have a function in providing environmental conditions in which mating can be successful.

Pearl organs and breeding tubercules in male cyprinids are induced by androgens. Androgens also induce changes in fin shape in males, including elongation of the dorsal fin in the three-spot gourami (*Trichopodus trichopterus*, Osphronemidae) and the extension of the caudal fin into the 'sword' that is characteristic of male swordtails, such as the green swordtail (*Xiphophorus helleri*, Poeciliidae) (Liley & Stacey, 1983). In male *Oncorhynchus* spp., changes in colouration and body shape are governed by androgens, especially 11-KT. A field study of male *O. tshawytscha* during the spawning season, which compared the large, dominant males, termed 'hooknoses', with smaller subordinate males that sneaked fertilisations, 'jacks' (see also Chapter 8), found that in hooknoses, plasma levels of 11-KT were positively correlated with the height of the dorsal hump, a secondary sexual trait (Butts *et al.*, 2012). In jacks, plasma levels of T and 11-KT were positively related to the length of the kype, an elongation of the jaw in mature male salmonids that is used in aggressive interactions.

Male *G. aculeatus* illustrate the importance of androgens, particularly 11-KT, for secondary sexual traits that are relevant to male–male and male–female interactions during the reproductive season, but also in inducing functional traits crucial for reproduction (Borg, 2007). The breeding colours of the male, the red throat and abdomen and blue eyes, are androgen-dependent and are lost in castrated males. These colours are relevant in male–male territorial interactions and in male–female courtship (Chapters 7 and 8). An important androgen-dependent secondary sexual trait of male *G. aculeatus* is the hypertrophy of the kidney, with kidney tubule cells synthesising spiggin, the glue the males uses to construct its nest. Spiggin mRNA is not found in the kidneys of females, but can be induced by treating females with androgens. The most potent androgen for inducing spiggin production is 11-KT and there is some evidence that *G. aculeatus* has an androgen receptor specific to 11-KT.

Effect of social interactions on reproductive endocrines

The reciprocal interaction between the endocrine system and the environment can involve social interactions. These are particularly important in the regulation of sex change in species that are sequential hermaphrodites (Chapter 10). However, even in gonochoristic species, social interactions can have important effects on the endocrine state of an individual.

Dominant male *A. burtoni* have higher GnRH receptor type 1 (GnRH-R1) gene expression in the pituitary, higher levels of circulating androgens, elevated expression of steroid receptors in the brain, and the cell bodies of the GnRH-1 neurons are larger than in subordinate, non-territorial males (Maruska *et al.*, 2011). When a subordinate male has the opportunity to improve its social status and becomes dominant, it shows territorial and reproductive behaviours within minutes. There are changes in gene expression in the POA and in GnRH-1 neurons. Within 30 minutes of a change in status, pituitary LH-β mRNA expression had increased and reached stable levels characteristic of dominant males by 72 h after the change. Pituitary FSH-β mRNA expression also increased rapidly to levels characteristic of dominant males. Levels of plasma LH and FSH also increased rapidly to those characteristic of dominant males, although levels of FSH were much lower than those of LH. In addition, plasma levels of 11-KT were positively correlated with the mRNA levels of FSH-β and LH-β. The rapid response to the change in the social environment suggests that the BPG axis of subordinate males is maintained at a sub-threshold level, allowing subordinate males to respond rapidly to a change in the social environment (see also Chapter 7).

In salmonids, social interactions are implicated in the regulation of the progestagen, 17α,20β-P. In male *O. mykiss*, the opportunity to court a nest-digging female resulted in an increase in plasma 17α,20β-P and a concomitant increase in sperm production. The production of the progestin is stimulated by a chemical in the urine of a female (Liley *et al.*, 1986; Scott *et al.*, 2010).

Role of pheromones in reproductive interactions

Chemicals released into the environment by an individual, as a result of metabolic and physiological processes, have the potential to play a role in social interactions, including the interactions between the sexes. Some chemicals may not be detected by another individual and so cannot convey any information. Some chemicals may be detected, but again transmit no useful information. A third possibility is that the chemical is detected and provides the recipient individual with useful information about the physiological state of the emitting individual, allowing the recipient to 'spy' on the emitter. However, this transfer of information is inadvertent. Finally, the chemical may act as a pheromone, acting as a signal whose function is to transmit information from the emitter to the recipient (Liley & Stacey, 1983; Stacey & Sorensen, 2008; Stacey, 2011). In practice, it may be difficult to distinguish between the third and fourth categories unless there is a structure whose function is unambiguously to produce a pheromone. Hormones, such as the sex steroids and their metabolites, clearly have the potential to act as pheromones, accurately signalling the reproductive state of the emitting individual.

The role of pheromones in the context of reproduction has been most fully studied in *C. auratus*, especially the female (Stacey & Sorensen, 2008; Stacey, 2011). A mature, female *C. auratus* releases pheromones in three succeeding contexts. Vitellogenic and post-vitellogenic but pre-ovulatory females release a 'recrudescent pheromone', which attracts males to the vicinity of the female and stimulates them to start nudging the female. This pheromone is induced by E_2, but its identity is not known. After the LH surge that initiates the process of ovulation, the female releases a mixture of steroids: androstenedione (AD), 17α,20β-P, both of which are released through the gills, and the metabolite 17α,20β-dihydroxy-4-pregnen-3-one 20β-sulphate (17α,20β-P-S), which is released in the urine. These three, the pre-ovulatory pheromones, reach peak levels successively. Initially, high levels of AD increase agonistic behaviour amongst males. Increasing 17α,20β-P triggers an LH increase in the attending males, at least partly by reducing the level of dopamine inhibition, which has the effect of stimulating sperm production and increasing sperm quality. The males show prolonged following and inspection of the female. With the increase in 17α,20β-P-S, there are brief but intense bouts of the male chasing the female and a further increase in LH. The 17α,20β-P also increases the frequency of ovulation in other females, but the function of this effect is obscure. After ovulation, the levels of the pre-ovulatory steroidal pheromones decline rapidly. Now, the movement of eggs into the gonoduct stimulates the synthesis of $PGF_{2α}$ and its metabolite 15-keto-prostaglandin $F_{2α}$ (15-K-$PGF_{2α}$), which act as a post-ovulatory pheromone. This pheromone immediately stimulates male courtship and also further enhances sperm production. $PGF_{2α}$ is released across the gills and in the urine, whereas 15-K-$PGF_{2α}$ is released in pulses in the urine. Pheromones released across the gills form a filament in the water as the female swims, whereas those released in the urine tend to form discrete patches in the water.

Male *C. auratus* also release a pheromone, mainly AD, but its role is not clear. Exposure to AD does inhibit sperm production in other males. This effect and the effect of the pheromones of the female on sperm production may be important in sperm competition between the males attending to the ovulating female (Chapter 8).

The picture that emerges is of sexually mature *C. auratus* living in complex social networks, mediated, at least partly, by pheromones. It is possible that PGF pheromones are found throughout the cypriniform fishes. By contrast, some cypriniforms, including the Cobitoidea (loaches) and Acheilognathinae (bitterlings), appear not to be able to detect sex steroids. Bitterlings deposit their eggs in the gill cavity of freshwater mussels (Chapters 7 and 8), so chemical detection of a suitable mussel may have priority over the detection of sex steroids, although a range of alternative olfactory cues may play a role in mating decisions (Smith *et al.*, 2004a).

The picture in salmonids is mixed. Male parr of *S. salar* are sensitive to water-borne PGFs, which have a priming effect, including increased levels of LH, steroids and sperm. By contrast, *Oncorhynchus* spp. seem to be insensitive to PGFs, although male *Oncorhynchus* spp. do respond to chemical cues from females that have ovulated (Stacey, 2011).

In some gobies, the odour of mature males attracts vitellogenic females and may help the female to find the male's nest site. The testes in species such as the black goby (*Gobius niger*, Gobiidae) have a specialised structure, the mesorchium (Chapter 4), which plays a role in the maturation of spermatozoa, but also synthesises steroids, including a metabolite etuiocholamine, which may act as a male pheromone (Stacey, 2011).

Endocrine-disrupting chemicals

The complexity of the reproductive axis in teleosts makes it potentially prone to disruption by environmental pollutants. Aquatic habitats, especially fresh and coastal waters, are particularly susceptible to pollution, because of the inflow of sewage and other drainage of waste products from domestic and commercial activities and, in rural areas, run-off from agricultural activities, including pesticides and herbicides. A wide range of pollutants have adverse effects on the reproductive system of teleosts in addition to their effects on other physiological processes (Kime, 1995). These pollutants include heavy metals and their salts, organochlorine pollutants, carbamate pesticides and a range of other industrial pollutants, including effluents from pulp mills and acid rain. Both males and females experience adverse effects on their reproductive performance. In extreme cases, this can lead to recruitment failure and the ultimate extinction of the population. Many lakes in northern Europe lost their fish populations in the latter half of the 20th century, most likely because of reproductive failure as acid precipitation (acid rain) progressively lowered the pH of lake waters (Hendriksen *et al.*, 1989).

Endocrine-disrupting chemicals (EDCs) are an important class of pollutants for the reproductive axis. An EDC is not produced endogenously by the individual fish, but has an exogenous source in the external environment. It causes endocrine changes in the individual, which have adverse effects on the health of the individual or its progeny (Olsson *et al.*, 2008; Vajda & Norris, 2011). Most attention has been paid to EDCs that adversely affect the reproductive axis.

There are several mechanisms by which EDCs may act (Scholz & Klüver, 2009; Vajda & Norris, 2011). An EDC may mimic an endogenous hormone and cause the activation of regular reproductive processes such as vitellogenesis. Conversely, an EDC may act as an antagonist of an endogenous hormone, thus blocking the activation of regular processes. This blocking may occur because the EDC interferes with the regulation of hormone receptor levels. An EDC may interfere with the synthesis, transport and metabolism of endogenous hormones. In the CNS, an EDC may influence the neuroendocrine system and the related feedback interactions. Interest in EDCs is currently high, because of concerns about the effects of EDCs on fertility in human populations, and teleosts are seen as potentially valuable indicators of the presence of EDCs at biologically relevant levels in the environment.

An indication that EDCs could have pervasive effects in the natural environment came from a study of the condition of intersex in the roach (*Rutilus rutilus*, Cyprinidae), a gonochoristic species that is widespread in northern European rivers (Tyler & Jobling, 2008). Intersex is the condition in which the gonads contain both male and female characteristic cells. In males, the testes contain oogonia and oocytes, while in females the ovaries contain spermatogonia and spermatocytes. The intersex condition reflects the bipotentiality of the teleostean gonad (Chapter 3). In natural populations of gonochoristic teleosts, the incidence of intersex is low, although it does vary between species.

The focus of attention was the incidence of intersex in *R. rutilus* living in rivers receiving effluents from waste water treatment plants (WWTPs) (Tyler & Jobling, 2008; Tyler & Filby, 2011). In rivers downstream of WWTP outflows, the percentage of intersex varies from 16% to 100%. At sites upstream of a WWTP, but perhaps receiving some effluents from WWTPs even further upstream, the percentages were lower, at 11–44%. At 51 river locations in the UK exposed to WWTP effluents, intersex *R. rutilus* were collected at 86% of the sites. The condition of the testes of intersex males ranged from the presence of a single oogonium or a small nest of oogonia embedded in an otherwise normal testis to extreme cases in which up to half of the testis was composed of ovarian tissue. In some intersex males, the testis efferent duct was missing and replaced by an ovarian cavity. The degree of feminisation of male *R. rutilus* was greater in fish sampled immediately downstream of a WWTP. Curiously, the carnivorous European perch (*Perca fluviatilis*, Percidae) and northern pike (*Esox lucius*, Esocidae) did not exhibit the severe sexual disruption found in the omnivorous *R. rutilus* living in the same waters.

Intersex in male *R. rutilus* reflects the oestrogenic properties of the effluents and is an example of an organisational effect, in that the pattern of differentiation of the gonads is disrupted (see also Chapter 3). An example of an activational effect of oestrogenic effluents was the presence of VTG in the blood plasma of male *O. mykiss* and *C. carpio* held in cages in, or close to, WWTP effluents.

Most of the oestrogenic activity of the WWTP effluents was caused by the presence of the natural sex steroids, E_2 and oestrone (E_1), together with the synthetic oestrogen, ethinyloestradiol (EE_2), which is a component of the oral contraceptive pill. In some effluents, other non-steroidal chemicals with oestrogenic properties, including alkyl phenols, phthalates, pesticides and insecticides, also contributed to the overall oestrogenic activity.

The presence of VTG in the blood plasma of male teleosts provides a biomarker for the presence of oestrogenic EDCs, but androgenic and anti-androgenic effects may also occur. Mosquitofish (*Gambusia affinis*, Poeciliidae) exposed to the effluent from pulp mills provided evidence that the effluent is androgenic. Females developed the gonopodium that is characteristic of males (see Chapter 8; Olsson *et al.*, 2008). Another useful biomarker for the presence of androgenic chemicals is the production of spiggin in the kidneys of female *G. aculeatus* (Katsiadaki, 2007).

There is experimental evidence of endocrine disruption and circumstantial evidence from studies of natural populations for sites in Europe, North America, India, Africa and Southeast Asia (Scholz & Klüver, 2009; Vajda & Norris, 2011). Nevertheless, there is still uncertainty about the ecological impact of EDCs on teleost populations. Does any reproductive disturbance they cause translate into adverse effects on population abundance and persistence over time? A field experiment in north-western Ontario (Canada) introduced EE_2 into a lake at levels that approximated oestrogenic levels associated with waters receiving WWTP effluents (Kidd *et al.*, 2007). A population of fathead minnow (*Pimephales promelas*, Cyprinidae) in the lake collapsed because

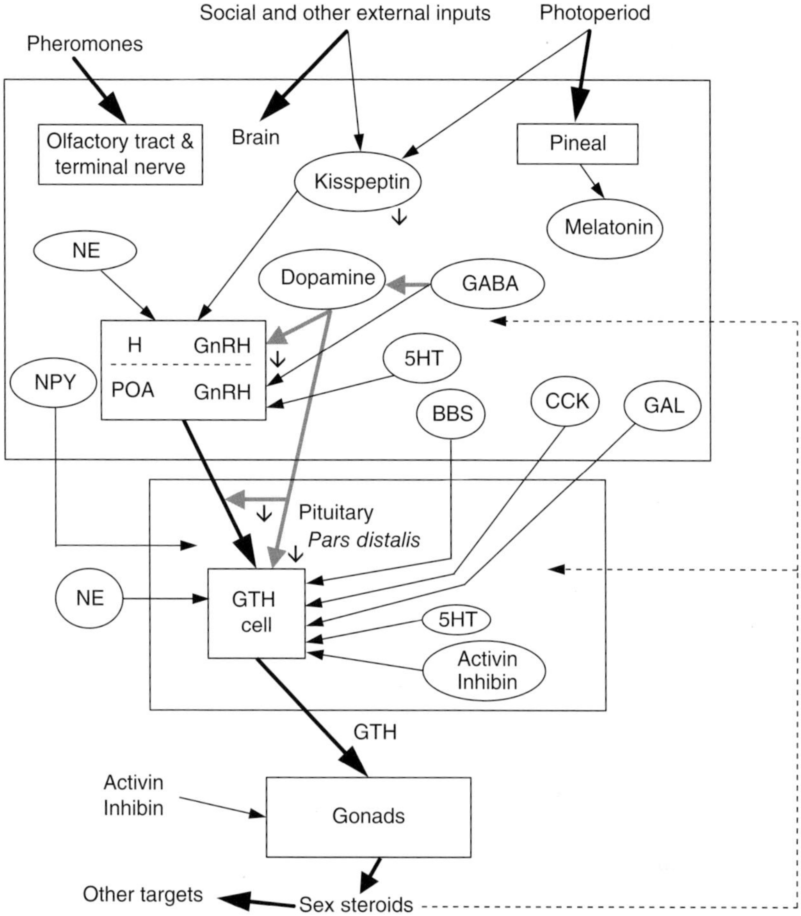

Figure 5.7 A provisional model of the physiological regulation of reproduction in teleost fishes in which dopamine has an inhibitory role. Well-established links are indicated by thick lines, but thickness does not imply quantitative effects. Inhibitory pathways are shown in grey. Positive and negative feedback pathways are shown as dashed lines. BBS, bombesin; CCK, cholecystokinin; GABA, γ-aminobutyric acid; GAL, galanin; GnRH, gonadotrophin-releasing hormone; GTH, gonadotrophin; 5HT, serotonin; NE, nornoradrenaline; NPY, neuropeptide Y; POA, pre-optic area; H, hypothalamus. ↓ indicates a negative effect. (Source: Kraak *et al.* 1998. Reproduced with permission of Taylor and Francis Group.)

of recruitment failure. The fish showed the typical symptoms of the effects of oestrogenic EDCs, including high levels of VTG and intersex in males. This study illustrates the potential of EDCs to have impacts at the population level.

Conclusions

The endocrine system, especially the BPG axis, is an exquisite system for the control of reproduction in teleost fishes (Fig. 5.7). It controls the timing of the onset of sexual maturation during puberty. It then controls the timing of the process of gametogenesis and, through its timing of the cycles of reproductive activity and inactivity, defines the life-history pattern in iteroparous species. In association with the endocrine systems controlling growth, development and metabolism, it regulates the allocation of resources between reproduction and somatic processes. It also regulates reproductive behaviour to ensure the required synchrony between the physiological state of the gonads and the behaviour of the individual. This synchrony is refined because of the sensitivity of the endocrine system to the social environment. The endocrine system is the physiological mechanism by which the life-history pattern characteristic of the species is achieved and is central to any understanding of the reproductive biology of the teleost fishes (see Chapter 11).

The lability and bipotentiality of the teleost gonad has made possible the development of techniques in aquaculture that allow control of the sex of fish that are to be grown to marketable size. The increased understanding of the physiological control of reproduction has allowed the development of techniques that overcome the inhibition of reproduction that is frequently associated with environments presented by aquaculture facilities (Chapter 12). The same lability also makes reproduction in teleosts, particularly freshwater species, prone to disruption by pollutants that act as endocrine disrupters. However, such disruption may act as an early warning of pollution that could also affect the reproductive health of human populations.

Chapter 6
Environmental control of reproduction

Introduction

Timing is a critical factor in the reproductive biology of teleosts. The geophysical characteristics of the Earth impose well-defined temporal cycles, which are relevant to the biology of teleosts (Ricklefs & Miller, 2000). The spinning of the Earth on its axis generates a daily (diurnal) cycle of light and dark. Superimposed on this is the circannual cycle as the Earth orbits the sun. Because of the axial tilt of the equator at 23.4°, this circannual cycle is characterised by seasonal cycles, which are more well defined the further a location is from the equator. In aquatic systems, particularly the seas, the rotation of the moon around the Earth generates tides because of the gravitational effects of the moon and sun on water bodies.

The changes in the environment, which are driven by these geophysical cycles, have implications for the survival, growth and reproductive success of teleosts. The operation of the physiological mechanisms that regulate reproduction (Chapter 5) should be timed to maximise the fitness of the individuals reproducing.

These geophysical cycles also influence the availability of food resources. Reproduction represents a costly investment for an individual (Wootton, 1998), so the timing of the reproductive cycle must ensure that the individual can obtain sufficient food to support that investment. The timing should also ensure that the progeny encounter adequate feeding conditions as they assume a free-living life.

Seasonal patterns of reproduction

The definition of the time period during the year when individuals in a population are sexually active can present technical problems (Hutchings & Myers, 1994; Lowerre-Barbieri *et al.*, 2011), but in many species reproduction is restricted to a discrete period of the year. These periods tend to be better defined at higher latitudes. Two methods have been used to define the stability of timing of the spawning seasons. The first is to use the time series of the dates of peak spawning in a population obtained over several years. Peak spawning is the date at which the highest frequency of individuals in a population is spawning. The mean date and the variance of the date can then be calculated (Cushing, 1975). The second is to identify the date of peak spawning in a given year and calculate the variance in the distribution of spawning times within a population for that year (Lowerre-Barbieri *et al.*, 2011). Although the date of peak spawning may not vary much from year to year, there may be significant variations across years in the distribution of the times of spawning within years.

Global seasonal patterns

The most striking and predictable environmental consequence of global geophysical factors is the change in the length of daylight (photoperiod) during the diurnal cycle of night and day. At the equator, day and night lengths are equal and this does not change over the year. Further and further from the equator, the seasonal change in photoperiod becomes greater. At high latitudes, the photoperiod is 24 hours in mid-summer (continuous daylight), but zero in mid-winter (continuous night). Away from the tropics, photoperiod increases after the winter solstice and decreases after the summer solstice, the increase and decrease depending on the distance from the equator. At the spring and autumn equinoxes, the day and night lengths are equal. Closely correlated with the seasonal changes in the photoperiod are seasonal changes in temperature. At the tropics, seasonal changes in temperature in aquatic bodies are small. At higher latitudes, the temperature changes are great, particularly in shallow water bodies, in some cases leading to ice covering the water during the winter. These seasonal changes in temperature are important because, with a few exceptions, teleosts are ectothermic and so their body temperature closely matches that of the water in which they live (Helfman *et al.*, 2009). Other seasonal changes, such as those in wind speed and direction, rainfall and food levels, are largely driven by these changes in photoperiod and temperature.

The tropics are defined as the region of the Earth where the sun is directly overhead at least once in the year. The tropics lie between latitudes 23.44°N and 23.44°S. Areas between 5°S and about 15°N tend to have two rainy seasons a year, each lasting approximately 2 months (Munro, 1990a). Further from the equator, there is usually a single rainy season and a single dry season. The significance of these seasonal changes in rainfall is their effects on water levels in streams, rivers and small lakes (Lowe-McConnell, 1979, 1987; Munro, 1990a). For coastal marine locations, the outflows of rivers can have local effects by bringing nutrients to waters close to the point at which they debouch. Tropical and subtropical coastal marine systems may also be seasonally enriched by upwellings of nutrient-rich cool waters originating in the deeper waters of the sea.

In seas and lakes at higher latitudes, the seasonal cycle of photoperiod and temperature causes cycles in the production of planktonic organisms at the base of the aquatic food chain (Cushing, 1975; Wetzel, 1983).

Environmental factors that show seasonal changes have two possible roles in determining the temporal pattern of reproduction in fishes. They can act as ultimate factors or as proximate factors (Lowerre-Barbieri *et al.*, 2011).

Ultimate factors

It is assumed, though rarely tested, that the timing of reproduction tends to maximise the individual fitness of the reproducing fishes. This effect on fitness will have two components. The first, and probably the most important, is the effect of timing on the survival of the progeny (Wingfield, 2008; Lowerre-Barbieri *et al.*, 2011). A second component, relevant only for iteroparous species, is the effect of timing on the future reproductive success of the reproducing adult. Ultimate factors act as selective agents operating on heritable components of timing (Lowerre-Barbieri *et al.*, 2011).

The important ultimate factors for the progeny are usually assumed to be availability of suitable food, the risk of predation and avoidance of adverse physical conditions. For some species, the timing of spawning may reflect a requirement for the growing season of the progeny to be sufficiently long to enable them to acquire the reserves they need to improve their chances of surviving the first winter of life (Conover, 1992). For adults, the effect of food levels on the timing of reproduction can be mediated by a cycle of storage of reserves in the form of lipids at times of good feeding, and the mobilisation of those reserves at a suitable time for reproduction (Wootton, 1979). Other ultimate factors that may be relevant for reproductive timing are predation on adults, the suitability of environmental conditions for behaviour related to reproduction and feeding biology (Robertson, 1991).

A complication in identifying ultimate factors is the presence in some species of populations that spawn at different times of the year, but which occur in the same general geographical area (Borg, 2010). The Atlantic herring (*Clupea harengus*, Clupeidae) has autumn and spring spawning populations off the coast of Newfoundland

and in the Baltic Sea. In Scandinavia, autumn and spring spawning vendace (*Coregonus albula*, Salmonidae) coexist within the same lakes. Thus, different spawning seasons may evolve in what seem to be similar environmental conditions. In some cases, these differences may reflect underlying genetic differences between populations as they adapt to the local conditions. In other cases, the differences may reflect phenotypic plasticity, with the same genetic background responding to different local conditions.

Proximate factors

These are factors that impact on the physiological processes that regulate reproduction. They act during the ontogeny of individuals in comparison with ultimate factors that act as selective agents driving heritable changes between generations. Proximate factors must be detectable by the sensory systems of the individual fish so the information can be passed to the central nervous system (CNS). Obvious candidates are photoperiod, temperature, the lunar and tidal cycles, the chemical composition of water and the flow rate of water. A problem with analysing proximate factors is that some of them are highly correlated, such as photoperiod and temperature at higher latitudes, lunar and tidal cycles in shallow marine waters, and water flow rates and chemical composition. The evolutionary history of species has taken place in the presence of these correlations.

Potentially, proximate factors can affect the timing of reproduction by their effects on the initiation of gametogenesis, the stimulation of vitellogenesis and the induction of the final maturation, ovulation and oviposition (Wang *et al.*, 2010)

Spawning seasonality in freshwater systems

For temperate and high latitude areas in the Northern Hemisphere, three basic seasonal spawning patterns occur (Billard, 1981; Wootton, 1998; Wang *et al.*, 2010). The first group, which consists largely of salmonids, including Atlantic salmon (*Salmo salar*, Salmonidae) and brown trout (*Salmo trutta*, Salmonidae), spawn mostly in autumn and early winter. They are typically total spawners (Chapters 1 and 4), so females spend only a relatively short period of time on the spawning grounds, although males tend to stay longer because they attempt to spawn with several females. The duration of the spawning period within a year reflects the individual variation in dates of spawning. At the low temperatures of winter, the eggs develop slowly so the progeny only emerge as free-living young in the following spring. The second group, which includes esocids such as northern pike (*Esox lucius*, Esocidae) and percids such as the European perch (*Perca fluviatilis*, Percidae), spawn in late winter or spring. These also tend to be total spawners. The third group consists of summer spawners, such as the cyprinids, the freshwater bream (*Abramis brama*, Cyprinidae) and the tench (*Tinca tinca*, Cyprinidae). These include both total and batch spawners. The eggs develop quickly at the higher water temperatures of summer and so hatch and take up independent lives in that summer. The effect of these patterns of breeding seasonality is that the young-of the-year fish all take up independent lives in the interval between spring and early autumn.

There are species that do not conform to these three patterns, most notably the burbot (*Lota lota*, Gadidae). This species is found in cold waters of both Eurasia and North America. It spawns in winter at water temperatures of 0.5–4 °C (Maitland & Campbell, 1992). This time and temperature of spawning might reflect phylogenetic relationships. *L. lota* is a freshwater representative of the predominantly marine codfishes (Gadidae). In Lake Oahe (North and South Dakota, USA), the mean date of peak spawning of *L. lota* was 14 January (June, 1977). The next species spawning within the year was *E. lucius*, with a peak spawning on 25 April.

As changes in photoperiod and temperature are less well defined in tropical and sub-tropical regions, seasonality in these regions relate more to the cycle of wet and dry seasons, particularly for freshwater systems. The consequence is a cycle of flooding, when the rivers burst their banks and water flows over the adjacent terrestrial habitats and then a period when the river is confined within its banks. The flooding results in a pulse of nutrients entering the water from the flooded land. In turn, this pulse supports a flush of production, which includes food suitable for young fish (Lowe-McConnell, 1979, 1987; Munro, 1990a).

The spawning of tropical riverine fishes is usually timed to coincide with the onset of the flood and consequent enhanced feeding opportunities. Large South American and African riverine characids show this spawning seasonality. The adults often show an upstream migration before spawning (Chapter 7). For some species, the migration is short, before the adults move laterally onto the floodplain to spawn. In other species, the upstream migration is long and may even end in the headwaters, where spawning takes place at the start of the rains. African cypriniform fishes show similar patterns. The riverine species are typically total spawners. After spawning, they remain on the floodplain, feeding and accumulating lipid reserves before being forced back to the rivers as the floods recede. The dry season, when fish are confined to the rivers, is often a period of relatively poor feeding (Lowe-McConnell, 1979; Welcomme, 1979).

In small tropical streams, there is considerable interspecific variation in the patterns of reproductive seasonality. In a small Panamanian stream, six species of small-bodied characids showed distinctive temporal patterns of spawning (Kramer, 1978). Two species spawned at the start of the rains, with the breeding season lasting only a few days. Two species spawned over 2–4 months, while two species spawned throughout the year (Fig. 6.1). Three species of viviparous poeciliids (Chapter 10) in a rainforest stream in Venezuela showed evidence of reproductive activity throughout the year, but

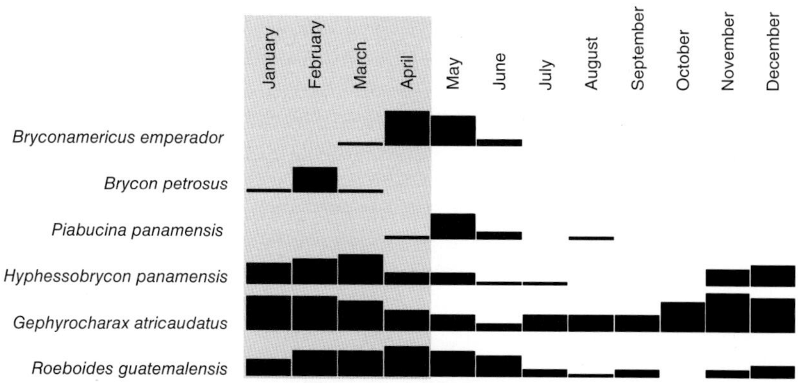

Figure 6.1 Relative frequency of sexually mature adults of six species of characiform fishes in streams on the Pipeline Road, Panama Canal Zone, based on samples collected between December 1973 and June 1975. Shading indicates the dry season. (Source: Adapted from Kramer 1978. Reproduced with permission of the Ecological Society of America.)

the proportion of gravid females peaked just prior to or during the early phases of the two wet seasons each year (Winemiller, 1993).

In large tropical lakes, many ostariophysan species enter rivers and migrate upstream to spawn at the time of floods (Chapter 7). However, endemic clupeid species in Lake Tanganyika spawn throughout the year, but do show spawning peaks when there is a seasonal upwelling of nutrients driven by wind-induced currents. Cichlids in Lakes George and Victoria spawn throughout most of the year, but with peaks tending to occur during wet seasons. Lake Sibaya (Sibhayi) in South Africa is situated close to the southern border of the tropics (approximately 27–32°S) and shows increased seasonality. The Mozambique tilapia (*Oreochromis mossambicus*, Cichlidae), has a well-defined breeding seasonality, thus reflecting the relationship between the intensity of the seasonality and the discreteness of the breeding season (Munro, 1990a).

Cichlid species in Lake Jiloá in Central America illustrate an interaction between seasonality and social behaviour in determining spawning seasonality (McKaye, 1977). Different species spawn in either the wet or the dry season, but interspecific competition for spawning sites is intense and this is reflected in the spawning seasonality (Fig. 6.2). In the wet season, the Midas cichlid (*Amphilophus citrinellum*, Cichlidae) aggressively excludes all but two species from breeding.

Spawning seasonality in marine species

The patterns of spawning seasonality in seas and oceans reflect, in some respects, the patterns in freshwater species. At high latitudes, the timing of peak spawning tends to be precise (Cushing, 1975). The peak spawning date for plaice (*Pleuronectes platessa*, Pleuronectidae) in the Southern Bight of the North Sea was in January, with a standard deviation of only 7 days. Similar precision is seen in other species such as *C. harengus* in Norwegian waters. At lower latitudes, the timing becomes less precise. This is well illustrated, intraspecifically, by the Atlantic silverside (*Menidia menidia*, Atherinopsidae) (Conover, 1992). The species is found down the eastern seaboard of North America. At high latitudes (Nova Scotia: 44°N) spawning only takes place in June, beginning at the lowest temperature that supports growth. Further south (South Carolina: 32°N), the spawning season extends from March to July. These differences may reflect the time required for the progeny to grow and store lipid reserves before the onset of adverse winter conditions, particularly at high latitudes (Conover, 1992).

At lower latitudes in South Asia, monsoons induce seasonal changes in current systems and small clupeids show increased breeding in response to a seasonal change in nutrient levels in up-welling areas (Lowe-McConnell, 1979). Species on continental shelves at low latitudes show a range of breeding seasonalities from those that spawn throughout the year, though with seasonal peaks, to those that have restricted breeding seasons.

There are different patterns of spawning that can produce an extended breeding season. A species may be a total spawner, but with different individuals spawning at different times, so a portion of the population is spawning at any given time. A species may be a batch spawner, with each individual spawning over an extended period (Lowerre-Barbieri *et al.*, 2011). Large individuals may spawn over a longer interval than smaller individuals (Reichard *et al.*, 2009a).

Even some species living in the abiotically benign conditions of coral reefs show seasonal patterns of breeding (Robertson, 1991). These patterns do not have the

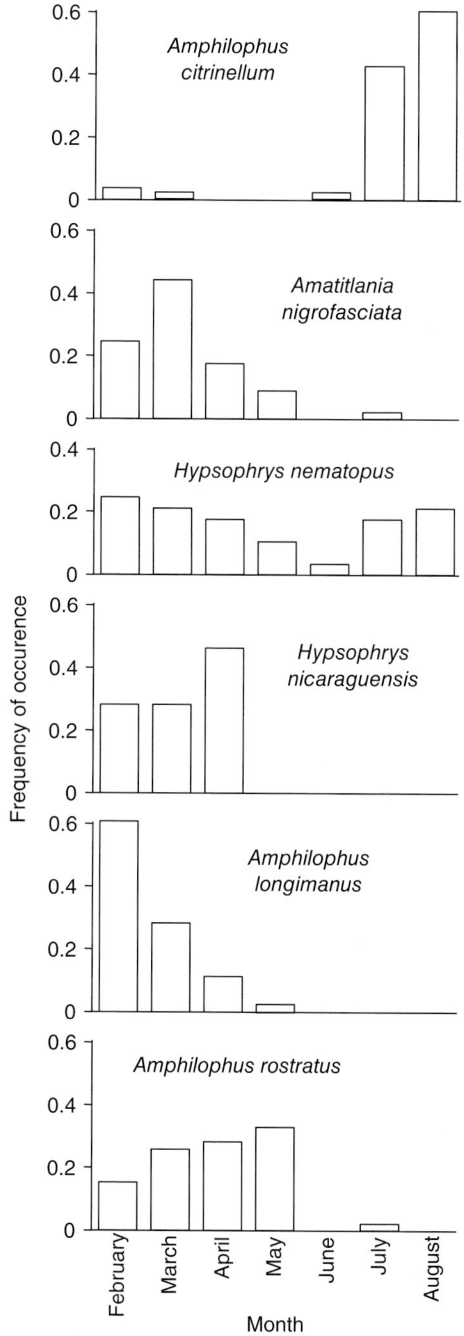

Figure 6.2 Relative frequency of breeding pairs of six cichlid species from February to August 1974 in Lake Jiloá, Nicaragua. (Source: Adapted from McKaye 1977. Reproduced with permission of the Ecological Society of America.)

discreteness of breeding seasons at higher latitudes and there is a trend for a decreasing strength of spawning seasonality the closer the reef is to the equator. On a given reef, even species in the same family spawn at different times. On the Great Barrier Reef, species in the Apogonidae (cardinalfishes), Pomacentridae (damselfishes) and Labridae (wrasses) include species that spawn at different times of the environmental seasonal cycle. Jamaican reef dwellers include species that are not seasonal spawners and species with a bimodal annual spawning pattern. The striped parrotfish (*Scarus iserti*, Scaridae) spawns in the summer months off the north coast of Jamaica, but has peaks of spawning in winter and spring months on the south coast of the island. The evidence is of idiosyncratic temporal patterns in the spawning of reef dwelling species, each responding in a unique way to the pattern of seasonal variation.

The explanation for these species patterns falls under two groups of hypotheses (Thresher, 1984; Robertson, 1991). The first suggests that the timing of spawning aids the dispersal of eggs or larvae away from the reef and helps to reduce predation on these early life-history stages. The second set of hypotheses suggests that seasonal timing of spawning relates to the biology of adults. One of the possibilities is that potentially adverse environmental conditions influence the timing of spawning. In Panama, there are two main seasons: one wet and calm, the other dry and windy. Spawning activity on the Caribbean and Pacific coasts during the dry season is reduced when physical conditions are particularly unfavourable (Robertson, 1991). Spawning seasonality may also provide a period of low reproductive activity when adults can recuperate from the expenditures associated with reproduction.

Many marine species living in the open oceans in the tropics have reproductively active adults at most times of the year, although individuals may be active at different times of the year (Cushing, 1975). Some species of tuna (Scombridae) have spatially extensive spawning areas and individual fish spawn batches of eggs at short time intervals (Helfman *et al.*, 2009). The difference in temporal patterns of spawning between tropical marine species and marine species at high latitudes may reflect the temporal pattern of the production cycle in the seas (Cushing, 1975).

Lunar-related reproductive cycles

Because of the close association between the lunar cycle and the tidal cycle, the spawning of estuarine and shallow water marine species is often closely related to the lunar cycle (Taylor, 1984; Takemura *et al.*, 2010). This lunar cycle of spawning may be superimposed on the seasonal variation in spawning activity. More surprisingly, lunar cycles in spawning also occur in a few freshwater species for which the tidal cycle is ecologically irrelevant.

Tidal heights are at a maximum at full and new moon spring tides and at a minimum at half moon neap tides, which occur at an interval of about 14 days. Thus, there are two geophysical cycles: the change in the intensity of moonlight, and the alternation of spring and neap tides.

There are two patterns of reproductive activity related to the lunar cycle (Takemura *et al.*, 2010). Some species have a lunar cycle, with spawning activity occurring at approximately 28 days, the length of the full lunar cycle. In species with a semi-lunar cycle, individuals spawn at approximately 14 days. However, a quasi semi-lunar cycle occurs when each individual spawns with a lunar periodicity, but individuals utilise different peaks of the lunar cycle. Two categories of marine fishes have well defined lunar or semi-lunar reproductive cycles (Taylor, 1984). There are species that spawn in the intertidal zone and depend on the tidal cycle to gain access to this zone. The second group consists of reef fishes, living in shallow, tropical waters.

Intertidal spawners

Species in the Salmoniformes, Atheriniformes and Tetraodontiformes have intertidal spawning, typically spawning on high tides (Taylor, 1984). The eggs are usually exposed to air for periods in their development, hatching when inundated by the following spring tide.

Atherinids provide good examples of intertidal spawning. The California grunion (*Leuresthes tenuis*, Atherinopsidae) takes advantage of unusually high tides associated with the full or new moon to lay eggs at night in sand above the surf zone. The Gulf grunion (*Leuresthes sardina*, Atherinopsidae) spawns on diurnal spring tides when they are higher than nocturnal high tides. The mummichog (*Fundulus heteroclitus*, Fundulidae) is one of several topminnows that spawn in the intertidal zone. It has a semi-lunar cycle, spawning for several days just before the spring tide. Striped killifish (*Fundulus majalis*, Fundulidae) spawn just after spring tides on the days of receding high tides. *M. menidia* also has a semi-lunar pattern, spawning eggs high up in the intertidal zone.

A feature common to these intertidal spawners is that they lay their eggs in the high intertidal zone that is only accessible to the fish at high tides. This mode of spawning may reduce predation on eggs, at least from aquatic predators (see also Chapter 7).

Coral reef spawners

Coral reef fishes commonly show lunar or semi-lunar spawning patterns (Johannes, 1978; Taylor, 1984; Robertson *et al.*, 1990; Robertson, 1991). Such lunar-related cycles tend to be more common in benthic spawners, whose larvae disperse after hatching, than in pelagic spawners (Thresher, 1984). A study observed Neotropical species at sites in both the Atlantic and Pacific oceans (Robertson *et al.*, 1990; Robertson, 1991). Among the species, 15 damselfish (Pomacentridae) and two blennies (Blenniidae), were benthic spawners. The diversity of spawning patterns observed illustrates the apparently idiosyncratic evolution of spawning periodicity among species. Ten species had a lunar cycle of spawning, but the time of peak spawning varied between species, so reproductive activity ranged over most of the lunar cycle. Four species had a semi-lunar pattern, with three species spawning during the week preceding the new and full moon, and a fourth species spawning almost at the new and full moon. Five species showed no indication of lunar cyclicity.

One proposed function of lunar-related spawning is that it facilitates the dispersal of eggs or larvae away from the reef (Johannes, 1978). However, this hypothesis only seems to be strongly supported by species that live in habitats in which dispersal is only possible at some lunar phases. Another possibility is that moonlight allows the larvae to move more rapidly into the water column and away from predators associated with the reef surface. Lunar-related cycles may also represent adaptations to adult rather than larval biology (Robertson, 1991). A lunar cycle will synchronise the breeding of adults and this may lead to enhanced defence of nest sites, because many breeding territories are defended at the same time. Some groupers (Serranidae) show lunar-related behaviour, migrating to and aggregating around the spawning grounds, with gametes released synchronously (Takemura *et al.*, 2010). A lunar-related cycle may also allow time between spawnings, during which the adults can recover from the costs of spawning.

Freshwater spawners

Although rare, lunar reproductive cycles occur in some freshwater cichlids. The most detailed descriptions have been of substrate-spawning cichlids from Lake Tanganyika (Nakai *et al.*, 1990; Desjardins *et al.*, 2011). In a study of eight species, the two species with the most well defined lunar cycle laid their eggs on an exposed or semi-exposed rock surface. Spawning took place in the second quarter of the lunar cycle (Nakai *et al.*, 1990). In a second study, dominant breeding males of *Neolamprologus pulcher* (Cichlidae) had high serum levels of testosterone (T) and 11 ketotestosterone (11-KT) in the first quarter moon, just before full moon (Desjardins *et al.*, 2011). These studies suggest that the lunar cycle means that young emerge when night-time light is highest at full moon. This would enhance parental protection from nocturnal predators of the young.

In assessing the adaptive significance of lunar-related reproductive cycles, it has to be determined whether the lunar cycle is being used merely as a timing device, or whether the changes in light intensity or tidal levels and associated changes in water currents are playing a causal role in reproductive success.

Diel reproductive cycles

The third geophysical factor that affects the timing of reproduction is the alternation of day and night in the diel cycle. In both temperate and tropical waters, some species show diel cycles of reproductive activity. In such species, the timing of spawning of egg release is related to the diel cycle.

A survey of 12 species of inshore marine fishes in the Peconic Bays, New York (Ferraro, 1980), showed that eight species, including the Atlantic menhaden (*Brevoortia tyrannus*, Clupeidae), spawned primarily in the evening or at night. Two species began spawning in the afternoon, with the spawning continuing into the night. One species, the scup (*Stenotomus chrysops*, Sparidae), spawned in the morning, while the Atlantic mackerel (*Scomber scombrus*, Scombridae) spawned throughout the day. Sciaenid species in the Gulf of Mexico spawned at dusk and in the evening (Holt *et al.*, 1985). Many coral reef fishes also spawn at dusk and at night (Roberston, 1991). Even species that are normally diurnal, such as *P. fluviatilis*, spawn at night. Other species spawn at dawn, including some pelagic-spawning, reef-dwelling surgeonfishes (Acanthuridae) (Thresher, 1984) or freshwater species, including the zebrafish (*Danio rerio*, Cyprinidae) (Spence *et al.*, 2008).

A common explanation for diel spawning is that it is timed to minimise predation on the eggs in a given environment. There are a number of possible mechanisms for a reduction in predation. The eggs may be released at a time when egg predators are inactive or cannot detect the eggs. The eggs may be spawned after the predators have become satiated by their previous feeding. A third possibility is that the release of eggs all at the same time swamps the egg predators. For coral reef fishes, another explanation is that diel spawning is timed to facilitate dispersal of the propagules (Robertson, 1991). Other explanations of diel spawning suggest that it relates to the biology of adults.

Multiple environmental cycles

In many species, both freshwater and marine, a diel cycle of spawning is superimposed on a seasonal reproductive cycle. In some species, there may be lunar-related and diel reproductive cycles superimposed on a seasonal cycle. This superimposition of cycles raises at least two questions.

The first is related to the role of the geophysical cycles as ultimate factors. How do the cycles interact to select the reproductive timing? One approach to this question is to assume that each cycle, diel, tidal, semi-lunar, lunar and seasonal cycles, has an independent effect either on its own or in combination with other cycles to generate a particular timing of reproductive activity (Yamahira, 2004). This model assumes that each cycle has a unique functional relationship between a fitness component and the timing of reproduction and that the individual fitness components combine multiplicatively to generate an overall fitness function. This model was tested by a study of the timing of intertidal spawning by the puffer, *Takifugu niphobles* (Tetraodontidae), on beaches in Japan. The results suggested that the timing of spawning was determined by the fitness components related to the tidal cycle and the diel cycle rather than the semi-lunar cycle of spring and neap tides (Yamahira, 2004). The seasonal cycle was not incorporated into this analysis.

The second question is related to the role of the geophysical factors as proximate factors. The physiological, regulatory processes that integrate the response of the reproductive system to multiple environmental cycles have still to be clarified. Indeed, the physiological effects of strong seasonal signals such as photoperiod are still not fully understood.

Endogenous cycles and reproductive activity

One possibility for the physiological control of reproductive activity by geophysical cycles is that a geophysical cue such as photoperiod has a direct physiological effect and that in the absence of the cue reproductive activity is not initiated. However, experimental evidence has suggested that at least some teleost species have endogenous cycles of reproductive activity, such that even in the absence of external cues, reproductive activity alternates with periods of reproductive quiescence. The criterion for an endogenous cycle is that it should free-run in the absence of relevant environmental cues, although experimentally it may be difficult to exclude all potentially relevant external cues. In teleosts, three types of endogenous reproductive cycles have been identified. A circannual cycle is characterised by the occurrence of reproductive activity at approximately 12-monthly intervals. A lunar-related endogenous cycle has reproductive activity at about 14- or 28-day intervals. A diel cycle has reproductive activity at approximately 24 hours. A characteristic of endogenous cycles is that, in the absence of relevant external cues, the period of the cycle (i.e. the time between successive events) does not exactly match the period of the geophysical cycle. For example, an endogenous diel cycle may have a period of 23 hours, rather than the 24 hours of the natural cycle. In practice, the endogenous cycle is synchronised, or entrained, with the natural cycle by cues provided by the natural cycle. This is the same process by which a clock that runs persistently slow can be corrected by adjusting the clock by reference to a reliable, external timing signal. The timing cues are called 'Zeitgebers', literally 'time givers'. Experimental studies in which adult fish have been maintained under constant conditions for long periods have provided evidence of circannual cycles.

The sub-tropical stinging catfish (*Heteropneustes fossilis*, Heteropneustidae), raised under continuous light (LL) or continuous darkness (DD) at 25 °C had an annual ovarian cycle comparable to that seen in *H. fossilis* collected from nature (Sundararaj, 1978). A fuller study of female rainbow trout (*Oncorhynchus mykiss*, Salmonidae) provided more evidence of the presence of a circannual cycle (Duston & Bromage, 1988; Borg, 2010). Females kept on a constant short photoperiod of 6:18 hours light:dark (LD 6:18) matured at intervals of about a year. Females kept on a long photoperiod (LD 18:6) or on continuous light (LL) matured at an interval of approximately 160 days. The study of *O. mykiss* suggested that the endogenous, circannual oscillator was dissociated from the neuroendocrine mechanisms controlling maturation. This raises the question of how the endogenous cycle and the physiological mechanisms integrate their effects.

Two studies of the three-spined stickleback (*Gasterosteus aculeatus*, Gasterosteidae) used populations from different latitudes. Sticklebacks collected in the Netherlands were maintained continuously on photoperiodic schedules of LD 16:8 or LD 8:16 at 20 °C (Baggerman, 1990). The fish on the short photoperiod (LD 8:16) had a short breeding season of about 7 days and thereafter failed to breed again. Fish on the long photoperiod (LD 16:8) initially bred for 91 days, followed by a non-breeding interval of 108 days. The next breeding period lasted a mean of 47 days followed by a non-breeding period of 149 days. Few fish survived to reach a third breeding period. These data give an endogenous period of 91 + 108 (=199) days and 47 + 149 (=196) days, both well short of a year. A comparable study of a more northerly, Baltic Sea population in Sweden (Bornestaf & Borg, 2000) used a constant temperature of 17 °C. Male *G. aculeatus* kept continuously on photoperiods of LD 18:6 or LD 12:12 had endogenous periods significantly shorter than a year, but there was no significant difference in male response to these different photoperiods. Under these conditions, some aberrant cycles occurred, with some males showing irregular spawning periodicities.

A study of *F. heteroclitus* provided evidence of a lunar-related endogenous cycle (Taylor, 1984). Fish maintained on a constant photoperiod of LD 14:10, with a low light intensity of 0.2 lux, showed an endogenous semi-lunar cycle. In the absence of external cues, this endogenous cycle drifted out of phase with the natural lunar cycle.

Diel endogenous cycles are almost ubiquitous in living organisms. It is likely that these cycles underlie the daily cycles in endocrine levels observed in fish that have a daily spawning (Chapter 5). The genetic mechanisms that underlie an endogenous cycle are partly understood and involve a series of clock genes (Albrecht, 2012). Clock genes have been identified in the brain of *S. salar* (Davie *et al.*, 2009), but their role in controlling endogenous diel cycles in teleosts is not fully understood.

Proximate factors

The timing of reproductive events in teleosts probably depends on interactions between endogenous cycles and external environmental factors acting as cues (Munro, 1990b). These environmental factors, the proximate factors, have several categories of effects on reproduction (Wingfield, 2008). An important role is when the proximate factor acts as a predictive cue (Munro, 1990b; Wingfield, 2008). A predictive cue allows an individual to make long-term predictions about the habitat and, as a consequence, regulate the timing of its reproduction. This long-term prediction is essential because gametogenesis takes time and so must be initiated well before the gametes are required. In salmonids, it can take about a year for the completion of gametogenesis (Chapter 4).

On a shorter timescale, fine-tuning of the timing of reproduction depends on local cues such as food availability, rainfall and temperature. This local information helps to ensure that the individual makes an appropriate response at the correct time. Social proximate factors can act as synchronising cues, stimulating final maturation and helping to integrate the mating behaviour of individuals (Chapters 5 and 8). Proximate factors may, on occasion, act as terminating cues, causing the regression of the gonads.

However, such regression may also result from an endogenous process, independent of external cues. Reproductive activity may be disrupted by proximate factors such as adverse weather, inadequate food, predators or pathogens, which cause the individual to switch from reproductive activity to an avoiding strategy, such as taking refuge, moving to another habitat or mounting an immune response (Wingfield, 2008). Most emphasis tends to be placed on proximate factors that act as predictive cues.

Photoperiod and temperature

The photoperiod is the most reliable of geophysical cues, and at temperate and higher latitudes frequently acts a predictive cue for teleost species, sometimes in association with water temperature. Throughout their evolutionary history, teleosts have experienced changes in photoperiod that are highly correlated with temperature, because both photoperiod and temperature depend on the energy input from the sun. The elucidation of the relative importance of photoperiod and temperature as proximate factors requires careful attention when designing experiments. A further complication is that whereas photoperiod can act only as a timing cue, temperature can act as both a timing cue and a modulating factor. It acts as the latter because teleosts are ectothermic, so their body temperature is close to the external environmental temperature. Temperature has direct effects on metabolic rates and consequently can modulate the rate at which all physiological processes can occur, including those related to reproduction (Wang *et al.*, 2010).

Studies on species from temperate and high latitudes led to the identification of three groups, defined by their response to photoperiod and temperature (Wang *et al.*, 2010). Group I primarily contains salmonid species. The photoperiodic control of reproduction in salmonids has received much attention because of the importance of some salmonids, such as *S. salar* and rainbow trout (*Oncorhynchus mykiss*, Salmonidae) in aquaculture. Salmonids naturally have an annual reproductive cycle, with breeding usually taking place in the autumn and early winter, but in some species this happens in late winter and early spring. The females are total spawners, so under natural conditions all the annual production of eggs takes place over a short period of time. The aquaculture industry prefers eggs to be produced over a more extended period of time, so photoperiodic manipulation has been used as a method to alter the timing of spawning. These experiments have also shone light on the role of the photoperiod in controlling the timing of reproduction in natural populations (Scott, 1990; Bromage *et al.*, 1993, 2001). Temperature has little effect on the timing of spawning as long as the range corresponds to that experienced in the natural habitat, typically 8–13 °C (Scott, 1990).

In salmonids, initial gonad recrudescence begins about a year before spawning, but by the following spring only the early stages of gametogenesis are present in males and females (Campbell *et al.*, 2003). Under the increasing photoperiod that characterises spring, gametogenesis accelerates and vitellogenic oocytes appear in the ovaries and spermatogenesis progresses. For autumn spawners, the decreasing photoperiod that follows the summer solstice accelerated the gonadal maturation, although this effect was not present in late winter spawners (Scott, 1990).

A study of 2-year old female *O. mykiss* exposed the fish to a long photoperiod (LD 18:6) from January and then separate groups were transferred to a short photoperiod (LD 8:16) at monthly intervals from March to June and then maintained on that short photoperiod until spawning (Duston & Bromage, 1988). All the females spawned much earlier than the natural spawning time of December, with the groups starting spawning on 31 July (March group), 13 August (April group), 30 August (May group) and 16 September (June group). From a series of such experiments, it was estimated that the change from long to short photoperiod to subsequent spawning took, on average, about 31 weeks (Scott, 1990). A feature of the photoperiodic response was that the response to a given photoperiod was dependent on the photoperiod experienced previously. Thus a photoperiod of LD 12:12 was interpreted as a long photoperiod if it followed a period of LD 8:16, but as a short photoperiod if it followed a regime of LD 14:10. The striking feature of the results was that the percentage of females spawning increased from 26% for females transferred in March, to 92% of females in June. This result led to the suggestion that there is a critical period in the endogenous circannual cycle when a 'gate' is open, which for the rest of the cycle is closed. The hypothesis suggests that fish will only go on to mature if, at the time the gate is open, they have achieved a threshold physiological state. This physiological state is thought to reflect the energy reserves or growth status so that the fish have sufficient resources to support reproduction.

The adaptability of the photoperiodic response of salmonids is demonstrated by their ability to continue to spawn in autumn even after transfer to the temperate Southern Hemisphere. In upland Kenya, close to the equator where annual photoperiodic change is small, spawning and spent fish were found throughout the year (Scott, 1990).

Gasterosteus aculeatus is a non-salmonid in which photoperiod is the major predictive, proximate cue (Baggerman, 1990; Borg, 2010). This species typically spawns in spring and early summer, but earlier at the lower latitudes in the species range (Baker, 1994). In males, spermatogenesis starts soon after the completion of the breeding season as the photoperiod starts to decrease, and by winter the testes contain viable spermatozoa. However, at this time, the males show no secondary sexual traits or sexual behaviour. After the completion of spawning by females, the ovary shows only slight development, so in winter the ovary contains oogonia and oocytes, some of which have reached the cortical alveoli stage (Chapter 4). In the cold temperatures of winter, the gonads of males and females show little change. With the onset of spring, the photoperiod starts to increase. Leydig cells in the testes of the male become active, releasing androgens, which stimulate the development of the secondary sexual characteristics and eventually the males start showing reproductive behaviours. In females, populations of oocytes become vitellogenic, so in spring the first batch of mature oocytes is ready for ovulation and spawning. Successive batches of oocytes become mature as the spawning season progresses. These changes largely reflect proximate responses to photoperiod.

When male and female *G. aculeatus* were exposed to a long photoperiod (LD 16:8) from November, they became sexually mature, but if exposed to a short photoperiod (LD 8:16) none became mature. As the annual changes in photoperiod occurred, the proportion of *G. aculeatus* exposed to LD 8:16 that matured increased, so by the following spring, all the fish exposed to LD 8:16 matured, although their breeding season was short. The onset of maturity tended to be accelerated by high temperatures, but a combination of a short photoperiod (LD 8:16) and high temperature, which is an unnatural combination, tended to cause an increase in the frequency of atretic oocytes in females, and a reduction in the kidney epithelium height in males, both symptoms of reproductive dysfunction.

An interpretation of these results for *G. aculeatus* is that there is a circannual endogenous cycle in photosensitivity. At a given time in the year, there is a photoperiodic threshold that has to be exceeded by the external photoperiod if a fish is going to mature. In November, for example, this threshold is lower than 16 hours, but higher than 8 hours, whereas in spring, the threshold is close to 8 hours. This circannual change in photosensitivity results from a daily, endogenous cycle in sensitivity to light, which determines the position of a photo-inducible phase (Baggerman, 1980, 1990; Borg, 2010). In these experiments, maturation was all or nothing: either a fish matured under the conditions it experienced or it failed to mature. Typically fish would mature in about 40 days.

The second group of temperate teleosts (group II) includes both marine and freshwater species, with examples from the perches (Percidae), temperate basses (Moronidae), seabream (Sparidae) and codfishes (Gadidae) (Wang *et al.*, 2010). In contrast to group I species, temperature plays a cuing role in the timing of reproduction. In most species in group II, a decrease in temperature and a decrease in photoperiod are required to initiate gonadal recrudescence. It is possible that all species in group II require a period of low temperatures for several months for gametogenesis, including vitellogenesis, to proceed to completion. Some group II species such as European seabass (*Dicentrarchus labrax*, Moronidae) can spawn during the period of cold water temperatures. For other species, such as *P. fluviatilis*, the increase in water temperature and perhaps photoperiod that occurs in spring is required for final oocyte maturation and spawning. The effect of this is to synchronise the spawning. This synchronisation is important for total spawners such as *P. fluviatilis*, because the female must have access to males when she is about to spawn.

Perca fluviatilis spawn in early spring, with the gonads forming about 20% of the total body weight. The initiation of gametogenesis during the previous autumn depends on a decline in the photoperiod and a decrease in temperature (Wang *et al.*, 2006). The length of the photoperiod is not relevant, it is the decrease that is important. Over winter, the water temperature must fall below 10 °C for the gametogenesis to continue (Fig. 6.3).

In Atlantic cod (*Gadus morhua*, Gadidae), the reduction in photoperiod in autumn initiates maturation, with spawning usually taking place in winter and early spring (Hansen *et al.*, 2001; Davie *et al.*, 2007). The modulating effect of temperature is indicated by the acceleration of gonadal maturation at higher temperatures that occurs in some populations of *G. morhua* off the east coast of Canada, leading to interannual variation in the date of peak spawning (Hutchings & Myers, 1994).

A third group of species (group III) includes mainly temperate cyprinids, which spawn in spring and summer, such as the European gudgeon (*Gobio gobio*, Cyprinidae) and *T. tinca* (Wang *et al.*, 2010). Initiation of gonadogenesis requires a decrease in temperature or photoperiod, but, in contrast to group II species, a decrease in both does not seem to be necessary. However, vitellogenesis and spermiogenesis require the increase in photoperiod and temperature that characterises spring. Ovulation and spawning do not take place until a critical temperature is reached, in the case of *T. tinca* at 18 °C.

There is variation amongst temperate cyprinids in the timing of gonadal recrudescence. The goldfish (*Carassius auratus*, Cyprinidae) starts recrudescence in autumn, with the gonads growing over the winter and gamete maturation taking place in spring. Spawning occurs if local proximate cues, such as temperature, the presence of vegetation and social synchronising cues, are present (Hontela & Stacey, 1990; see Chapter 5). Common carp (*Cyprinus carpio*, Cyprinidae) show a different pattern. Gonadal recrudescence occurs immediately after spawning in summer and is completed in about 2 months (Hontela & Stacey, 1990). The gonads then remain quiescent over winter, with final maturation occurring in spring. *T. tinca* does not initiate recrudescence until spring and spawns in summer. No gonadal development takes place below 10 °C and the date of spawning is related to the cumulative number of days that the water temperature is above 10 °C (Breton *et al.*, 1980).

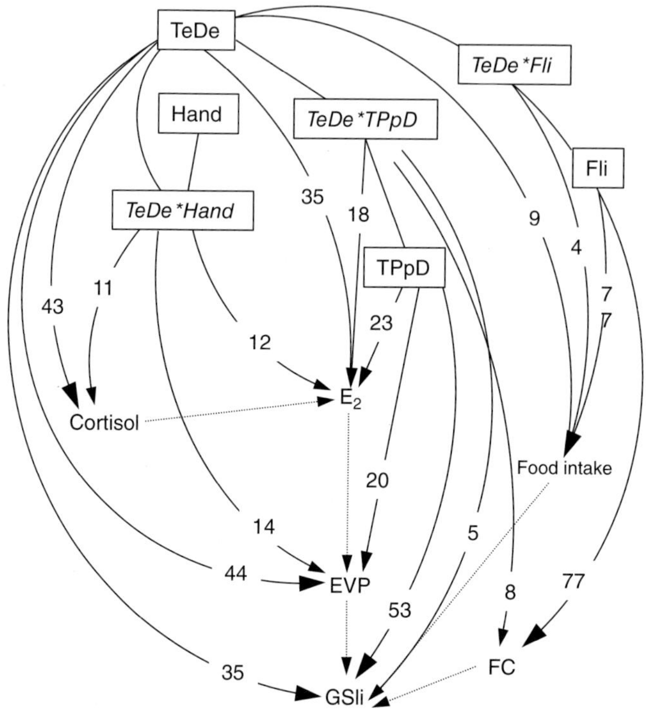

Figure 6.3 Model of induction of the reproductive cycle of female European perch (*Perca fluviatilis*). Solid lines indicate effects, with relative contribution expressed as a percentage of the total variance. E_2, oestradiol-17β; EVP, female in exogenous vitellogenesis; FC, mesenteric fat consumption (initial – final fat index); Fli, Initial fat index; GSIi, increase in gonadosomatic index; Hand, breeding stress associated with handling; TeDe, amplitude of temperature decrease; TPpD, time of photoperiod decrease. Interactions are italicised. (Source: Wang *et al.* 2006. Reproduced with permission of Elsevier.)

These three groups are not all-inclusive for temperate species. Some species, such as the common dab (*Limanda limanda*, Pleuronectidae), initiate gonadal recrudescence and complete gonadogenesis, ovulation and spawning in response to increasing temperature and photoperiod (Wang *et al.*, 2010).

Research on the European minnow (*Phoxinus phoxinus*, Cyprinidae) provides a cautionary tale about the interpretation of results obtained under simplified laboratory and aquaculture conditions, which may not accurately reflect the role of photoperiod and temperature (Scott, 1979). A laboratory experiment that exposed *P. phoxinus* to a naturally increasing photoperiod at about 15 °C evoked some increase in ovarian size and oocyte development, but they failed to reach levels characteristic of the natural population. By contrast, an experiment at the same temperature but in which *P. phoxinus* were exposed to the seemingly unnatural condition of a rapid change in photoperiod from LD 8:16 to LD 16:8, led to ovarian size and oocyte composition similar to that found in the natural population. The key to this anomaly came from a study of the behaviour of *P. phoxinus* in Loch Walton, Scotland. In winter, *P. phoxinus* spent the short daylight hours concealed under stones in relative darkness, only emerging after dark. In spring, when water temperature exceeded about 8 °C, the behaviour of *P. phoxinus* changed and the fish became active during daytime. Thus it is the behaviour of *P. phoxinus* in relation to water temperature that causes an abrupt change in the photoperiod experienced in natural conditions. However, even in this experiment, the *P. phoxinus* did not spawn, indicating that critical local physical and social cues were missing. Individual fish are not simply passive receptors of external conditions, but by their behaviour can adjust their exposure to external cues. Thus, fish may actively move from one temperature or light regime to another.

In addition to predictive, proximate cues that initiate gametogenesis, and any local or synchronising cues that are required for ovulation and spawning, there may also be terminating cues, which bring spawning to a close. In some species, the end of spawning may occur even in constant environmental conditions because of an endogenous process. This is seen in both male and female *G. aculeatus* (Baggerman, 1990). However, in other species, an environmental cue may cause the termination of spawning (Munro, 1990b). In the rose bitterling (*Rhodeus ocellatus*, Cyprinidae), decreasing photoperiod and high water temperatures cause gonadal regression, while in the spring-spawning tabira bitterling (*Acheilognathus tabira*, Cyprinidae), a temperature greater than 25 °C causes gonadal regression (Hontela & Stacey, 1990).

Such gonadal regression is frequently followed by a refractory period (Baggerman, 1990). In this period, even the usual initiating proximate cues have no effect on gonadal recrudescence. Only at the end to the refractory period can proximate cues initiate gonadal recrudescence. This refractory period provides a basic pattern of timing on the reproductive cycle, and probably prevents recrudescence at times unfavourable for reproduction.

Lunar-related and tidal cycles

After the completion of gonadal recrudescence, some species show, within the breeding seasons, cycles of spawning related to the lunar cycle. *F. heteroclitus* exhibits both a seasonal reproductive cycle and lunar-related cycles within a reproductive season (Day & Taylor, 1984; Hsiao *et al.*, 1994). A period of short days and low temperatures is required for the initiation of oogenesis, but then long photoperiods and high temperatures stimulate vitellogenesis. Once the ovaries contain vitellogenic oocytes, the individual females go through several cycles of semi-lunar spawning before entering gonadal regression. Each cycle involves the recruitment of a cohort of vitellogenic oocytes, which undergo final maturation and ovulation for spawning at the high tide associated with a full and new moon. The proximate cues associated with each cycle are unclear, but may be the change in light intensity at night as the moon waxes and wanes, or the changes in tidal height and current speeds associated with the high tide – the ebb tide cycle.

Rainfall, flooding and other cues

In the tropics, the annual changes in photoperiod and temperatures are much smaller than at temperate and high latitudes. Some tropical species seem able to respond to small changes in photoperiod. Two viviparous poeciliid species, the porthole livebearer (*Poeciliopsis gracilis*, Poeciliidae) and molly (*Poecilia sphenops*, Poeciliidae), in El Salvador (approximately 14°N) showed increases in fecundity at a photoperiod of 12.9 hours of light compared with fecundity when exposed to 11.5 hours of light (Burns, 1985). However, the role of photoperiod as a predictive cue for tropical species is unclear.

In many freshwater tropical species, spawning is related to the flood cycle, which raises the question of what predictive cues initiate gonadal recrudescence to ensure spawning at an appropriate stage of the flood cycle. The main cues that have been suggested are rainfall, water conductivity and water level (Munro, 1990a). In several African elaphantfish species (Mormyridae) and neotropical knifefish species (Gymnitiformes), a decrease in water conductivity initiated gonadal recrudescence (Schugardt & Kirschbaum, 2004; Kirschbaum *et al.*, 2008). Both of these groups have electro-receptors and this may make them sensitive to water conductivity. In characin (Characiformes) species studied, a decrease in conductivity did not induce gonadal recrudescence. Characins do not have electro-receptors.

Anecdotal and some experimental studies have implicated rainfall as a predictive proximate cue or as a local cue (Munro, 1990a; Kirschbaum *et al.*, 2008) for some species, including the Indonesian silver barb (*Barbonymus gonionotus*, Cyprinidae) and three-striped African glass catfish (*Pareutropius buffei*, Schilbeidae). Combinations of rainfall, conductivity decrease and water level may also act to initiate gonadal maturation. There is a need for studies of tropical species that dissociate the relevant predictive proximate cues and the local cues involved in the reproductive cycle.

Social factors as cues

Social interactions can play a major role as synchronising cues, ensuring that male and female fish are ready for successful mating (Chapters 5 and 8). The social induction of gonadal maturation has been reported for a few species. In experimental studies on a species of clown goby (*Gobiodon erythrospilus*, Gobiidae), a coral reef species, single juveniles did not mature, but maturation was quickly induced by the presence of a sexually mature individual (Hobbs *et al.*, 2004). This species is also a serial sex-changer (Chapter 10) and the sex of the juvenile is determined by the sex of the adult. Social interactions can also inhibit maturation. In the spotty wrasse (*Notolabrus celidotus*, Labridae), a reef fish of temperate waters around New Zealand, larger females inhibited the maturation of smaller individuals (Jones & Thompson, 1980).

Stress and reproduction

Labile perturbation or modifying factors are unpredictable events in the environment (Wingfield, 2008). If the perturbation is of a sufficient magnitude, it stimulates the expression of an emergency life-history stage. Such perturbations are stressors, evoking a stress response in individuals experiencing the perturbation. A key component of the stress response is a reallocation of resources away from functions such as reproduction, which are not essential for immediate survival, to activities that contribute to survival (Fuzzen *et al.*, 2011).

In teleosts, long-term exposure to stressors can affect several components of reproductive activity. Adverse effects include failure of maturation, reduced gonadal mass, decreased levels of plasma gonadotrophins (GTHs) and gonadal steroids, decreased concentrations of vitellogenin, reduced size of oocytes and increased frequency of atresia of oocytes in the ovary (Leatherland *et al.*, 2010). A wide range of environmental factors can act as stressors, with adverse effects on reproduction (Billard *et al.*, 1981).

Water quality can be degraded by anthropogenic effects, with compromised water quality having effects on components of reproduction. In freshwater, low pH has been associated with acidic precipitation. In parts of North America and Europe, the resulting low pH has caused the decline or extinction of teleost populations because of recruitment failure. Experimental studies show that low pH decreases the fecundity and egg fertility in some species, including the fathead minnow (*Pimephales promelas*, Cyprinidae) and brook trout (*Salvelinus fontinalis*, Salmonidae) (Billard *et al.*, 1981). Low dissolved oxygen concentrations, which can occur either naturally or as a result of pollution by organic wastes, inhibited spawning by *P. promelas*. Pollutants from industrial activities, such as selenium, copper and mercury, also interfere with components of reproduction. The adverse effects of pollutants that act as endocrine disrupters are described in Chapter 5.

The social environment can also lead to disruption of reproduction. Dominant adults may inhibit the reproductive activity of subordinates (Jones & Thompson, 1980). If fish are overcrowded, reproduction can be inhibited, as can occur in aquaculture facilities. Other aspects of aquaculture can also disrupt reproduction, including transportation, handling and holding conditions (Billard *et al.*, 1981). Aquaculture facilities often lack the local or synchronising cues that are required for the final maturation, ovulation and spawning of eggs (Chapter 12).

To evoke a stress response, the relevant environmental stimuli must be transduced into neural and neuroendocrine events. The sensory systems provide a rapid and complex input, which is processed and the responses coordinated by the CNS (Leatherland *et al.*, 2010). There are two main responses.

The first is an activation of the neurons of the sympathetic component of the autonomic nervous system. These neurons release a catecholamine hormone, noradrenaline (norepinephrine), from their synapses in many organ systems. The sympathetic component also activates chromaffin cells, some of which are located in clusters in the inter-renal tissue of the head kidney. The stimulated chromaffin cells release the catecholamines, adrenaline (epinephrine) and noradrenaline, into the blood system. These neurohormones play a role in the release of glucose from glycogen reserves in the liver and muscle, increasing the levels of glucose available to fuel any required activity. This pathway is particularly evoked when the individual experiences a life-threatening event such as an attack by a predator. If the stress is mild or moderate, the levels of circulating catecholamines do not greatly change. The role of these catecholamines in the regulation of the brain–pituitary–gonad (BPG) reproductive axis is poorly understood (Leatherland *et al.*, 2010; Fuzzen *et al.*, 2011).

The second response to stress is an increased release of corticotrophin-releasing hormone (CRH) by neurons that originate in the pre-optic nucleus and terminate in the adenohypophysis (Leatherland *et al.*, 2010; Fuzzen *et al.*, 2011). The CRH evokes an increase in the secretion of adrenocorticotrophic hormone (ACTH) by the corticotrophs of the pituitary. The ACTH is transported in the blood to its main target cells, the steroidogenic cells of the inter-renal gland. Under the influence of ACTH, these inter-renal cells synthesise and release corticosteroids, particularly cortisol. This axis, the brain pituitary inter-renal (BPI) axis, has many parallels to the BPG axis. The anatomical locations of the neurons producing CRH are close to the neurons producing gonadotrophin-releasing hormone (GnRH), and the pituitary cells producing ACTH are close to cells producing the GTHs (Chapter 5). This may provide opportunities for significant cross-talk between the BPI and the BPG axes.

In a stressed individual, an action of cortisol is to redistribute resources away from gonadal growth and maturation into processes contributing to the response to the stressor. This redistribution forms part of the allostatic load, which is the physiological cost of chronic neural or neuroendocrine stress responses (Leatherland *et al.*, 2010).

An effect of corticosteroids, such as cortisol, may be a reduction in the activity of neurons producing GnRH. When adult male Nile tilapia (*Oreochromis niloticus*, Cichlidae) were exposed to chronic social stress, they became reproductively inactive. Levels of mRNA for GnRH-I and GnRH-II were reduced in the brain. Other neurotransmitters associated with

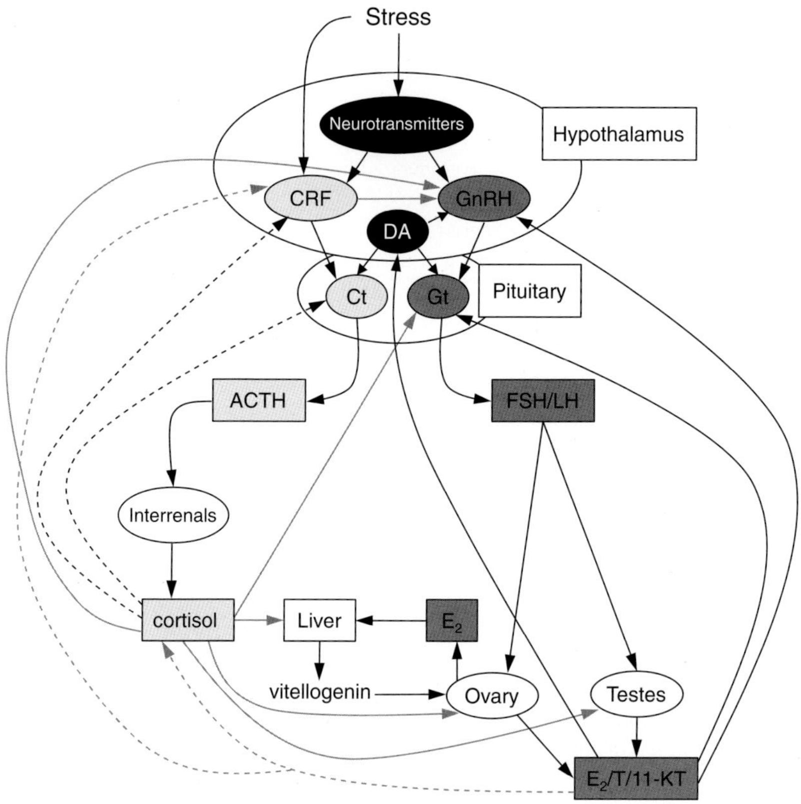

Figure 6.4 Model of the interactions between the brain–pituitary–inter-renal (BPI) stress axis (heavy shading) and brain–pituitary–gonadal (BPG) reproductive axis (light shading) in teleosts. Neurotransmitters (black) regulate both these axes. Solid black arrows indicate stimulation. Dashed black arrows indicate inhibition. Solid grey arrows indicate potential effects of cortisol and corticotrophin-releasing factor (CRF) on the BPG axis. Dashed grey arrows indicate potential effects of sex steroids on the BPI axis. 11-KT, 11 ketotestosterone; ACTH, corticotrophin; Ct, corticotrophs; DA, dopamine; E_2, oestradiol-17β; FSH, follicle-stimulating hormone; GnRH, gonadotrophin-releasing hormone; Gt, gonadotrophs; LH, luteinising hormone; T, testosterone.
(Source: Fuzzen *et al.* 2011. Reproduced with permission of Elsevier.)

the neural circuits that regulate the stress response may also be involved in the regulation of GnRH neurons (Fig. 6.4; Fuzzen *et al.*, 2011; see Chapter 5).

The effects of cortisol on the production of GTHs by the pituitary are unclear. This may reflect interspecific differences or differences in the experimental protocols used.

The main resource-demanding process in the reproducing adult female is the loading of the oocytes with vitellogenin that is processed into yolk (Chapter 4). The vitellogenin is synthesised in the liver under the influence of the ovarian steroid oestradiol-17β (E_2). Not surprisingly, an action of cortisol is to decrease vitellogenesis by the liver as part of the redistribution of resources in the stress response. A study of the effect of cortisol on female *O. mykiss* suggested that cortisol was interfering with the activity of the oestrogen receptor (ER). Even when some vitellogenesis does occur, cortisol can be transferred from the female to the developing oocyte, accumulating in the yolk. In this way, the effects of stress may be transmitted intergenerationally if it is found that there are adverse effects of the cortisol on fertilisation rates or embryonic development.

The relationship between the reproductive and stress axes may not always be inhibitory. In the semelparous Pacific salmon, *Oncorhynchus* spp., there is a chronic activation of the BPI axis that accompanies sexual maturation and spawning. This activation of the BPI axis does not suppress the activity of the BPG axis. It is likely that the capacity of cortisol to mobilise energy is important for the upstream migration and spawning activities of the salmon, which have ceased to feed on entering freshwater (Fuzzen *et al.*, 2011). In a sense, the stressful nature of reproduction in these species can be judged by the lack of any post-spawning survival. The mobilisation of reserves is so effective that over 70% of bodily reserves are consumed in the upstream migration and spawning (Glebe & Leggett, 1981; see Chapter 7).

Evidence is accumulating that cortisol has multiple sites of action in the BPG axis and that the inter-relationships between the BPG and the BPI axes are complex (Fig. 6.4). A major challenge in the study of the reproductive physiology of teleosts is to understand how the various axes are regulated by the brain–pituitary system, notably how the reproductive, stress, growth and metabolic axes interact and mutually regulate each other in the context of the environmental stimuli being processed by the CNS.

Neuroendocrine mechanisms mediating environmental effects

Because of its effect on the rate of chemical reactions, temperature can have a direct effect on physiological processes such as reproduction, as well as potentially acting as a predictive cue. However, cues such as photoperiod or rainfall can only affect the reproductive system by some transduction process (Scott, 1979). The only predictive or local cue that has been studied in detail is the photoperiodic cycle. How do changes in the duration of daylight affect the neuroendocrine BPG reproductive axis? In teleosts, light reception takes place at three sites: the retina of the eye, and extra-retinally in the pineal organ and in deep brain structures in the diencephalon (Borg, 2010; Falcón *et al.*, 2010).

The pineal may be the key element in the photoperiodic response (Falcón *et al.*, 2010). In teleosts, the pineal is a vesicle attached to the roof of the diencephalon by a slender stalk. The pineal usually lies below a window in the skull through which light enters. The epithelium of the pineal consists of true cone-like photoreceptors and ependymal epithelial cells. Neural connections link the pineal with the rest of the brain, with some neurons conveying information from the pineal to the rest of the brain (efferent neurons) and some neurons bringing information to the pineal (afferent neurons). Crucially, the pineal also produces melatonin, an indoleamine hormone (Bromage *et al.*, 2001). Melatonin is also produced by the retina, but retinal melatonin seems to have mainly local effects. By contrast, the pineal is primarily responsible for the levels of melatonin in the cerebrospinal fluid and the bloodstream and acts on specific targets through melatonin receptors (Fig. 6.5).

The critical feature of melatonin production by the pineal gland is its relation to the daily changes in light levels. During the dark phase of the diel cycle (scotophase), the pineal produces melatonin and blood levels of melatonin are high. During the light phase (photophase), the pineal ceases producing melatonin, and levels in the bloodstream decline to low levels. Thus, the pineal effectively transduces light levels into levels of circulating melatonin (Borg, 2010; Falcón *et al.*, 2010).

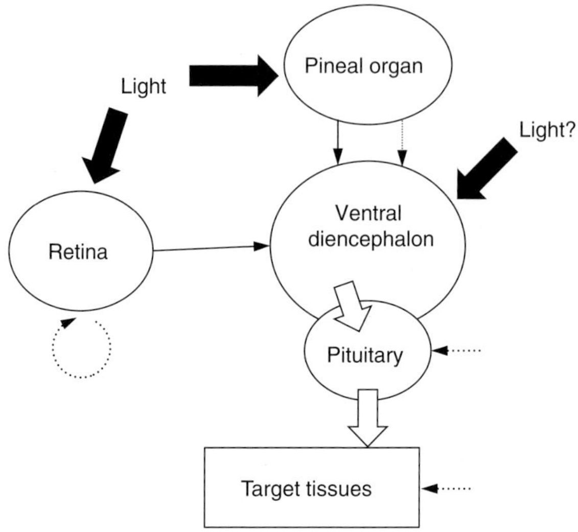

Figure 6.5 Photoneuroendocrine regulation in teleosts. Light acts on photoreceptor cells of the pineal organ and retina, thereby permitting synchronisation of their internal molecular clocks. Light might also impact on other photosensitive and circadian structures in the ventral diencephalon and peripheral organs (indicated by '?'). Photoperiodic information elicits neural and hormonal responses from the retina and pineal organ. The neural response (solid black arrows) from the retina and pineal organ reach the ventral diencephalon through the retino-hypothalamic and the pineal tracts, respectively. This information indicates day length and variation in ambient illumination. Hormonal information is relayed by melatonin (dashed arrows), providing information on day length and season. In the retina, melatonin only has local effects. By contrast, pineal melatonin is released into the cerebrospinal fluid and blood, and acts on specific targets through melatonin receptors. Melatonin may act directly on the pituitary gland itself to modulate the production of hormones, or on target tissues (e.g. the gonads). Melatonin may also exert effects via the hypothalamus, which receives nervous input from the pineal organ and retina. (Source: Adapted from Falcón *et al.* 2007. Reproduced with permission of Elsevier.)

Species differ in the precise pattern of melatonin production by the pineal during the scotophase, but the adaptive relevance of this interspecific variation is not known (Migaud *et al.*, 2010). In species such as *G. morhua*, there is a delay at the start of the dark phase before melatonin levels rise to a peak towards the end of the dark phase (Fig. 6.6A). A second pattern, shown for example by the cichlid *O. niloticus*, has a discrete peak in the middle of the dark phase (Fig. 6.6B). The commonest pattern in teleosts is a rapid increase in melatonin that immediately follows the onset of the dark phase. Migratory salmonids typically show this pattern (Fig. 6.6C).

In salmonids, there is a direct relationship between light levels and melatonin levels. However, in the majority of teleosts studied, there is evidence of an endogenous diel cycle of melatonin production by the pineal, which free-runs in the absence of a diel cycle and which is entrained by the diel cycle (Bromage *et al.*, 2001).

For temperate and high-latitude teleost species, a key feature is that the duration of the elevated levels of melatonin is directly related to the duration of the dark phase. In this way, seasonal changes in photoperiod are directly mapped into seasonal changes in the duration of elevated melatonin levels (Falcón *et al.*, 2010; Migaud *et al.*, 2010).

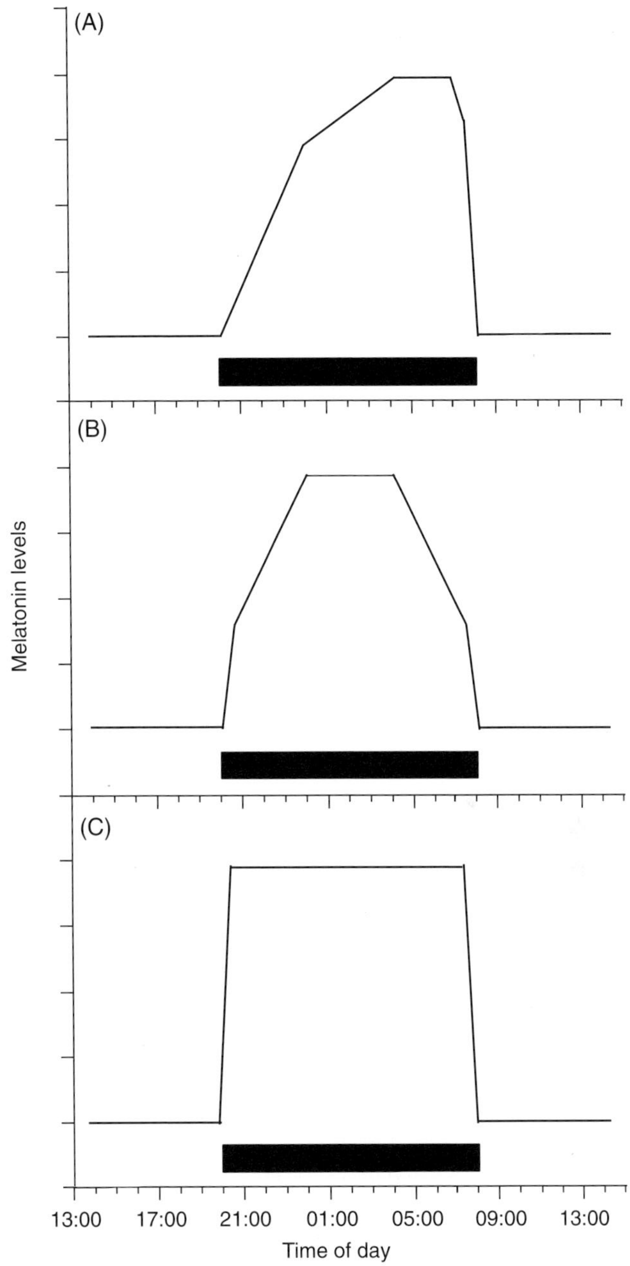

Figure 6.6 Summary of different melatonin profiles recorded in teleosts. The horizontal black bar denotes a dark period. (A) In these species there is a delay at the start of the dark phase before melatonin levels rise to a peak towards the end of the dark phase. Examples include Atlantic cod (*Gadus morhua*) and haddock (*Melanogrammus aeglefinus*). (B) A discrete peak in the middle of the dark phase, a pattern shown by the Nile tilapia (*Oreochromis niloticus*). (C) The most common pattern in teleosts is a rapid increase in melatonin that immediately follows the onset of the dark phase and rapid decrease at its conclusion. Examples include Atlantic salmon (*Salmo salar*), rainbow trout (*Oncorhynchus mykiss*) and Atlantic halibut (*Hippoglossus hippoglossus*).
(Source: Falcón *et al*. 2010. Reproduced with permission of Elsevier.)

The level of melatonin in the dark phase is also sensitive to temperature, with levels reduced at lower temperatures and elevated at higher temperatures.

Changes in light levels during the night as a consequence of the lunar cycle may also be transduced into a melatonin signal. The goldlined spinefoot (*Siganus guttatus*, Siganidae) is a tropical reef-dwelling species with a lunar spawning cycle during the reproductive season, typically spawning around the moon's first quarter. The blood melatonin level at midnight during the new moon period (dark night) was higher than the level during the full moon (brightly lit night) (Takemura *et al.*, 2010).

Melatonin receptors belong to the family of G-protein-coupled receptors, with their characteristic transmembrane domains (Chapter 4). The receptors are found in the brain, pituitary and peripheral tissues, including the gonads (Falcón *et al.*, 2010). Evidence for *D. rerio* has implicated melatonin in the maturation of ovarian follicles (Carnevali *et al.*, 2011).

However, the evidence that melatonin has a major physiological role in mediating the photoperiodic effect in teleosts is still limited (Borg, 2010). Melatonin may play a significant role in coordinating mechanisms such as feedback control of gonadotrophin signalling (Migaud *et al.*, 2010). Levels of melatonin reflecting the light–dark diel cycle could feed into the kisspeptin system (Chapter 5) to form a key link in the photoperiodic control of reproduction through its effect on neuroendocrine mechanisms (Migaud *et al.*, 2010, 2012).

Another thread in the story of the photoendocrine control of reproduction comes from a study of male *G. aculeatus* (O'Brien *et al.*, 2012). When males were switched from a 'short' photoperiod of LD 10:14 to a 'long' photoperiod of LD 17:7, one day of exposure to the long photoperiod caused an increase in levels of the TSH-β1 subunit of the thyroid-stimulating hormone (TSH) in the pituitary. This suggests a rapid activation of the brain–pituitary–thyroid axis, preceding a response by the BPG axis to the photoperiod change. However, the relationship between these two axes in the control of reproduction in teleosts is still unclear.

The mechanisms by which other predictive cues such as rainfall are transduced into neuroendocrine signals are unknown. There is clearly a need for explicit quantitative models of the mechanisms by which photoperiodic and other cues have effects on the reproductive axis.

Food as a proximate factor

A crucial modifying environmental factor for teleost species is the availability of suitable food at appropriate times during the reproductive cycle. Reproduction requires a high expenditure of resources because of the outlays on egg and sperm production (Chapter 4) and the behavioural activities associated with reproduction (Chapters 7–9). One impact of food availability that is experienced by most teleost species is independent of details of their reproductive biology. High food levels support the growth of fishes, leading to larger body size (Wootton, 1998). In female teleosts, fecundity is highly correlated with body size (Chapter 11), so high food levels lead, indirectly through their effect on growth, to high fecundities. Any effects of food availability that are mediated through the BPG axis are dependent on the details of the reproductive biology of the species.

A characteristic of some teleost species is that there is a cessation or reduction in feeding in the run-up to, or during, the spawning season. A consequence of this curtailment in food intake is that expenditures on the later stages of gametogenesis and behavioural activities associated with reproduction must be met from reserves built up when the fish were actively feeding. In other teleost species, feeding continues actively throughout the reproductive cycle and the expenditures on reproduction can be met by this feeding activity.

This difference between species provides an example of the distinction between capital breeders and income breeders (Houston *et al.*, 2006; Stephens *et al.*, 2009). In capital breeders, compensatory feeding occurs in advance of breeding, with the expenditure on reproduction met from the accumulated reserves of lipid and protein. Income breeders meet expenditure on reproduction from the current intake of resources provided by feeding behaviour. Capital breeding and income breeding represent end points of a continuum. In pure capital breeders, reserves of lipids and protein cover 100% of the expenditure on reproduction, whereas in pure income breeders, 0% of the expenditure is covered from stored reserves. Pure capital and pure income breeding represent different forms of linkage between the individual and the environment. Capital breeding allows a temporal separation of feeding and reproductive activities, so they can be carried out at times that are optimal for both. The temporal alternation of somatic and reproductive expenditures is achieved by a cycle of storage and depletion of reserves (Wootton, 1979; Reznick & Braun, 1987). An obvious advantage of capital breeding is that reproductive success is insulated from unfavourable feeding conditions during the reproductive season. By contrast, an advantage of income breeding is that individuals can take advantage of favourable feeding conditions during the breeding season.

Feeding and capital breeders

Most studies on the effect of food availability on capital breeders have focused on species of high commercial interest, especially salmonids (Salmonidae), *C. harengus* (Clupeidae) and gadoids (Gadidae). Some tropical species, which are total spawners, reproducing in synchrony with the flood cycle of rivers, may also be capital breeders (Munro, 1990a). A question that arises for capital breeders is whether there is a critical window during the feeding phase, when the physiological condition of the fish determines whether it will reproduce in the next spawning season.

Salmonids, such as *S. salar*, are extreme capital breeders. Anadromous salmonids cease feeding before entering freshwater to commence their upstream migration to the breeding grounds, so all subsequent expenditure on reproduction, physiological or behavioural, must be met from reserves (Thorpe *et al.*, 1998; Thorpe, 2007). The cessation of feeding seems to depend on individual salmon reaching a threshold lean body mass (Kadri *et al.*, 1995). Semelparous species such as the Pacific salmon, *Oncorhynchus* spp., expend more than 70% of their body reserves on the migration, completion of gametogenesis and reproductive behaviour (Glebe & Leggett, 1981; see Chapter 7). Even in iteroparous species, such as *S. salar*, expenditure can approach 70% of body reserves, but at high levels of expenditure, post-breeding survival is low.

Experimental and field studies on *S. salar* suggest that there are two critical windows (or gates), when the physiological condition of the individuals, possibly indicated by the level of lipid reserves or the rate of change of lipid levels, determines whether that individual will subsequently mature and reproduce in the next spawning season (Thorpe *et al.*, 1998; Thorpe, 2007). The model for *S. salar* may generalise to other salmonid species.

The first window is in autumn, a full year before the spawning season. Only if the physiological condition of the individual exceeds a genetically determined threshold will the individual initiate the early stages of gametogenesis, which are relatively low in their demand for resources. A second and critical window opens in the following spring, about 6 months before the expected time of spawning in the autumn. Again, only if the physiological condition of the individual exceeds a genetically determined threshold does gametogenesis start its later stages, especially vitellogenesis in the females. The length of the critical window in spring was 3–4 weeks. Photoperiodic cues are essential for the onset of maturity, but the process is inhibited unless the thresholds are surpassed during the critical periods.

The feeding success of the individual during winter and especially in spring will determine whether this spring threshold will be surpassed. High growth rates in winter increase the probability of maturing the following autumn (Jonsson *et al.*, 2013). In spring, individuals that are going to mature have higher growth rates and are in better physiological condition than immature fish (Kadri *et al.*, 1995). Later in the summer, maturing fish become anorexic and their growth declines, whereas immature fish continue to feed. A consequence of these processes is that under aquaculture conditions, maturation, which is usually an unwanted process, can be inhibited by the application of appropriate feeding regimes. For example, the proportion of first-year male *S. salar* that become precocially mature can be limited by restricted feeding, which reduces their lipid reserves in spring at the time of the critical window (Rowe & Thorpe, 1990a,b).

The presence of critical windows may be important for anadromous salmonids because their upstream migration takes them into habitats that provide poor feeding opportunities for fish of their size. Individuals that entered freshwater with inadequate reserves would have had no possibility of replenishing them, even if feeding could be resumed.

A contrast is provided by the marine *C. harengus*. This species is an iteroparous extreme capital breeder, with total spawning. A feeding phase is followed by a reproductive phase during which feeding ceases. Lipid reserves are used to meet metabolic demands, while protein reserves support gametogenesis (Iles, 1984). An experimental study of individuals from the Norwegian spring spawning stock of *C. harengus* subjected fish to periods of food deprivation at different times during the feeding phase in summer to determine the presence of critical windows (Kennedy *et al.*, 2010). However, the timing of the acquisition of food had no effect on maturation, suggesting the absence of a critical window during feeding. Maturation seemed to depend on individuals reaching a threshold physiological condition, but not on when in the feeding phase they reached that condition. Some female *C. harengus* in relatively poor condition did initiate vitellogenesis, but they reabsorbed their oocytes by pre-ovulatory atresia (Chapter 4). The food of *C. harengus* is zooplankton, which is spatially and temporally unpredictable in its abundance. It is probably advantageous for individuals not to have a restricted critical window, so they can exploit high-density patches of zooplankton encountered at any time in their feeding phase and achieve the appropriate physiological threshold required for maturation.

Gadus morhua (Gadidae), plaice (*P. platessa*, Pleuronectidae) and turbot (*Scophthalmus maximus*, Scophthalmidae) are marine species of the North Atlantic and connected waters. All have high commercial value. Like *C. harengus*, they are all iteroparous species with determinate fecundity. In all three species, maturation and fecundity are a function of both the level of stored reserves and the intensity of feeding leading up to the initiation of spawning. They are capital breeders, but with some of the characteristics of income breeders.

An experimental study of 1-year-old *G. morhua* subjected fish to different periods of starvation in the period from July to November, when the number of vitellogenic oocytes is being determined (Karlsen *et al.*, 1995). All fish were starved from late January to May, i.e. over the spawning season. The length of the period of starvation had no effect on the proportion of fish maturing. This suggests that, if there is a critical threshold, even the starved fish exceeded that threshold. The best-fed fish had the highest fecundity, but this was entirely explained by body size. The longer the period of starvation, the smaller the fish were at maturity. The number of vitellogenic oocytes was adjusted to reflect body size. Between January and May, females lost more of their somatic and liver mass than males, which suggests that females invested more in reproduction than males. However, in repeat spawning, *G. morhua* potential fecundity was related to the food availability in the autumn, when the oocytes were vitellogenic, as well as to body size (Kjesbu *et al.*, 1991).

An experiment on *S. maximus* that were approaching maturity provided groups of fish with similar amounts of food over a year, but the intensity of feeding was varied at different times of the year (Bromley *et al.*, 2000). If the feeding of females was restricted in the period of oocyte recruitment (October to January), and especially during the vitellogenic phase (February to May), fewer females matured and the growth of vitellogenic oocytes was slowed. However, the number of oocytes recruited to vitellogenesis was not affected in those females that did mature. In the experiment, some *S. maximus* continued to feed, providing the possibility of an income contribution to the process of vitellogenesis. For males, the pattern of feeding had no effect on the proportion of males maturing.

A problem facing iteroparous capital or partial capital breeders is the regulation of investment in reproduction in the current breeding season and the retention of stored reserves that can be used to promote survival in the post-breeding period (Bromley *et al.*, 2000). In species with determinate fecundity, vitellogenesis before the spawning period defines the maximum number of eggs that can be spawned during the breeding season. Fecundity can then be regulated in relation to available stored resources and any current income by pre-ovulatory atresia (Chapter 4). Atresia returns resources into general circulation that had been invested in reproduction.

Oncorhynchus mykiss is a non-anadromous, iteroparous, total spawner widely used in aquaculture. When fed high or low rations over a full year, the fish on high rations had a higher fecundity than fish on low rations (Springate *et al.*, 1986). This difference was entirely accounted for by differences in body size at spawning. The pattern of feeding is relevant. Fecundity, even after adjustment for body size, was higher in *O. mykiss* fed a high ration between April and August, when vitellogenesis was occurring, than when the high ration was fed from August to November. Reducing the ration over these months had no detrimental effect on fecundity (Bromage & Jones, 1991). Subsequently, when feeding was stopped or reduced, fecundity was adjusted by pre-ovulatory atresia.

In *C. harengus*, *G. morhua*, *S. maximus* and *P. platessa*, atresia during the vitellogenic phases seems to be the mechanism for adjusting the potential fecundity (i.e. the number of vitellogenic oocytes established) to generate the realised fecundity (i.e. the total number of eggs spawned over a breeding season). In *C. harengus*, atresia seems to play a major role in the adjustment of potential fecundity (Kennedy *et al.*, 2010). In *P. platessa*, which feed from the end of one spawning season to the beginning of the next, levels of pre-ovulatory atresia vary from 10% to 30%. The rates are highest in the earlier phases of vitellogenesis, but decrease towards spawning (Kennedy *et al.*, 2008). About 50% of the gonadal growth in *P. platessa* is supported from stored reserves. *G. morhua* can have losses of up to 36% of vitellogenic oocytes through atresia during spawning. *S. maximus* had lower rates of atresia and it was not clear whether this was the result of a down-regulation of fecundity or the removal of oocytes that were not going to complete vitellogenesis successfully (Bromley *et al.*, 2000).

Feeding and income breeders

Species that have indeterminate fecundity, so that vitellogenic oocytes can be recruited during the breeding season, and are batch spawners are likely to have the total number of eggs spawned over a breeding season determined by their current intake of food. This has been most clearly shown by female *G. aculeatus*. The species spawns in spring and summer and can have a breeding season extending over several weeks. They are small-bodied fish, which means their ability to store reserves of lipid and protein is restricted. The number of spawnings during a breeding season is related to the food available during the breeding season, with a strong correlation between ration size and number and frequency of spawnings (Wootton, 1977; Fletcher & Wootton, 1995; Wootton & Fletcher, 2009). However, the batch fecundity, the number of eggs laid at a given spawning, is largely determined by body size. The inter-spawning interval, the time between successive spawnings, is inversely correlated with ration, so at high rations females can spawn at intervals as short as 3 or 4 days. The least variable response to differences in food availability is egg quality, as assessed by egg size and lipid content. At lower rations, the body mass of the female declined, suggesting that body reserves are contributing to egg production, but the main adjustment to a decline in current income from feeding was a reduction in the number of spawnings.

Can an income breeder adjust its reproductive performance to changes in food availability within a breeding season? The question was addressed in the oviparous *G. aculeatus* and in the lecithotropic, viviparous guppy (*Poecilia reticulata*, Poeciliidae) by experiments in which the levels of food availability were varied over three successive spawnings (Reznick & Yang, 1993; Ali & Wootton, 1999a). Two ration levels were used, high (H) and low (L) and so individual fish experienced one of the following feeding protocols, H-H, H-L, L-L, L-H, with the ration changed after a spawning. The species differ in that, in *G. aculeatus*, a female develops a batch of eggs in the inter-spawning interval, which typically last 4–8 days, whereas *P. reticulata* gestates a clutch of developing embryos up to birth, a gestation period of about 28 days (Chapter 10). Female *G. aculeatus* showed an increase in the inter-spawning interval when the ration was low (Fig. 6.7A). When a period of low ration was followed by a high ration, the interval to the next spawning was still longer than if the high rations had been present in both periods. Egg characteristics and, to a lesser extent, batch fecundity were insensitive to the current rate of food consumption (Fig. 6.7B). In *P. reticulata,* the effect of ration levels on the interval between successive births was similar to that in *G. aculeatus* (Fig. 6.7C). By contrast, in *P. reticulata*, a low ration reduced the litter size (Fig. 6.7D), although the young were larger. The increase in offspring size was largely a consequence of a high lipid content of the progeny. However, in the matrotrophic least killifish (*Heterandria formosa*, Poeciliidae), females that experienced periods of limited feeding produced smaller young (Reznick *et al.*, 1996a).

Experimental studies on *G. aculeatus* have also allowed the evaluation of the effects of variation in food availability before the start of the breeding season (Ali & Wootton, 1999b; Inness & Metcalfe, 2008). Although the experimental protocols differed, the studies showed that if food availability is sufficiently high during periods of feeding, then periods of starvation lasting as long as 5 days did not significantly affect the reproductive performance of the fish when they started spawning. The fish were able to compensate for periods of starvation by their feeding rate when food was present, albeit there was evidence that these fish had a reduced life span (Inness & Metcalfe, 2008). If a low rate of food is fed continuously, few fish mature (Wootton, 1973; Inness & Metcalfe, 2008). This suggests there may be a threshold condition that has to be surpassed for *G. aculeatus* to initiate vitellogenesis.

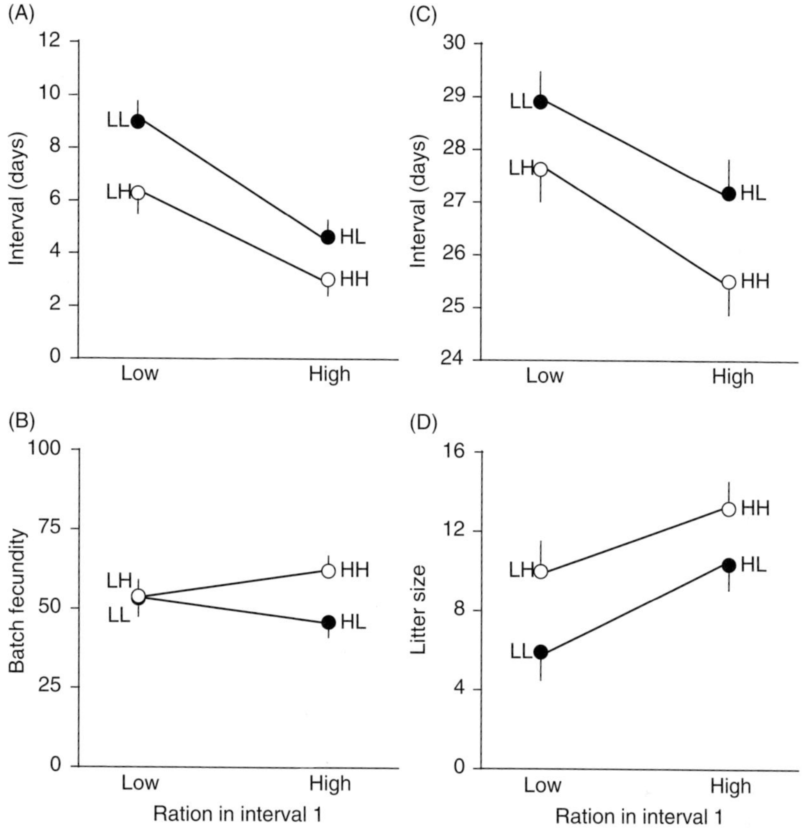

Figure 6.7 Effect of ration on reproductive variables. Levels of food availability were varied over three successive spawnings, with two ration levels used, high (H) and low (L), and so individual fish experienced one of the following feeding protocols, H-H, H-L, L-L, L-H, with the ration changed after a spawning. (A) Interval between spawnings for female three-spined sticklebacks (*Gasterosteus aculeatus*), an oviparous species; (B) batch fecundity of female *G. aculeatus*; (C) interval between broods for female guppies (*Poecilia reticulata*), a lecithotropic, viviparous species; (D) litter size of female *P. reticulata*. The responses of the two species to ration were similar for the interval between spawning and parturition, but diverged in fecundity/brood size. Fecundity was relatively insensitive to ration in *G. aculeatus*, but litter size in *P. reticulata* was reduced in response to a low ration. [Sources: (A, B) Adapted from Ali & Wootton 1999a (reproduced with permission of John Wiley & Sons); (C,D) Adapted from Reznick & Yang 1993 (reproduced with permission of the Ecological Society of America.)]

These results demonstrate the diversity of responses of income breeders to changes in food availability, but too few species have so far been studied to develop any general model of outcomes.

Feeding and skipped spawnings

Low levels of food availability may lead to the phenomenon of skipped spawnings. The estimation of the total egg production by a population of fish during a spawning season is an important component of the data required to manage a commercial fishery (Chapter 12). Skipped spawning is a factor that can affect the accuracy of such estimates (Rideout *et al.*, 2005; Jørgensen *et al.*, 2006; Rideout & Tomkiewicz, 2011). Individual fish can mature only once in their lifetime with the transition from the juvenile to the adult phases. Juvenile fish, by definition, do not spawn. A sexually mature fish exhibits a skipped spawning when it fails to spawn in a given breeding season having spawned in a previous spawning season. After one or more skipped spawnings, the individual will resume spawning.

Skipped spawnings have been observed in about 31 teleost species from a variety of families, but most studies have been on the females of species with determinate fecundity (Rideout & Tomkiewicz, 2011). Skipped spawning can take one of three forms: retaining, reabsorbing and resting. Retaining skipped spawners are fish that have produced fully developed gametes, but these are not released. This is typically because the fish have not encountered the relevant local and synchronising cues that are required for successful spawning. Reabsorbing skipped spawners have initiated gametogenesis, but this is aborted before vitellogenesis is completed. The vitellogenic oocytes are then reabsorbed by pre-ovulatory atresia. Resting skipped spawners are sexually mature fish that do not enter the later stages of gametogenesis, such as vitellogenesis. In females, the oocytes are all in the primary growth phase (Chapter 4).

Modelling studies have suggested that skipped spawnings might occur in relation to contrasting levels of food availability (Jørgensen *et al.*, 2006). At high food levels, the growth a female fish can achieve over a year, if it skips spawning, can lead to an increase in fecundity in succeeding years that more than compensates for the loss in the skipped year. This effect is because of the strong positive correlation between body size and fecundity (Chapter 11). However, there is no good evidence from natural populations that this is an adaptive explanation for skipped spawnings. A potential fitness cost of this strategy is if death occurs prior to spawning, in which case the potential fecundity benefit of a skipped spawning is lost.

Most, but not all, of the evidence suggests that skipped spawning in the form of reabsorbing or resting modes is a consequence of poor feeding conditions and the failure of individuals to acquire sufficient reserves to support the subsequent spawning. Skipped spawning by females associated with poor feeding conditions has been described for *G. morhua* and the winter flounder (*Pseudopleuronectes americanus*, Pleuronectidae). In *G. morhua* females that are going to skip a spawning, the levels of the ovarian steroid hormone $E_{2,}$ which is required for vitellogenesis, fail to show the increase that occurs in females beginning vitellogensis (Chapter 5). Experimental studies on *P. americanus* suggested that there is a critical period immediately after spawning. If females do not receive sufficient food in this period, they fail to spawn in the following year (Burton, 1994). In salmonids, species living in northern lakes with low productivity may spawn only every 2 years. This occurs in the anadromous Arctic char (*Salvelinus alpinus*, Salmonidae) and the non-anadromous lake trout (*Salvelinus namaycush*, Salmonidae).

Skipped spawning provides a mechanism by which fish can adjust their reproductive activity to the prevailing feeding conditions. An alternative mechanism is the adjustment of fecundity by pre-ovulatory atresia after vitellogenesis has started.

Physiological basis for the effect of feeding on reproduction

The effect of feeding on reproduction is usually mediated by its effect on the physiological condition of the individual fish. This raises the question of what indicators of physiological condition provide information to the reproductive axis. Possible indicators are the level of lipid reserves, the rate of acquisition of reserves or the growth rate (Chapter 5). A second problem is how critical thresholds, if present, are determined. The mechanisms by which feeding conditions control reproduction have yet to be elucidated.

Conclusions

The environment forms the ecological arena in which the success of the reproductive biology of a species is tested. The environment provides the food that is required for the fundamental Darwinian transformation of food into progeny. It also provides cues that can be used to ensure that reproduction takes place at the appropriate time. Thus, several questions arise and, as yet, there are only partial answers to those questions. At the evolutionary, or ultimate, level, there is the question of the adaptiveness of the reproductive traits in relation to the environment experienced by populations of a species (Chapter 11). At the proximate level, there is the physiological question of how the individual receives the cues provided by the environment and converts them into signals that contribute to the regulation of reproduction in relation to the resources available. The overarching question is how the individual integrates the information provided by the environment with the information provided internally to calibrate the reproductive effort that the individual should make under the prevailing conditions to maximise its reproductive success. Once the physiological 'decision' is made to engage in reproduction, a fish must navigate its way to the site of reproduction, and in some cases secure and defend that site from rivals. In some species, fish also prepare the site for spawning and subsequent care of the offspring.

Chapter 7

Migration, territoriality and spawning site selection

Reproductive Biology of Teleost Fishes, First Edition. Robert J. Wootton and Carl Smith.

Introduction

While gametogenesis (Chapter 4) is an essential prelude to reproduction, a series of additional steps is typically necessary to place sexually mature fish in the right location to ensure that reproduction is successful. These steps can include spawning migrations, establishment of a territory, and selection and preparation of sites for spawning. The control of these behaviours is mediated hormonally (Chapter 5) and cued by the environment (Chapter 6).

Migration

The periodic mass movements of fishes, sometimes on a spectacular scale, have long been recognised and fish migratory behaviour is a relatively well-researched field of fish ecology (Harden Jones, 1968; Northcote, 1978; McKeown, 1984; Dingle, 1996). Many species engage in long-distance migrations, including economically important species such as Atlantic cod (*Gadus morhua*, Gadidae) (Rose, 1993), Atlantic herring (*Clupea harengus*, Clupeidae) (Slotte & Fiksen, 2000), anchoveta (*Engraulis ringens*, Engraulidae) (Cury *et al.*, 2000) and chub mackerel (*Scomber japonicus*, Scombridae) (Hiyama *et al.*, 2002). Some fish movements are of short periodicity, such as on a daily cycle, and are associated with feeding, predator avoidance or control of energy expenditure. Reproductive migrations tend to be on a larger scale, are horizontal rather than vertical, and are often strongly seasonal.

The spatial scale over which reproductive migrations occur can vary from a few hundred metres to thousands of kilometres. In the northern pike (*Esox lucius*), lake populations perform small-scale spawning migrations, of no more than a few hundred metres in some cases, to shallow lake margins in early spring. As water levels decline after spawning, these areas serve as nurseries for the young stages (Bregazzi & Kennedy, 1980). By contrast, the European eel (*Anguilla anguilla*, Anguillidae) migrates 5000–6000 km from the coasts of Europe to its spawning grounds in the Sargasso Sea, a region of the North Atlantic Ocean east of Bermuda and north of the Antilles (Tesch, 1977; Deelder, 1984). Adult eels confined to landlocked water bodies, or which meet migration barriers, are capable of moving briefly over land during migration. After spawning, the specialised leptocephali larvae are transported by the northern gyre of the Gulf Stream, and the unpigmented juvenile eels, termed glass eels, arrive in European coastal waters 11 months to 3 years later. Following further development into pigmented elvers, the juveniles migrate up West Atlantic river systems from Scandinavia to North Africa, including rivers of the Baltic and Mediterranean Seas.

Comparable migrations are made by other anguillid eels. The American eel (*Anguilla rostrata*, Anguillidae) also spawns in the Sargasso Sea, possibly west of the area used by *A. anguilla*. Juvenile *A. rostrata* drift in the southern gyre of the Gulf Stream, eventually entering river systems on the Atlantic coast of North America, but as far south as Panama, and throughout much of the West Indies. In the Pacific Ocean, adult Japanese eels (*Anguilla japonica*, Anguillidae) undertake a migration from freshwater sites in Japan and eastern Asia to spawning grounds west of the Mariana Islands. Like *A. anguilla* and *A. rostrata*, the leptocephali larvae of the *A. japonica* rely on major oceanic currents to complete their migration, in this case utilising the North Equatorial Current to carry them west and north to the Japanese islands and Asian mainland (Tsukamoto, 1992).

The periodicity of spawning migrations similarly varies greatly. Semelparous species make only a single migration during their lifetime and die following reproduction (Chapter 11). Anguillid eels characterise this mode of reproductive migration, as do many populations of Pacific salmon (*Oncorhynchus* spp.) and capelin (*Mallotus villosus*, Osmeridae). In some species, spawning migrations are annual, corresponding to changes in photoperiod (sticklebacks), temperature (temperate marine flatfish) or water level (tropical cyprinids) (Chapter 6). In other species, spawning migrations occur with relatively high frequency. Male melon butterflyfish (*Chaetodon trifasciatus*, Chaetodontidae) undertake small-scale migrations of 100–400 m to temporary spawning territories (Yabuta, 1997). Spawning occurs only at dusk and with lunar periodicity, although not always in every lunar cycle.

Terms and definitions

A suitable definition of migration is those movements that result in an alternation between two or more discrete habitats, which often occurs with regular periodicity and involves a large proportion of the population. The goal of reproductive migration is mating. A key element of migratory behaviour, which distinguishes it from other movements, is the 'undistractability' of migrants (Dingle, 1996); movements are persistent and migration is not motivated directly by resources, at least on a proximate scale, although ultimately resources are being sought.

Fish may migrate entirely within the typical environment of the adult. Species that migrate within the marine environment are oceanadromous, those within freshwater potamodromous (Fig. 7.1). Diadromous migrants may move between environments with different salinity regimes. Anadromous species are those in which adults migrate from the marine environment into freshwater to reproduce. The embryos of anadromous teleosts complete development in freshwater, and after a period feeding, the juveniles migrate back to saline water to grow and mature. Examples include migratory trout and salmon (Salmonidae), shads (Clupeidae), sturgeons (Acipenseridae) and sea catfish (Ariidae). Catadromous species do the reverse: adults migrate from a freshwater environment to a marine one, with juveniles developing in a saline environment before migrating back into freshwater to mature as adults. Examples of catadromous species include the anguillid eels (Anguillidae), species of mullet (Mugilidae), snooks (Centropomidae) and flounders (Pleuronectidae). A variation of diadromy is amphidromy. Here spawning occurs in either freshwater (freshwater amphidromy), with larval stages moving promptly out to sea at an early stage of development, or seawater (marine amphidromy) in which larval stages move into freshwater. With freshwater amphidromy, development is completed in seawater, after which the juveniles migrate back into freshwater. Maturity is achieved in freshwater, where spawning occurs. The

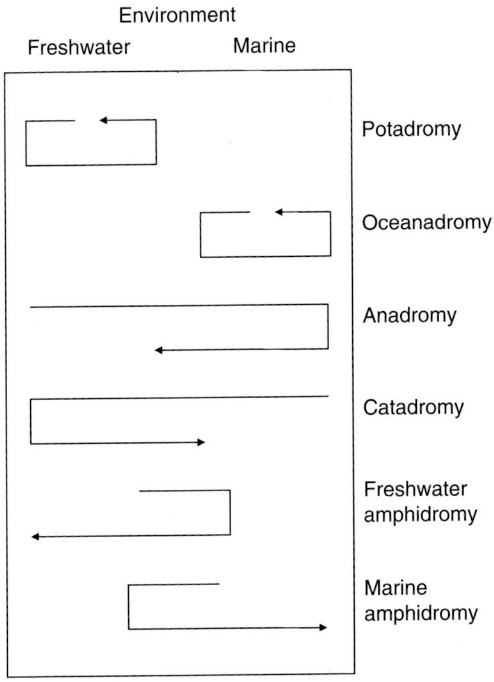

Figure 7.1 Different forms of spawning migration in freshwater and marine environments. In potadromous species, migrations occur entirely in freshwater, and in oceanadromous species, they occur entirely in the marine environment. Anadromous species spawn in freshwater, and the juveniles, after a period of residence in freshwater, migrate into seawater. Catadromous species do the opposite, with adults migrating into seawater to spawn, and juveniles eventually migrating back into freshwater. In freshwater amphidromy, adults spawn in freshwater and the young stages migrate promptly into the marine environment, but return to freshwater to complete maturation. In marine amphidromy, spawning occurs in seawater and the young stages migrate temporarily into freshwater, but return to sea to grow and mature.

reverse occurs in marine amphidromy (Fig. 7.1). In amphidromy, migration is not an immediate prelude to reproduction. Amphidromous species are found among the galaxiids (Galaxiidae), smelts (Osmeridae) and sleepers (Eleotridae). However, simple classification of migration in this way does not capture the remarkable variability and complexity of fish spawning migrations, and a more meaningful approach is to consider the patterns underlying the evolution of migratory behaviour and their adaptive significance. Thus, Watanabe *et al.* (2013) recognised a distinction between temperate and tropical amphidromy, which entail different levels of dependence on the ocean phase of migration. The distinction between these two categories of migration reflects fundamental differences in the life histories of the taxa that display them.

Migratory patterns associated with reproduction

The body size of teleosts is often orders of magnitude greater than their egg, embryo and larval stages (Wootton, 1998). A consequence is that the habitat and dietary requirements of adult fishes, as well as their range of potential predators, are distinctly different from those of their offspring. One way to view migration, then, is as a set of

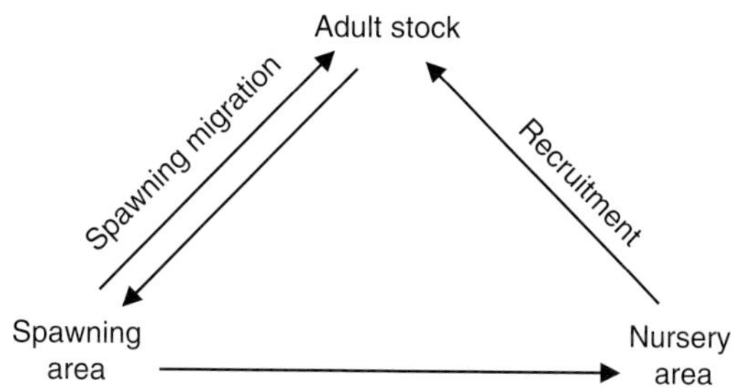

Figure 7.2 A conceptual model of the interrelationships between different life-history stages and spawning migrations and recruitment. Migration permits an individual to maximise its fitness in a specific habitat type.

movements between specific habitat types to maximise individual fitness by providing the optimal conditions for the individual and its offspring at each life-history stage (Fig. 7.2). This approach to understanding migration emphasises a trade-off between optimal conditions for the adult and those for offspring; migration is a potentially expensive behaviour (see later), but one that can provide substantial fitness returns through enhanced offspring growth and survival.

The three key functions of the environment in the life history of fish are for reproduction, and to provide refuge and food. The choice of suitable sites for spawning, which maximise the success of egg and larval development, is discussed in detail towards the end of this chapter (see the section 'Spawning site selection'). Migration to suitable spawning habitats is exemplified in the salmonids. In this group, eggs are deposited in river and occasionally lake beds with a coarse gravel substratum and with water flow that provides the developing stages with sufficient oxygenation so they are not smothered, but that is sufficiently fine to prevent eggs and embryos from being washed out during periods of high water flow. Female salmonids are selective of the sites in which they dig spawning nests, termed 'redds', and will sometimes dig up the previously deposited eggs of other females in a process referred to as 'superposition'.

For the early life stages, avoidance of predation and access to suitable nursery or shelter areas, from which predators are absent or limited, are key determinants of survival. In *E. lucius*, eggs are spawned in the shallow flooded margins of lakes, rivers and ponds, which the young stages subsequently use as nursery areas and where fish predators, especially older conspecifics, are absent. In the marine migratory *M. villosus*, eggs are spawned intertidally on sand or gravel beaches, sometimes close to the tide line and often so high up the beach that the adults, males in particular, become stranded and die. A consequence is that egg mortality is limited because the eggs are effectively out of the reach of many potential egg predators. Interestingly, some capelin spawn offshore in deeper waters, and this variation in spawning life history may be linked to variation in the degree of semelparity/iteroparity (Christiansen *et al.*, 2008; see Chapter 11). Capelin that spawn offshore show a semelparous life history in both sexes, whereas intertidally spawning populations are iteroparous. These differences in life-history strategy appear to relate to differences in juvenile mortality rates, although the ecological basis to these differences is not clear (Christiansen *et al.*, 2008). A number of other species also spawn intertidally, and in all cases this spawning strategy appears to have evolved as a means of limiting predation on young stages. Examples include California grunion (*Leuresthes tenuis*, Atherinopsidae), sharpnose sculpin (*Clinocottus acuticeps*, Cottidae), the midshipman (*Porichthys notatus*, Batrachoididae) and some populations of anadromous three-spined sticklebacks (*Gasterosteus aculeatus*, Gasterosteidae). The reproductive biology of beach-spawning fishes is reviewed by Martin *et al.* (2004) (see the section 'Terrestrial spawning sites').

Another key variable for young stages is the availability of food. Starvation is a significant cause of mortality for early developmental stages, which are unable to withstand prolonged starvation as they possess limited energy reserves (Lasker, 1981). The timing and location of spawning migrations of many oceanic marine fishes appear to have been selected to ensure the co-occurrence of egg hatching with blooms of suitable planktonic food items (Lough & Manning, 2001). For example, the spawning migration of bluefin tuna (*Thunnus thynnus*, Scombridae) to the Gulf of Mexico places larval stages in relatively cool but fertile water masses that maximise their growth, which would not be the case if eggs were deposited in surrounding warmer but less fertile waters (McGowan & Richards, 1989).

A further goal of reproductive migration may be to obtain a mating partner. In some species, mate acquisition may be the principal goal of migration. In lek breeding cichlids (see Chapter 8), females migrate to the sites of leks simply to mate; no other resources are sought, and after mating females return to areas suitable for rearing offspring (Barlow, 1991).

Adaptive significance of reproductive migrations

One model for understanding the evolution of reproductive migration is through comparing the fitness advantage that adults can obtain by moving between habitats that provide the resources for their own maintenance and survival, as well as those of their offspring, with a life history without migration. Migration, then, represents a set of movements between habitats that maximise the survival, energy intake, growth and mating success of individuals at each life-history stage (Chapter 11). The fitness benefits that accrue from migratory behaviour, however, must be traded off against the fitness costs imposed by migration. Thus, migration decisions are condition-dependent and provide scope for variation in migratory behaviour among individuals in the same population with direct consequences for reproductive success (Dingle, 1996).

Comparison of life-history strategies of migratory and non-migratory members of the same species, and even of the same population, can provide insights into the selective pressures leading to the evolution of migration (Kaitala *et al.*, 1993). The term 'partial migration' is used for populations subdivided into individuals that are either migratory or resident. Partially migrant populations often show polymorphism; migratory forms are typically larger and more robust than non-migrants. Examples include anadromous and non-migratory salmonids, and populations of *G. morhua* that comprise coastal residents and migratory Arctic fish (Møller, 1969). In some species, migratory and non-migratory individuals show no colour, size or morphological differences; examples include *C. harengus* and smelt (*Osmerus eperlanus*, Osmeridae) (Jonsson & Jonsson, 1993).

A proximate explanation for the larger size of migrants is that they obtain access to more abundant food resources, with the result that they achieve substantially larger body sizes than non-migratory individuals (Dingle, 1980; Northcote, 1992). Anadromous male sea trout (*Salmo trutta*, Salmonidae) were 1.7 times larger than non-migratory males in the River Voss system in Norway (Hindar *et al.*, 1991). Similarly, a migratory population of *G. aculeatus* in California showed higher growth rates and achieved larger body sizes than a related non-migratory population, despite developing from eggs of comparable size (Fig. 7.3). Size differences in this species had a significant genetic component (Snyder, 1991a).

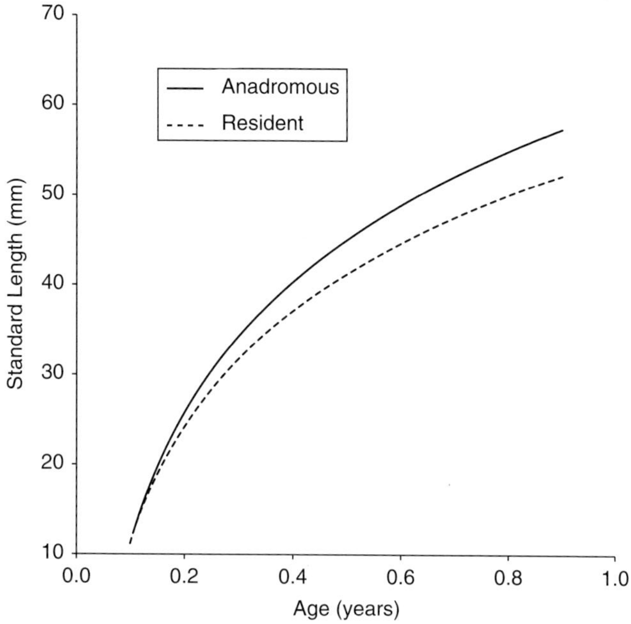

Figure 7.3 Von Bertalanffy growth curves for migratory anadromous (solid line) and resident freshwater (dashed line) populations of the three-spined stickleback (*Gasterosteus aculeatus*) from the Navarro River, California. (Source: Snyder 1991b. Reproduced with permission of Springer Science + Business Media.)

Migration can confer increased reproductive potential as a consequence of larger body size and higher gamete production. In anadromous salmonids, female fecundity is greater than in non-anadromous individuals (Hutchings & Morris, 1985). The mean fecundity of anadromous female *S. trutta* in Lake Vangsvatnet, Norway, was 1790 eggs, while that of residents was 330 eggs (Jonsson, 1985). Investment in gonads per unit body mass was 20% greater in anadromous females than resident females, although there was no difference among males (Jonsson & Jonsson, 2006). A comparable pattern has been observed in Arctic charr (*Salvelinus alpinus*, Salmonidae) (Nordeng, 1983) and *G. aculeatus* (Wootton, 1976), although the reverse was observed in the coporo (*Prochilodus mariae*, Prochilodontidae), a South American characin in which resident females invest up to five times more energy in egg production than do migratory females (Saldaña & Venables, 1983).

Migrants may also have improved competitive ability during spawning as a result of their larger body size. Large males and females are more dominant, better able to acquire high-quality territories for spawning and egg development, and are more successful in nest defence and parental care than smaller individuals (van den Berghe & Gross 1989; Maekawa *et al.*, 1994). Larger males and females may also be more attractive as mates (Hutchings *et al.*, 1999; Rowland, 1989; Sargent *et al.*, 1986; see Chapter 8).

In species with partial migration, it is typically males that display migratory and non-migratory forms, whereas females more commonly tend to be fully migratory (Jonsson & Jonsson, 1993). This disparity between the sexes demonstrates differences between males and females in the benefits of migratory behaviour, and therefore in the strength of selection acting on them (Chapter 8). Female egg production is strongly size-dependent in teleosts; the number and also quality of eggs a female produces are often directly correlated with female body size (Bagenal, 1973; Wootton, 1979; see Chapters 4 and 11), and thus females experience strong natural selection to achieve a large body size. Although male reproductive success may also be correlated with size, e.g. through competition for resources and mates, smaller males can achieve high reproductive success through alternative mating tactics, which are favoured through sexual selection (Chapter 8). Consequently, while male reproductive success may be size-related, natural selection on females for growth to a large size will typically be stronger than in males.

Energetic costs of migration

The benefits of migration must be traded off against costs associated with the behaviour. A potentially significant cost of migration is a substantial increase in energy expenditure. Long-distance oceanic migrations and ascent of rivers are likely to elevate energy expenditure, while energy intake may be constrained or may cease altogether during migration (Chapter 6).

Migrating 2-year-old female sockeye salmon (*Oncorhynchus nerka*) enter the Skeena River in British Columbia at a weight of approximately 2270 g with a total energy content of 17 620 kJ. During the 500-km upstream migration, the fish do not feed, and arrive at the spawning grounds in Babine Lake with an energy content reduced to an average of 7950 kJ, having expended over 50% (Brett, 1986). Spawning reduces their energy content further to 3890 kJ, after which all fish die. For *S. salar* migrating in Norwegian rivers, the energetic costs of migration and spawning were lower than for *O. nerka*, both males and females expending 60–70% of their energy content. This species is iteroparous and not all individuals die after spawning.

In anadromous populations of *S. trutta*, while the benefits of migration manifest themselves in the form of enhanced growth, there are also energetic costs associated with the return to freshwater to spawn. Anadromous and resident populations of *S. trutta* in Swedish rivers at different altitudes showed a distinctive pattern. The density of juvenile migratory *S. trutta* was highest at low altitude sites, where the costs of migration were assumed to be lowest. At higher altitude sites, the density of juvenile migratory *S. trutta* was lower than that of resident *S. trutta*. At the highest altitude sites, migratory fish were absent (Bohlin *et al.*, 2001). This result suggests the migratory strategy may be constrained by the energetic costs of migration.

American shad (*Alosa sapidissima*, Clupeidae) are anadromous, with a distribution extending down the east coast of North America. Migrants do not feed when they re-enter freshwater to spawn. In rivers in Florida, in the southern part of their distribution, ambient water temperatures are relatively high (approximately 19 °C) and fish show a concomitant elevation in metabolic rates. Here, migrating fish utilise 70–80% of their energy reserves during migration and are semelparous (Glebe & Leggett, 1981). In cooler, northern New England rivers, the ambient temperature is lower (around 14 °C). As a consequence, the fish use only 40–60% of their energy reserves during migration and show a degree of iteroparity, with repeat spawners in the population (Glebe & Leggett, 1981). The energetic costs of migration in this species are not equal between the sexes; females incur a significantly greater energy cost than males, which may limit the scale of upstream movement (Leonard & McCormick, 1999). The costs of migration may be evident not only in terms of depleted energy reserves, but also in terms of reproductive investment. There was a somatic energetic cost of migration in Chinook salmon (*Oncorhynchus tshawytscha*, Salmonidae), but also a decrease in ovarian investment, primarily through a reduction in egg size (Kinnison *et al.*, 2001).

Survival costs of migration

Migration may substantially increase the risk of mortality, through either predation or possibly exposure to disease. The annual survival rates of *S. alpinus* were 0.68 for resident fish, but substantially lower, at 0.52, for anadromous fish (Nordeng, 1983). In *S. salar*, 20–25% of seagoing juvenile smolts in two Norwegian rivers were eaten by predators during their first week at sea (Hvidsten & Mokkelgierd, 1987; Hvidsten & Lund, 1988). In *S. trutta*, figures are comparable, with mortality rates of juveniles at sea estimated to be between 40% and 56%, while for non-migratory fish estimates vary between 16% and 47%, with most estimates lower than 40% (Jonsson & Jonsson, 2006). The extent of energy investment on migration correlates negatively with post-spawning survival within and among populations. In a comparison of 12 Norwegian populations of *S. salar*, the estimated energy investment on migration and spawning was negatively correlated with subsequent survival (Jonsson *et al.*, 1997).

Physiological adaptations associated with migration

For diadromous species, a significant challenge in undertaking a spawning migration is in developing the physiological capacity to osmoregulate in both fresh and saltwater. This challenge applies both to adults that undertake spawning migrations, and to the juvenile stages that complete the diadromous life cycle by returning either to freshwater, in the case of catadromous species, or to seawater in anadromous species.

Adult salmonids that make a spawning migration from seawater to freshwater undergo profound morphological and physiological changes. These changes involve shifting energy reserves to the gonads, particularly in females, as well as physiological preparation for reproduction, including the development of sexually selected traits such as bright colouration and a modified jaw (termed a kype) for fighting with rival males, as well as traits associated with migration itself, such as accumulation of fat reserves. Juvenile salmonids that migrate from freshwater to seawater must undergo a process termed smoltification, which enables them to make the transition from a freshwater phase to a marine one. Smoltification is a complex process that is mediated hormonally and involves changes in body pigmentation, visual pigments, carbohydrate and lipid metabolism and osmoregulation. Environmental cues that activate smoltification include photoperiod (Sigholt *et al.*, 1995), and possibly also temperature and feeding rate. In metamorphosing from a parr to a smolt, juveniles develop a silvery appearance, the result of an increase in the deposition of guanine and hypoxanthine platelets under the scales and in the dermis, which has an adaptive value in a marine environment through camouflage by reflection (Bone & Moore, 2008). Exposure of fish to thyroid hormones promotes silvering of fish, demonstrating that this is a hormonally mediated process that is probably controlled by temperature and photoperiod. Body composition also changes during smoltification, with a marked decrease in body lipid composition. Tissue lipid comprises a higher proportion of polyunsaturated fatty acids (PUFAs), which is typical of marine fishes. Smolts show a phase of rapid growth that results in a relative lengthening of the tail region, with the result that their body shape becomes more streamlined (Fig. 7.4). Gill function also changes, with a marked increase in the number of specialised chloride cells and ion permeability, both key adaptations to enable the move from hyposmotic freshwater to hyperosmotic seawater. In freshwater, the parr face a net ion influx, whereas in the sea, smolts face a net ion efflux. Further adaptations to contend with a changed salinity regime are shown in the gut, kidney and urinary bladder. Many of these physiological changes occur while fish are still resident in freshwater and before they commence their downstream migration to the sea (Aas *et al.*, 2010).

Figure 7.4 Atlantic salmon (*Salmo salar*) parr (top) and smolt (bottom), illustrating the change in appearance following transformation. (Source: © Stephen McCormick, University of Massachusetts Amherst, USA. Reproduced with permission of Stephen McCormick.)

Mechanisms of orientation and homing during spawning migrations

Spawning migrations can take place over enormous distances, often through seemingly featureless environments such as open ocean. Yet teleosts are able to navigate, like many migratory vertebrates, with great accuracy. Given the substantial costs of migration, there should be strong selection on the sensory mechanisms that serve to guide migrants and enable them to locate sites of reproduction. The cues that may play a role in migration are diverse, varying among taxa and habitat types. Research on the mechanisms by which fish are able to orientate themselves and home to sites of reproduction is probably best understood in salmonids fishes. Young *O. nerka* migrating to a marine or lake habitat have been shown to utilise a combination of polarised light and rheotaxic cues to orientate themselves, as well as cues from the Earth's magnetic field (Quinn & Brannon, 1982). On their return journey as adults to spawn, the mechanism by which adults navigate the open ocean to converge on their natal rivers is poorly understood, but the few data available are consistent with the adoption of geomagnetic cues combined with some form of cognitive map (Quinn, 1984; Wiltschko & Wiltschko, 2005). By contrast, the cues used by *O. nerka* to locate natal rivers are predominantly olfactory. The smolts imprint on chemical cues from their natal streams before departing the river (Dittman & Quinn, 1996).

In the hardhead catfish (*Arius felis*, Ariidae), the onset of its oceanadromous migration is determined by an endogenous clock that is linked to changes in day length. Orientation during migration to spawning grounds may be based on a sun-compass mechanism to maintain a preferred direction once migration is under way (Steele, 1985). A sun-compass mechanism requires an appreciation of the passage of time, requiring an internal clock, the angle of the sun and a compass direction. The adoption of sun-compass orientation in fishes has been shown in several fishes from marine, estuarine, lacustrine and riverine habitats (Goodyear & Ferguson, 1969; Smith, 1985; Levin *et al.*, 1992).

Yellowfin tuna (*Thunnus albacares*, Scombridae) are able to discriminate between different Earth-strength magnetic fields, which they use in navigation (Walker *et al.*, 1984). Several fish taxa that migrate in open oceans, including several tuna species, salmonids and eels, possess magnetite in their skull that may permit magnetoreception (Walker *et al.*, 1997). The otoliths of the teleost's inner ear are an alternative potential seat of geomagnetic sensing, which are primarily concerned with balance, but have been implicated as having a role in the detection of magnetic fields (Harada *et al.*, 2001).

For riverine teleosts, but also for some marine species, the direction of water movement provides a cue for orientation during spawning migration. Anadromous migrating adult salmonids show positive rheotaxis (i.e. a tendency to swim against flowing water) during upstream migration to spawn, while downstream migrating smolts show negative rheotaxis. Water flow may determine the speed, timing, duration and direction of migration. High rates of water flow may be required to stimulate migratory fish to begin ascent of a river. In migratory shads (Clupeidae) and salmon, migration into rivers appears to be initiated by high river discharge rates. Downstream migration of young stages is also often associated with passive drift during periods of high flow (Reichard *et al.*, 2002).

In the marine environment, water currents may provide cues for orientation as well as facilitating movement. The use of electronic data storage tags has provided evidence that plaice (*Pleuronectes platessa*, Pleuronectidae) in the North Sea utilise tidal movements to assist with migration to and from spawning grounds. The adoption of tidal stream transport involves a pattern of vertical movement that matches tidal cycles. Fish leave the seabed as tidal currents flow in their direction of migration. Once the tide turns, they settle on the seabed and wait until the tide again turns in the direction in which they are migrating. *P. platessa* are able to travel over several kilometres rapidly, while reducing their energy expenditure by 20–40% (Weihs, 1978; Metcalfe *et al.*, 2006).

Chemical stimuli play a key role in the orientation and homing of some teleosts during their reproductive migration. A significant role for olfactory cues in particular has been demonstrated in the migration of salmonid species. During their freshwater phase, prior to downstream migration, young salmon imprint on the characteristic odours of their home river. The memory of these odours later guides their migration with remarkable accuracy. Although a role for olfactory cues in salmon migration was proposed well over a century ago (Hara, 1994), their role was properly demonstrated in a series of ingenious studies in which hatchery-reared salmon were exposed to synthetic chemicals, such as morpholine and benzenethanol (Hasler & Scholz, 1983; Dittman & Quinn, 1996). Fish were exposed to these chemicals in a hatchery during smoltification and then released. As adults, these fish later homed in to artificial streams containing the specific odour with which they had been imprinted. These studies demonstrated that salmon imprint as smolts during a brief period prior to seaward migration. The transition from parr to smolt is associated in salmonids with increased levels of the hormone thyroxine, and this hormone may function in olfactory learning (Hasler & Scholz, 1983). Artificially increasing thyroxine levels in salmonids has the effect of increasing their long-term memories of an odour, and elevated plasma thyroxine is associated with an ability to imprint on odour in several salmonid species (Morin & Doving, 1992). The mechanism by which thyroxin mediates imprinting in salmon appears to be through its influence on neural development in the olfactory system, which thereby facilitates imprinting (Nevitt *et al.*, 1994; Dittman & Quinn, 1996; see Fig. 7.5). There is some evidence that salmon imprint on odours at a series of stages during their downstream migration, which they later track back in sequence. Such a

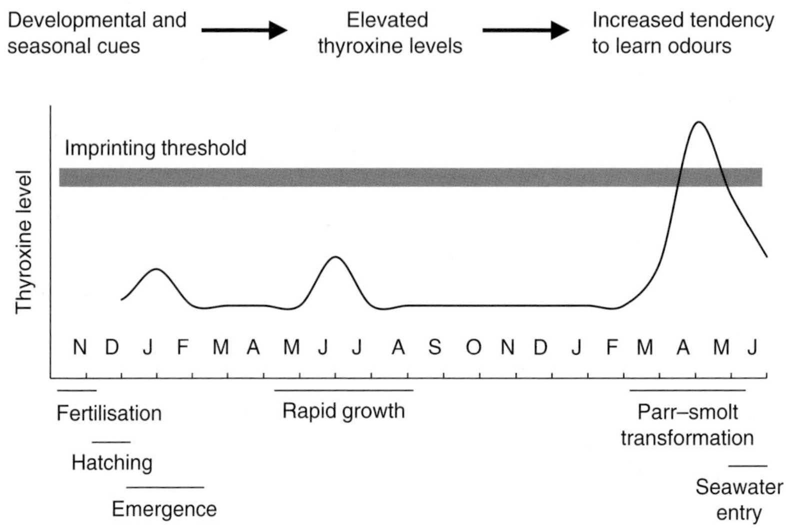

Figure 7.5 Hypothetical relationship among thyroxine levels, olfactory imprinting and migration in hatchery-reared coho salmon (*Oncorhynchus kisutch*) over a 20-month period. (Source: Adapted from Dittman & Quinn 1996. Reproduced with permission of The Company of Biologists Ltd.)

mechanism would involve transitory increases in thyroxine to facilitate imprinting at specific points during migration (Dittman & Quinn, 1996). The odours on which salmonids imprint during migration are a subject of debate. While site-specific odours derived from the underlying rock, soil, bacteria and flora appear to play a role (Hasler & Scholz, 1983), the odour of conspecifics belonging to the same population may also influence homing (Nordeng, 1977; Courtenay *et al.*, 1997). Support for the idea that salmonids recognise members of their own population comes from transplanted fish, which have a tendency to home to their ancestral rivers without having previously been exposed to them (Pascual & Quinn, 1994).

The mechanism by which salmon are able to imprint and home to natal rivers may involve odorant receptors in the peripheral nervous system. These comprise two main classes: olfactory receptors (ORs), which are involved in perception of environmental odours; and vomeronasal receptors (VNRs), involved in the perception of pheromones. A study of *S. salar* parr undergoing transformation to smolts showed a transient, 50-fold increase in the expression of odorant receptor transcripts (Dukes *et al.*, 2004). Notably, both olfactory and vomeronasal receptors were involved, implying that both environmental odours and pheromones play a role in imprinting. A later study on one non-anadromous and two anadromous populations of the same species identified significant differential expression of seven olfactory receptor genes in anadromous fish, but not in fish from the non-anadromous population (Johnstone *et al.*, 2011). These findings strongly implicate a role for specific olfactory receptors in the imprinting and homing of anadromous *S. salar*.

While learning through imprinting has been best studied in salmonids, there is also evidence for learning in spawning migrations in other species. A learned component to teleost orientation and homing permits an individual greater flexibility in response to rapid changes in the environment (Odling-Smee & Braithwaite, 2003). The Indo-Pacific brown surgeonfish (*Acanthurus nigrofuscus*, Acanthuridae) relies on landmarks in the environment during spawning migrations. Landmarks can be natural or man-made, but tend to be large (at least 1 m long and 0.5 m high) and therefore, presumably, obvious. Experimental displacement of landmarks influenced migration, although if landmarks were displaced relatively long distances (over 6 m) they were no longer used, suggesting that a series of landmarks are used simultaneously with no single feature of the environment serving solely for orientation (Mazeroll & Montgomery, 1998).

In such a phylogenetically diverse group as the teleosts, in which migration may have evolved on numerous occasions, there does not appear to be any single prevalent mechanism by which fish orientate themselves during migration. In some of the best-studied species, a combination of cues appears to be used, with the reliance on each changing at different stages of the migration, e.g. with rheotaxic cues being used during downstream migration of salmonid smolts and olfactory cues on the return migration.

Control of timing of migration

Spawning migrations are tightly linked to seasonality in reproduction (Chapter 6). Reproduction tends to be seasonal, even in relatively constant environments, presumably because gamete production is expensive and gamete release is selected to occur at the optimal period for current versus future reproduction (Roff 1992; Wootton 1998; see Chapters 6 and 11). Consequently, migration is predicted to be timed so that reproductive success is optimised. Migratory behaviour, like other aspects of teleost reproduction and maturation, is mediated hormonally (Chapter 5).

Water temperature functions as a cue for the onset of gametogenesis in many fishes (Chapters 3 and 6), but can also serve as a cue for the initiation of migration and in orientation during migration. Water temperature can initiate both upstream and downstream migration of teleosts in river habitats (Jonsson, 1991). In years with higher than normal temperatures, fish may begin migration earlier than in cooler years. The initiation of migration may be linked directly to gonad maturation and gametogenesis, but may have adaptive significance in matching reproduction with temperature-dependent seasonal peaks in productivity (Cushing, 1990; Sims *et al.*, 2004), as illustrated by *M. villosus* (Carscadden *et al.*, 1997), Pacific herring (*Clupea pallasi*, Clupeidae) and flounder (*Platicthys flesus*, Pleuronectidae) (Sims *et al.*, 2004).

One of the most detailed studies of the control of timing of migration in a fish is that of Baggerman (1957) on *G. aculeatus*. Migratory *G. aculeatus* winter in the sea or in brackish water and migrate into freshwater in spring. At the end of the spawning season, they migrate back to sea. Baggerman (1957) demonstrated that under exposure to a short day length and low temperature during winter, the secretion of thyroid hormone (principally thyroxine) in sticklebacks is suppressed, with the result that sticklebacks have a preference for seawater. In spring the photoperiod increases, which stimulates the production of both gonadotrophic hormones and thyroxine under the control of the anterior pituitary. If temperatures during early spring are low, the production of hormone is reduced but not suppressed. Conversely, warm spring temperatures will tend to elevate hormone production. The effect of thyroxine on the fish is to alter their salt water preference to a freshwater preference, with the fish showing an increasing motivation to enter rivers and begin reproduction. Thyroxine may also play a role in regulating the metabolic changes associated with increased locomotion necessary for upstream migration. The cues used by sticklebacks to terminate migration are not known, although it is notable that in some populations the fish cease migration when only a short distance inland (tens of metres at most), while others may migrate for 30 km or more before starting to breed. Downstream migration occurs when day length is shortening and water temperatures are declining.

Although seasonal migrations are often cued by temperature and photoperiod, other signals may also play a role. The timing of entry into rivers by salmonids, either from oceanic or lake habitats, is probably associated with photoperiodic, tidal and lunar cues (Dittman & Quinn, 1996; DeVries *et al.*, 2004). In species inhabiting tropical latitudes, changes in water level associated with patterns of rainfall typically serve as a cue to migration and reproduction rather than temperature and photoperiod (Lowe-McConnell, 1979). In the Amazon system, a range of teleost taxa make spawning migrations between main river channels and associated floodplain lakes in response to changes in water level, and the same is true of tropical African river systems (Welcomme, 2000).

There may be sex differences in the patterns of spawning migration. Electromyogram (EMG) telemetry was used to investigate the swimming speeds of *O. nerka* migrating to their spawning grounds through the Fraser and Nechako Rivers in British Columbia. Migrating male *O. nerka* expended more energy than females, although they did not tend to arrive at the spawning grounds earlier than females (Hinch & Rand, 1998). The greater energetic expenditure on migration by males may relate to the development of secondary sexual characteristics, including a pronounced humped back and kype jaw (see also Chapter 8), during migration that are not displayed by females and which make males less streamlined.

Human impacts on migration

Migratory fishes appear especially susceptible to human impacts, either through direct mortalities associated with harvesting, or indirectly through river engineering or habitat degradation. As a result, the activities of humans have been, and continue to be, detrimental to populations of migratory teleosts.

Fish undertaking spawning migrations often concentrate in large numbers at predictable times and locations and, consequently, have long been the target of fisheries. For example, migrating *T. thynnus* have been intercepted using the traditional *tonnara* trap system off the coast of Sicily in the Mediterranean Sea for centuries (Fromentin & Powers, 2005). Similarly, migrating salmonids and eels have been captured using nets and traps in European waters since prehistoric times. In some cases, migratory species have become severely depleted through overfishing. For example, the anadromous migratory hilsa shad (*Tenualosa ilisha*, Clupeidae) is an important commercial species of estuarine fisheries in South Asia (Blaber, 1997). This fish enters large rivers and migrates as much as 1200 km upstream to spawn. There are extensive and valuable, though largely unregulated, gillnet fisheries for migrating *T. ilisha*, as well as high seas fisheries that target stocks in the open ocean. Gillnets are deployed to intercept the fish as they enter estuaries, and the fish are also netted throughout the river systems through which they migrate. These fisheries have seen steady declines in catch as a result of overfishing (Hoq, 2007).

Another significant impact on migratory fishes comes from engineering schemes, particularly the construction of dams and weirs, but also the regulation of rivers for flood defence. The effect of these schemes is to modify the flow regimes of rivers, and to present obstacles to migrating fish which often result, predictably, in rapid and severe declines in the population size of migratory species. Some dams completely block the river channel, preventing fish from reaching spawning sites. Other dams may permit fish to pass to varying degrees, but so modify the environment (e.g. by creating reservoirs) that migration success is reduced. Dams can also cause changes to the thermal regime and oxygen conditions of rivers. Water released from dams can be depleted of oxygen, with devastating effects on fish populations downstream. Alternatively water can be supersaturated with oxygen, which is also damaging to some fishes. On the downstream migration from spawning sites, adults, and later juveniles, may be forced to pass through the turbines of hydroelectric generators that can kill or injure them. There are numerous examples of negative impacts of dams on migratory fish populations. On the Snake River, situated in the River Columbia Basin in the Pacific Northwest of the USA, the construction of hydroelectric dams has led to the virtual extinction of migratory *O. tshawytscha* (Karieva *et al.*, 2000). The same pattern in other fish taxa is repeated worldwide with depressing regularity (see reviews in Northcote, 1998; Lucas & Baras, 2001).

The negative impacts of dam building can be partially mitigated. Mitigation schemes can involve the construction of fish passes, of which there are a variety of forms (Lucas & Baras, 2001). Fish ladders comprise a stepped series of pools through which fish can pass, either by swimming over a weir or notch in the wall linking the pools or through a submerged pipe or opening. Fish locks operate by allowing migratory fish to accumulate in a holding pool, which is then isolated, and, like a conventional boat lock, flooded and the fish released into the upstream section of the river above the obstruction. Fish elevators similarly allow migratory fish to accumulate in a holding area from which they are lifted in either cages or baskets over the obstruction. More sophisticated elevators involve lifting fish in large water-filled tanks.

The impacts of habitat degradation on migratory fish can also be considerable. Boulder and gravel removal from riverbeds, siltation of spawning sites, water abstraction, and organic and inorganic pollution may all be detrimental to migratory fish at all life stages, but are particularly pronounced if they affect reproduction. Thermal pollution, either local or global, can affect the timing and pattern of fish migrations if the behaviour is associated with temperature cues. The large-scale impacts of climatic warming (whether human-induced or otherwise) on fish migration are difficult to predict, but are potentially profound in some species (Genner *et al.*, 2004).

Territoriality

Territorial behaviour is performed by a broad range of animal taxa. However, the concept of territoriality arose and continues to be developed primarily from observations of teleost and bird behaviour (Grant, 1997). Dominance and territoriality are linked. Dominance by certain individuals over others enables dominant individuals to obtain a disproportionate share of resources. Territoriality occurs when dominance is associated with attachment to a specific site (Wootton, 1998). A territory is functionally distinct from a home range. A home range represents an area used by an individual in its daily activities, but which is not defended (Helfman *et al.*, 2009). A territory may be contained within a home range, or a fish may temporarily abandon or restrict its home range while it engages in territoriality. In some cases, the home range and territory may correspond exactly, at least for the period for which a territory is defended. However, in most fish species, a territory owner utilises a home range, but with most of its activities concentrated within its territory.

Territoriality is of interest because it is a key element of the mating systems of some species (Hixon, 1987; see Chapter 8), and can play a role in regulating population size through limiting the number of individuals able to breed (Elliot, 1990; Both & Visser, 2003). Territoriality is defined differently by various authors; Maher & Lott (1995) identify

48 definitions. This situation has probably arisen because territories serve different purposes in different species (see Kaufmann, 1983 for discussion).

The concept of territoriality and definitions

One of the earliest and most simple definitions is that of Noble (1937), who defined territoriality as the defence of any area. Later modifications include the defence of any excluded area (Schoener, 1968) and any fixed exclusive area (Brown & Orians, 1970). Kaufmann (1971) provided a broad definition, comparable to a home range that represented an exclusively occupied area that might be maintained by passive avoidance. A 'core area' within the larger and loosely demarcated territory could represent a region of maximum utilisation from which conspecifics might be actively excluded, e.g. in the immediate proximity to a nest site. Davies (1978) provided a statistical definition of territoriality in terms of its effects on the distribution of animals. He considered territoriality as occurring whenever individuals are spaced out more than would be expected from a random distribution of suitable habitats. His definition ignores the manner in which spacing occurs, and does not necessarily imply exclusive use of areas. However, it has the advantage of focusing on the consequences of spacing out, equating nearest neighbour distances and territoriality.

The concept of territoriality is closely associated with intraspecific aggression, and is primarily associated with reproduction in fishes, although sometimes it is associated with feeding and occasionally with both. Interspecific interactions also need to be considered, as some fish species defend concentric or serial territories, i.e. a series of territory boundaries held against different species. Concentric territories have long been recognised as a feature of territorial coral reef fishes (Low, 1971; Myrberg & Thresher, 1974; Sale, 1975).

Adaptive significance of territoriality

Successful establishment of a reproductive territory, usually by males, is a key determinant of fitness in territorial species. Efforts to understand the adaptive significance of territoriality have focused on the various requisites of survival and reproduction (Verner, 1977). Resources secured through territoriality include spawning sites, cover and materials for nest building, as well as food for adults and offspring. The majority of studies of territoriality have addressed competition for feeding territories and, as a consequence, the theory of feeding territoriality is better developed than for reproductive territoriality. For simplicity, theorists often ignore the fact that territories can have multiple functions (but see Hixon, 1987). For example, the territory of *G. aculeatus*, defended during the breeding season, houses its nest and serves as an area in which to court females, but is also the area in which the male forages (Wootton, 1976). By defending a space around their nests, males obtain energy, minimise courtship interruptions and prevent rivals from stealing fertilisations and later their eggs (van den Assem, 1967).

Cost–benefit analysis and evolutionary stable strategy

The two most important conceptual advances in understanding the function of territoriality are the development of the 'economic defensibility' (Brown, 1964) and evolutionary stable strategy (ESS) models (Maynard-Smith, 1974). These approaches assume that individuals are territorial because it enhances their inclusive fitness and that maximisation of individual fitness is the ultimate driver that has resulted in the evolution of territoriality. Brown's (1964) argument is that animals only defend territories when the costs of territoriality are less than the benefits. His model is useful for making predictions about when animals should defend territories and about the size of the territory they should defend. It is also a powerful approach to understanding how different components of territoriality, e.g. territory size, might change in relation to environmental variables. Models of this type assume that natural selection will favour the size of territory that maximises an individual's reproductive success. If individual reproductive success is an additive function of gross benefits and costs, then the optimal territory size will be the one that returns the maximum benefits in relation to territorial costs. This type of model is applicable if the costs of territory holding are frequency-independent, relating simply to the energetic expenditure of patrolling and establishing boundaries (Parker & Knowlton, 1980; Schoener, 1987). An assumption of economic models is that animals will be able to alter their territory size without restriction (Hixon, 1980; Schoener, 1983; Grant, 1997; Adams, 2001). However, expansion of contiguous territories may be severely limited by the adjoining territories of neighbours, with the result that observed territory size will be smaller than predicted (Maynard Smith, 1974; Hixon, 1980; Adams, 2001).

Simple cost–benefit models ignore frequency-dependent selection, whereby the success of a given behavioural strategy is contingent on the number of other individuals engaging in the same or an alternative strategy. Frequency-dependent effects can often arise, and in these cases ESS models are more appropriate. An ESS is a strategy that, once fixed in a population, cannot be invaded and replaced by an alternative strategy. A powerful approach to understanding frequency-dependent selection is game theory. In a game theory context, an ESS is the strategy that enables an individual to achieve maximum success against the strategies of all other individuals. A game theory approach can be used to understand territoriality in terms of a sequential assessment game in which an individual engages in territorial contests with neighbours. Through a series of contests, an individual makes a judgment about its own fighting ability relative to that of its neighbours and thereby delineates the boundaries of its territory. As the territories of teleosts are frequently contiguous (Keenleyside, 1979), a game theory approach to modelling territoriality may offer a better representation of reality than optimisation.

Aggression and territoriality

Teleosts defend their territories through a combination of threat displays, chasing and direct aggression. In the European bitterling (*Rhodeus amarus*, Cyprinidae), males establish territories around freshwater mussels, which are used by females as spawning sites. Males are highly aggressive to rivals and express a suite of behaviours in response to intruding males. Responses range from displays involving fin spreading, through vigorous parallel swimming, to direct attacks involving biting rivals, slapping them with their tail fins and headbutting (Wiepkema 1961; Smith *et al.*, 2004a). However, the level of aggressiveness displayed by fishes can vary widely among individuals of the same sex, between sexes, among populations of the same species, and among closely related species.

There is a large and disparate literature on aggression, dominance and territoriality in teleosts. While each of these aspects of behaviour is related, the relationships between the different manifestations of aggression are complex. Most of these studies have been carried out in the laboratory. Some caution must be observed in generalising from laboratory studies to field situations, as environmental variables such as habitat heterogeneity and adult density can strongly determine social organisation and the intensity of aggression shown. In van den Assem's (1967) classic study of territoriality in *G. aculeatus*, it was shown that when a group of sexually mature males are placed in an aquarium with a homogenous substrate and limited cover, only some are able to establish territories. The males with the biggest territories were the most aggressive and females spawned most eggs with them. Furthermore, territory size was positively correlated with male success at hatching the eggs. Aggression was defined as the number of attacks initiated by the territory owner towards rival males. Van den Assem (1967) proposed a scenario for the evolution of territoriality. He argued that if variation in the ability to obtain a territory has a genetic basis, and if high levels of aggression result in bigger territories and males with bigger territories enjoy high reproductive success, then both aggressiveness and territory size will be under directional selection. However, while aggressiveness in this species has been shown to have a genetic component (Bakker, 1994a), other laboratory studies have failed to find a correlation between territory size and reproductive success (Sargent & Gebler, 1980; Sargent, 1982). The main difference between the studies of van den Assem (1967) and Sargent (1982) was that the latter provided the males with a more complex environment, giving rise to territories of varying quality. In the field, negative correlations between inter-nest distance (a measure of territory size) and aggression, and between aggression and the number of eggs females laid in male's nests were detected (FitzGerald, 1983). Furthermore, no correlation was found between a male's level of aggressiveness and his ability to obtain a territory in artificial pools housing potential rivals (FitzGerald & Kedney, 1987). There was also no correlation between a male's ability to win fights in dyadic encounters and his ability to obtain territories in more complex situations. This example indicates the difficulty of obtaining a relevant scale for quantifying behaviour and mitigates against generalisations about such complex phenomena as territoriality. Obviously an animal cannot obtain and maintain a territory without having the capacity to display aggression to intruders, but it is far from clear that the most aggressive and dominant males enjoy the highest reproductive fitness. One explanation for this complexity might be through a conflict between female mating decisions and male territorial dominance. It may not always be in a female's interest to mate with the most dominant and aggressive male (Qvarnström & Forsgren, 1998), though in some circumstances it may be, and in some instances females may be unable to avoid such matings (Casalini *et al.*, 2009).

Thresher (1978) suggested that the size of territory defended and aggression are independent. In a study of the threespot damselfish (*Pomacentrus tripunctatus*, Pomacentridae), a species in which males defend a territory both for feeding and reproduction, he found that seasonal adjustments to these two parameters differed and that the size of the area defended correlated poorly with attack rates against intruders. Territoriality in *P. tripunctatus* appears to be the result of two causally independent mechanisms, one dealing strictly with the amount of space defended and the second with the vigour with which such an area is defended. Thresher (1978) usefully introduced the terms 'attack readiness' and 'spatial commitment' to distinguish between the two manifestations of territorial aggression.

Patterns of occurrence of territoriality

A minimum of 266 teleost species belonging to 57 families, of the 515 extant families listed by Nelson (2006), show territoriality (Table 7.1), although this figure is probably a considerable underestimate. At least 128 species maintain a territory for the primary

Table 7.1 Occurrence of territoriality in teleosts with respect to territory function, environmental regime, adult feeding guild (modified from Keenleyside 1979), reproductive guild (from Balon 1975) and geographical region. Entries refer to the number of species in each category, these being mutually exclusive for any single category. In cases where categories do not sum to the number of species, information is incomplete.

		Territory function				Environmental regime				Adult feeding guild					Reproductive guild			Region	
Family	No. of species	Reproduction	Feeding	Both	Not known	Marine	Brackish	Anadromous	Freshwater	Grazer	Scavenger	Benthivore	Planktivore	Piscivore	Nonguarder	Guarder	Bearer	Tropical	Temperate
Lepisosteidae	1		1						1					1	1			1	
Anguillidae	1				1			1			1				1				1
Cyprinidae	10	8			2				10			5	5		9	1		1	9
Catastomidae	1	1							1	1					1				1
Characidae	1				1				1				1		1			1	
Lebiasinidae	1				1				1				1		1			1	
Callichthyidae	1				1				1			1				1		1	
Ictaluridae	2	2							2			2				2		2	
Malapteruridae	1				1				1			1						1	
Sternopygidae	1				1				1									1	
Umbridae	1	1							1			1				1			1
Osmeridae	1		1					1		1					1			1	
Salmonidae	11	1	10					8	3			1		10	11				11
Batrachoididae	3	1			2	3						3				3			3
Adrianichthyidae	2	1	1				2						2				2	2	
Poeciliidae	3	1			2				3				3				3	3	
Cyprinodontidae	7	2			5		7					7				7		7	
Gasterosteidae	6	6				1			5			6				6			6

(*Continued*)

Table 7.1 (Continued)

Family	No. of species	Territory function				Environmental regime				Adult feeding guild					Reproductive guild			Region	
		Reproduction	Feeding	Both	Not known	Marine	Brackish	Anadromous	Freshwater	Grazer	Scavenger	Benthivore	Planktivore	Piscivore	Nonguarder	Guarder	Bearer	Tropical	Temperate
Scorpaenidae	5	2		2	1	5								5			5		5
Hexagrammidae	4	3			1	4						4				4			4
Cottidae	8	3			5	4			4			8				8			8
Opistognathidae	1				1	1						1					1	1	
Centrarchidae	7	6	1						7			6		1		7			7
Percidae	3	3							3			3				3			3
Apogonidae	3	3				3							3				3	3	
Malacanthidae	1	1				1												1	
Gerreidae	1				1	1						1						1	
Sparidae	2		2			2								2	2			2	
Chaetodontidae	6	1	1		4	6						6			6			6	
Pomacanthidae	6	3			3	6				3		3				6		6	
Cirrhitidae	1	1				1						1						1	
Oplegnathidae	1				1	1						1			1			1	
Cepolidae	1				1	1							1						1
Cichlidae	41	39	1		1				41	15		19	4			25	14	41	
Embiotocidae	1				1	1						1					1	1	
Pomacentridae	29	7	9	1	12	29				22			7			29		29	
Labridae	19	8	1		10	19						19			3	3		19	

Scaridae	6	1	3	1	1	6				6					6			6	
Stichaeidae	3	2			1	3										1		3	
Zoarcidae	1	1				1						1					1		1
Pholidae	1	1				1						1				1			1
Anarhichadidae	2	1			1	2						2				2			2
Harpagiferidae	1	1				1													1
Tripterygiidae	5				5	5						5						5	
Labrisomidae	2				2							2						2	
Clinidae	1	1				1						1				1			1
Chaenopsidae	2				2	2							2			2		2	
Blenniidae	14	10		2	2	14						14				14		8	6
Callionymidae	1	1				1						1			1				1
Odontobutidae	1				1				1			1				1		1	
Gobiidae	9	4		2	3	7			2			9				9		3	6
Acanthuridae	4		4			4				4						4		4	
Belontiidae	4	3			1				4				4			4		4	
Balistidae	2			2		2						2				2		2	
Monocanthidae	1	1				1								1		1		1	
Ostraciidae	4				4	4						4			4			4	
Tetraodontidae	2	2				2						2				2		2	

purpose of reproduction, either for obtaining mates or raising young, or both. Territoriality exclusively for feeding, though well studied, is relatively infrequent, particularly among freshwater teleosts (Barlow, 1993). While this survey does not provide a comprehensive review of territoriality among taxa, it does highlight some salient points. First, reproductive territoriality appears to be broadly associated with species, from a range of environmental conditions, that guard nesting sites or provide parental care, whereas feeding territories are associated with non-guarders that are predominantly benthic feeders or grazers. A similar pattern to territoriality was observed by Smith & Wootton (1995a) and Grant (1997), who proposed that the predominance of reproductive territoriality in fishes compared with feeding territoriality may reflect the greater defendability of mates, nest sites and progeny compared with food resources, which tend to be less spatially and temporally predictable. Pelagic species are not associated with territoriality, presumably because the behaviour has a strong spatial focus, which is not relevant in the pelagic environment.

A further tentative conclusion that can be drawn from Table 7.1 is that feeding and reproductive territoriality appear to be distinct, and that few species appear to maintain territories for both purposes, though there are cases, e.g. some blenny (Blenniidae), goby (Gobiidae), damselfish (Pomacentridae) and cichlid (Cichlidae) species. Notably, feeding territories of fishes tend to be larger than reproductive territories (Grant 1997). This difference in territory size may be a consequence of the lower spatial clumping of food resources, or it may reflect the greater seasonality of reproductive territories; feeding territories, which are often permanent, may need to be larger to buffer the territory holder against long-term changes in resource availability (Grant, 1997). In those species in which territories are established predominantly for feeding, territoriality may still shape or influence breeding systems, as the extent to which male and female feeding territories overlap will strongly influence the potential for polygyny (Hixon, 1987).

Despite acknowledging what appear to be consistent patterns in the distribution of territoriality in fishes, a reflection of phylogenetic and adaptive effects, some caution needs to be exercised. These patterns may simply reflect the species most studied by fish ecologists. For example, it is notable that there are few records of territoriality among deep-water marine fish families, and this may simply reflect the paucity of basic data for these species, not an absence of the behaviour (Breder & Rosen, 1966).

Establishment and maintenance of territories

Reproductive territories serve as sites for courtship and mating, oviposition and parental care. The suitability of a territory for these functions may depend on the location of the territory and the resources within it. Several variables determine whether a fish should attempt to defend a territory. These factors are primarily environmental, and include the density of resources (it pays to defend resources at high densities), its spatial and temporal patchiness (it may pay to defend a spatially patchy resource that is always available), and its spatial and temporal predictability (it may pay to defend a territory that is spatially unpredictable, but temporally predictable). A further factor is the abundance of predators, as the value of a territory to an individual might be predicted to decline with an increasing risk of predation. The availability of refuges from predators influenced territory quality and female mating decisions in the tailspot wrasse (*Halichoeres melanurus*, Labridae) (Karino *et al.*, 2000). However, the impact of predators may not always be predictable. In the Lake Tanganyika cichlid *Cyprichromis leptosoma*, males establish territories in proximity to the predatory cichlid

Lepidiolamprologus profundicola. When engaged in parental care, *L. profundicola* are territorial and limit the risk of predation on *C. leptosoma* by driving away other predatory cichlids (Watanabe, 2000).

A further factor that may determine whether a fish should attempt to defend a territory is competitor density. The key role of density in reproductive territoriality is illustrated in a study by the bluehead wrasse (*Thalassoma bifasciatum*, Labridae), a protogynous hermaphrodite (Warner & Hoffman, 1980; see Chapter 10). In this species when population densities were high, territorial male mating success correlated negatively with density, but not with the availability of females, suggesting that territorial defence was the principal determinant of mating success. The way in which density affected territorial male mating success was through the additional time and energy males were forced to devote to territorial defence.

Prior residence of a territory can influence an individual's success in establishing and holding a territory. A prior residence advantage in contests for territories, such that residents outcompete intruders, can arise when residents have better knowledge of a territory or place greater value on it than intruders, and has been termed the payoff asymmetry hypothesis (Bradbury & Vehrenkamp, 1998). There is good evidence to support the payoff asymmetry hypothesis in teleosts. For example, Chellappa *et al.* (1999) experimentally investigated territorial interactions between pairs of sexually mature males of the freshwater angelfish (*Pterophyllum scalare*, Cichlidae). They showed that in all contests over breeding territories with a resident–intruder asymmetry, the resident fish won the encounter, regardless of their relative size.

There is a clear relationship between the level of territorial aggressiveness and the value of the resource defended. Male *G. aculeatus* protecting nests containing eggs are generally more aggressive than those with empty nests (FitzGerald & van Havre, 1985). Pressley (1981) found that male *G. aculeatus* were more aggressive in defending their nests if they contained more and older eggs. Territorial fish appear to evaluate the perceived threat that intruders pose to the defended resource using intruder characteristics such as the degree of colouration, body size and shape of the intruder. Thus, the territory holder can distinguish among conspecifics and heterospecifics, juveniles, rival males and females. After the nature of the intruder has been recognised, the level of territorial defence may be further adjusted to the behaviour shown by the opponent, such as movements, body orientation, and aggression. This effect is most pronounced in species that defend concentric territories. The aggressive response of territorial male beaugregory damselfish (*Stegastes leucostictus*, Pomacentridae), a small, coral reef species with paternal care, varied among potential egg predators, with the most aggressive response directed at *T. bifasciatum*, which was the principal predator of eggs, and a relatively weaker response to ophiuroid egg predators that posed less of a threat (Payne *et al.*, 2003). In the Brazilian damsel (*Stegastes fuscus*, Pomacentridae), conspecifics were egg cannibals and elicited the strongest territorial response from males (Osório *et al.*, 2006).

Body size is an important determinant of dominance in fishes and is a key factor, other variables being equal, in territorial contests. Larger individuals tend to occupy territories when space or resources are limiting. There was a highly significant effect of body size on dominance and territoriality in the rose bitterling (*Rhodeus ocellatus*, Cyprinidae). In this species, monopoly of territories was also correlated with reproductive success measured using paternity analysis (Casalini *et al.*, 2009). Male mating tactics in this species, like many others, is determined by male size (Chapter 8). Size, not age, determined male success in securing a territory in *G. aculeatus* (Dufresne *et al.*, 1990). In this species, there was a positive correlation between male body size and territory size

(Candolin & Voigt, 2001). Larger males also reduced their territory size less than smaller males as competition for territories intensified or risk of predation increased, suggesting that larger males may be able to occupy larger and more open territories both because of superior competitive ability and either their lower susceptibility to predation or a tendency to engage in more risky behaviour.

Physiological basis of territoriality

Our understanding of the physiological changes associated with territoriality in fish owes much to several decades of research on the African cichlid, *Astatotilapia burtoni* (Fernald, 2009, 2012; Huntingford, 2012; see also Chapter 5). This species has a lek mating system (Chapter 8) with maternal care of the offspring (Chapter 9). Males express two alternative phenotypes, depending on their social and reproductive status. Territorial males are brightly coloured and display a distinctive black (lachrymal) eye bar, forehead stripe, vertical black bars on the body, a black spot on the posterior tip of the opercula, and a large red-orange humeral patch above the pectoral fins (Fig. 7.6). Territorial males dig spawning pits, show social dominance and are reproductively active (Fernald, 2012). Non-territorial males lack the bars and spots of territorial males and, like females, are cryptic in colouration. They are socially subordinate and reproductively suppressed. Instead of preparing and defending a spawning site, non-territorial males tend to school with females.

Male *A. burtoni* can switch reversibly between the territorial and non-territorial phenotype as territory sites become available or disappear, through the death of territorial males or disturbance to spawning areas from extreme weather events and the incursions of terrestrial animals (Fernald & Hirata, 1977; Huntingford, 2012). When a

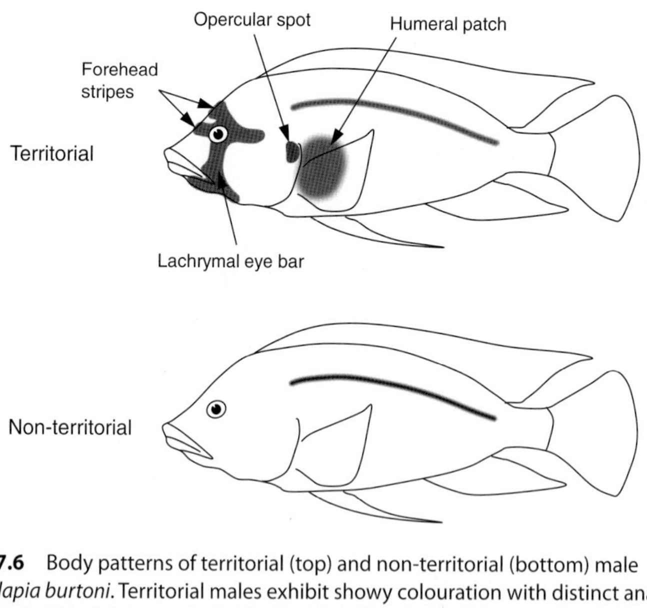

Figure 7.6 Body patterns of territorial (top) and non-territorial (bottom) male *Astatotilapia burtoni*. Territorial males exhibit showy colouration with distinct anal fin spots, vertical black bars on the body, an opercula spot, dark (lachrymal) eye-bar stripes and a red-orange humeral patch above the pectoral fins. Non-territorial males lack the bars of their territorial counterparts and are cryptically coloured like females. (Source: Fernald 2009. Reproduced with permission of Elsevier.)

male occupies a territory that has become available, he undergoes a marked change after only a few minutes. Males quickly display the lachrymal eye bar, body bars, and humeral and opercula spots that characterise territorial males, and rapidly increase their level of aggression towards rivals. Males also begin to express reproductive behaviour towards gravid females (Fernald, 2009). After a delay of a few days, new territory holders also express profound changes to their reproductive physiology. Their testes mature and begin producing androgens, and also grow, with their gonadosomatic index doubling within days of acquiring a territory. These changes are tightly coupled to changes in the hypothalamic–pituitary–gonadal (HPG) axis (Chapter 5). In the hypothalamus, neurons release gonadotrophin-releasing hormone 1 (GnRH-1), which promotes the release of gonadotrophins from the pituitary, stimulating gonad maturation and the production of the androgens testosterone (T) and 11 ketotestosterone (11-KT). Expression of GnRH receptors, particularly GnRH-R1, is up-regulated, with territorial males displaying larger GnRH1 neurons in the preoptic area of the brain (Davis & Fernald, 1990; Maruska & Fernald, 2011; Maruska *et al.*, 2011; Fernald 2012). Growth of GnRH neurons is rapid, with substantial increases observable within a day of a change in territorial status and the process complete in less than 1 week (White *et al.*, 2002). Full territorial physiological status is achieved after approximately 1.5 weeks.

Loss of a territory results in a reversal of these physiological changes. A male that is displaced from its territory by a rival quickly changes colour and alters its behaviour, becoming less aggressive and ceasing courtship. Complete reversal of the changes in brain structure and gonads seen in socially ascending males is completed in approximately 3 weeks (Fernald, 2009; Fig. 7.7). The significance of the difference in the interval in acquiring (1 week) and losing (3 weeks) structural changes in reproductive physiology may relate to the ecology of *A. burtoni*, which occupies a physically and

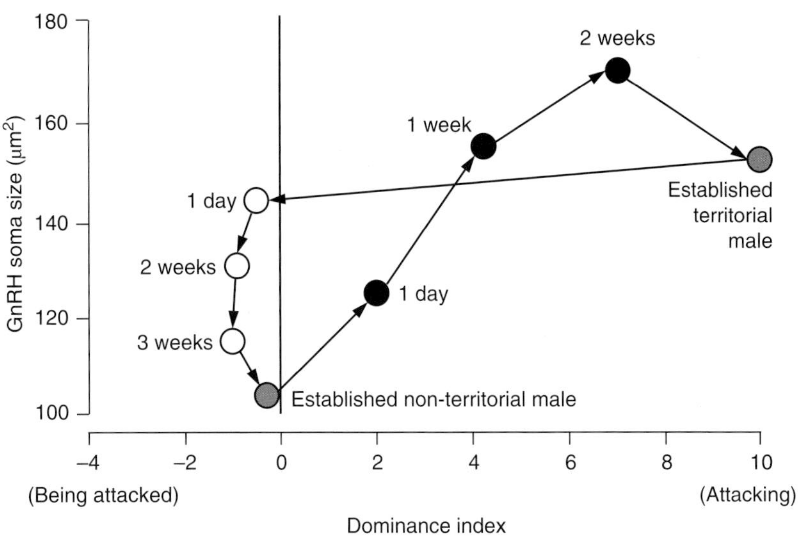

Figure 7.7 Illustration of immunoreactive gonadotrophin-releasing hormone (GnRH-ir) neuronal soma size in response to changes in male *Astatotilapia burtoni* in the frequency of aggressive and submissive behaviours expressed as a dominance index. Black circles indicate the fish is gaining territorial status and white circles indicate it is losing territorial status. Grey circles indicate a transition from one status to another. (Source: Fernald 2009. Reproduced with permission of Elsevier.)

socially unstable environment. Switching to a subordinate appearance and behaviour immediately after the loss of a territory probably reduces the risk of injuries to a losing male. However, retention of a fully functional reproductive system for a longer period may permit a male to quickly reoccupy a territory and resume reproduction, should a territory become available in the short term (Fernald, 2009).

Energetic costs of territoriality

Where they have been measured, the costs of territorial defence are substantial. Variation in the degree of territorial defence by an individual is related to their resource holding potential (RHP), which in turn is related to food availability. For a given density of male *G. aculeatus*, the proportion of males that established territories and built nests increased with ration size (Stanley & Wootton, 1986). When food was distributed so that some males regularly obtained more food than others, these males tended to increase the size of their territories, suggesting there may be benefits to defending bigger territories in some circumstances.

It is likely that individuals will have their own optimum territory size. One of the physiological factors that may differ among individuals is the level of internal energy reserves, reflected by the level of hunger. A hungry individual may place higher priority on feeding than on defence and consequently set its level of defence and the territorial boundaries lower than one in a better energy state. Small individuals may incur higher energetic costs than larger individuals (FitzGerald *et al.*, 1989) and may be forced to defend larger territories than bigger individuals. If the costs of defending large territories are too high, these smaller individuals may have to adopt alternative reproductive tactics (Chapter 8) or defer reproduction until they are larger.

An estimate of the expenditure of energy on territoriality has been made for the Amaragosa pupfish (*Cyprinodon nevadensis*, Cyprinodontidae) (Feldmeth, 1983). Male pupfish attempt to guard small territories within which females spawn. Dominant territorial males spend most of their time during daylight defending their territories against intruding males that attempt to sneak fertilisations. The energetic cost of defending a territory was approximately 20 J h^{-1}, double that of simply holding position within the territory. Interestingly, the energetic costs to males of engaging in alternative mating tactics were equal to or higher than the costs to territorial males (Feldmeth, 1983). Other evidence that territoriality represents a costly behaviour is demonstrated by territorial *G. aculeatus*, which spend protracted periods guarding their territory (van den Assem 1967), with consequent reduced growth (Wootton, 1994a), depleted energy reserves (Chellappa & Huntingford, 1989; Smith & Wootton, 1995b) and increased mortality (Chellappa *et al.*, 1989).

Genetic basis of territoriality

There are relatively few studies on the genetics of territoriality in fish (Noakes, 1986; Bakker, 1994b). Evidence that levels of territorial aggressiveness have a genetic basis can be deduced from the substantial inter-population differences in aggressiveness within a species and among closely related species. Territorial aggression can be quantified in a number of ways. A commonly used test to measure the aggressive motivation of territorial males is that developed by van Iersel (1958) to study aggression in *G. aculeatus*. In the test, a rival male, confined in a glass tube, is introduced into the territory of an isolated experimental male at a certain distance from its nest and the number of bites against the intruder are counted during a period of 3–5 minutes.

Variants on this method have become a standard technique for measuring the aggression levels of territorial fish (Sevenster, 1961; Itzkowitz & Slocum, 1995; Le Comber *et al.*, 2003). Other methods involve the use of dead specimens of rivals (Helfman, 1983), models (Rowland, 1999), video playback (Balshine-Earn & Lotem, 1998) and the recording of aggressive behaviours of isolated territorial males separated from neighbours by a barrier (Tavolga, 1956).

Gaudreault & FitzGerald (1985) used van Iersel's method to compare levels of aggressiveness toward conspecifics by territorial males of three related species of stickleback that breed sympatrically. *G. aculeatus* were more aggressive than the closely related blackspotted stickleback (*Gasterosteus wheatlandi*, Gasterosteidae) and nine-spined stickleback (*Pungitius pungitius*, Gasterosteidae), with the reproductive success of *G. wheatlandi* significantly reduced when in sympatry with *G. aculeatus*.

Significant differences in levels of aggression among related species have been reported in other taxa. The Chinese bitterling (*Rhodeus sinensis*, Cyprinidae) aggressively defend territories around freshwater mussels, which they use for spawning. The closely related *R. ocellatus* and European bitterling (*Rhodeus amarus*, Cyprinidae) are less aggressive (Pateman-Jones *et al.*, 2011). *R. sinensis* are mussel specialists, using a limited number of species, whereas *R. ocellatus* and *R. amarus* are generalists, making monopolisation of mussel spawning sites through aggression more difficult. As a consequence, *R. ocellatus* and *R. amarus* tend to compete for matings through (post-copulatory) sperm competition (Chapter 8), whereas *R. sinensis* compete through pre-copulatory territorial aggression (Fig. 7.8). Further evidence for a genetic basis to territorial aggression in fish comes from bidirectional selection experiments (Francis, 1984; Bakker, 1985). Selection experiments for and against aggression with *G. aculeatus* and paradise fish (*Macropodus opercularis*, Osphronemidae) were successful in demonstrating changes in aggression levels that differed from unselected control lines.

Territoriality and population dynamics

Territoriality is a significant component of reproduction in a number of fish species and also has implications for population size and regulation. Historically there has been considerable debate about the relative importance of density-dependent and -independent variables in regulating animal populations (Sinclair, 1989). Territoriality can act as a density-dependent factor in population regulation if there is a minimum territory size; if space is limited, territoriality will have the effect of limiting local population density. Thus, a population-level effect of territoriality could arise if individuals that are unsuccessful in obtaining a territory move elsewhere or are more vulnerable to predation. There may also be a population effect of territoriality if non-territorial individuals defer reproduction, are unable to reproduce at all, or have a lower reproductive success than territorial individuals.

Fretwell & Lucas (1970) proposed the 'ideal free distribution' to describe the pattern of settlement and distribution of individuals with an ideal knowledge of resource distribution and the ability to move freely between patches of resource, a concept that stemmed from the polygyny threshold model of Orians (1969). They proposed that, under these conditions, individuals are expected to distribute themselves so that the average fitness in all resource patches is the same. However, if access to resources is restricted by the territorial behaviour of other individuals, density in the best habitats can be lower than expected with an ideal free distribution. The problems of settlement in patches already occupied by territory holders led to the development of the 'ideal despotic distribution' (Fretwell, 1972), which has proven a helpful concept for

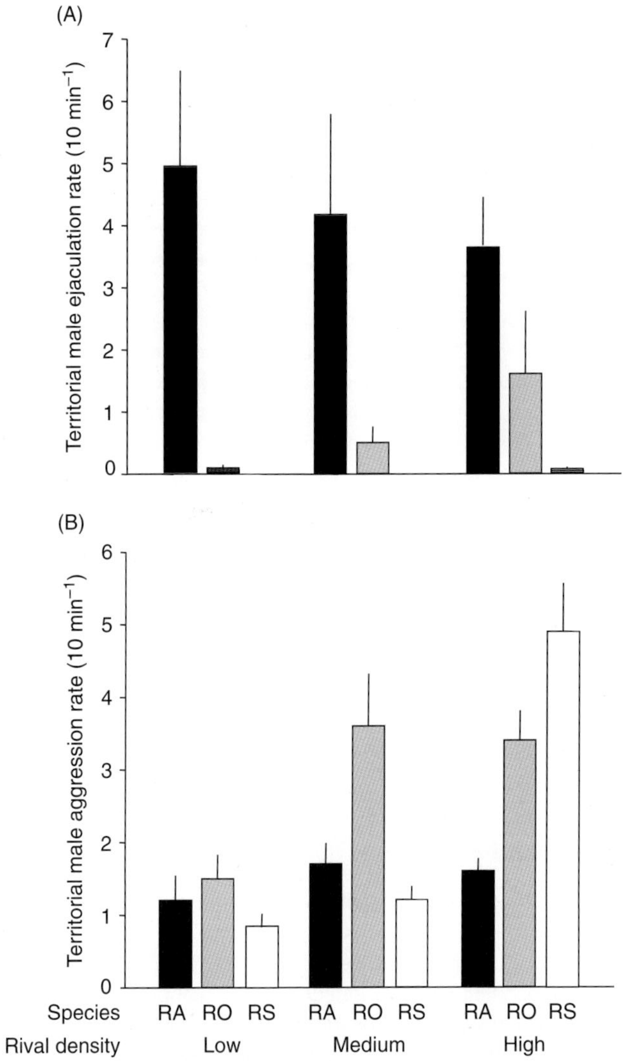

Figure 7.8 (A) Mean (+SEM) rate of ejaculation by territorial males of three species of bitterling [RA, *Rhodeus amarus* (black bars); RO, *R. ocellatus* (grey bars); RS, *R. sinensis* (white bars)], at three levels of rival density: low, one rival; medium, two rivals; high, four rivals. (B) Mean (+SEM) rate of aggression by territorial males of three species of bitterling at three levels of rival density. (Source: Adapted from Pateman-Jones *et al.* 2011.)

understanding the consequences of territoriality for density dependence and thereby its effect on population dynamics.

Individuals that are prevented from establishing a territory face two possibilities: exclusion from the territory of a dominant individual, or subordination within the territory. In teleosts there are few examples of subordination, except under laboratory conditions, where dominance hierarchies frequently establish but may simply be an artefact of confined laboratory conditions. Individuals that are unable to establish territories or remain in the territories of others often become 'floaters'. In some cases, floaters may be relatively inconspicuous and wait for a territory vacancy, which can occur with the death of a territory owner. In other species, floaters may continually challenge for ownership of a territory. Floaters have been described in the lek mating Mozambique tilapia (*Oreochromis mossambicus*, Cichlidae). In this species, males adopt one of three mating tactics: territoriality, sneaking and floating (Oliveira & Almada, 1998). Floaters lack the distinctive markings of territorial males, which appeared to limit aggression directed towards them; they are sometimes even courted by territorials. Floaters may be able to maximise their reproductive success by postponing reproduction and devoting energy to growth and survival. Van den Berghe (1992) showed that floating male peacock wrasse (*Symphodus tinca*, Labridae) did not lose weight during the breeding season as do territory owners, and were consequently more likely to survive the winter.

Dunham *et al.* (1995) used a population model to explore the floating strategies of fighting with a territory owner or waiting until a territory owner dies and then taking its place. Waiting was predicted to be adaptive when the injury rate during contests for territories, reproductive rate of non-territorials, and mortality rate of territory owners were high. Conversely, floating was predicted not to be an adaptive tactic when the rate of successful territory takeover, death rate of non-territorials and reproductive rate of territory owners was high. While it is useful for making predictions about the evolution of a floating tactic, this model did not consider the potential benefits of adopting alternative mating tactics by floating males. In *R. ocellatus*, floating males are able to reproduce by adopting alternative tactics, although their reproductive success is frequently lower than that of territory holders (Kanoh, 2000), though that need not always be the case (Chapter 8).

The 'buffer effect' was defined by Brown (1969) as the tendency of individuals to occupy the best sites at low densities, while at higher densities poorer sites are expected to be occupied. Patterson (1985) demonstrated a buffer effect in the convict cichlid (*Amatitlania nigrofasciata*, Cichlidae), a biparental, territorial, cave-nesting species. *A. nigrofasciata* were stocked in experimental pools with nesting areas of two types: caves and stones. Caves were preferred by fish and were occupied at a higher rate than stones. The stone areas were colonised later and by fewer fish. Defence of cave nest sites was significantly more aggressive than that of nests in the stone area. The study demonstrated that at a low population density, individuals occupied good-quality territories, while at higher densities, some individuals were forced into lower-quality sites. Reproductive success was not measured in the study. However, cave sites, which were less exposed to the risk of predation, may have conferred benefits to parents through higher survival of young stages.

For birds, territorial behaviour has been shown to play a key role in determining both breeding density and number of young produced under natural conditions (Brown, 1975). For teleosts, studies on *G. aculeatus* are particularly informative with respect to the role of territoriality in constraining reproductive output. Wootton (1985), drawing on van den Assem (1967) and his own long-term observations on a population of sticklebacks from the oligotrophic River Rheidol in Wales (Wootton & Smith, 2000), outlined a hypothesis of population limitation based on the idea that food is limiting. At low food levels, the number of males reaching sexual maturity is too low to saturate the available space for territories. Egg production is also low because some females fail to mature while those that do mature produce few eggs. Thus, low food levels lead to fewer offspring being produced. When conditions change and food levels are high, more males establish territories until males are establishing territories of the minimal acceptable size. When this point is reached no more territorial settlement is possible and the size of the male breeding population is limited by territoriality. Even though females are able to produce more eggs at high food levels, the capacity of males to fertilise and care for them is restricted, with the result that recruitment is limited by the density of males. However, as Wootton (1985) acknowledged, mortality during the first few months of life of a cohort can also determine population abundance in teleosts. In a study of an anadromous population of *G. aculeatus*, Whoriskey & FitzGerald (1987) argued that abiotic factors were probably more important than density-dependent biotic factors in limiting population growth at their study site, a highly productive saltmarsh in which it was unlikely that food was ever limiting.

Spawning site selection

The eggs of fishes are rarely deposited haphazardly; more usually they are placed at preselected sites that afford protection or a better-quality environment for development. Even species with buoyant or pelagic eggs that complete development in the water column may discriminate with respect to areas for spawning to maximise offspring development and feeding. Mechanisms for site selection are relatively poorly understood, although recent studies have shed light on this aspect of reproductive behaviour.

Sites for reproduction

Sites for oviposition may comprise anything from an unmodified substrate to specially constructed nests or bowers that serve as receptacles for eggs, as well as signals of parental quality that function in mate choice. Sophisticated site preparation and nest building are almost always associated with parental care (Chapter 9). The principal roles of site selection, modification and nest construction in teleosts are to ensure the availability of oxygen to eggs and young stages and to limit their predation. In some cases, access to an abundance of food during development may also play a role. Balon (1975) devised a classification of fish reproductive styles (termed 'reproductive guilds'), based on the site of reproduction and mode of parental care, which summarises the basic divisions among fishes (Table 1.2).

Site selection without habitat modification

In many cases, the sites for oviposition may be selected by an individual, but the environment in which spawning takes place is not modified by the fish. Many demersal-spawning fishes are broadly categorised as scatter spawners, whereby pairs or groups of fish shed eggs and sperm over a bare substratum or vegetation, seemingly haphazardly, and no protection or parental care is provided. The degree to which spawning sites are selected in these cases is rarely known. However, some studies have demonstrated a surprising degree of selectivity by teleosts that conform to this mode of spawning. In the zebrafish (*Danio rerio*, Cyprinidae), a small cyprinid native to the floodplain rivers of north-east India and Bangladesh, eggs are selectively deposited by females on a coarse substrate and in association with aquatic vegetation. When eggs were experimentally incubated in different substrate types, survival was greatest on the preferred substratum (Spence *et al.*, 2007). Similarly, in species that lay strings or clumps of eggs that, superficially at least, appear to be deposited indiscriminately, eggs are often deposited in locations and at water depths that tend to maximise development rate and survival. In the European perch (*Perca fluviatilis*, Percidae), a species that lays long ribbons of eggs, oviposition occurs on submerged vegetation of a particular structure and at specific depths (Gillet & Dubois, 1995). Eggs deposited on the lake bed, or that were dislodged from vegetation, had higher mortality rates than those on vegetation (Smith *et al.*, 2001a), possibly through suffocation or infection by microorganisms.

In many pelagic species that spawn in the open ocean, spawning site selection would appear irrelevant, although this may not be the case in all species. There is some evidence that the spatial, as well as the temporal, distribution of spawning is tailored to the requirements of early life stages (Robertson, 1991). The timing of spawning of pelagic fishes often corresponds with the period of peak food abundance for first feeding larvae (Leggett & DeBlois, 1994). There is also evidence that the location of spawning of fishes with pelagic early life stages may favour larval feeding (Robertson, 1991), and also perhaps avoidance of predators (Johannes, 1978; Robertson ,1991). In other species, the location and timing of spawning appear to be determined by cues that enable spawning among adults to be coordinated or to limit their risk of predation, rather than for any direct benefit to offspring (Colin & Clavijo, 1988; Robertson, 1991).

Site preparation

An extension to the selection of a location for oviposition is the active preparation of sites for egg deposition. Many species lay their eggs on a hard surface and often engage in cleaning activity to prepare the site for their eggs. The effect of cleaning is to remove algae, vegetation and sediment. The value of cleaning may be in ensuring adhesive eggs attach properly to the chosen surface and to remove material that may suffocate the eggs. Cleaning behaviour often involves a fish picking or biting at the substratum, and carrying material a short distance away. In some species of damselfish (Pomacentridae), individuals lie with their ventral side close to the substrate and briefly undulate their body, which may assist in clearing debris from the site. Pairs of whitetail major (*Stegastes leucorus*, Pomacentridae) prepare oviposition sites by scouring their intended spawning substrate. They pick up mouthfuls of sand and repeatedly spit them out against the site they have selected. After 'sanding' the substrate, they fan the site with their pectoral fins, presumably to remove the sand that has accumulated (Breder & Rosen, 1966). The preparation of spawning sites by cleaning is also common in substrate-spawning cichlids. One or both parents may engage in site preparation, although when both males and females are involved, it is often one sex that performs a greater share of the workload. Cleaning of a spawning site is usually directed at a hard inert surface, such as a rock. Some species also use living substrates (see also later). *P. scalare* cleans the leaves of aquatic plants as a site for egg deposition. The process of leaf cleaning in this species, and perhaps others, may be functional in preparing the site for receiving eggs, as well as playing a role in courtship and the establishment of a pair bond between prospective parents (Keenleyside, 1979; see Chapter 8).

Site excavation

In some species, a pit is excavated in which the eggs are deposited. The eggs may subsequently remain within the pit guarded by one or both parents. Alternatively, the pit may serve only as a spawning site, with the eggs removed and brooded elsewhere. This mode of spawning site preparation is frequently encountered with substrate-brooding cichlids, but also in North American sunfishes (Centrarchidae). In sunfishes, males exclusively excavate nests and these serve as receptacles for eggs during spawning, but also as sites for subsequent parental care. Nests are dug by males principally using their tails and vary in size and shape depending on species. Large objects, such as vegetation and woody debris, are removed by the males pulling, pushing or lifting the items away from the nest site in their mouths (Keenleyside, 1979). Nests are often clustered together, along river margins or in lakes, which may provide group defence for nesting males from egg cannibalism by females and attempts to sneak fertilisations by other males (Keenleyside, 1972; Chapter 8).

In substrate-brooding cichlids, nest digging can be extensive and protracted, sometimes lasting days and involving the displacement of large amounts of sand and gravel, vegetation and occasionally large pebbles and rocks in larger species. Nest sites are frequently excavated under natural or artificial structures, such as large rocks, submerged woody debris or discarded man-made objects. Artificial nest sites have been designed by fish biologists to facilitate the study of fish behaviour. In some cases, these artificial nest sites appear superior to natural sites. Artificial nest sites comprising sections of plastic piping attached to a heavy base have been used extensively to study the courtship and parental behaviour of *S. leucostictus* (Itzkowitz & Makie, 1986; Payne *et al.*, 2002). These artificial nests are usually occupied within hours of being deployed and are used in preference to natural nest sites. The reproductive success of males that use them is also superior to those using natural nest sites (Itzkowitz *et al.*, 1995).

Site preparation and young brooded elsewhere

In some cases, males may prepare a site that is used only for spawning, with eggs removed once they are fertilised. Examples include the spawning bowers of African mouthbrooding cichlids and a marine pufferfish. Bowers are constructed by excavating and heaping the substratum, usually sand and gravel, into distinctive mounds that are highly variable and often species-specific. Females may use features of bowers in their mating decisions, and bower construction is often associated with lek mating systems (Chapter 8). After spawning in the bower, the non-adhesive eggs are quickly picked up by the female (in the case of maternal mouthbrooders, such as *Oreochromis* spp.), by both parents (e.g. mango tilapia, *Sarotherodon galilaeus*) or only by the male (e.g. blackchin tilapia, *Sarotherodon melanotheron*). Extravagant bowers are also constructed by non-cichlids, including male pufferfish (Tetradontidae) (Kawase *et al.*, 2013; see Fig. 7.9).

Figure 7.9 Pufferfish bower nest. (Source: Kawase *et al.* 2013. Reproduced with permission of Macmillan Publishers.)

Site preparation and eggs buried

Where parental care is not provided, a spawning pit may be excavated and, after spawning, the eggs are buried, presumably for their protection during development. There are several examples of this mode of spawning site preparation, and it is exemplified by the salmonids, but also by some cyprinids.

In salmon and trout, the spawning pits, termed 'redds', are prepared by females in coarse gravel in rivers and sometimes lake beds. Redds are located in clean gravel exposed to running water. Fish that occupy lakes typically spawn in streams that empty into the lake, while river populations tend to prepare redds in headwaters or smaller tributaries. Some lake populations of trout and salmon do not have ready access to flowing water, and in these cases the fish spawn on stony wave-washed lake shorelines (Elliott, 1994). To prepare a redd, the female lies on her side and undulates her tail and body with powerful strokes, thereby displacing gravel and creating a shallow depression. As nest preparation progresses, the female performs 'probing behaviour' to test its shape and depth. She does this by flexing her tail upwards and lowering her anal fin into the gravel. As the nest gets deeper, the angle of the female's body during probing increases. Once the angle of her body reaches approximately 20°, digging ceases (Esteve, 2005). After spawning her eggs, which are immediately fertilised by one or more males, she repeats the digging process slightly upstream of her first redd, with the result that gravel is carried over the eggs, burying them. Females may spawn a series of egg batches, with the excavation of each new redd serving to bury the eggs from the previous spawning. Within the gravel, the large eggs of trout and salmon develop into an alevin (larval stage with a yolk sac) before emerging from the gravel as a juvenile. Selection of an appropriate substratum for a redd by the female is crucial if development of the eggs and alevins is to be successful. Gravel must be sufficiently coarse to allow proper ventilation of the eggs, but also sufficiently fine to prevent the eggs and alevins from being swept out of the gravel interstices during periods of high water flow. Preferred gravel size varies among salmonid species, but also within species, with larger individuals tending to prefer coarser gravel. The cues that stimulate females to choose a particular site for digging a redd are not completely understood. However, the rate of water flow, its oxygen content and the rate of groundwater seepage through the substratum all appear to play a role. The degree of hyporheic flow through the gravel may indicate to a female the extent to which eggs laid at a particular site will be exposed to water flow that will oxygenate them, thereby serving as a reliable indicator of likely egg and alevin survival.

Several riverine chub species of the genus *Nocomis* (Cyprinidae) prepare pebble nests and bury eggs in a fashion analogous to that of salmonids, though with notable differences. In this group, males use their mouths to first excavate a saucer-shaped pit in a riffle. Pebbles removed from the pit are heaped on either side of the pit in a crescent, creating a central trough. The next step is the construction of a platform in the pit. The platform is composed of pebbles carried back from the margin of the pit and deposited in the pit, initially in the centre and downstream end of the pit, then the upstream end until the entire pit is re-filled with pebbles. On top of the platform, the male constructs a spawning mound using pebbles collected from up to 25 m away. The completed nest comprises a circular mound of pebbles with a central trough on its upstream side (Fig. 7.10). If undisturbed, a male is capable of constructing a nest in a single day (Maurakis *et al.*, 1991). The functional value of first removing and then replacing the heap of pebbles appears to be to create a well aerated spawning surface that is free of sand and silt through which water can percolate. In the hornyhead chub (*Nocomis biguttatus*, Cyprinidae), spawning occurs in the central trough of the nest. After several spawnings, the male drops pebbles over the eggs and moves to a new site

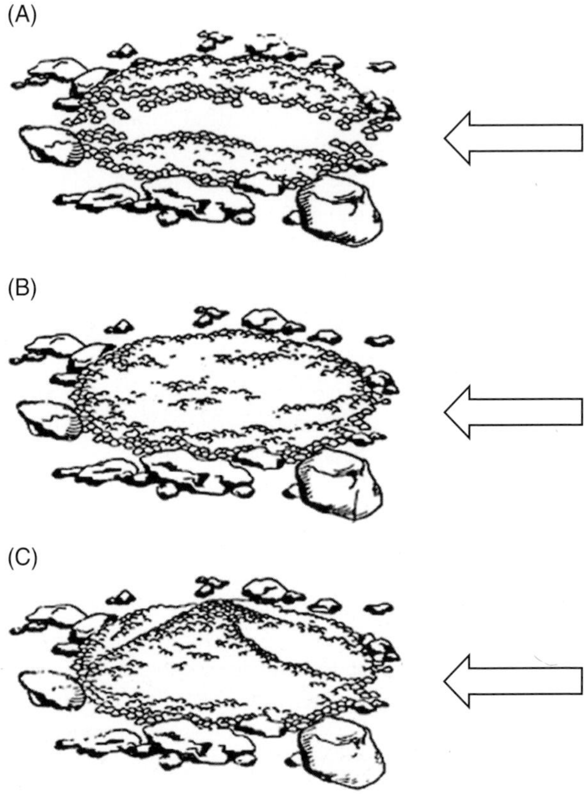

Figure 7.10 The stages in the construction of a pebble nest by *Nocomis* species: (A) concavity; (B) platform; (C) mound. Arrows indicate the direction of water flow. (Source: Maurakis *et al.* 1991. Reproduced with permission of American Midland Naturalist.)

to build another nest. In other species, the male excavates a depression in the spawning mound in which the pair spawn, with the eggs then covered by the male. Nest construction and spawning behaviour vary among chub species, and in some species the eggs are not actively buried by the male, e.g. river chub, *N. micropogon* (Cyprinidae), although the activities of other fishes using the nest for spawning tend to bury the eggs. Nest construction in other nest-building cyprinids, including the stonerollers (*Campostoma* spp., Cyprinidae) and cut-lips minnow (*Exoglossum maxillingua*, Cyprinidae), broadly match those of the nocomid chubs (van Duzer, 1939; Miller, 1962).

Excavation of shelters and tunnels

A wide range of teleosts deposit their eggs in shelters that they partially excavate or prepare entirely themselves. Spawning sites that are concealed in tunnels or caves may limit the risk of predation on eggs and young stages, but also suffer an attendant increase in the risk of suffocation or siltation (Keenleyside, 1979). Consequently, shelters and tunnels are often associated with parental care (Chapter 9). Many fish excavate shelters, often in or under submerged objects such as rocks or branches. The behaviour is seen in goby, blenny, cichlid, loricariid and cottid species, but is also a feature of many other families. In some species, the size and extent of excavations can be impressive. In the Orinoco sailfin catfish

(*Pterygoplichthys multiradiatus*, Loricariidae), breeding tunnels can be 1.3 m in length and 45 cm wide, and the tunnelling activities can undermine riverbanks (Nico *et al.*, 2009).

Some fishes adopt ready-made shelters for spawning in the form of old gastropod and bivalve shells. Male sand gobies (*Pomatoschistus minutus*, Gobiidae), which inhabit the shallow coastal waters of north-west Europe, typically excavate a nest cavity under the valves of dead bivalves, although they readily adopt other objects for this purpose. Several species of lamprologine cichlid nest inside empty gastropod shells. These fish often occur in association with 'shell beds', accumulations of empty gastropod shells. Pairs, or sometimes groups of fish, guard excavated shells and use them as spawning sites. Eggs laid deep within the shell, while protected from predators, risk suffocation and so are aerated ('fanned') by the parent (Chapter 9) to oxygenate them (Schradin & Lamprecht, 2000).

In some cases, nesting tunnels are prepared in association with other species. In a remarkable association, some species of goby establish cooperative relationships with tropical alpheid shrimps. The association between shrimp and goby may be obligate or facultative, and there may be as many as 130 species of goby belonging to 12 genera that form these mutualistic associations (Karplus & Thompson, 2011). The shrimp digs a tunnel in which both animals seek refuge and where reproduction in both species takes place (Karplus, 1987, 1992). Tunnels can be extensive and require constant maintenance to prevent their collapse. Shrimps are either blind or have poor eyesight and rely on the goby to alert them to predators. While outside the burrow, the shrimp keeps one of its antennae in contact with the goby. If a potential predator approaches the pair the goby flicks its tail, which serves as a signal to the shrimp to retreat to the burrow (Preston, 1978). Because shrimps have poor eyesight, they locate a suitable goby partner using olfactory cues, while gobies locate shrimps using visual cues (Karplus, 1987, 1992).

Nest construction

Some of the most sophisticated forms of nest construction by fish involve the use of specific materials selected and brought to the nest site or produced by the fish themselves. The construction of a nest in nest-building species represents a key component of reproduction, with nest attributes having direct implications for reproductive success. Nests can be considered as 'extended' phenotypic traits (Dawkins, 1982), which may have a significant genetic component, under both direct and indirect selection, as well as an environmental component that involves learning and cultural transmission. Perhaps the best-studied nests are constructed by the sticklebacks (Gasterosteidae). Male sticklebacks construct a nest with the aid of a glycoprotein glue secreted by the kidney (termed 'spiggin') to stick pieces of vegetation together (van Iersel, 1953). The nest is a relatively complex structure, and there is variation among species, and populations within species, in the form nests take. Nest construction is also variable within populations, and individuals may tailor nests to local conditions, for example in response to flowing or still-water habitats (Rushbrook & Barber, 2008). Nest construction varies among stickleback species. In *G. aculeatus*, the nest is sited at the bottom of a shallow pit excavated by the male or between stones, and the same is true of the closely related *G. wheatlandi*. In other species of stickleback, the nests are typically built among the branches of aquatic vegetation, although *P. pungitius* occasionally builds its nest on the substrate (Wootton, 1984a). The nest of a stickleback provides the focus for courtship, a receptacle for the eggs and a shelter for developing embryos. In addition, the nest may function in mate choice; variation in nest construction may be a reflection of male 'quality' (Barber *et al.*, 2001a; see Fig. 7.11) and may be used by females in mate choice decisions (Östlund-Nilsson, 2001; see Chapter 8).

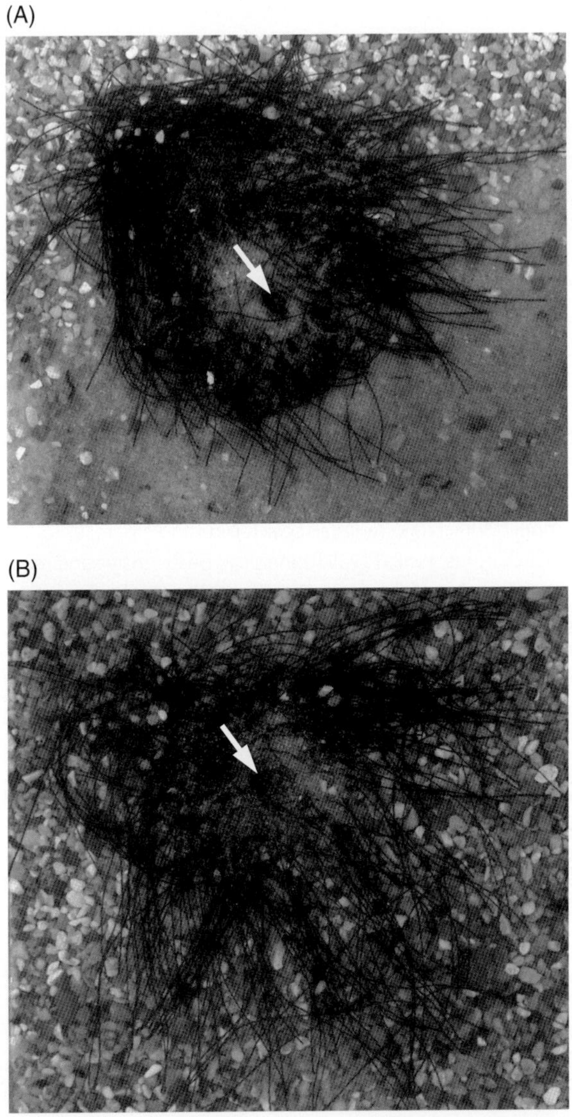

Figure 7.11 Photographs of the nests of male three-spined sticklebacks (*Gasterosteus aculeatus*) showing different degrees of 'neatness' and 'compactness'. Arrows indicate nest entrance. (A) A highly compact and neat nest; (B) a nest of low compactness and neatness. (Source: Iain Barber, University of Leicester, LE1 7RH, UK. Reproduced with permission of Iain Barber.)

Bubble nests are a highly specialised form of nest produced by several taxa of fishes, but are best known in air-breathing anabantoid fishes, although leaf fishes (Nandidae), callichthyid catfishes and synbranchid eels also raise their young in them. Bubble nests comprise a floating raft of mucus foam produced from a special gland in the mouth, usually by the male parent. Pieces of vegetation may be incorporated into a bubble nest for strength, and may be anchored to rooted plants. Bubble nests are associated with tropical fish living in oxygen-poor conditions, and their function appears to be to oxygenate developing eggs and embryos by raising them above the water surface, while also protecting them from desiccation. In the atipa (*Hoplosternum littorale*, Callichthyidae), a bubble-nesting armoured catfish, the bubble nest is heaped with vegetation which may help to limit temperature variation in the nest (Hostache & Mol, 1998; see Fig. 7.12).

Terrestrial spawning sites

In a few rare cases, fish spawn their eggs out of water (Martin & Carter, 2013). In the splash tetra (*Copella arnoldi*, Lebiasinidae), adhesive eggs are laid on the underside of leaves hanging above the water surface. To spawn, a male and female leap on to overhanging vegetation simultaneously, where they adhere through surface tension, aided by their broad pectoral fins. A small batch of eggs is deposited before the pair fall back into the water, the entire process lasting just a few seconds. These spawning leaps are repeated until approximately 50 fertilised eggs have been deposited on the vegetation (Krekorian & Dunham, 1972). The fish are capable of jumping to a height of 120 mm, though the eggs are typically 70–90 mm above the water level. After egg deposition, the male flicks water at the eggs using his tail fin to prevent them becoming dehydrated, at a rate of approximately once per minute. The eggs hatch after 3 days and the larvae are washed off the vegetation by the male's splashing behaviour (Krekorian, 1976).

A number of fish species deliberately leave the water to spawn. In the characin, *Brycon petrosus* (Characidae), which inhabits Neotropical streams, males and females beach themselves on damp river margins to spawn before returning to the river (Kramer, 1978). The eggs complete development amongst damp gravel and vegetation, with larval stages finding their way back to water after hatching. During spawning, both sexes appear to produce an unusual quantity of mucus, which may assist them in moving on land and avoiding desiccation (Kramer, 1978). Similarly, in some populations of *M. villosus*, fish spawn predominantly on gravel beaches in the intertidal zone (Nakashima & Wheeler, 2002). The eggs, which are negatively buoyant, stick to beach sediment and often become buried through wave action. The developing eggs are incubated in the substrate, with the hatched larvae eventually dispersing into open water when wave action eventually frees them (Taggart & Leggett, 1987). In *L. tenuis*, spawning is cued by the highest tides of the lunar cycle (Martin *et al.*, 2011; Chapter 6). Adults allow themselves to be carried onto beaches on a falling spring tide. Once on the beach, the adults actively crawl on the wet sand, with the females burying themselves tail first and depositing 1000–3000 eggs, which are fertilised by males that associate with them. The eggs complete development in damp sand above the water line, and hatch in response to the wave action of the subsequent spring tide, typically 10–14 days later. Hatching can be delayed for as long as 36 days post-fertilisation if the tide fails to reach the site where the eggs are incubating (Smyder & Martin, 2002; Moravek & Martin, 2011).

The habit of terrestrial spawning is also seen in cyprinodonts, galaxiids, tetraodontids and silversides. The eggs of many terrestrial spawning fish show adaptations to resist desiccation and also delayed hatching until the eggs are immersed (Martin, 1999). The adaptive advantages of terrestrial spawning, while not always evident, are probably linked to evading egg predation and possibly also to ensure adequate oxygenation or more rapid rates of development from exposure to elevated temperatures (Sayer & Davenport, 1991).

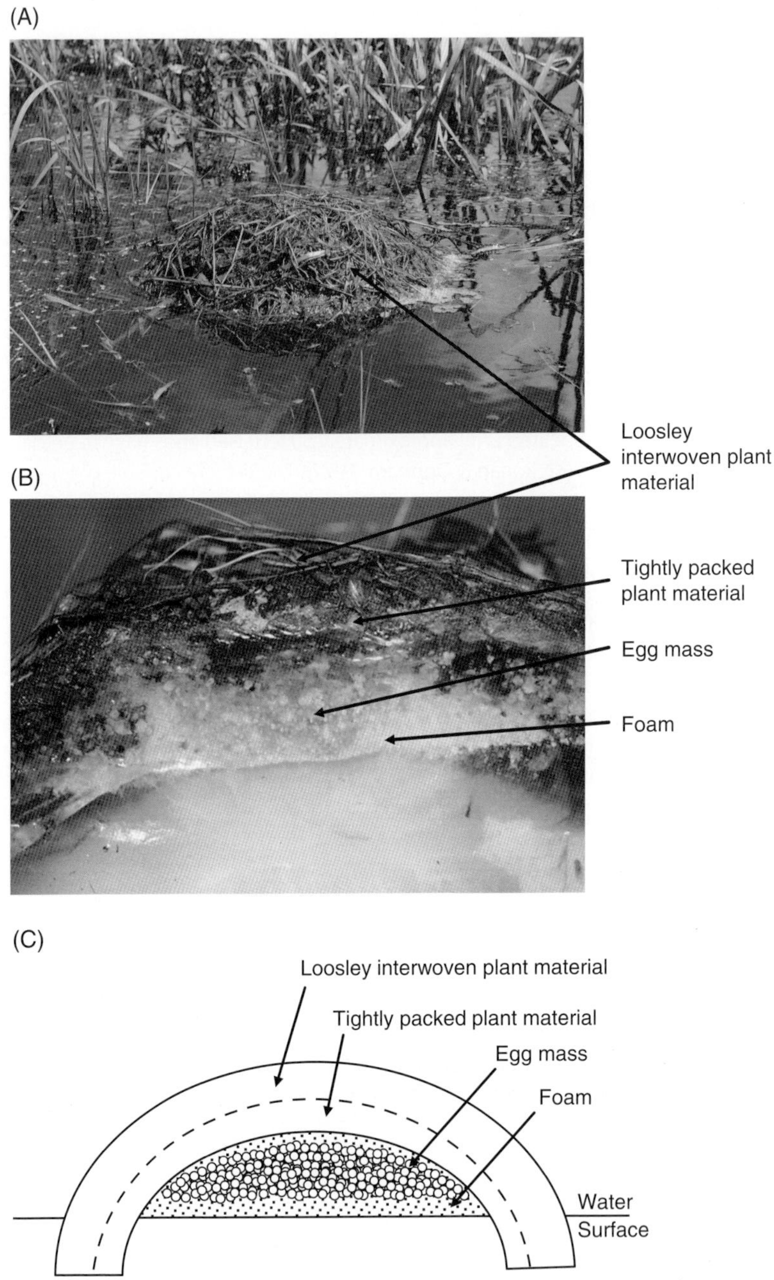

Figure 7.12 A cross-section of the bubble nest of the atipa (*Hoplosternum littorale*), showing layers of vegetation heaped above the eggs, with a mass of foam providing buoyancy. [Sources: (A, B) Jan H. Mol, University of Suriname/CELOS, PO Box 9212, Paramaribo, Suriname. Reproduced with permission of Jan H. Mol; (C) Adapted from Hostache & Mol 1998. Reproduced with permission of Elsevier.]

Cues used in spawning site selection

The adaptive basis to spawning site selection is to promote development and survival of early life stages, and the cues used for the selection of spawning sites reflect this, although the mechanisms of spawning site selection in fishes are generally poorly understood. Key variables in selecting spawning sites appear to be oxygen availability and, related to this, substratum type. In some cases, the degree of protection afforded by a site, as well as temperature, may play a role.

Oxygen availability plays a key role in fish behaviour (Kramer, 1987), including oviposition decisions. In the beaugregory damselfish (*Stegastes leucostictus*, Pomacentridae), oxygen availability determines the rate of development and survival of embryos, and spawning sites that maximise these parameters are preferentially selected, presumably using dissolved oxygen as a cue (Payne *et al.*, 2002). Similarly, in *R. amarus*, which uses the gills of living unionid mussels as a spawning site, females use the concentration of dissolved oxygen emerging from the exhalant siphon of a mussel to make oviposition decisions (Smith *et al.*, 2001b). Mussels vary in the efficiency of their uptake of dissolved oxygen. In addition, the presence of other bitterling eggs and embryos in a mussel's gills reduces available oxygen. Thus, by using oxygen as a cue for oviposition choice, female bitterlings place their eggs in mussels in which their progeny avoid competition with conspecifics (Spence & Smith, 2013).

In salmonids, females appear to 'test' the suitability of the substratum prior to digging spawning redds and depositing their eggs (Esteve, 2005). There is a link between female salmonid substratum preference, substratum permeability and the survival of embryos (Chapman, 1988; Bernier-Bourgault & Magnan, 2002; Brabrand *et al.*, 2002). These studies suggest a direct link between female spawning site preferences and reproductive success. A similar relationship has been found in other species, including European grayling (*Thymallus thymallus*, Salmonidae) (Gönczi, 1989), yellow perch (*Perca flavescens*, Percidae) (Robillard & Marsden, 2001), and rainbow smelt (*Osmerus mordax*, Osmeridae) (Lischka & Magnuson, 2006). In *M. villosus*, spawning site decisions appear to be related to both substratum type and temperature (Nakashima & Wheeler, 2002).

Spawning symbioses

Some fishes are unusual in using other animals as sites for depositing their eggs. These spawning relationships present the evolutionary biologist with some intriguing questions, although in all cases further research is needed to better understand the coevolutionary dynamics of these associations (Karplus, 2014).

One such association is that between protandrous anenomefishes (clownfishes) and sea anemones (Actiniaria). At least 25 species in the genus *Amphiprion* (Pomacentridae) and the spinecheek anemonefish (*Premnas biaculeatus*, Pomacentridae) associate with approximately 10 species of sea anemone in the Indo-West Pacific region (Fautin, 1991). Anemonefish guard an anemone in small groups, usually comprising a single dominant female in association with a harem of smaller males. The eggs are laid close to the anemone, which affords them protection through the stinging capsules (nematocysts) on the anemone's tentacles. Although stings from nematocysts are lethal to most fishes, anemonefish are unharmed. The mechanism by which anemonefish avoid being stung appears to relate to their skin mucus, which provides protection from nematocyst discharge, although the fish are susceptible to the toxin released by nematocysts (Mebs, 1994). Whether the mucus contains compounds that inhibit nematocyst discharge or simply lack compounds that stimulate discharge has still not been established (Mebs, 2009). As well as an innate protective mechanism, there also appears to be a role for conditioning. Anemonefish embryos appear to imprint on their host anemone and home to the same species when they settle after a planktonic phase. The association between fish and cnidarian is best considered a commensalism, although in some cases it appears to be a true mutualism, as anemonefish obtain protection for themselves and their eggs, and anemones, at least in some cases, benefit from aggressive defence of the anemone from fishes that feed on them (Godwin & Fautin, 1992). Anemones may also obtain a nutritional benefit from the fish through their excretion of ammonia, sulphur and phosphorous, which are utilised by the anemone's zooxanthellae (photosynthetic dinoflagellates contained in specialised cells of the anemone that provide it with energy; Porat & Chadwick-Furman, 2004).

Another fish that utilises a living invertebrate as a spawning site is the Japanese tubesnout (*Aulichthys japonicus*, Aulorhynchidae). Females place their eggs in the peribranchial cavity of a sea squirt (Ascidiacea) (Sasaki, 1977). Details of their breeding biology are poorly known, but it appears that there may be internal fertilisation in this species, although this is yet to be confirmed (Akagawa *et al.*, 2008). Females appear to carefully select a sea squirt, though it is not clear what cues they use in making this choice. The female releases a clutch of approximately 50 eggs in the water column, which they catch in their mouths. The female then places her eggs in the peribranchial cavity of the sea squirt (Akagawa *et al.*, 2004, 2008). The eggs are adhesive and remain fixed where they are placed (Breder & Rosen, 1966). An unrelated species, the sunrise (*Pseudoblennius percoides*, Cottidae) is also reported to spawn inside ascidians, although details of its biology have not been described (Munehara, 1991). There are reports of sculpin (Cottidae) and filefish species (Monacanthidae) that use marine sponges as sites for oviposition, although the nature of these associations are not well understood either and are incompletely described (Munehara, 1991; Akagawa *et al.*, 1995).

Perhaps the best understood of these spawning relationships is that between bitterling fishes (Acheilognathinae, Cyprinidae) and unionid and margaritiferid mussels. These mussels occupy benthic freshwater habitats and filter water through their gills to obtain food and oxygen and remove metabolic waste. Water enters the mussel gill through an inhalant siphon leading to the mantle cavity. Water is passed through the gills to their inner surface and is channelled via the water tubes to the suprabranchial cavity from which it is expelled through the exhalant siphon (Bauer & Wächtler, 2000). Bitterling comprise approximately 80 species and subspecies in three genera. All bitterling must lay their eggs in the gills of living freshwater mussels, in which their eggs and embryos are incubated. During the breeding season, males typically develop bright nuptial colouration and compete for territories around freshwater mussels. Female bitterling develop an unusually long ovipositor that they use to deposit their eggs deep inside the gill cavity of a mussel through its exhalant siphon (Fig. 7.13). Female oviposition decisions are based on both male and mussel quality (Casalini *et al.*, 2009; Agbali *et al.*, 2010). Clutch size varies among species, from one to six eggs per spawning in the European bitterling (*Rhodeus amarus*), to over 200 in a single spawning in the Itasenpara bitterling (*Acheilognathus longipinnis*) (Ogawa *et al.*, 2000). Males fertilise the eggs by releasing sperm over the inhalant siphon of the mussel.

Mussels often contain eggs and embryos from multiple ovipositions. The eggs and embryos of *R. amarus*, the only species so far studied in detail, can inflict significant fitness costs on mussels (Reichard *et al.*, 2006, 2007a) through damage to gill epithelium, competition with mussels for oxygen and disruption of water circulation over the gills (Stadnichenko & Stadnichenko, 1980; Smith *et al.*, 2001b). Bitterling embryos also compete with their host mussel for nutrients (Spence & Smith, 2013). Mussel species

Figure 7.13 A pair of European bitterling (*Rhodeus amarus*) about to spawn in a swollen mussel (*Unio tumidus*). The female *R. amarus* has an extended ovipositor and is examining the exhalant siphon of the mussel prior to oviposition. (Source: Carl Smith.)

differ in the anatomy of their gills, oxygen consumption and flow rate of water circulating through their gills (Smith *et al.*, 2001b; Liu *et al.*, 2006). Differences among mussels are important for the survival of bitterling eggs and embryos, as the most important cause of their mortality is ejection from the mussel gills and suffocation during incubation (Kitamura, 2005).

A rarely studied group that show similar oviposition behaviour to the bitterling are the lazy gudgeons (*Sarcocheilichthys* spp., Cyprinidae). This is a group of approximately 10 species of small freshwater fishes, inhabiting middle or lower reaches of rivers and lakes in East Asia (Zhang *et al.*, 2008). They, like bitterling, use unionid mussels as a spawning substrate, although instead of the female placing her eggs in the gills of the mussel through the exhalant siphon, the eggs are laid through the inhalant siphon into the mantle cavity (Ba⊠na⊠rescu & Nalbant, 1973; Luo *et al.*, 1977). Females possess a short ovipositor and the eggs are unusually large, approximately 4 mm in diameter, increasing to 7.5 mm prior to hatching. These fish, like bitterling, appear to show selectivity in the mussels they use (Barabanshchikov, 2004).

Another group that uses other animals for oviposition are some members of the marine snailfishes (Liparidae). Species in the genera *Allocareproctus* and *Careproctus* deposit their eggs in the gill cavities of lithodid king crabs. Liparid egg clutches are large, over 800 eggs, and more than a single clutch may be laid in the gills of a crab. The relationship appears to be a wholly parasitic one. Snailfish eggs compress the crab's gills, causing gill necrosis and, in some cases, entire loss of the gills from one side of the body (Love & Shirley, 1993; Somerton & Donaldson, 1998). Mortality as a result of parasitism by snailfish is frequent, with approximately one-third of infected crabs dying as a result of parasitism (Somerton & Donaldson, 1998). The impact of snailfish on host populations is unclear, but in some regions the prevalence of infection by snailfish can be substantial (Poltev & Mukhametov, 2009).

Female snailfish possess a pronounced ovipositor, though how they deposit their eggs in the branchial cavity of a host crab is unclear, as is the means by which the eggs are fertilised by the male. Snailfish are highly selective of crabs for oviposition, with those that have recently moulted exclusively selected. The reason for this preference is that if the crab moults, the eggs of the snailfish risk being lost from the host crab's gill. Snailfish egg development is shorter than the length of the inter-moulting period of the crab. Thus, eggs deposited on recently moulted crabs complete development before the moult occurs. The mechanism by which snailfish identify whether a crab has recently moulted is unclear, but it may be based on olfactory cues linked to moulting hormones. Male crabs use this mechanism to identify pre-moult females for mating (Love & Shirley, 1993). The mechanistic and evolutionary basis to this parasitic relationship is intriguing but, sadly, remains poorly understood.

In all cases where fish have used other species as spawning sites, the benefits to the fish appear to be through protection, strikingly so in those that use mussels and crabs for oviposition, and also improved egg ventilation. Notably the sites for egg deposition are often gill chambers where a continuous current of water will supply the eggs with oxygen and remove metabolic waste.

Conclusions

While behavioural ecologists have developed a sound understanding of the adaptive basis to the behavioural aspects of migration, territoriality and site selection, this has largely been accomplished in isolation from other aspects of teleost reproduction. The challenge remains to combine the behavioural components of reproduction, which often have a demonstrable link to reproductive success, with the endocrine and neurophysiological mechanisms that control them. Notably, research with teleost models, including *A. burtoni* and *G. aculeatus*, has gone some way to integrating these discrete fields, although our overall understanding remains superficial. After completing migration, establishment of a territory, and selection and preparation of sites for spawning, a fish must accomplish mating, a process summarised by its mating system.

Chapter 8

Mating systems and sexual selection

Reproductive Biology of Teleost Fishes, First Edition. Robert J. Wootton and Carl Smith.

Introduction

The way in which mating is accomplished is summarised by a species' mating system. Selection that acts on differences in fitness among individuals due to the number and identity of their mates is termed sexual selection, and it is a pervasive and powerful form of selection that shapes the evolution of mating systems (Jennions & Kokko, 2010). Sexual selection tends to act more strongly on males, a consequence of males producing numerous small gametes (spermatozoa), while females, by definition, produce a relatively smaller number of large gametes (eggs), (Chapter 4). The concept of sexual selection was formulated by Darwin (1871) in his attempt to apply natural selection to humans, specifically to explain differences among human races. In doing so, he identified a framework for explaining the evolution of intrasexual competition and intersexual mate choice. Sexual selection is intimately related to the evolution of mating systems, because mating systems strongly influence – and are influenced by – the strength of sexual selection. Parental care is widespread in fish and comprises an important component of mating systems (see Chapter 9 for a separate discussion).

Sexual selection differs qualitatively from natural selection, which acts independently of the number and identity of an individual's mates, because it is analogous to a zero sum game: one individual that is favoured by sexual selection gains a fitness advantage at a cost to a rival. A result is that sexual selection tends to be strongly directional, in contrast to natural selection, which more frequently tends to result in stabilising selection (Jennions & Kokko, 2010). There are several hypotheses to explain the mechanisms by which sexual selection occurs. The field of sexual selection theory is an active one and studies on teleosts have contributed substantially to its development.

Classification of mating systems

A mating system defines the processes by which fertilisation is achieved, and includes courtship (including coercion and forced copulations), mate choice (both pre- and post-copulatory), pair-bond formation, fertilisation and parental care. Mating systems are a prominent feature of many teleosts and are characterised by striking variability among species and plasticity within species.

Mating systems have been categorised on the basis of the number of partners with which the members of a given sex typically mate. In species with parental care, these categories correspond with which sex provides parental care and which deserts its partner after mating (Maynard Smith, 1977). This approach generates four main classes of mating system: monogamous (one male mating with one female, with both sexes providing parental care), polygynous (one male mating with two or more females, with males typically deserting), polyandrous (one female mating with two or more males, with females typically deserting) and promiscuous (both sexes mating with two or more mates and neither providing parental care). A form of promiscuity, but with the possibility of mate selection, is termed polygynandry (Table 1.1).

An alternative approach is to classify mating systems by the degree of relatedness between mating partners. Thus, mating may be random with respect to genetic relatedness, or may show negative or positive assortative mating (Shuster & Wade, 2003). A classification based on genetic relationships has most relevance to plants, the mating systems of which are dominated by the fitness consequences of self-fertilisation and outcrossing, and how these are influenced by pollen transmission (Barrett & Harder, 1996).

Emlen & Oring (1977) made one of the first attempts to systematically categorise mating systems by recognising the influence of ecological factors. In particular, they identified male monopolisation of resources and mates as fundamental to the evolution of mating systems. Mating systems were categorised on the basis of the extent to which males can directly monopolise access to females or the resources necessary for reproduction. For example, female (or harem) defence polygyny is a system in which males are able to mate with more than a single female by guarding groups of females and preventing other males from mating with them, while resource defence polygyny involves males monopolising key resources, such as breeding sites, that give the male exclusive access to females. Under this classification, mating system evolution is an outcome of the spatial and temporal distribution of mates and resources. The major failing of this approach to classifying mating systems is that it overlooks the role that females play in sexual selection, assuming a male-biased causal framework and ignoring the role of female life history, behaviour and distribution, and the role of cryptic female choice.

An attempt to accommodate female role has been made by classifying mating systems from the standpoint of each sex separately. Hence, a promiscuous mating system, in which both sexes mate with several partners, can be classed as polyandrous from the female perspective, and polygynous from the male perspective. The shortcomings of this approach are that a polyandrous system, in which a single female mates with several males, while the males mate with only a single female, would be classed simultaneously as male monogamy and female polyandry. Thus, from the female perspective, classic polyandrous and promiscuous mating systems are indistinguishable (both are classed as female polyandry). Yet the strength of sexual selection on females in a classic polyandrous mating system is predicted to be stronger than in a promiscuous system (if the sex ratio is at unity), because in classic polyandry females compete for access to males, which is not normally the case in a promiscuous mating system.

An additional oversight in attempts to classify mating systems is the potential role of density. Competition for mates, as with competition for any other resource, may be sensitive to density (Kokko & Rankin, 2006), yet density is rarely acknowledged as a variable that shapes mating system evolution. The impact of density might be through encounter rates, both with rivals and potential mates, with direct outcomes for sexual conflict, mating rates, mate choice decisions and parental care (Kokko & Rankin, 2006).

A further advance in the classification of mating systems was proposed by Shuster & Wade (2003). Their approach was to explicitly examine the process rather than the outcome of mating system evolution, which more closely corresponds with the view of Bateman (1948), who argued that mating systems are a product of sexual selection, in contrast to the view of Darwin (1871), who held that mating systems were primarily responsible for generating sexual selection. Shuster & Wade (2003) incorporated the spatial and temporal distribution of females ready to mate, in addition to features of male and female life-history and reproductive behaviour, in understanding how the 'opportunity for selection' acts on the sexes. The opportunity for selection is a standardised measure of the intensity of sexual selection, which typically acts most strongly on males as a result of the higher variance in male reproductive success. Using this model of mating system evolution, Shuster & Wade (2003) identified 12 major categories of mating system, and over 40 subcategories. The model of mating system evolution proposed by Shuster & Wade (2003) has proved controversial (e.g. Gwynne & Kelly, 2004), but represents the only fundamental innovation in our understanding of mating systems since Emlen & Oring (1977).

Ultimately, our understanding of mating system evolution may not be furthered by attempting to construct discrete categories, because by doing so, an implicit assumption is that a species expresses a particular class of mating system that is more or less fixed. In reality, the way sexual selection acts on males and females varies continuously within broad limits, both within and among species. Thus the mating system that emerges in a given population will be sensitive to variation in the abiotic and biotic variables (including density) that impinge on variance in reproductive success in males and females. A better representation of a mating system comes from quantifying the Bateman gradient for each sex, i.e. measuring the relationship between mating success and reproductive success. Identifying how selection operates can then be understood by examining the mechanisms that generate covariance between trait values and mating success (Jones & Ratterman, 2009).

Sexual selection and theories of mate choice

Mate choice has the effect of imposing sexual selection on the opposite sex. It is a key aspect of mating system evolution and is responsible for the evolution of elaborate ornaments and behaviours, typically in males, that influence female mating patterns. Throughout this chapter it will be assumed that females are the choosy sex.

There are exceptions to this general role, and they are discussed where appropriate. The term 'mate choice' implies active mate discrimination, but also encompasses what are termed 'mating biases'. Thus, females may resist the attentions of males and only mate with those that are able to overcome their reluctance. Female choice might also be cryptic, taking place after mating and biasing the fertilisation success of particular males (Kokko *et al.*, 2003). Cryptic female mate choice may particularly be a feature of teleosts with internal fertilisation (Eberhard, 1996; see Chapter 10). Thus mate choice, mate preference and mating bias are used interchangeably, and imply simply that, when expressed, they generate a higher mating success for a subset of males. Mate choice does not necessarily imply active sampling or discrimination among prospective mates, although this aspect of mate choice is also under sexual selection.

Direct intersexual selection

Sexual selection may act directly on female mate choice for male traits that increase her lifetime reproductive output. Thus, if males provide a resource that yields a fitness benefit to the female or her young, selection will act on the female tendency to mate with males that provide those resources. In teleosts, mating biases towards males with high fertility, with high-quality territories or nest sites for oviposition, or that can provide effective parental care to offspring, can potentially undergo direct selection.

There is good evidence for direct mate choice in teleosts. In the river bullhead (*Cottus gobio*, Cottidae), males dig a cavity nest, often at the base of a rock, where they guard and aerate a series of egg clutches spawned in the nest by females. Parental care of a clutch of eggs lasts approximately 4 weeks and the male fans the eggs throughout this period using his pectoral fins. Males perform courtship to females, and females are choosy about the males with which they mate (Morris, 1955). Females had a significant mating preference for large males, though they avoided males 150% larger than their own body size, possibly because of the risk of cannibalism by the male during courtship (Bisazza & Marconato, 1988). This mating preference was proposed as providing a direct fitness benefit to females, because larger males were more successful at defending their offspring from nest predators, particularly other *C. gobio*. Larger males may also be more effective at fanning the developing young.

Another example is the European bitterling (*Rhodeus amarus*, Cyprinidae), a species in which females lay their eggs in the gills of living freshwater mussels (Chapter 7). Bitterling eggs and embryos complete development inside the mussel gill cavity without parental care, which takes approximately 1 month (Smith *et al.*, 2004a). The survival and development rates of bitterling eggs is density dependent; at high densities survival is low and development rate are reduced, and females prefer spawning in mussels that contains few eggs and embryos (Smith *et al.*, 2001b). Male bitterling guard territories around freshwater mussels and court females, leading them to a mussel within their territory. Females mate with males that are able to successfully defend a mussel from rivals. Only the largest and most dominant males in a population, which are often the oldest individuals, have possession of mussels (Smith *et al.*, 2002a). Consequently, females obtain a direct fitness benefit from mating with larger older males, and there is direct selection by females, through mussel preference, of males with large body size and longevity. Other examples of direct fitness benefits to females from mate choice include territory quality in three-spined sticklebacks

(*Gasterosteus aculeatus*, Gasterosteidae) (Sargent, 1982) and male parental quality in the sand goby (*Pomatoschistus minutus*, Gobiidae) (Forsgren, 1997).

Elaborate male ornaments and behaviours are often linked with benefits to females through indirect selection (see the section 'Indirect intersexual selection'). However, they may also signal direct benefits to females. In the bicolor damselfish (*Stegastes partitus*, Pomacentridae), male courtship accurately signalled male parental quality (Knapp & Kovach, 1991). In a population of *P. minutus*, male body size signalled how much a male subsequently fanned eggs in his nest (Lindström & Hellström, 1993), although this pattern was not consistent in another population (Forsgren 1997). In the guppy (*Poecilia reticulata*, Poeciliidae), males with large areas of orange colouration had faster-swimming and more viable spermatozoa than males with less pigmentation (Locatello *et al.*, 2006). In this species, male colouration signalled fertility, with direct selection for this trait enhancing female reproductive output.

Despite several studies providing convincing evidence for direct fitness benefits of mate choice, the challenge remains to design studies that unambiguously distinguish whether male traits function primarily as signals of direct or indirect benefits to females.

Indirect intersexual selection

In some species, males provide no resources to females in the form of nest sites, parental care or protection. Indeed, the only male contribution to the female in some cases is sperm to fertilise her eggs. In these instances, an explanation based on direct benefits of mate choice cannot be invoked to explain female mate choice. An extreme example is seen in lekking species. Here males gather in a 'lek' (derived from a Swedish word meaning 'to play'), in which males defend small territories without resources and where they display to females and engage in aggressive or ritualised contests with other males. Females visit leks to mate with males, but appear to obtain no other benefit. An evolutionary conundrum is why females express such strong mate preferences without receiving any obvious benefit from doing so; this difficulty has been termed 'the paradox of the lek' (Kirkpatrick & Ryan, 1991). A possibility is that they obtain indirect benefits from mate choice, although an alternative is that mate preferences are arbitrary (see the section 'Arbitrary mate choice'). Examples of lek mating teleosts include many mouthbrooding cichlids from the African Great Lakes. In these fishes, males often construct sand and gravel nests, termed 'bowers', which are used as spawning sites (Chapter 7). Bower shape and position, as well as male courtship behaviour, may play a role in female mating decisions (Fryer & Isles, 1972). Females typically visit a series of males, but spawn only with a single male, or, in some species, a limited number. Once the female has completed spawning, she departs from the lek with a mouthful of fertilised eggs to begin solitary parental care, with no further contribution from the male parent (Barlow, 1991).

Indirect benefits of mate choice are obtained in the generation succeeding that in which mate choice is performed and assume a direct relationship between a male trait and his viability. Theories of indirect benefits present the evolutionary biologist with some difficulties and are therefore controversial. Indirect benefits have also proven more difficult to demonstrate than direct benefits, and there are fewer persuasive examples.

There are three models to explain indirect benefits of mate choice: good genes, compatible genes and antagonistic genes.

Good genes

Under a 'good genes' (more correctly 'good allele') model, the prediction is that female mate preferences are under selection through the choice of males that father offspring of superior viability. Females are able to assess male viability using male ornaments or behaviour that accurately, or 'honestly', signals male genetic superiority. Good genes models of mate choice are exemplified by the much criticised 'handicap principle' (Zahavi, 1975, 1977). Here an honest signal is one that is costly to the male. As ornaments are effectively a handicap, only genetically superior males are able to express them fully. A more widely accepted derivation of the handicap principle was developed by Hamilton & Zuk (1982), who argued that ornaments might serve as indicators of parasite resistance.

While good genes models of sexual selection are intuitively appealing, it has proved challenging to find evidence to support them and they raise theoretical difficulties. An implicit assumption of good genes models is that male traits under selection have additive genetic effects. Thus, when a male mates with a female, irrespective of female genetic background, offspring have greater viability as a consequence of the male contribution. This assumption has rarely been demonstrated. Another problem with this hypothesis is that additive genetic variance is predicted to deplete under selection, and it has proved challenging to explain how female mate choice could be maintained as a consequence. Hamilton & Zuk's (1982) proposal for parasite-driven selection does offer a mechanism for maintaining genetic variance, through the cycling of host and parasite resistance and virulence, which would tend to maintain heritability for resistance genes, and thereby the ornaments that signal those genes, and female mate preferences for them. A weakness of many studies that purport to show good genes effects is that they often fail to exclude direct selection on male traits; simply because a male possesses an exaggerated ornament and a female chooses to mate with that male is not robust evidence for good genes selection in operation.

Indirect selection for good genes has been demonstrated in *G. aculeatus*. Male *G. aculeatus* develop red colouration on their throats during the breeding season. In some populations, female sticklebacks spawn more frequently with males that show the greatest intensity of red colouration. Barber *et al.* (2001b) showed for a Scottish population that offspring fathered by males with the greatest intensity of red had greater resistance to a cestode parasite (*Schistocephalus solidus*; see Fig. 8.1). Interestingly, this study also showed that the offspring of redder males suffered a cost of lower growth rates. This cost of mating with redder males offers an explanation for why genetic variance for male colour is not depleted in the population, as fluctuations in food availability, size-selective predation and rates of parasitism by *S. solidus* would have the effect of varying the strength of selection on female preference for redder males. Thus, Barber *et al.*'s (2001b) study conforms with Hamilton & Zuk's (1982) model of selection for good genes through fluctuating resistance to parasites. This study also illustrates the contradiction in the term 'good genes' to describe this form of indirect selection, as a gene (or suite of genes) that confers an indirect fitness benefit through female mate choice in one environment may be detrimental to fitness (i.e. represent 'bad genes') under different conditions.

Compatible genes

Another mechanism for indirect selection on mate choice is through genetic compatibility (more correctly 'compatible allele'). In this model, a female enhances the viability of her offspring by mating with a genetically compatible male. In contrast to good

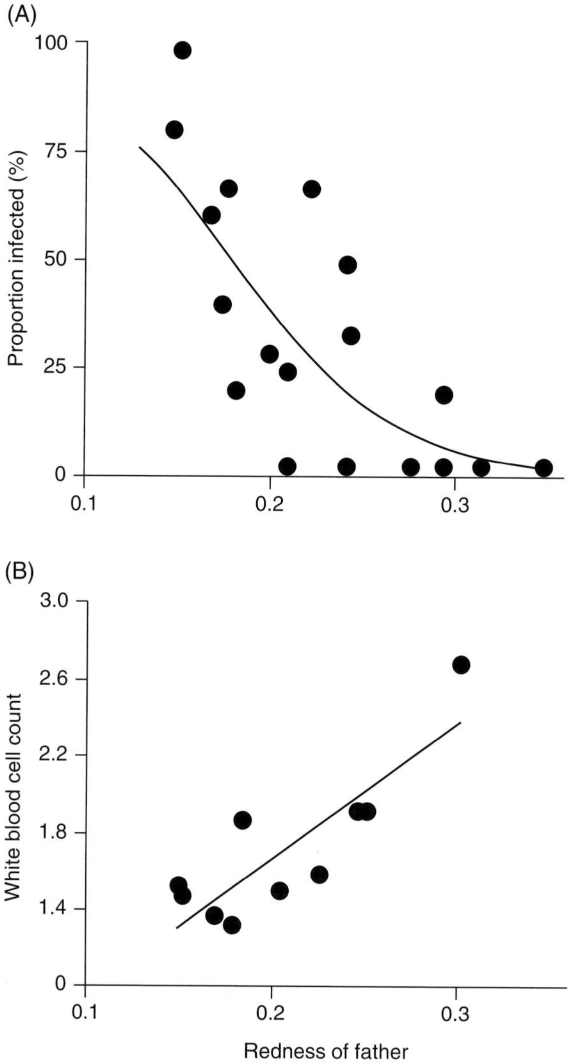

Figure 8.1 (A) The relationship between infection by the cestode parasite *Schistocephalus solidus* and paternal redness score in three-spined sticklebacks (*Gasterosteus aculeatus*). (B) The relationship between white blood cell count and paternal redness score in *G. aculeatus* that became infected with *S. solidus*. (Source: Barber *et al.* 2001b. Reproduced with permission of the Royal Society.)

genes models, offspring viability is enhanced only when particular male and female genotypes are combined (Zeh & Zeh, 1996), and thus compatible genes are predicted to show non-additive variation (Neff & Pitcher, 2005). Where a mating system is underpinned by compatible gene effects, females will not necessarily share mate preferences, which is the opposite prediction of the good genes model.

The genetic compatibility paradigm also raises conceptual difficulties as to how it operates. The functional basis to compatibility is unclear, but may be related to advantages associated with enhanced heterozygosity. The process also demands that a female is able to recognise her own genotype, or at least components of her genotype, as well as those of potential mates in making mate choice decisions. How a female is able to match her genotype to that of a compatible male is unclear. For genetic compatibility to be able to function as the basis to a mate choice system, it is likely to be limited to specific genetic systems. One such genetic system comprises the genes of the major histocompatibility complex (MHC), for which there is evidence of a role in mate choice. The MHC is a family of highly polymorphic genes that encode a set of transmembrane proteins. The function of these proteins is to distinguish between self and non-self antigen and present foreign peptides to T-cell receptors, thereby playing a key role in resistance to infectious and autoimmune disease (Hill, 1991). Many studies have implicated MHC genes in influencing mate choice, including in fish (Forsberg *et al.*, 2007; Consuegra & Garcia de Leaniz, 2008; Yeates *et al.*, 2009). MHC genes appear to influence odour, and thereby have the potential to function in an odour-based mate recognition system, enabling females to identify mates that can provide good genes genetic benefits through enhanced immunocompetence, or to identify genetically compatible mates (Penn & Potts, 1999; Tregenza & Wedell, 2000; Penn, 2002).

One promising hypothesis is the proposal that an intermediate, rather than a maximal, level of MHC sequence variability will be optimal for offspring viability (Reusch *et al.*, 2001; Milinski *et al.*, 2005; Forsberg *et al.*, 2007; Kalbe *et al.*, 2009). This situation only applies to species with duplicated MHC genes (Chapter 1). Intermediate MHC dissimilarity may be important in mating between animals from different populations or races, where outbreeding depression can result from genetic incompatibilities, possibly the result of disruption to co-adapted gene complexes (Turelli & Orr, 2000; Barton, 2001; Jiggins *et al.*, 2001).

An example of genetic compatibility in fish is shown by the rose bitterling (*Rhodeus ocellatus*, Cyprinidae). In this species, offspring survival and growth are largely determined by egg size (an additive female effect), but also the genetic compatibility of males and females. Females show a mating preference for genetically compatible males, and this preference correlates with the degree of functional dissimilarity of male and female MHC genes (Agbali *et al.*, 2010; Reichard *et al.*, 2012). The cues used by females to assess male compatibility have yet to be determined, but are almost certainly olfactory (Casalini *et al.*, 2009).

Antagonistic genes

This hypothesis claims that female preference for male traits has a negative effect on female fitness, thereby generating an antagonistic arms race between the sexes (Holland & Rice, 1998). Males are predicted to express traits that give them a fitness advantage in mating with females, such as through being better able to attract females for mating or through fertilising their eggs more efficiently. A key element of this model is that when females mate with males that express these traits, they suffer reduced fitness. The result is that females evolve resistance to the trait, which elicits selection in males on the trait to overcome female resistance alleles. The outcome is cyclical antagonistic coevolution that has been termed 'chase-away' selection, which results in exaggerated male traits and strong mate discrimination in females. Potential mating costs to females associated with chase-away selection include exposure to toxins in seminal fluid, manipulation of their endocrine system, increased exposure to parasites and predators, and elevated time and energy expenditure in avoiding males (reviewed by Arnqvist & Rowe, 2005).

In teleosts, an example of chase-away selection is found in *P. reticulata*, which is viviparous (Chapter 10). Male *P. reticulata* persistently court and attempt to mate with females, which substantially disrupts female foraging. Male harassment was estimated to reduce food intake by 25% (Magurran & Seghers, 1994), representing a potential loss in fecundity. Female *P. reticulata* show a tendency to avoid males (Magurran & Nowak, 1991; Magurran & Seghers, 1994), and because they can use stored sperm to fertilise ova for up to 8 months (Winge, 1937), they potentially do not need to routinely associate with males. The evolution of female resistance to male courtship appears to be a general feature of live-bearing poeciliid fishes (Plath *et al.*, 2007). Another example is the barbed spermatophores of the Malabar ricefish (*Oryzias setnai*, Adrianichthyidae) and the protective 'genital pads' that have evolved in response in females (Grier, 1984; see the section 'Sexual conflict').

Arbitrary mate choice

Mate choice decisions may not always generate a benefit for the choosing sex. Two models of intersexual selection that do not invoke fitness benefits of mate choice are Fisherian selection and selection through sensory bias.

Fisherian selection

This model of mate choice, developed by Fisher (1930), is based on the idea that female mate preference is influenced by male ornaments or behaviour, but without these male traits conveying any information about male quality. Thus, an ornament is attractive to females for an arbitrary reason, and the greater the elaboration of the ornament, the stronger the attraction it has for females. An assumption of this mechanism of mate choice is that the alleles controlling the ornament under selection in males are genetically coupled to the alleles that control female preference for the ornament. This linkage disequilibrium between male trait and female preference is maintained through sexual selection, with the result that the male trait tends to become more elaborate and female preference for the trait stronger over time – hence the name for this type of selection as Fisher's 'runaway' process. Fisherian selection is predicted to select for ever greater elaboration of male traits until a point is reached when natural selection, e.g. through predation, constrains sexual selection on the male trait (Andersson, 1994). Conditions for Fisherian selection are found in *G. aculeatus*. In this species, male red nuptial colouration and female preference for red are genetically correlated (Bakker 1993), suggesting the operation of Fisherian selection. Interestingly, more than one mechanism of mate choice has been suggested for this species, including good genes (Barber *et al.*, 2001b), Fisherian selection (Bakker 1993) and sensory bias (Smith *et al.*, 2004b; see the section 'Sensory bias'), and the mating system of this species appears to vary among populations (Smith & Spence, 2013).

Fisher (1930, 1958) additionally argued that the origin of runaway sexual selection would be initiated through female preference for a trait that conferred a fitness benefit, through either direct or indirect selection. Later models have suggested these starting conditions need not be in place to initiate Fisherian selection (Lande, 1981; Kirkpatrick, 1987). The Fisherian model of intersexual selection most closely reflects Darwin's (1871) original formulation of how sexual selection through mate choice might operate. The reformulation of Fisherian sexual selection by Lande (1981) and Kirkpatrick (1987) (the so-called Lande–Kirkpatrick model, or simply the LK model) can be considered as a null model for intersexual selection (Kirkpatrick & Ryan, 1991; Prum, 2010). Thus, while few examples of Fisherian selection have been demonstrated in fishes, an alternative view is that all cases of intersexual selection should be viewed as arising through this process, unless additional evidence demonstrates otherwise (Prum, 2010). This approach has the advantage of providing a null hypothesis against which alternative ideas of indirect selection can be tested.

Sensory bias

Female preferences for male traits may exist because females have existing predispositions in their sensory systems for certain stimuli, whether visual, auditory or olfactory (Ryan, 1990; Fuller *et al.*, 2005). Male traits that stimulate those existing biases may subsequently be selected through sexual selection. Female sensory biases may be arbitrary. Alternatively, they may have evolved under natural selection, e.g. in the context of prey detection. Female mating preferences that arise through a sensory bias, such as Fisherian selection, do not assume a fitness benefit for the female from mate choice decisions. The key element of the various sensory bias hypotheses is that the preference evolves first, with the favoured male trait evolving later as a consequence of how the bias shapes the evolution of the trait. Another feature of the hypothesis is that males as well as females should express a sensory bias, although females may subsequently evolve a stronger or more refined response than males through sexual selection. An appealing aspect of the sensory bias hypothesis is the prediction that a pre-evolved preference for a male trait will be universal, with the result that all females will share the preference for a male trait when it evolves. From a conceptual viewpoint, then, this model presents no difficulties in explaining how an arbitrary male ornament could be elaborated through sexual selection. This model of mate choice is not fundamentally different from Fisherian selection and is sometimes categorised with it (Shuster & Wade, 2003).

There are several examples where a sensory bias has been invoked to explain the elaboration of male traits in fishes. One is in the livebearing swordtails and platyfishes (*Xiphororus* spp., Poeciliidae), which possess a pronounced sword-like extension to the lower lobe of their caudal fin. Female *Xiphororus* show a mating preference for males with swords, even in species in which males do not exhibit the trait. Female preference for swords appears, at least in some analyses, to pre-date the evolution of swords, which may therefore have evolved through a female sensory bias (Basolo, 1990). Subsequent phylogenetic studies have questioned this conclusion (Meyer, 1997), and female preference for swords in *Xiphororus* spp. may simply reflect a preference for large male body size (Rosenthal & Evans, 1998).

Another proposed example of sensory bias is in *P. reticulata*. In this species, females have a mating preference for males with orange spots, although the strength of this preference varies among populations. Both male and female *P. reticulata* also have a preference for orange food items, with the strength of variation in attraction to orange food items explaining most variation in female preference for orange in a mating context (Rodd *et al.*, 2002). In *G. aculeatus*, both males and females were shown to have a preference for red in a foraging context, while females had a preference for males with red nuptial colouration. A red preference was also shown in the nine-spined stickleback (*Pungitius pungitius*, Gasterosteidae), a species without red male nuptial colouration. A phylogenetic analysis showed that the preference for red must therefore have evolved prior to red nuptial colouration (Fig. 8.2), a key prediction of the sensory bias hypothesis (Smith *et al.*, 2004b).

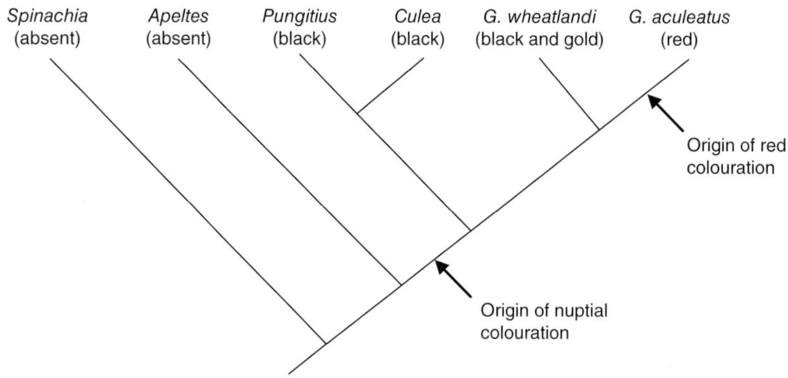

Figure 8.2 A phylogenetic tree of the family Gasterosteidae. Colours in parentheses under each species name indicate male nuptial colouration. Arrows indicate the evolution of male nuptial colouration and red colouration. (Source: Smith *et al.* 2004b. Reproduced with permission of The Royal Society.)

Male mate choice and female ornaments

While it is generally the case that females are the choosy sex, under certain circumstances a role for male mate choice can be predicted from sexual selection theory. Male mate choice is expected when males are constrained in the number of females with which they can mate, and when females show variability in the number and viability of their offspring (Sargent *et al.*, 1986). Being choosy over mates, however, may carry a cost in terms of increased time and energy devoted to seeking a mate (Parker, 1983), greater risk of predation (Forsgren, 1992; Houde, 1997; Herdman *et al.*, 2004), and loss of potential future matings (Johnstone, 1997). Therefore, males, like females, must trade off the benefits of increased reproductive success of choosing females against the costs associated with choosiness. If the cost of choice in terms of the probability of finding an alternative mate and assessing their quality does not exceed the benefit resulting from an improvement in their fitness, then male mate choice is predicted to evolve (Johnstone, 1997; Bonduriansky, 2001).

Because males are most frequently the care-giving sex in teleosts (Chapter 9), they may experience constraints with regard the number of offspring for which they can simultaneously provide care. In some cases, this constraint may be because caring for multiple broods of offspring at different stages of development makes behavioural demands on a male that are incompatible. Thus, eggs may require regular fanning to aerate them, while free-swimming larval or juvenile stages need to be actively protected from predators. In other cases, there may be a physical constraint on the number of young a male can accommodate. Such a constraint is present in species that brood young in specialised pouches, such as seahorses (Syngnathidae), in their mouths, such as gouramies (Anabantidae), cardinalfishes (Apogonidae), cichlids (Cichlidae), jawfishes (Opistognathidae) and bony tongues (Osteoglossidae), or on their body surface, such as pipefish and seadragons (Syngnathidae) and frogfishes (Antennariidae and Tetrabrachiidae). The rate at which parental males are able complete a parental cycle in relation to that rate at which females can generate new batches of eggs can also play a role in the evolution of male mate choice. In some cases, the constraints placed on males by parental care result in sex role reversal, whereby the reproductive rate of females exceeds that of males (see Chapter 9). In sex role-reversed species, females are competitive in mating with males and show conspicuous colouration that may function in courtship, while males are the choosy sex. However, sex role reversal is not necessary for male mate choice to evolve, suggesting that potential rates of reproduction are a poor predictor of mate choice. In some species, males preferentially court larger females. Larger females usually produce bigger clutches of eggs or young, which may also be of larger size and greater viability (Chapter 4). In *P. reticulata*, males direct more displays at large females, unless females are monopolised by other males, in which case, additional males tend to switch courtship to smaller females (Houde, 1997). In pairwise tests, males preferred the bigger of two females, and the strength of male preference was stronger for females of large absolute size (Dosen & Montgomerie, 2004). Other examples of male preference for larger females include the sailfin molly (*Poecilia latipinna*, Poeciliidae) (Ptacek & Travis, 1997), *G. aculeatus* (Kraak & Bakker, 1998) and Pacific blue-eye (*Pseudomugil signifer*, Pseudomugilidae) (Wong & Jennions, 2003).

Male choosiness may also occur in species in which males experience a significant cost of courtship and mating, particularly for those individuals that experience high mating success. Male preference linked to the costs of mating occurred in the lek-breeding mouthbrooder, *Astatotilapia flaviijosephi* (Cichlidae), a species endemic to Israel in which female size is correlated with fecundity (Werner & Lotem, 2003). Males were shown to be more likely to court larger females, and courted larger females for longer periods. In this species, there are no costs of parental care for the male or costs associated with searching for females, as the females visit the lek to mate. Thus, it appears that the costs associated with courtship and sperm production alone were sufficient to drive male mate preference. Courtship was potentially costly through energy expended on behaviour and the elevated risk of predation on courting males (Werner & Lotem 2003).

In some species, females exhibit ornaments comparable to those of males, and these may have evolved under sexual selection. In the two-spotted goby (*Gobiusculus flavescens*, Gobiidae), Amundsen & Forsgren (2001) demonstrated a significant preference by males for the bright orange bellies of females (Fig. 8.3). The orange colour is due to the pigmented eggs visible through the female's skin, and orange-red chromatophores in the abdominal skin (Fig. 8.4). Females perform courtship displays in this species, swimming close to a male and exposing their bellies. The benefits to males of choosing brightly coloured females are not clear. One possibility is that female colour may correlate with egg viability, although that has yet to be demonstrated. Notably, small males in this species were indiscriminate in choosing females, whereas large males actively chose colourful females (Amundsen & Forsgren, 2003). The difference in preference with male size may be because small males rarely have the opportunity to mate and the cost of mate choice by small males may outweigh any fitness benefits that accrue from it.

The colouration and behaviour of female *G. flavescens* are not unique. In the Central African cichlid *Nanochromis transvestitus* (Cichlidae), females have a striking red-maroon belly when sexually mature that they also display to males, which are drab. Details of the mating system of this fascinating fish have yet to be fully investigated (Stewart & Roberts, 1984). Comparable examples of female ornaments and female display exist in other cichlids, e.g. in the Topaz cichlid (*Amatitlania myrnae*, Cichlidae) (Tobler, 2007) and the rainbow krib (*Pelvicachromis pulcher*, Cichlidae) (Barlow, 2000).

In the convict cichlid (*Amatitlania nigrofasciata*, Cichlidae), females possess gold scales on their belly, a trait largely absent in males. Experimental tests by Beeching *et al.* (1998) failed to show male preference for more brightly ornamented females. However, in contests with other females, colour did play a role, with brightly coloured females eliciting stronger displays and aggression from other females. Thus, this trait may have evolved

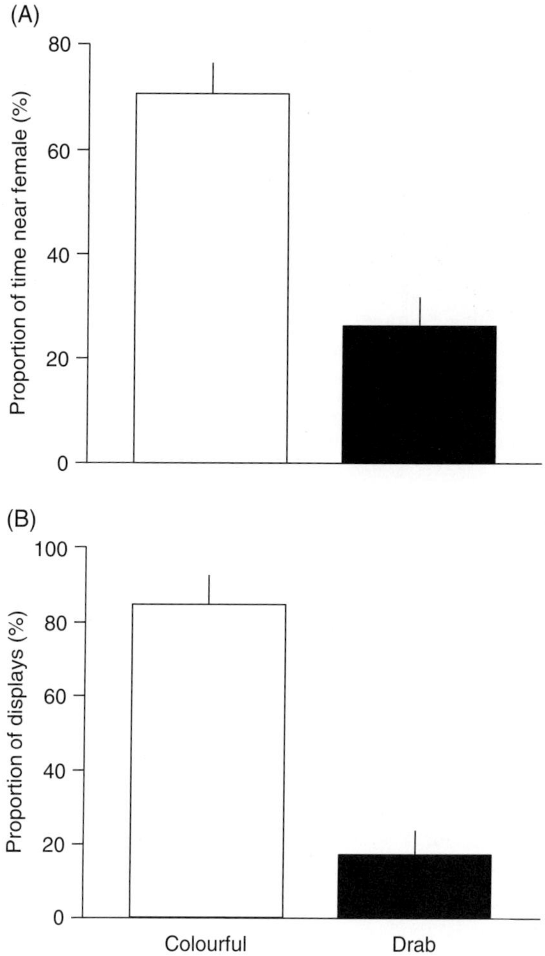

Figure 8.3 Mate preferences of male two-spotted gobies (*Gobiusculus flavescens*) for females varying in experimentally manipulated colouration. (A) Mean (+1 SE) proportion of time (%) spent in association with colourful and drab females. (B) Mean (+1 SE) proportion of displays (%) directed at colourful and drab females. (Source: Adapted from Amundsen & Forsgren 2001. Reproduced with permission of National Academy of Sciences, USA.)

through intrasexual, rather than intersexual, selection acting on females, perhaps in competition over males (Beeching *et al.*, 1998). How intrasexual selection operates on females in this species, and what information female colour conveys remain unclear.

Mutual mate choice and the 'double process' of sexual selection

Darwin (1871) identified a role for mutual mate choice in his theory of sexual selection, which he termed the 'double process' of sexual selection. Remarkably, given the huge interest in sexual selection, this aspect of Darwin's otherwise influential ideas has

Figure 8.4 Courtship display by female two-spotted goby (*Gobiusculus flavescens*), with the female displaying her bright orange belly. Note that a parasitic copepod is attached to the female's first dorsal fin. (Source: Amundsen & Forsgren 2001. Reproduced with permission of National Academy of Sciences, USA.)

received relatively little attention. Darwin identified two forms of mutual mate choice. In the first, males and females are attracted by the same traits in each sex. In the other, the sexes prefer different traits. Thus, intersexual selection can take one of four forms:

(1) Females choose ornamented males and females are not ornamented (male 'monodecus'), which represents the conventional view of intersexual selection.
(2) Males choose ornamented females, and males are not ornamented (female 'monodecus').
(3) Males and females mutually choose mates using the same ornaments as the basis to choice ('homodecus').
(4) Males and females mutually choose mates using different ornaments as the basis to choice ('heterodecus').

In the case of forms 3 and 4, both sexes simultaneously undergo intersexual selection. In the case of form 3, the implication is that the same mechanism of mate choice operates in both sexes, and in the case of form 4 the mechanisms may differ.

The conditions under which mutual mate choice is predicted to evolve were considered in a model by Kokko & Johnstone (2002), which represented a refinement of a model by Kokko & Monaghan (2001). The model predicted that mutual mate choice would be most likely in species with biparental care, particularly in which the impact of two parents on offspring survival was greater than double that of a single parent. Other conditions for the evolution of mutual mate choice predicted by the model are when there is variation in quality within both sexes, when parental care by both sexes is substantial, when extra-pair matings are rare, and when mate searching costs are low for both sexes. Notably, Kokko & Johnstone's (2002) model did not predict a pivotal role for the operational sex ratio (OSR), the ratio of females and males ready to mate, a factor that is frequently implicated in the evolution of competition for matings and mate choice.

Despite its neglect, there is ample evidence for what we have termed homo- and heterodecus mate choice in fishes, with the conditions predicted by Kokko & Johnstone (2002) for the evolution of mutual mate choice met in some taxa, most notably in biparental substrate-brooding cichlids (Axelrod, 1993; Barlow, 2000; Jackson, 2001).

However, studies that explicitly examine the difference between different forms of mutual mate choice are currently lacking. An implication of Darwin's double process is that the highest 'quality' mates seek each other out, presumably generating above-average 'quality' offspring, and thereby driving variance in fitness.

Mate choice copying

While a common assumption is that mate choice is based solely on cues associated with a potential mate, social cues have also been implicated in playing a role (Westneat *et al.*, 2000). Mate choice copying involves an individual (usually a female) making a mate choice decision based partly or wholly on the responses of other individuals to a prospective mate. Thus, if a female observes another female select a male for mating, the prediction, if mate choice copying operates, is that she will show a greater probability of choosing to mate with the same male. Evidence for mate choice copying comes largely from studies on *P. reticulata* (Dugatkin, 1992, 1998; Dugatkin & Godin, 1993), but has been recognised in other fish species, including medaka (*Oryzias latipes*, Adrianichthyidae) (Grant & Green, 1996), blackbarred limia (*Limia nigrofasciata*, Poeciliidae) (Munger *et al.*, 2004), *G. aculeatus* (Frommen *et al.*, 2009), ocellated wrasse (*Symphodus ocellatus*, Labridae) (Alonzo, 2008), and *P. latipinna* (Witte & Ryan, 1998, 2002).

The benefits of mate choice copying come from avoiding costs associated with finding potential mates and assessing their suitability. Therefore, the evolution of this behaviour is predicted to vary not only with the costs associated with mate choice, but also the benefits of mate choice decisions, and the ability of an individual to discriminate among potential mates that vary in quality (Stöhr, 1998). Mate choice copying is most likely to evolve in species without parental care or with maternal care, and in polygynous or promiscuous mating systems (Witte & Nöbel, 2011). Mate choice copying is not predicted in species with a significant role for compatible genes as the mechanism for indirect selection on mate choice, as females will not necessarily be genetically compatible with males selected by other females. An outcome of mate copying will be increased variance among males in reproductive success, with consequent stronger sexual selection on males.

Intrasexual selection

Some degree of direct competition among males for mates and matings is a feature of almost all mating systems. From a conceptual viewpoint, competition among males for access to females is straightforward and relatively easy to demonstrate in field and laboratory studies. Intrasexual selection is associated with the evolution of morphological traits. One of the best known examples in fish is the overdevelopment of the jaw in male salmonids, termed a 'kype', which coincides with the onset of reproduction (Fig. 8.5). The kype is a weapon that serves the male during fighting, while also precluding feeding. In cyprinids, males develop epidermal tubercles during the breeding season. These may have a function as sexual ornaments, but in some species may function as weapons. In bitterling fishes, males ram rivals with their snouts, which possess clusters of robust and abrasive tubercles (Smith *et al.*, 2004a; Fig. 8.6). In many cichlid species, males develop enlarged jaws and unicuspid teeth at sexual maturity, and these features may have evolved under intrasexual selection through contests over mates and matings. Perhaps the most common male trait to have undergone sexual selection is body size, and many fish species show size dimorphism between the

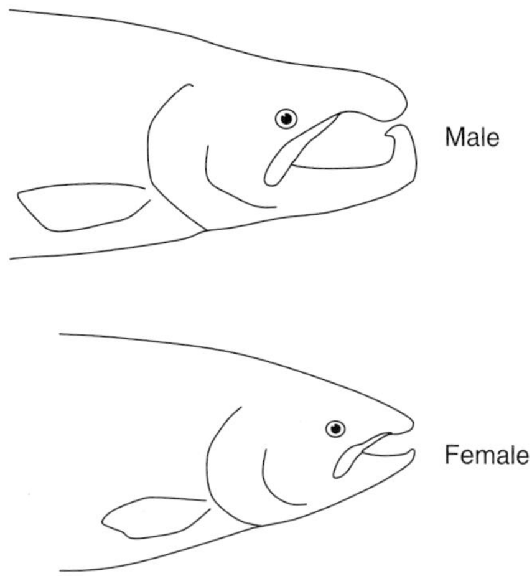

Figure 8.5 Jaw development in sexually mature Atlantic salmon (*Salmo salar*). The male exhibits overdevelopment of the jaw, termed a 'kype', which is used in fights with other males during spawning.

Figure 8.6 A pair of sexually mature rose bitterling (*Rhodeus ocellatus*). The male (right) displays epidermal tubercles on the snout and around the eyes (highlighted with arrows). Tubercles are not displayed by the female (left). (Source: Carl Smith.)

sexes. Male body size frequently correlates positively with a male's ability to secure and hold resources necessary for reproduction against other males through combat, such as territories and nest sites (Parker, 1992).

Male ornaments, although often associated with mate choice, may also signal competitive ability to other males. Body size, bright colouration, sounds, electrical discharge, and morphological traits may all accurately signal male fighting ability, as well as readiness to engage in combat. In many cases, these signals are combined with body postures and behaviours that enhance the strength of the signal to rivals (Maynard Smith & Harper, 2003). In the golden dwarf cichlid (*Nannacara anomala*, Cichlidae), males engage in sometimes lengthy contests with other males for nest sites, which are essential for reproduction (Enquist *et al.*, 1990). Males display to each other, changing the colour and banding pattern on their bodies as the contest between them escalates. Visual displays eventually give way to physical contact, including tail beating, mouth wrestling and eventually circling, in which males attempt to tear each other's fins. These contests can end at any stage, with one male signalling submission by closing their fins and changing colour (Enquist *et al.*, 1990). The winner of these contests is usually the heavier male, and the duration of contests tends to be longer when the weight difference between rivals is small (Enquist & Leimar, 1987).

Intrasexual selection among females also occurs in fishes, although it is less common than in males. In sex role-reversed species, such as the broad-nosed pipefish (*Syngnathus typhle*, Syngnathidae), female–female competition occurs over mates (Berglund, 1991). Females may also compete with each other when nest sites are limiting, e.g. in the peacock blenny (*Salaria pavo*, Blenniidae) (Almada *et al.*, 1995).

The interaction of intra- and intersexual selection

The effects of intra- and intersexual selection can potentially interact. Thus, females may use male contests to measure potential mate quality. The outcome of male contests would tend to facilitate female mate choice where the benefits of mate choice are direct, e.g. if males compete for nest sites or territories that serve for rearing offspring. Where mate choice benefits are indirect, the role of male contests on female mate choice is less clearly defined. In the case that male competitive success correlates with male quality, in accordance with a good genes model of mate choice, male contests will reliably indicate male viability as a mate. In this situation, it would pay females to instigate male contests, and there is some evidence that this occurs in teleosts (Wong & Candolin, 2005). However, where the mechanism of mate choice is based on compatible genes, male success in intrasexual contests will not signal male quality as a mate. In this case, dominance by males through aggressive contests may conflict with female mate preference through intersexual selection, possibly leading to sexual conflict (see the section 'Sexual conflict'). Where mate choice is based on arbitrary mechanisms (such as Fisherian selection and sensory bias), male intrasexual contests will indicate nothing about mate quality to females.

The strength of sexual selection and the operational sex ratio

The OSR is the ratio of the relative numbers of males and females in the population that are ready to mate. The OSR is distinct from the population sex ratio, or even the sex ratio of sexually mature individuals, and is usually strongly male-biased. The OSR expresses the relative number of competing males per reproducing female. The greater the number of competing males in relation to available females, the higher the OSR, which has often been used as an index of the strength of sexual selection (Kvarnemo & Ahnesjö, 1996). Several studies have linked the OSR to mate competition in fishes. In *P. reticulata*, *G. aculeatus* and zebrafish (*Danio rerio*, Cyprinidae), males modulate courtship behaviour in response to changes in the OSR, with lower rates of courtship when the OSR is more male-biased. In the case of *G. aculeatus* and *D. rerio*, males also showed elevated levels of aggression in response to an increase in the male bias of the OSR (Jirotkul, 1999; Le Comber *et al.*, 2003; Spence & Smith, 2005).

There are limitations to the usefulness of the OSR as an index of sexual selection. It assumes a linear relationship between OSR and strength of sexual selection, which is overly simplistic and unlikely to be justified in most situations, especially at low population densities (Klug *et al.*, 2010). For example, Head *et al.* (2008) found no changes to the strength of sexual selection on male nuptial colouration, courtship behaviour or body and sperm characteristics in response to changes in the OSR in *P. reticulata*. Other relevant parameters, including the potential reproductive rate (PRR; Clutton-Brock & Parker, 1992) and breeding sex ratio (BSR; Arnold & Duvall, 1994) suffer from similar limitations (Shuster & Wade, 2003). An alternative to the OSR is the Bateman gradient (BG) – the number of offspring an individual produces as a function of its mating success (Bateman, 1948; Jones & Ratterman, 2009). The BG itself suffers a limitation in that it fails to measure how mate choice alters reproductive success, but is a more direct measure of the strength and direction of sexual selection than the OSR (Jones & Ratterman, 2009; Jennions & Kokko, 2010). Reliable estimates of BG for fishes under natural conditions are currently lacking and the challenge is to show that this parameter, which is considerably harder to measure than the OSR, can provide more comprehensive insights into how sexual selection shapes mating system evolution.

Courtship

Courtship in fishes can be relatively simple, little more than a demonstration of the readiness to mate. It can also comprise a complex suite of behaviours that signal key features of mate quality with a key role in mate choice decisions. Thus the importance of courtship for teleosts varies widely among taxa. Ultimately the function of courtship is to ensure that mature conspecifics are able to identify each other and mate successfully. Courtship is most elaborate in species that display parental care and which establish long-term pair bonds.

Functions of courtship

Courtship in fishes can have a variety of functions. These include mate identification and attraction, mate stimulation, mating synchrony and maintenance of pair bonds in biparental species (Keenleyside, 1979). In some species, courtship may have the function of limiting aggressive interactions between potential mates, particularly those that are strongly territorial (Wootton, 1998). Tinbergen (1951) classed the functions of courtship as orientation (attracting the attention of potential mates), persuasion (influencing mate choice decisions), synchronisation and reproductive isolation.

Mate identification and attraction

In its most simple form, courtship may identify that an individual belongs to the same species and is the appropriate sex. In species in which males and females look similar, e.g. in many cichlid species, courtship can have the function of simply enabling the sexes to recognise each other (Barlow, 2000). However, it is important not to overlook the fact that teleost fishes occupy a sensory world quite different from our own, and while the sexes may appear superficially similar to our visual systems, that is not necessarily the case for the fish.

Courtship may also transmit information about an individual that indicates its suitability as a mate, though the function of courtship will depend on the mechanism of mate choice. If courtship behaviour is energetically expensive, the vigour or duration of courtship may honestly indicate some aspect of mate quality. Where females obtain direct benefits from mate choice, the vigour of male courtship may be under selection as an indicator of the quality of those benefits. Male courtship may accurately signal male parental ability (Knapp & Kovach, 1991; Lindström & Hellström, 1993). If a good genes mechanism operates, courtship may signal (or enhance a signal of) some aspect of male phenotype that correlates with genetic quality. In the case of mate choice for genetic compatibility, cues that are used to measure compatibility, such as olfactory cues released from the body surface, in urine, gonadal fluids or from the gills, may be disseminated by courtship behaviour. Where the mechanism of mate choice is arbitrary, such as through Fisherian selection or sensory bias, courtship is not predicted to convey any information about male quality and simply to display or highlight the trait that is under selection (Searcy & Nowicki, 2005).

Mate stimulation, mating synchrony and appeasement

The role of courtship may be to elicit an appropriate response from a prospective mate, with courtship comprising a series of reciprocated behavioural steps that synchronise male and female behaviour and facilitate mating. A well-known example is the 'zig-zag dance' of *G. aculeatus* (Tinbergen, 1951). The zig-zag dance comprises a sequence of steps, starting with an initial approach by the male to a gravid female and involving the male jumping sideways towards and away from the female. Gravid females react to this behaviour by adopting a head-up posture. The male responds by swimming quickly to his nest with the female following. The male then 'shows' the nest entrance by poking his snout into the entrance while turning on his side. In response, the female swims past the male and forces her way into the nest. The male then performs 'quivering', in which he touches the flank of the female with his snout while appearing to tremble or quiver. This stimulus is necessary to induce the female to spawn. Once the female deposits her eggs, she swims out of the nest and the male passes through it and fertilises the eggs (Wootton, 1976). If at any stage of this process either the male or female fails to respond in the appropriate way, courtship will tend to break down and spawning will fail to take place. Therefore, in this species, and probably in many others, courtship serves as a mechanism that inhibits interspecific matings with sympatric species. In the case of *G. aculeatus*, males risk hybridisation with sympatric populations of the blackspotted stickleback (*Gasterosteus wheatlandi*, Gasterosteidae) and *P. pungitius*.

In some cases, courtship may serve to limit the aggressive responses from a potential mate. In *C. gobio*, females show a mating preference for males larger than themselves. Nest-guarding territorial males are aggressive towards conspecifics and females risk injury or even cannibalism when approaching a nesting male. Females perform appeasement behaviours as part of their courtship of males (Morris, 1955). Appeasement of male aggression was proposed by Tinbergen (1953) as an explanation for the evolution of female courtship behaviour.

Maintenance of pair bonds

In species with biparental care, especially those with prolonged care and stable within-pair relationships, courtship may have the function of reinforcing the pair bond. Pairs of biparental cichlids perform elements of their initial elaborate and prolonged courtship behaviour as a 'greeting ceremony' whenever one returns to the other after separation. These greeting behaviours include flaring the opercles, quivering and lateral displays and are readily observed in many cichlid species (Baerends & Baerends van Roon, 1950). The role of these 'greeting' displays in maintaining the pair bond is unclear, athough their ubiquity and frequency suggest there is one. In the dwarf seahorse (*Hippocampus zosterae*, Syngnathidae), pairs are monogamous and exhibit four distinct phases to their courtship over a period of 2–3 days (Fig. 8.7). The initial phase of courtship is characterised by reciprocal quivering, which takes place 1 or 2 days prior to copulation. On the day of copulation itself, courtship is performed for longer and is greatly elaborated, culminating in the pair rising in the water column with mating occurring in midwater, with the female depositing her eggs in the male's brood pouch (Masonjones & Lewis, 1996). Several species of monogamous seahorse engage in elaborate daily courtship rituals. The function of this behaviour appears to be in facilitating reproductive synchronisation.

Ornaments and signals in courtship

Females tend to prefer to mate with more ornamented males. In species in which females are ornamented, males may also prefer ornamented females. Fish utilise a variety of morphological and behavioural traits during courtship that are directed at

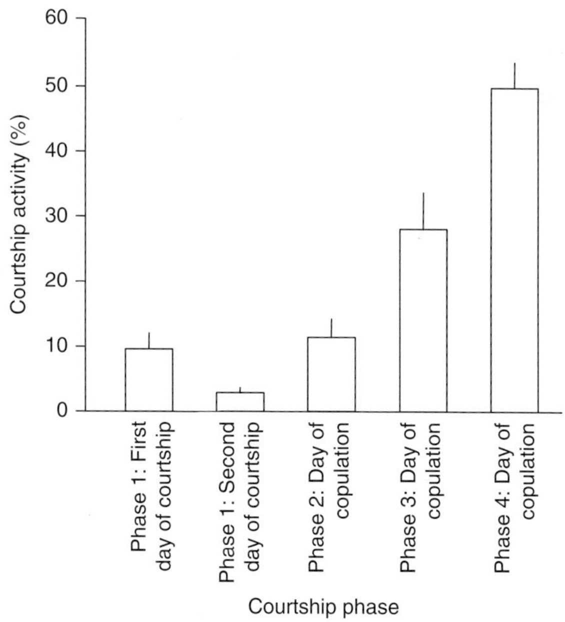

Figure 8.7 Mean (+1 SE) proportion of time spent on courtship activity by dwarf seahorses (*Hippocampus zosterae*) observed across different courtship phases. (Source: Masonjones & Lewis 1996. Reproduced with permission of American Society of Ichthyologists and Herpetologists.)

the visual, olfactory, auditory, tactile and even electrical senses of potential mates. Courtship typically involves behavioural displays that draw attention to the displaying individual and which may accentuate the impact of ornaments. More than one mating signal may be used during courtship, which may comprise a series of multicomponent signals directed at different sensory modalities.

Visual

Visual signals are common in fish and the visual ecology of teleosts is relatively well studied, especially in coral reef fishes (e.g. Marshall, 2000). Light transmission in aquatic habitats is highly variable and images observed in water depend on water depth and what is suspended in the water. Consequently, visual signals used in courtship are often tailored to the photic environment in which species occur. Vision in teleosts is broadly comparable to that in terrestrial vertebrates, but is more variable, reflecting the greater variety of light regimes they experience in different aquatic environments (Nicol, 1989). Fish visual pigments comprise the protein opsin, which is bound to a vitamin A-based aldehyde. These pigments have the capacity to convert light to nervous impulses, with different pigments being sensitive to different light wavelengths (Shichida & Matsuyama, 2009). Most teleost fishes possess good trichromatic colour vision and in some cases tetrachromatic, with some species also able to see ultraviolet (UV) light (Losey *et al.*, 1999; Bone & Moore, 2008). Because ultraviolet light is absorbed by water, ultraviolet sensitivity is associated with species found in shallow or surface waters (Guthrie & Muntz, 1993). In the brown trout (*Salmo trutta*, Salmonidae), sensitivity to UV light changes with age. Young *S. trutta* occupy shallow water and possess ultraviolet sensitive visual pigments, while older fish occupy deeper water and lose the capacity to see short wavelength light (Bowmaker & Kunz, 1987).

Body size

A key visual trait, known to function in both intra- and intersexual selection, is body size. A preference for large male body size by females has been demonstrated in a number of species, although many studies purporting to show size effects may be confounded with differences in the genetic background of males used in trials, or other traits that may covary with size. One study that controlled for genetic background was that of Magellan *et al.* (2005) using *P. reticulata*. Magellan *et al.* (2005) presented female *P. reticulata* with males that were full siblings, but raised at different temperatures so that they grew at different rates and were of different sizes during preference trials. Females significantly preferred larger males, despite the limited genetic difference between males, suggesting a role for body size in female mate choice (Fig. 8.8).

Males also appear to demonstrate a preference for mating with larger females in some species. Male preference for larger females may be linked to the general correlation between female fecundity and body size in fishes (Chapter 11; Wootton, 1998), which means that directing courtship at larger females may be more profitable. In *A. flaviijosephi*, males spent more time courting images of larger females and also stayed longer in their vicinity (Werner & Lotem, 2003). Herdman *et al.* (2004) showed a preference for larger females in *P. reticulata*. In this case, male preference was modified by other cues that indicated female mating status, with males less likely to mate with large females that had a higher probability of having mated previously.

In sex role-reversed pipefish species, females prefer larger males (Rosenqvist & Berglund, 2011). Males also appear to prefer larger females, at least in the straight-nosed pipefish (*Nerophis ophidion*, Syngnathidae). Reproductive success is correlated

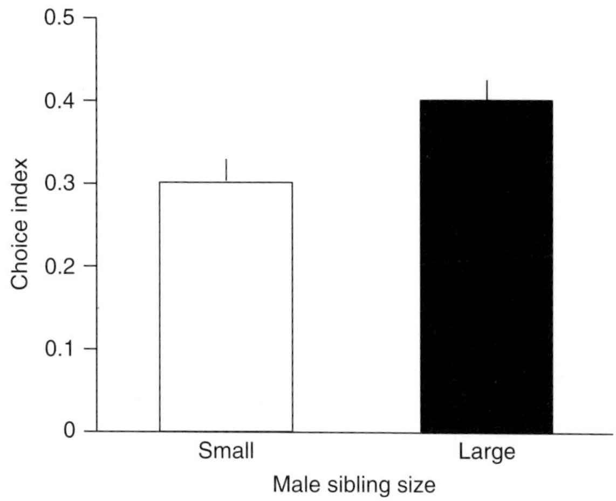

Figure 8.8 Mean mate choice index (+1 SE) by female guppies (*Poecilia reticulata*), scored as the ratio of time a female spent oriented towards a male to the time she spent in that male's proximity. 'Small' denotes smaller male sibling, 'Large' denotes larger male sibling. (Source: Magellan *et al.* 2005. Reproduced with permission of Springer Science + Business Media.)

with body size in both sexes; larger males can brood larger clutches, and larger females produce larger clutches. A similar pattern has been described in the paternally mouth-brooding Banggai cardinalfish (*Pterapogon kauderni*, Apogonidae). In this species, larger females produce larger eggs, which are of higher quality, while larger males are able to successfully brood larger clutches (Kolm, 2002). Both are examples of mutual mate choice.

Large body size may be preferred by females because it signals some aspect of male quality. Male body size may correlate with direct benefits to females, e.g. territory size and quality, or parental ability. Body size might also indicate additive genetic benefits. Large body size may be an honest indication of resistance to parasites, foraging success or longevity and may have a heritable component (Andersson, 1994).

Morphological trait size

In addition to body size, the size of specific traits that function as ornaments may be important in courtship. In many teleost species, fin size and shape may play a role that is independent of body size. In some populations of the green swordtail (*Xiphophorus helleri*, Poeciliidae), females prefer males with longer extensions to their caudal fin (termed a 'sword'), which the male displays to the female during courtship (Basolo, 1990). Sword size and body size correlate in this species, but where predators are abundant, sword size is relatively smaller, presumably because of constraints placed on sword evolution by predation (Basolo & Wagner, 2004). In *P. latipinna*, dorsal fin size can influence female mate preference by increasing the apparent size of a male (MacLaren *et al.*, 2004). Other ornaments include head crests, e.g. in blennies (Neat & Lengkeek, 2009), and breeding tubercles, which are keratin-based epidermal structures that sexually mature males (and sometimes females) develop during the breeding season in a number of fish families (Collette, 1977; Wedekind *et al.*, 2008). Like body size, the presence and size of these ornaments may signal direct or indirect benefits to potential mates, or they may simply represent arbitrary cues.

Colour

Many teleosts exhibit striking colouration during reproduction, and colour represents a significant visual ornament. The range of colours and patterns exhibited by fish is enormous. The rich colour patterns of teleosts are generated by the interaction of numerous pigment cells in the skin, termed chromatophores, which can alter the colour pattern by contracting and dilating under nervous control (Baerends & Baerends-van Roon, 1950). The costs of colour production have been considered trivial (Zahavi & Zahavi, 1997), although this may not always be the case. In particular, carotenoid-based colouration is probably costly to produce and many colour signals associated with courtship and mate choice utilise carotenoids. Carotenoids are complex molecules that selectively absorb short-wavelength light and consequently appear yellow or red. Carotenoids are obtained entirely in the diet by vertebrates and have a number of critical physiological functions, making them vital, and possibly limiting, components of the diets of teleosts. These features make carotenoid-based colouration a potentially honest indication of individual condition and foraging ability (Olson & Owens, 1998). There is evidence that red carotenoid-based pigments are considerably rarer in the environment than yellow carotenoids (Goodwin, 1980, 1984). While vertebrates have a limited capacity to synthesise red carotenoids from yellow, this appears to be an energetically costly process, and carotenoids are principally utilised for pigmentation in the form they are obtained in the diet (Hill, 1996).

One of the best-known examples of nuptial colouration is that displayed by male *G. aculeatus*. In this species, breeding males typically develop a red-orange throat, blue eyes and gray sides to their body, and a metallic blue sheen on the dorsal surface. The 'red' pigmentation of male sticklebacks is highly variable and is based on more than one carotenoid pigment, which may explain why male sticklebacks vary in hue so widely among populations (McLennan, 2007). Different carotenoid pigments may also convey different information to potential mates or rivals (McLennan, 2007), although this has yet to be tested systematically. Not all *G. aculeatus* populations develop red nuptial colouration. Some populations in North America are described as melanic, developing black not red colouration (Reimchen, 1989), which is the most likely ancestral state for male nuptial colour in this species (McLennan, 1996). On the Atlantic coast of North America, there are also populations that develop white nuptial colouration (Jamieson *et al.*, 1992). Nuptial colouration in *G. aculeatus* develops at the onset of the breeding season under hormonal control (Chapter 5) and is believed to have a role in male–male contests (Tinbergen, 1951), but also in influencing female mating decisions as an indication of direct mating benefits (Semler, 1971; Milinski & Bakker, 1992), through indirect additive genetic benefits (Barber *et al.*, 2001b), and as a sensory bias (Smith *et al.*, 2004b). Thus, even in a relatively well-studied system like *G. aculeatus*, the expression and function of colour as a visual cue appear to be unusually complex (Smith & Spence, 2013). A caveat to these and other studies is that, while many types of visual ornament can be measured quite readily, colour is notoriously difficult to quantify accurately (Endler, 1990), especially when colours of different hue are compared (Hill, 1996). Making the proper measurement of colour patterns in fish is an exacting field of research in its own right.

UV reflectance

Because humans are blind to UV light, the significance of UV reflectance in the colour patterns of teleosts can frequently be overlooked (Losey *et al.*, 1999), although there is evidence that UV-reflecting colouration plays a significant role in courtship in some

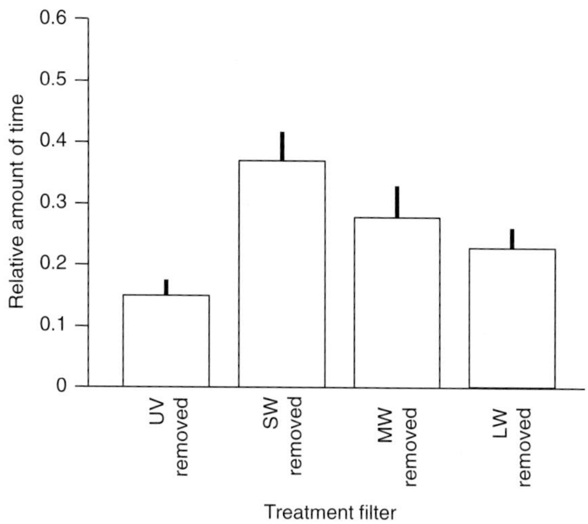

Figure 8.9 Mean relative time (+1 SE) spent by female three-spined sticklebacks (*Gasterosteus aculeatus*) in front of males viewed through filters that removed one of the following: UV wavelength light, short-wavelength light (i.e. blue light; SW); medium-wavelength light (green; MW); and long-wavelength light (red; LW). (Source: Adapted from Rick & Bakker 2008.)

species. In *P. reticulata*, UV plays a role in mating decisions. By presenting males to females placed behind screens that either blocked or allowed the passage of UV light, Smith *et al.* (2002b) demonstrated that UV reflectance could influence female mating decisions, although there was no significant effect of female UV reflectance on male responses to females. In the Panuco swordtail (*Xiphophorus nigrensis*, Poeciliidae), UV ornamentation may have evolved because it enables brightly coloured males to evade their chief predator, the Mexican tetra (*Astyanax mexicanus*, Characidae), which has limited UV sensitivity (Cummings *et al.*, 2003).

Using *G. aculeatus*, Rick & Bakker (2008) presented females that were ready to spawn with males that they viewed through filters that removed either UV wavelength light, short-wavelength (i.e. blue light), medium-wavelength (green) and long-wavelength (red) light. As predicted, there was a significant effect of the removal of long-wavelength light on female response to males. It is well established that female *G. aculeatus* in many populations show a mating preference for red males. However, Rick & Bakker's (2008) results also showed that the effect of removing UV wavelength light was comparable to filtering out red light (Fig. 8.9). This result implies that UV reflectance may be of equal importance to red colouration in some aspects of the mating system of *G. aculeatus*.

Display behaviour

Courtship often involves conspicuous display behaviour. The courtship of different species is highly variable among taxa, although related species often show similar patterns of courtship. Courtship displays are typically more elaborate in species that exhibit territoriality, nest-building and parental care (Wootton, 1998). Courtship behaviour often involves rhythmic or repetitive actions, rapid movements towards or away from a prospective mate, the extension of the fins or opercula, and rapid undulation of the body or fins. The duration and frequency of courtship behaviours may function in

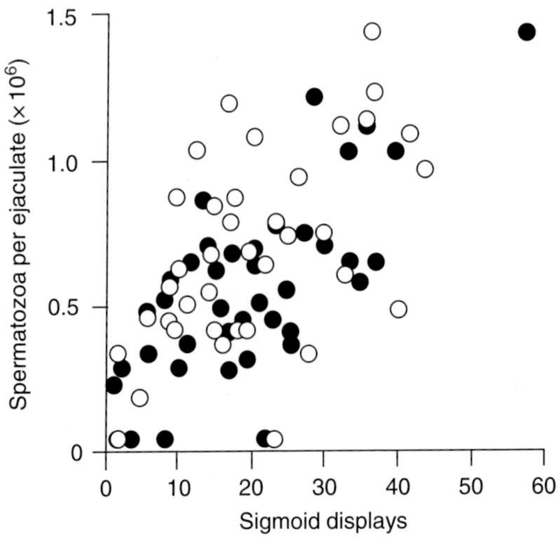

Figure 8.10 The relationship between numbers of spermatozoa per ejaculate and male sigmoid display behaviour for two populations of guppies (*Poecilia reticulata*) from Trinidad. Open circles, Tacarigua; closed circles, Tunapuna. (Source: Matthews 1999. Reproduced with permission of I.M. Matthews.)

mate choice, either to simply attract the attention of potential mates or to signal some aspect of individual quality as a mate.

In some cases, courtship displays may indicate mate quality. In *S. partitus*, courtship frequency advertised male parental ability and was correlated with the male's lipid reserves (Knapp & Kovach, 1991). In the Amur goby (*Rhinogobius brunneus*, Gobiidae), only the most vigorous males are able to perform courtship displays in fast-flowing water currents, and these are preferred by females as mates (Takahasi & Kohda, 2001). Males that are able to display in fast-flowing water were found to be more successful at raising eggs (Takahasi & Kohda, 2004). Direct benefits associated with courtship may also come from high rates of fertility. In two populations of *P. reticulata*, the rate of male sigmoid courtship displays was correlated with the number of spermatozoa in ejaculates (Matthews *et al.*, 1997; see Fig. 8.10).

If courtship vigour honestly indicates male quality, and therefore direct benefits to females during mate choice, an assumption is that courtship is an energetically expensive activity. Despite this key assumption, the energetic costs of courtship have never been competently measured in a teleost. Courtship may also attract the attention of predators and thereby carry a cost of an increased mortality rate. Courtship behaviour by males in wild populations of *P. reticulata* varied in relation to their risk of predation in nature (Magurran & Seghers, 1990). Males from the Upper Aripo in Trinidad, a site with few predators, performed conspicuous sigmoid courtship displays in the presence of a natural predator. Those from the Lower Aripo, by contrast, performed a lower proportion of sigmoid displays when exposed to predators, suggesting that courtship was risk-sensitive where there had been an evolutionary history of predation.

Auditory

Sound plays an important role in the lives of many fishes, including their reproduction (Myrberg, 1981; Hawkins, 1993; Amorim, 2006). Water transmits sound faster than in air and with a lower rate of attenuation. The gas-filled swim bladders of some species possess a connection to the ear that accentuates their sensitivity to underwater sounds (Hawkins, 1993). Fish produce sound using a variety of means, although the principal mechanism is by vibrations of the swim bladder initiated by contraction of sonic muscles (Ladich & Fine, 2006). Sounds tend to be of low frequency and low amplitude, but the types of sound generated by fish vary enormously among and within taxa (Amorim, 2006). Sound production and hearing tend to be correlated in fishes (Maruska *et al.*, 2007).

A role for auditory signals in territory establishment and courtship has been demonstrated in several groups. The Mozambique tilapia (*Oreochromis mossambicus*, Cichlidae) shows a correlation between sound production by males and courtship behaviour and subsequent spawning (Amorim *et al.*, 2003). Sound played a role in signalling the presence of a territorial male and might also play a role during spawning in synchronising sperm and egg release.

Among related species of Lake Malawi cichlid, male acoustic signals showed interspecific differences, which may play a role in facilitating species recognition during courtship (Amorim *et al.*, 2008). In this study, it was notable that components of the sounds produced by males varied among species, but also among individuals within species, raising the possibly of individual recognition based on acoustic signals.

In the sand goby (*Pomatoschistus minutus*, Gobiidae) and the related and sympatric painted goby (*Pomatoschistus pictus*, Gobiidae), males utilise low-frequency pulsed acoustic signals in their courtship. The two species generate distinct patterns of sound that permit them to be separated unambiguously (Pedroso *et al.*, 2013). In *P. minutus*, sound amplitude predicted male size, while calling rate predicted fat reserves. These findings imply that the frequency and temporal pattern of sound generated by males encoded species identity, while sound amplitude and call rate signalled male size and fat reserves (Pedroso *et al.*, 2013).

A role for sound signals in the courtship of fishes has been recognised in a wide range of unrelated fish taxa, including Gadidae, Pomacentridae, Centrarchidae, Sciaenidae, Mochokidae, Gobiidae, Cichlidae, Serranidae, Osphronemidae and Batrachoididae. Despite its widespread occurrence, the role of acoustic communication in mate choice and courtship in teleosts has hitherto received relatively little attention.

Olfaction and gustation

Fish can detect dissolved chemicals by smell (olfaction) and taste (gustation), and both provide a potential means of chemosensory communication in courtship and mating decisions. Olfactory cues can play a role in mate choice decisions in many fish, including poeciliids (McLennan & Ryan, 1997) and *D. rerio* (Spence *et al.*, 2008). In *G. aculeatus*, odour cues linked to MHC genes may play a key role in identifying compatible mates (Milinski, 2006), and the same appears to be the case in *R. ocellatus* (Agbali *et al.*, 2010; Reichard *et al.*, 2012).

Chemical cues in fish may be released from specialised exocrine glands (Chapter 5), or as water-soluble compounds from the gills or in the urine (Rosenthal & Lobel, 2006). The release of chemical compounds by fish is often associated with fin or body movements associated with courtship displays, possibly because the diffusion of compounds in water is relatively slow (Atema, 1996). Olfactory cues may also indicate readiness to mate. Male *P. reticulata* courted non-pregnant females more vigorously than pregnant females, presumably because non-pregnant females have a higher reproductive value. Males can probably assess female reproductive status using olfactory cues (Chapter 5).

Odours may also play a role in species recognition. Here the divergence of chemosensory traits that play a role in mating decisions in response to a variety of factors can potentially drive the evolution of prezygotic barriers (Smadja & Butlin, 2011). In some species there is sexual dimorphism in the development of chemoreception. Males of the deep-water bristlemouth (*Cyclothone* sp., Gonostomatidae), show greater development of the olfactory organ, which may function in mate finding (Marshall, 1967).

Tactile, electrical and other cues

Although relatively poorly researched, there is some evidence that tactile cues play a role in the courtship and mating decisions of some fishes. In the Mexican molly (*Poecilia sphenops*, Poeciliidae), males are polymorphic for the presence of moustache-like rostral filaments. These appear to have no sensory function, though females have a significant preference for males with the filaments (Schlupp *et al.*, 2010). Courtship and mating in *P. sphenops* involve contact between the male's snout and the female genital region. The filaments appear to convey tactile signals to the female (Schlupp *et al.*, 2010) and may have evolved through a sensory bias in this species (McCoy *et al.*, 2011).

Some nocturnal fishes, and others that inhabit turbid waters, possess the capacity for electroreception. This can be either passive or active. Passive electroreception involves detection of electrical fields in the environment, e.g. the muscular activity of living animals. In active electroreception, an individual measures the distortions to an electric field it produces itself with specialised electric organs. The electrical discharges of weakly electric elephantfishes of the genus *Campylomormyrus* (Mormyridae) serve in mate detection and discrimination, resulting in 'electric' mate recognition (Feulner *et al.*, 2009).

In an intriguing study, Shohet & Watt (2009) demonstrated that female *P. reticulata* preferred to associate with males that were most successful at completing a spatial learning task. Females showed no significant preference for males with greater orange saturation, contrary to other studies with this species (Magurran, 2005). There was also no correlation between male spatial cognition and orange saturation. While females preferred males with the most rapid rates of learning, the cues that females used to select males were uncertain, although they may have been correlated with courtship rates. An explicit link between cognition and reproductive success has yet to be demonstrated in fish, although there is some evidence in other vertebrate taxa (Boogert *et al.*, 2011).

Multiple cues

Males of many species exhibit several ornaments simultaneously. If all visual, auditory, olfactory/gustatory, tactile, electrical cues and courtship displays are considered, it is clear that fish display multiple cues that potentially play a role in mate choice. Consequently, the focus of many previous mate choice studies on single traits alone may be overly simplistic. Even those that deal with more than one trait often combine information on sensory inputs from just two modalities, such as olfactory and visual cues. The question of why many species exhibit multiple cues that play a role in mate choice presents an intriguing evolutionary question. Møller & Pomiankowski (1993) offered three hypotheses to explain the evolution of multiple ornaments. The 'multiple message hypothesis' argues that each ornament separately reflects a different aspect of the quality of an individual. The 'redundant signal hypothesis' posits that each signal gives a partial indication of condition, such that all traits taken together give an estimate of overall quality. Finally, the 'unreliable signal hypothesis' proposes that some ornaments are unreliable indicators of quality, but are maintained because they are not costly to produce and may influence mating decisions, though only marginally in some cases. Møller & Pomiankowski (1993) predicted that this last hypothesis should be most common where the strength of sexual selection is greatest. Under circumstances where conditions for the unreliable signal hypothesis are met, the relative importance of traits in female mate choice may depend more on their 'detectability' than on their reliability as signals (Schluter & Price, 1993). An additional hypothesis is the 'back-up hypothesis' (Johnstone, 1996), whereby each additional ornament serves to enhance the accuracy with which an individual's quality can be accurately assessed. Studies that examine multiple cues in mate choice studies in fish are rare. Some studies have demonstrated that female mating preferences change when different cues are presented to them experimentally, providing support for the multiple signal hypothesis. In the Lake Malawi cichlid *Pseudotropheus emmiltos* (Cichlidae), female tendency to associate with males differed if the females were presented only with visual cues compared with visual and olfactory cues (Plenderleith *et al.*, 2005). In a study that provides support for the back-up hypothesis, Hankinson & Morris (2002) showed that female pygmy swordtails (*Xiphophorus pygmaeus*, Poeciliidae) risk hybridisation with the related delicate swordtail (*Xiphophorus cortezi*, Poeciliidae). The ability of female *X. pygmaeus* to discern conspecifics was enhanced with visual and olfactory cues compared with each cue in isolation.

Probably the most comprehensive attempt to date to understand the action and role of multiple cues in mate choice is that of McLennan (2007) in *G. aculeatus*. In this species, a series of visual and olfactory cues play a key role in mate choice decisions by females. Different cues, from the same or different sensory modalities, appear to indicate different aspects of male quality or suitability as a mate. Olfactory cues may transmit information about sex, species, reproductive state and degree of genetic compatibility, while visual cues may indicate to females direct and indirect benefits of male quality. McLennan (2007) also recognised the potentially hierarchical nature of the way in which mate choice cues might operate. Thus, gravid females may initially be attracted to sites where breeding males are found by olfactory cues that indicate the presence of males with nests. Once on the breeding grounds, females may selectively attend to the most brightly coloured males, but then select to mate only with those brightly coloured males with the optimal degree of genetic dissimilarity measured through MHC-based olfactory cues. The way in which individuals integrate information based on multiple cues and use this information for making mate choice decisions remains a key question in understanding teleost mating systems.

A final consideration is the way in which changes to the sensory environment of fishes impinge on mating decisions. Anthropogenic effects in particular can be rapid, and aquatic environments appear to be particularly susceptible to change. Relatively few studies have tackled this question (van der Sluijs *et al.*, 2011), although it is an increasingly pressing problem with implications for mating system evolution and the maintenance of reproductive isolating mechanisms (Wong *et al.*, 2007).

Sexual signals and reproductive isolation

The evolution of sexual signals can potentially result in reproductive isolation, thereby driving cladogenesis, the process of speciation. Mating preferences can create reproductive barriers through one of two mechanisms. Sexual isolation arises through divergent mate preferences, with the result that individuals from one population are less likely to mate with those from a second population (Rundle & Boughman, 2010). Sexual selection against hybrids occurs when the hybrids of two populations are unsuccessful in obtaining matings because they fail to attract mates (Schluter, 2000). These mechanisms can result in allopatric speciation, but also sympatric speciation, a controversial idea that has taxed evolutionary biologists for decades. An example of sexual selection against hybrids is demonstrated by 'species pairs' or ecotypes of *G. aculeatus* in North America. In a small number of lakes, sticklebacks have diverged into two races: benthic and limnetic. Benthic sticklebacks are relatively deep-bodied, with less bony armouring and fewer, shorter gill rakers. Benthic *G. aculeatus* tend to forage on the substrate. By contrast, limnetic *G. aculeatus* are more slender-bodied, possess bony armour and have numerous long gill rakers, which adapts them to feeding on zooplankton in the water column (McPhail, 1993). The divergence of these forms dates from the last ice age, approximately 13 000 years ago, and probably arose through independent invasions of the same habitat on different occasions. The two forms are largely reproductively isolated, although they produce viable hybrids in the laboratory. The maintenance of the two ecotypes under natural conditions appears to be through selection against F_1 hybrids in mating competition. Male F_1 hybrids of benthic and limnetic parents from the same lake had limited reproductive success in competition with male limnetics for matings with female limnetics, suggesting a role for sexual selection against hybrids (Vamosi & Schluter, 1999).

Another example of sexual isolation is found with cichlid fishes from the African Great Lakes. In Lakes Victoria and Malawi, the cichlid faunas have diverged relatively recently at an unusually rapid rate, with more than 1500 species endemic to the region (Ribbink, 1991; Turner, 1999) and with endemic species within lakes often sharing a common ancestor (Meyer, 1993). One hypothesis to account for this rapid radiation is through strong female assortative mating for males from the same population and for particular colour forms. Female preferences are heritable and involve relatively few genes (Haesler & Seehausen, 2005; Barson *et al.*, 2007). The effect of strong female mating preferences may be reinforced by intrasexual competition between males for nesting sites. Field and laboratory studies suggest that males exclude conspecific males, or those belonging to closely related species of the same nuptial colouration, resulting in negative frequency-dependent selection and enabling rare male colour morphs to invade against the predominant female preference. This effect could then impose positive selection on rare female preference genotypes with a preference for the rare male colour, as the rare morph will be associated with above-average fitness (Seehausen & Schluter, 2004). Thus, sexual selection, rather than ecological speciation (*sensu* Rundle & Nosil, 2005), is proposed as the mechanism driving the rapid radiation of cichlids. Notably, different cichlid species and morphs appear to differ more in colour and other secondary sexual traits than in morphological traits linked to resource use. However, this view may be overly simplistic. Resource differences may be responsible for driving divergence in different environments that are difficult to observe or recreate in the laboratory, e.g. through predation or interference competition. The interaction of sexual and natural selection is also frequently ignored in attempts to link sexual selection and speciation (Ritchie, 2007; Maan & Seehausen, 2011) and, though frequently proposed as an important mechanism of speciation, wholly convincing evidence for a role of sexual selection as a mechanism of reproductive isolation is wanting (Ritchie, 2007). Sexual selection has the capacity both to promote reproductive isolation, e.g. through strong female assortative mating, and to erode reproductive isolation, e.g. through alternative male mating tactics (see later). A key question is to identify the order in which isolating mechanisms develop, and whether sexual selection plays a role in directly generating reproductive isolation, rather than acting as a secondary force alongside ecological selection (Coyne & Orr, 2004; Ritchie, 2007).

Cichlid and stickleback diversification is not unique among fishes, and several taxa also show striking rates of radiation and speciation, including Lake Baikal sculpins (Cottidae) (Sherbakov, 1999), Lake Tana *Labeobarbus* spp. (Cyprinidae) (de Graaf *et al.*, 2008), Lake Lanao barbs (*Puntius* spp., Cyprinidae) (Meyers, 1960) and Lake Titicaca pupfish (*Orestias* spp., Cyprinodontidae) (Northcote, 2000). These rapid radiations are, conspicuously, associated with lacustrine habitats. Understanding the extent to which these speciose groups have evolved through selection acting on sexual signals represents a challenge for fish evolutionary biologists.

Fertilisation

The interaction of spermatozoon and egg constitutes the process of fertilisation (Longo, 1987; see Chapter 4 for a fuller treatment). Fertilisation is completed when a single spermatozoa penetrates the egg plasma membrane via a microscopic cone-shaped aperture termed the micropyle (Chapter 4). Polyspermy, penetration of multiple spermatozoa, is efficiently inhibited in teleost fishes, although the process does occur in other fish taxa, but in these cases only a single spermatozoon nucleus fuses with that of the egg. Teleosts exhibit both external and internal fertilisation, as well as modes of fertilisation that are intermediate between the two (Chapter 10). Fish also possess specialised accessory sexual characters to facilitate fertilisation. Many adaptations for fertilisation are underpinned by competition among males in the form of sperm competition.

Sperm competition

A key insight to understanding the evolution of mating systems came with the recognition that males can engage in post-copulatory (in external fertilisers, post-ejaculatory) contests for fertilisations through the process of sperm competition (Birkhead & Montgomerie, 2009). Sperm competition occurs when the spermatozoa of different males compete for the fertilisation of a set of eggs (Parker, 1970). A rich body of evolutionary theory has been developed to account for morphological and behavioural traits associated with sperm competition (Ball & Parker, 1997; Parker *et al.*, 1997; Parker, 1998), some explicitly for fishes (Parker *et al.*, 1996; Stockley *et al.*, 1997; Petersen & Warner, 1998), which offer outstanding material for research on this subject.

Cost of sperm production

Although the energetic costs of spermatogenesis have not been quantified in fish, circumstantial evidence suggests that male reproductive success can be limited by sperm production. In the Lemon tetra (*Hyphessobrycon pulchripinnis*, Characidae), Nakatsuru & Kramer (1982) allowed males to spawn repeatedly with receptive females over a series of daily repeated spawning acts. Eggs were collected in small batches and incubated to measure the number of eggs that had been successfully fertilised. The proportion of eggs fertilised declined as a function of the number of spawning acts performed by the male, implying that males had experienced a decline in the quantity or quality of sperm released at successive spawnings.

Smith *et al.* (2009) performed a comparable study using *R. amarus*, a species that spawns in the gills of freshwater mussels. Here males were presented with a female and a mussel to which several cannulas were attached, which enabled entire ejaculates released by a male to be collected. The number of spermatozoa released in a series of ejaculates could thereby be quantified. Ejaculate size was measured for five successive ejaculations during a mating in the morning, at midday and in the late afternoon. In addition, half the males were allowed to see a rival male. The number of spermatozoa per ejaculate declined over successive ejaculations in each mating, and overall spermatozoa number declined over the course of a day of matings. Male *R. amarus* responded to the presence of a rival male by increasing the size of ejaculates. However, over the course of a day of 'competing' with a rival, males suffered a steeper decline in ejaculate size than solitary males (Fig. 8.11).

These and other studies on teleosts, and in other taxa, suggest that sperm production is limiting to male reproduction and that males can become sperm-depleted (Wedell *et al.*, 2002). Whether males become sperm-limited because sperm is costly to produce or because sperm storage is limiting is not clear. An important corollary is that females potentially risk a limitation to their fertility through male sperm depletion, especially in species with external fertilisation (Levitan, 1998), with consequences for the ways in which females distribute eggs among males and males distribute sperm among females.

Risk and intensity of sperm competition

Sperm competition theory seeks to predict optimal sperm allocation strategies for males in relation to the risk and intensity of sperm competition. The risk of sperm competition is the probability that a male's sperm will compete with the sperm of other males. The intensity of sperm competition is the extent of overlap of the ejaculates of different males, and is therefore a function of the number of males that engage in sperm competition and the quantity of sperm they contribute to a mating. The risk and intensity of sperm competition vary greatly among fish species, from those in which sperm competition is rare, such as pipefish and seahorses (Syngnathidae), to promiscuous group-spawning species in which males and females shed gametes in large spawning aggregations, as seen in many clupeids. Across species, the prediction is that with increased competition among the ejaculates of rival males, selection will tend to increase investment in sperm competition. An assumption of this model of sperm competition is that sperm from different males has an equal probability of fertilising a female's eggs, a state often referred to as a 'fair raffle' (Parker, 1998). An outcome of a fair raffle is that those males that release larger ejaculates will fertilise a greater share of eggs, with the consequence that sperm competition will select for greater sperm production. In fishes there is evidence to support this prediction. Stockley *et al.* (1997) categorised freshwater fishes on an ordinal scale with respect to the sperm competition intensity they experience and, controlling for phylogenetic history, examined the association with a number of reproductive traits. They demonstrated a correlation between intensity of sperm competition and relative testis size and an index of sperm quantity.

In contrast to among-species comparisons, within species the predictions of sperm competition theory with regard to the intensity of sperm competition are different and less intuitive. Males are predicted to maximise their expenditure on fertilisation in competition with a single competitor, but to reduce expenditure as the number of rivals exceeds one, as the probability of fertilising eggs diminishes with the number of competing males and increasing sperm competition intensity (Parker *et al.*, 1996; see Fig. 8.12). A number of studies provide support for this model in teleosts.

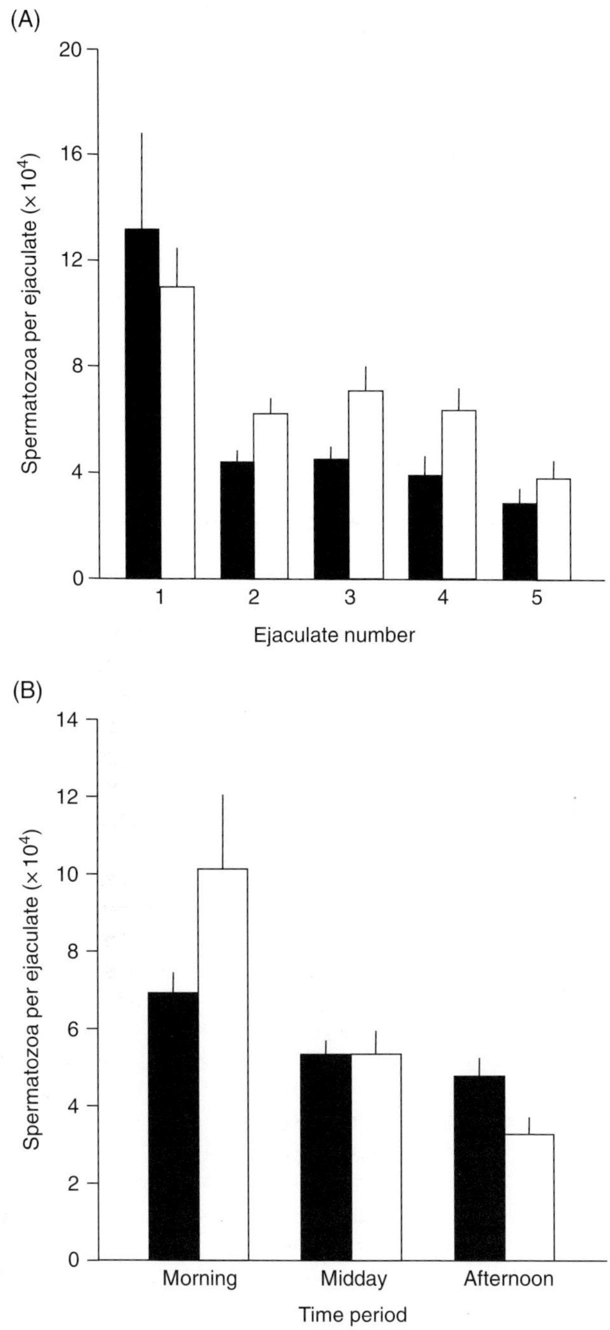

Figure 8.11 Sperm depletion in territorial male European bitterling (*Rhodeus amarus*) (A) Mean (+1 SE) numbers of spermatozoa per ejaculate in five consecutive ejaculates by territorial males at low (black bars) and high (white bars) levels of sperm competition risk. (B) Mean (+1 SE) numbers of spermatozoa per ejaculate in the morning (09.00–10.00), at midday (12.00–13.00), and in the afternoon (15.00–17.00) of territorial male *R. amarus* at low (black bars) and high (white bars) levels of sperm competition risk. (Source: Smith *et al.* 2009. Reproduced with permission of Elsevier.)

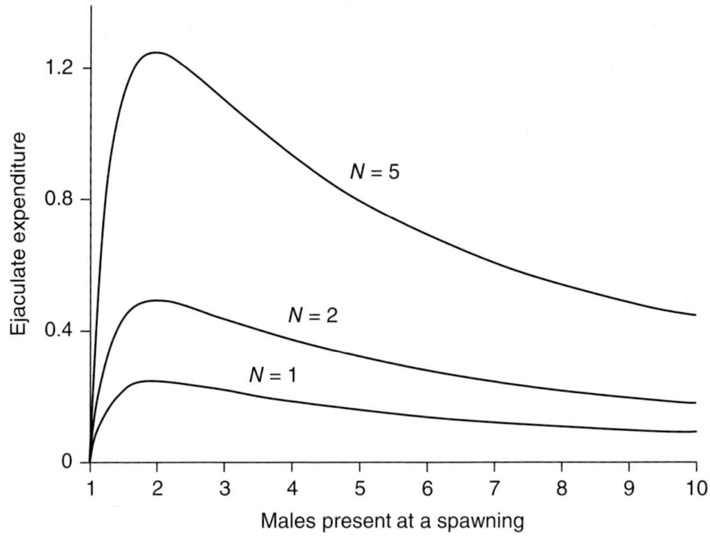

Figure 8.12 Predicted male sperm allocation when an ejaculate faces sperm competition from several other ejaculates (expressed relative to the average total reproductive effort per spawning) in relation to the number of males competing at a spawning. The curves are for different populations or species with different mean numbers of males present at a spawning. Spermatozoa are expected to compete in a 'fair raffle' and males are considered able to assess the number of competitors at the time of ejaculation. (Source: Wedell *et al.* 2002. Reproduced with permission of Elsevier.)

Sperm competition avoidance

Males sometimes display adaptations for forestalling sperm competition. In *R. amarus*, males actively lead females to freshwater mussels, which they use as spawning sites, without other males in close proximity or in which sperm has not recently been released (Smith *et al.*, 2002a). For dominant males in territorial species, an alternative, or additional, response to sperm competition is aggression (Petersen & Warner, 1998). Aggressive defence of females or sites of reproduction is common in many taxa, and represents a mechanism for avoiding or reducing the intensity of sperm competition. The conditions that determine whether an adaptive response to sperm competition is through aggression or adaptations for sperm competition, such as increased teste size, are uncertain (Petersen & Warner, 1998). Although not mutually exclusive, each makes different demands on males, which may result in behavioural, physiological or morphological trade-offs.

Pre-oviposition ejaculation

In some teleosts with external fertilisation, gamete release is asynchronous, with males releasing spermatozoa before females deposit their eggs. This behaviour is termed pre-oviposition ejaculation and has been recorded in *G. aculeatus* (Sevenster-Bol, 1962) and two bitterling species (Cyprinidae), *R. amarus* and *R. ocellatus* (Kanoh, 2000; Smith *et al.*, 2004a). Pre-oviposition ejaculation is also a feature of those species that produce mucous seminal fluids (see the section 'Seminal fluid composition'). Male *G. aculeatus* release sperm into their nest while performing a component of their courtship behaviour termed 'creeping through' (Sevenster-Bol, 1962), which involves the male passing through their nest prior to the female entering to deposit her eggs (Wootton, 1976). In bitterling, males perform pre-oviposition ejaculations over mussels, both in the role of a territorial male and as 'sneaker' males (see the section 'Male mating polymorphisms'). Using paternity analysis in a captive but free-ranging population of *R. amarus*, Reichard *et al.* (2004a) demonstrated that sperm released by a male a minimum of 14 minutes prior to egg deposition by a female resulted in successful fertilisation. Pre-oviposition ejaculation may be advantageous in ensuring precedence in fertilisation in species with predictable spawning sites in which fertilisation occurs in a microenvironment, such as a mussel (bitterling) or nest (stickleback), that limits the dispersal of sperm (Smith & Reichard, 2013). Despite being recorded in few species, pre-oviposition may be an overlooked feature of many fish species that spawn in confined spaces and experience sperm competition.

Seminal fluid composition

The chief components of seminal fluid are lipids, proteins and free amino acids. Seminal fluid also exhibits phosphatase, β-glucuronidase and protease activity. Monosaccharides are present and function in spermatozoa nutrition in species with sperm storage (Lahnsteiner *et al.*, 1993). An additional component of seminal fluid in some species is a sialoglycoprotein-rich fluid, termed sialomucins or just mucins (Chapter 4). The production of mucin-rich seminal fluids is not common in fishes, although it is reported in some gobies (Gobiidae), toadfish (Batrachoididae), cottids (Cottidae), blennies (Blenniidae), bitterling (Cyprinidae) and several families of catfish (Siluriformes) (Miller, 1984; Mazzoldi, 2001; Pateman-Jones *et al.*, 2011).

The role of mucins in fertilisation is to slowly release active spermatozoa over an extended period after ejaculation, which is accomplished by the gradual dissolution of the mucin in water (Marconato *et al.*, 1996; Scaggiante *et al.*, 1999). Mucins greatly extend the functional longevity of sperm and, consequently, the period over which fertilisation can occur. The period over which spermatozoa are released is dependent on the mucin content of the seminal fluid. Seminal fluid mucins play a role in fertilisation efficiency by minimising the amount of sperm that is released, thereby reducing the costs of sperm production. The role of mucins is best studied in gobies and blennies (Marconato *et al.*, 1996; Mazzoldi, 1999; Giacomello *et al.*, 2008). Male gobies, which typically nest in holes or small caves, deposit mucus bands, termed 'sperm trails', inside their nests during spawning (see also Chapter 4). Female egg deposition can take several hours and sperm trails release spermatozoa throughout the period of oviposition, thereby ensuring efficient fertilisation. By releasing spermatozoa in sperm trails, males are also liberated for mate and nest guarding during oviposition (Marconato *et al.*, 1996). In species that experience a high risk of sperm competition, mucus secretion and the structures associated with its production, such as 'sperm-duct' glands (Chapter 4), are well developed. The presence of mucins in seminal fluids is not confined to external fertilisers and they are often also highly developed in some internally fertilising species.

Sperm capacitation and motility

The capacitation of spermatozoa after ejaculation is associated with changes in pH, and ionic and oxygen concentration, and, in some species, with the presence of seminal fluid proteins (Morisawa *et al.*, 1983, Morisawa, 1985; Lahnsteiner *et al.*, 2004; Alavi &

Cosson, 2005; Schulz *et al.*, 2010; see Chapter 4). In marine species, spermatozoa typically activate on exposure to an increase in osmolarity, with the reverse occurring in freshwater (Kinsey *et al.*, 2007). After activation, the spermatozoa of most fish species swim for a short period of time, in the range of seconds to several minutes, with motility typically maintained for longer in marine than in freshwater species (see Chapter 4 for a fuller discussion). After this period, spermatozoa cease movement and effectively lose their capacity to fertilise eggs. However, there is evidence that sperm motility can be revived in some species. The mechanism by which fish spermatozoa can resume motility may be dependent on an altered external osmolarity leading to changes in their internal ionic composition, and ultimately flagellar activity. In *G. aculeatus*, Le Comber *et al.* (2004a) showed that the capacity of spermatozoa to fertilise eggs correlated positively with their motility. After 5–10 minutes, spermatozoa motility declined to half that immediately after release, and fertilisation rates were negligible. However, the addition of a saline solution returned sperm motility to its original rate. Interestingly, the ovarian fluid of female *G. aculeatus*, which is released by the female during oviposition, has the same effect on the spermatozoa, extending its viability from minutes to hours (Elofsson *et al.*, 2006). This means that sperm deposition by male *G. aculeatus* in the nest prior to spawning is an effective mating tactic, as oviposition by the female will have the effect of reactivating spermatozoa in the nest.

Spermatozoa motility and longevity play a key role in sperm competition (Pizzari & Parker, 2009). Faster-swimming sperm and those that were motile for longer had a greater probability of finding and fertilising an egg in Atlantic salmon (*Salmo salar*, Salmonidae) (Gage *et al.*, 2004) and *P. reticulata* (Boschetto *et al.*, 2011). Spermatozoa longevity and motility are potentially negatively correlated, especially if rapidly swimming spermatozoa deplete their energy resources quickly (Levitan, 2000). However, there is no evidence in *S. salar* that these parameters are negatively correlated (Gage *et al.*, 2002), and a positive correlation between spermatozoa longevity and motility has been detected in some cases (Rudolfsen *et al.*, 2008).

While spermatozoon length has sometimes been equated with swimming speed (Pitcher *et al.*, 2007), this may not always be the case. Instead, the ratio of spermatozoon flagellum length to head length was proposed as an index of sperm swimming speed (Humphries *et al.*, 2008). Estimates of this ratio correlated with predicted differences in sperm competition intensity among three bitterling species (Pateman-Jones *et al.*, 2011). An additional parameter that will influence male competitiveness is the proportion of sperm motile in an ejaculate, which appears to show wide variation but has so far received little attention (Snook, 2005).

Ejaculate size and ejaculation frequency

Ejaculation rates are high in some species. Male *R. amarus* can perform over 250 ejaculations each day during the spawning season (Smith *et al.*, 2009). Male *R. amarus* are able to ejaculate at this rate because they display semi-cystic spermatogenesis (Chapter 4), which permits them to produce numerous small ejaculates. Economical partitioning of ejaculates is also a feature of several blenny species (Blenniidae), which exhibit the same mode of spermatogenesis (Giacomello *et al.*, 2008; see Chapter 4).

There is evidence that male fishes of some species are able to modulate ejaculation size. Male bluehead wrasse (*Thalassoma bifasciatum*, Labridae) produced larger ejaculates in pair spawnings with bigger females that release more eggs. They also produced larger ejaculates during group spawnings when the intensity of sperm competition was greater (Shapiro *et al.*, 1994). Males possess precise muscular control over sperm release, which permits them to modulate ejaculate size (Rasotto & Shapiro, 1998). Males of the grass goby (*Zosterisessor ophiocephalus*, Gobiidae) and black goby (*Gobius niger*, Gobiidae) both adjust ejaculate size in response to sperm competition (Pilastro *et al.*, 2002), and the same response was observed in *R. amarus* (Smith *et al.*, 2002a). A comparable mechanism for controlling ejaculate size to that seen in *T. bifasciatum* appears to operate in these species (Pilastro *et al.*, 2002; Pateman-Jones *et al.*, 2011).

External fertilisation

External fertilisation is widespread in fishes and differs in crucial ways from internal fertilisation, with implications for mating system evolution. For external fertilisers, the typical pattern of spawning is that females oviposit in the water column or on a substrate and males shed sperm over them, which have a limited interval in which to fertilise the eggs (Chapter 1).

An important feature of external fertilisation that differs from internal fertilisation (Chapter 10) is that female fertility can be limited by sperm availability in external fertilisers. While it is generally considered that, because many more spermatozoa than eggs are produced, fertility rates should not be limiting to females, this does not always appear to be the case in externally fertilising species. In *T. bifasciatum*, clutches of eggs were collected after natural spawning and the proportion that had been fertilised was estimated (Petersen *et al.*, 1992). The median fertilisation rate was 76.5%, and in some cases less than 10% of egg clutches were successfully fertilised (Fig. 8.13). Variables that appeared to influence fertilisation success included proximity of the male and female when gametes were released, and the velocity of water movements, with rougher water significantly reducing fertilisation success. A similar pattern was observed in another wrasse, the slippery dick (*Halichoeres bivattatus*, Labridae) (Petersen, 1991). In this species, fertilisation success was increased when an additional 'streaking' male participated in spawning. A consequence is that male traits often considered as adaptations

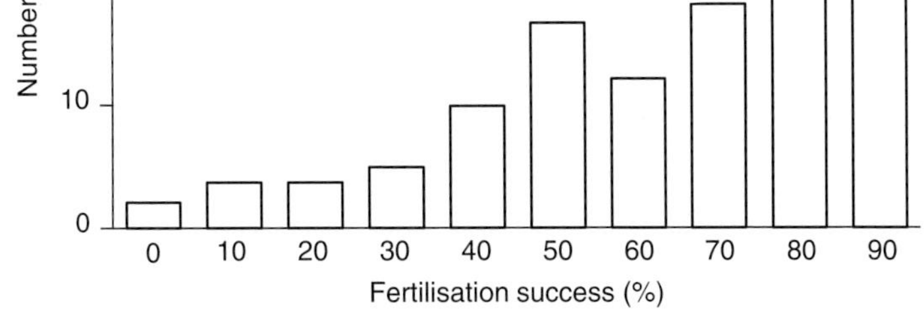

Figure 8.13 Frequency distribution of fertilisation success for 138 natural spawnings by bluehead wrasse (*Thalassoma bifasciatum*) with 20 or more eggs collected. (Source: Petersen *et al.* 1992. Reproduced with permission of the Ecological Society of America.)

for sperm competition, such as testis size, ejaculate size, ejaculation frequency and sperm motility, can also evolve in response to sperm limitation. Consequently, species with external fertilisation, in which fertilisation efficiency tends to be relatively lower than for internal fertilisers, typically possess relatively bigger testes and produce larger ejaculates than taxa with internal fertilisation (Levitan, 1998). These traits are not necessarily adaptations for sperm competition.

Another key difference between external and internal fertilisation is that female external fertilisers are able to exert little post-copulatory choice over which male fertilises their eggs. Conversely, external fertilisation enables males to exercise control over a mate, and extended mate guarding to protect paternity is unnecessary, as males need only be vigilant during oviposition. Male mating polymorphisms are still common in external fertilisers (DeWoody & Avise, 2001), but are usually more apparent in internal fertilisers. A possible evolutionary outcome of these conditions is the widespread occurrence of paternal care in externally fertilising fishes and its absence among internal fertilisers in which male control of paternity is limited (Chapter 9).

Internal fertilisation

Internal fertilisation involves fusion of the gametes inside the female reproductive tract. This mode of fertilisation is associated with the evolution of viviparity and internal bearing, although not exclusively. A fuller discussion of internal fertilisation is presented in Chapter 10. Points specifically relevant to mating system evolution are presented here.

To accomplish the transfer of sperm from males to females, fish have evolved a variety of copulatory structures. In teleosts, male intromittent organs are found in a number of families and have evolved independently on several occasions. These copulatory structures are highly variable in form, derived from a variety of tissues, and often bear different names depending on the taxonomic group in which they are found (for a systematic review, see Evans & Meisner, 2009; see also Chapter 10). Male copulatory structures appear to evolve rapidly and have diverged in fishes, and other taxa, to an unusual degree. They are clearly linked to male fertilisation success and, as such, appear to experience strong sexual selection, although the mechanism (whether it be through sperm competition, female choice or sexual conflict) is not always clear (Hosken & Stockley, 2004).

In species with internal fertilisation, spermatozoa are often organised into bundles termed spermatophores (Chapter 10). In others, they are arranged into unencapsulated sperm bundles termed spermatozeugmata, e.g. in viviparous kelp blennies (Clinidae) (Fishelson *et al.*, 2006) and *P. reticulata* (Magurran, 2005). While associated with external fertilisation, mucins also occur in the ejaculates of internal fertilisers, where they perform the same function seen in external fertilisers of increasing fertilisation efficiency by extending the period over which motile spermatozoa are released (Ginzburg, 1968).

Sperm storage occurs in some internal fertilisers. The significance of stored sperm for the mating system of *P. reticulata* has long been recognised (reviewed by Magurran, 2005). Females can produce broods in the absence of males for up to 8 months in this species, although the capacity to store sperm appears to vary among populations. Stored spermatozoa are used to fertilise eggs even if the female mates with additional males, but the fertilisation success of stored spermatozoa is typically lower than that from recently deposited sperm. The degree to which the ejaculates of different males, deposited at different times, interact remains to be fully elucidated in livebearers

(Magurran, 2005), although variables including male density (Soucy & Travis, 2003), relatedness (Ala-Honkola *et al.*, 2009) and maternal provisioning (Leips *et al.*, 2009) may all play a role. Sperm storage has the effect of greatly increasing the intensity of sperm competition. It also potentially enables a female to exercise control over which male fertilises her eggs. Cryptic female choice, which is recognised in other taxa, especially insects (Eberhard, 1996), is poorly understood in teleosts.

At least 18 teleost families representing nine orders exhibit internal fertilisation. In addition to the livebearers (Poeciliidae), other important teleost groups that display internal fertilisation are found in the New World rivulines (Rivulidae), goodeids (Goodeidae), four-eyed fishes (Anablepidae), kelp blennies (Clinidae), characids (Characidae), halfbeaks (Hemiramphidae) and sculpins (Cottidae).

Quasi-internal fertilisation

In some species, fertilisation is nominally external, but takes place in a site so restricted that it is analogous to internal fertilisation. One example is buccal fertilisation, seen in many mouth-brooding cichlids. In these fishes, the eggs are deposited by the female on the substrate, but are immediately picked up in her mouth. In many species, males possess coloured spots on their anal or pelvic fins, termed 'egg spots', which resemble the female's egg in size and colour (Fryer & Isles, 1972). Female cichlids respond to egg spots by nipping at the male's fins, and the male releases sperm into her buccal cavity where the eggs are fertilised (Fryer & Isles, 1972; Mrowka, 1987). Some cichlid species without well-developed egg spots also show this mode of fertilisation, so this trait is not a prerequisite for buccal fertilisation.

Sperm drinking is also a feature of the mating system of the bronze corydoras catfish (*Corydoras aeneus*, Callichthyidae). The female collects sperm in her mouth from the male's genital opening. The live spermatozoa pass rapidly through the length of her digestive tract and are discharged along with her eggs, which she releases into a 'pouch' created by her paired pelvic fins (Kohda *et al.*, 1995). The rapid passage of spermatozoa through the female's digestive tract is possible because this fish exhibits intestinal breathing, whereby air is gulped at the surface and passes through the stomach to the intestine where it is absorbed. This adaptation appears to facilitate the rapid passage of water and spermatozoa through the intestine and to the eggs (Kohda *et al.*, 1995). This mode of fertilisation may permit the female to exercise choice over which male fertilises her eggs.

In the bitterling fishes (Cyprinidae), fertilisation occurs in the gill chambers of living freshwater mussels. Female bitterling deposit their eggs into the gill chamber of a mussel with an unusually long ovipositor and males fertilise the eggs by releasing sperm into the inhalant siphon of the mussel, so that water filtered by the mussel carries the sperm to the eggs (Smith *et al.*, 2004a). The mussel is a functional extension of the female reproductive tract and bitterling spermatozoa exhibit specialised adaptations to enable them to function in this environment (Pateman-Jones *et al.*, 2011)

In all seahorses and pipefishes of the genus *Syngnathus* (Syngnathidae), males possess pouches in which eggs and larvae are brooded and where fertilisation occurs. Spawning is accomplished by the female using her ovipositor to deposit the eggs inside the pouch, with egg transfer taking only a few seconds, after which the pouch is closed. Sperm is released inside the pouch to fertilise the eggs (Chapter 10).

In several sculpin species, including the scalyhead sculpin (*Artedius harringtoni*, Cottidae) and possibly also the padded sculpin (*Artedius fenestralis*, Cottidae) and smoothhead sculpin (*Artedius lateralis*, Cottidae), males possess an intromittent organ and sperm are deposited internally but fertilisation occurs externally (Munehara *et al.*, 1989; Petersen *et al.*, 2005). Sperm atozoa associate with ovulated eggs in the ovary, but either do not enter the micropyle or penetrate the egg ooplasm. Fertilisation of the egg is not completed until the egg is laid and exposed to seawater (Petersen *et al.*, 2005). In the related *Alcichthys alcicornis* (Cottidae), sperm enter the micropyle of ovulated eggs, but fail to penetrate into the egg ooplasm until oviposition (Munehara *et al.* 1989). This pattern of fertilisation is termed internal gamete association (IGA). Paternal care is performed in species with IGA, but estimates of paternity in these fishes indicate that the care-giving male often has not fathered all the eggs under his care. Petersen *et al.* (2005) proposed that paternal care is able to persist in this mating system because males with clutches of eggs in their nest have a significant mating advantage through mate choice, and possibly also egg survival through a dilution effect (Chapter 10).

Male mating polymorphisms

Sexual selection is driven by variation in mating success and results in the evolution of a mating strategy, within the constraints of a given mating system, which optimises reproductive success. In some species, there may be more than a single successful mating strategy or 'mating niche', resulting in the evolution of alternative mating phenotypes. Mating polymorphisms are associated with strong sexual selection; natural selection probably plays only a trivial role in their evolution (Shuster, 2010). Because sexual selection almost always acts more strongly on males than on females, mating polymorphisms are typically associated with males.

Male mating polymorphisms often manifest themselves in the form of discontinuities in traits under sexual selection. In most cases, two male mating phenotypes occur together. Typically, one phenotype is large-bodied, territorial, aggressive and colourful, and actively courts females and drives away rivals. Other males in the same population tend to be small (possibly as a function of age), passive and cryptic, and frequently reproduce by attempting to fertilise the eggs of females they have not courted. Alternative mating phenotypes may also be predominantly behavioural. Behavioural polymorphisms may be underpinned by hormonally mediated differences in physiology, including brain chemistry.

The mechanism by which multiple mating phenotypes are maintained in the same population is probably through negative frequency-dependent selection. As the frequency of one male mating morph increases, its fitness declines relative to other morphs, while rare mating morphs show elevated fitness. Because sexual selection on traits closely linked to fitness is usually strong, changes in male mating morph frequencies are likely to be rapid. At a population level, different mating strategies and tactics are predicted to generate equivalent fitness payoffs (Shuster, 2010).

The terms reproductive 'strategy' and 'tactics' are frequently used in the context of male mating polymorphisms. These terms are not synonyms when used in the context of life-history evolution (Wootton, 1984b). For a given reproductive strategy, which is largely fixed, there are a range of mating tactics that offer minor, or possibly quite substantial, adjustments to behaviour, morphology and physiology, within the constraints imposed by the broader strategy. Thus, males may show territoriality (a reproductive strategy) as part of their mating system, but flexibility in the size of territory they defend in response to the prevailing environment or their condition (reproductive tactics). Although it can sometimes be difficult to draw a distinction between reproductive strategies and tactics (Taborsky, 2008), that does not mean that the distinction should be ignored. A discussion of male mating polymorphisms as alternative life-history strategies is considered in Chapter 11.

Male mating polymorphisms can be fixed over an individual's lifetime and may have a substantial genetic component. Alternatively, polymorphisms may show substantial plasticity. In some cases, an individual may exhibit different mating phenotypes sequentially, which may be linked to their age or size. In others, the transition from one mating strategy to another may be reversible over their lifetime. Probably the most common pattern of mating polymorphism in teleosts is one of displaying alternative mating phenotypes concurrently, with males switching rapidly between mating tactics, sometimes over the course of a single mating.

Fixed alternative male mating phenotypes are displayed by anadromous salmonids, *S. salar* being the best studied. Male *S. salar* either mature as small 'parr' or transform into smolts and migrate to sea, later returning as large, mature adults (Chapter 7). On the spawning grounds, large, migratory males compete aggressively for access to females, with larger males more successful in these contests. Mature parr reproduce by swimming quickly into a spawning nest, termed a 'redd', and releasing sperm over the eggs as they are deposited by the female and before the anadromous male can fertilise them. Mature male parr do not engage in prolonged fights, relying instead on their small size and relative inconspicuousness to achieve reproductive success. Mature parr additionally possess relatively larger testes than migratory males and produce spermatozoa that show adaptations that enhance fertilisation success, including faster swimming speed and adenosine triphosphate (ATP) content (Vladić & Järvi, 2001). Males that mature as parr are those that show fast early growth, which may relate to paternally inherited additive genetic differences among males. Male *S. salar* adopting these different mating strategies also display divergent brain gene-expression profiles (Aubin-Horth *et al.*, 2005).

Sequential mating phenotypes are described in *T. bifasciatum*. In this species, young males mature as an initial phase (IP) phenotype and spawn in groups. Older males, and also some females (as *T. bifasciatum* exhibit sequential protogynous hermaphroditism; see Chapter 10), subsequently develop as terminal phase (TP) males. TP males guard spawning sites, court females and spawn singly with a female. IP males perform 'streaking' behaviour, rushing to join a TP male and female during spawning and competing for fertilisations through sperm competition (Warner, 2001).

Bluegill sunfish (*Lepomis macrochirus*, Centrarchidae) display a combination of fixed and sequential mating phenotypes. Some mature at an age of approximately 7 years, as large 'parental' males. Parentals construct nests, court females and perform parental care. 'Cuckolder' males, by contrast, mature at 2–3 years and 'sneak' fertilisations. To do so they furtively approach the nest of a spawning parental male and attempt to reproduce by dashing into the nest and releasing sperm. As they grow older and larger, cuckolder males adopt a 'satellite' tactic, whereby they express female colour and behaviour and enter the nest of a spawning parental male to attempt egg fertilisation as a 'female mimic'.

Concurrent, flexible mating tactics are exemplified by *R. amarus*. In this species, which spawns in the gills of living freshwater mussels, the male mating tactic is not fixed, although larger, and usually older, individuals tend to guard mussels while actively searching for females and leading them back to their territories. Some males are unable, or do not attempt, to guard mussels, but instead occupy the territory of a mussel-guarding male and release sperm into mussels, often before a female has deposited her eggs. Bitterling sperm is mucin-rich and potentially able to fertilise eggs deposited in mussels for approximately 15 minutes after ejaculation. Other males associate with groups of females and participate in the spawnings of females belonging to the group.

Finally, a subset of territorial males both guard a territory and also aggressively invade the territories of neighbouring territory holders. These males either participate in spawning with their neighbours or disrupt courtship and attempt to lead females away to their own territories. The mating system is highly dynamic, with males capable of switching among tactics many times over the course of a single day (Reichard *et al.*, 2004a,b; Smith *et al.*, 2002a, 2003, 2004a).

Causal factors

The conditions that underpin male mating polymorphisms are relatively poorly understood, although instructive case studies exist. Environmental cues that elicit a switch between male mating phenotypes include density in *R. amarus* (Reichard *et al.* 2004a). At low male densities, the territorial tactic yields the highest reproductive success. As male density increases, the mean success of territorial males is eroded as more non-territorial rivals attempt to participate in matings. At a density at which the reproductive success of territorial behaviour is equivalent to non-territorial males, territoriality breaks down and the mating system switches to one of group spawning, with limited aggression among males (Reichard *et al.*, 2004b).

Interestingly, density may have an impact on male mating tactics in the related *R. ocellatus*, but by a different mechanism. In this species, embryo density during the first few weeks of development, at the stage the embryos are hosted by a mussel, has a significant effect on age and size at maturity. At low incubation densities, fish mature later and are larger at maturity than those developing at a high density (Agbali & Smith, 2012). This epigenetic effect may be adaptive, as egg densities in mussels correlate with bitterling population size, and male mating success is greater in large males at low population densities. The role of epigenetics, in both the loose sense of adaptive plasticity and the strict sense of inherited alterations to phenotype (Gluckman *et al.* 2011), has hitherto been overlooked as a mechanism for the evolution of male mating polymorphisms but warrants further investigation.

Because the pattern of growth in fishes is indeterminate, growth rate and body size are key intrinsic factors that dictate life-history evolution (Roff 1992; Wootton 1998), with direct implications for male mating phenotype (Taborsky 2008; Chapter 11). Larger and usually older males typically adopt a dominant role in which they attempt to monopolise females or the resources they require for reproduction. Smaller and often younger males attempt matings without investing in territory acquisition, courtship and parental care. Smaller body size may also lend itself to this mode of reproduction (Taborsky, 2008). The major role played by growth rate in determining male mating phenotype in salmonids has long been recognised (Fleming, 1996). Growth rate also underpins male phenotype in *S. ocellatus*. In this species, some males mature after 1 year and sneak fertilisations. Others delay maturation and grow to a larger body size, subsequently acting as territorial, nest-building males that court females (Alonzo *et al.*, 2000).

In many species, male mating tactics are probably size-dependent, with territorial males comprising older individuals within a single growth trajectory, and males that engage in alternative mating tactics being simply smaller, younger individuals. This conditional pattern of male mating status is probably widespread. Alternatively, males may express distinct growth patterns that dictate their mating role. This scenario appears to be the case in *S. ocellatus* (Alonzo *et al.* 2000), and perhaps in other species, though divergent patterns of growth are rarely demonstrated. Thresholds for adopting different mating phenotypes may also vary among populations (Aubin-Horth & Dodson, 2004; Piché *et al.*, 2008; see Chapter 11).

Floaters, sneakers, streakers, pirates, parasites, fighters and the bourgeoisie

A bewildering number of terms are used in the literature to label different types of male mating polymorphism. In some cases, these describe the same phenotype in different species, whereas in others there is a marginal functional difference. The terms available to label male phenotypes are often pejorative and sometimes misleading. 'Sneakers' are defined as males that attempt to mate with females furtively. In fact, the 'sneaker' tactic is often anything but furtive and appears to be highly visible to females and other males alike. 'Streakers' are males that rapidly enter the nest or spawning site of others males and attempt to fertilise the eggs of females they have not courted. The distinction between 'sneakers' and 'streakers' is often not apparent; functionally they appear identical as both reproduce primarily through sperm competition and without engaging in courtship of the female. Sneakers and streakers may also be 'floaters' and 'satellites'. Applying the term 'parasitic male' to the same males is unhelpful. Use of this term arises from the notion that these males parasitise the investment of dominant males in attracting and courting females. In cases where the 'parasitic male' fertilises eggs that are subsequently cared for by another male, this does represent parasitism, but in the form of brood parasitism. The term is particularly misleading when viewed from a female perspective. In some cases, females are constrained in their mate choice, or choice of oviposition site, by dominant or territorial males and may actually have a mating preference for males expressing the alternative mating phenotype (Reichard *et al.*, 2007b). In this case, it is arguably the dominant male that is performing the role of a 'parasite'.

Nest-guarding or territorial males are described by turns as adopting a 'fighter', 'parental', 'dominant', 'conventional' or 'bourgeois' mating phenotype, terms that unsatisfactorily express male mating roles. Although it is often asserted that male mating polymorphisms display enormous diversity, there are really only two main classes of male mating polymorphism. Larger and usually older males typically adopt a dominant role in which they attempt to monopolise females or the resources they require for reproduction. This contrasts with smaller, subordinate and often younger males that endeavour to mate without investing in territory acquisition, courtship or parental care, and frequently engage in sperm competition. Thus, the categorisation of the male mating role as either 'guarder' or 'non-guarder' appears adequate to unambiguously describe typical male mating polymorphisms. Despite this, the term 'sneaker' is deeply embedded in the literature, and a guarder–sneaker dichotomy is probably universally acceptable to describe these major polymorphisms. One additional, albeit rare, alternative mating phenotype is the 'female mimic', exemplified by the bluegill sunfish 'satellite' morph, but even this phenotype can be classed as a non-guarding sneaker tactic and is also associated with sperm competition, though endocrine control of female mimics may differ from that of sneakers (Mank & Avise, 2006).

Hormonal control of male mating polymorphisms is not yet well understood, with the most comprehensive insights coming from studies of the endocrine status of sequential hermaphrodites (Chapter 10). A proposed working hypothesis is that male mating polymorphisms in different species share the same endocrine regulatory processes, which have evolved convergently under common selection pressures (Mank & Avise, 2006). There might also be evolutionary progression from some male mating polymorphisms that require few specialised adaptions, such as sneaking, to more derived polymorphisms, such as female mimicry, in which the expression of female behaviour and phenotype have become decoupled from sex differentiation and gametogenesis. There is some evidence for this from comparative phylogenetic analysis, although the direct evolution of complex mating phenotypes, such as male mimicry, is also possible (Mank & Avise, 2006). A fuller appreciation of the regulatory processes and evolutionary pathways of male mating polymorphisms will come from a greater understanding of the neuroendocrine and endocrine control of reproduction in different phenotypes. The selection pressures that generate male mating polymorphisms can also be understood in the context of life-history theory (Chapter 11).

Male mating polymorphism and sperm competition

Some male mating polymorphisms are strongly associated with sperm competition, particularly the sneaker mating phenotype. These males tend to experience a higher risk of sperm competition because they usually mate in the presence of at least one competitor, usually the guarder male and often other sneakers. An outcome is that sneakers are predicted to show adaptations for sperm competition, such as relatively (or even absolutely) larger testis size, larger ejaculates and faster-swimming spermatozoa (reviewed by Montgomerie & Fitzpatrick, 2009).

Guarder male response to non-guarders

The response of guarder males to attempts by sneakers to mate is generally one of aggression (Petersen & Warner, 1998). In some cases, guarders may be able to assess the intensity of sperm competition and act adaptively. In *L. macrochirus*, guarder males assess the paternity of their offspring on the basis of the number of sneakers adjacent to their nest when spawning occurred, and adjust the level of their parental care appropriately (Neff, 2003). After the eggs hatch, guarder males were able to measure the extent of their paternity on the basis of olfactory cues and further adjust the level of parental care they provided to the developing offspring, relative to the proportion of offspring they had fathered. In the related pumpkinseed sunfish (*Lepomis gibbosus*, Centrarchidae), the paternity of guarder males was positively correlated with nest defence effort (Rios-Cardenas & Webster, 2005). Similarly, simulated intrusions by sneaking males led to an increase in filial cannibalism and a decrease in parental effort by guarder male scissortail sergeant (*Abudefduf sexfasciatus*, Pomacentridae) (Manica, 2004).

Female response to alternative male mating phenotypes

While non-guarder males always undermine the fitness of guarding males, their impact on the fitness of females is more ambiguous, although few studies have addressed female responses to alternative male mating phenotypes (reviewed by Reichard *et al.*, 2007b). An often unqualified assumption is that guarder males, which are those individuals that are successful in intrasexual competition for resources, are also expected to confer superior genetic quality to females. In many cases, females will obtain a fitness benefit from mating with guarder males, primarily through direct benefits, because guarders control resources of superior quality or provide parental care (Andersson, 1994). However, male dominance and control of resources may be poor overall predictors of the benefits that mating with a guarder male brings, especially if the benefits of mating are primarily indirect (Qvarnström & Forsgren, 1998; Moore & Moore, 1999; Reichard *et al.*, 2007b). An advantage to mating with non-guarder males can come to females through additive genetic benefits provided by non-guarders. In *S. salar*, offspring sired by mature parr had higher viability and grew faster than progeny of migratory guarder males, implying a higher additive genetic quality of non-guarder males (Garant *et al.*, 2005). A non-additive benefit of mating with non-guarders can similarly be favoured under sexual selection if a female is coerced into mating with a guarder that is genetically incompatible (Neff & Pitcher, 2005). Thus, female *S. salar* with a higher number of mates tended to produce more outbred offspring and had greater reproductive success through enhanced offspring survival (Garant *et al.*, 2005).

A benefit of mating with non-guarders can also come to females through enhanced genetic variability among offspring ('genetic bet-hedging'), whereby females increase the genetic diversity among their offspring (Watson, 1991; Jennions & Petrie, 2000). The benefits to females of multiple paternity by guarders and non-guarders may derive from spatially and temporally unpredictable gene–environment interactions. Females might also benefit through fluctuations in environmental conditions favouring different phenotypes between generations. Females whose eggs have been fertilised by both guarder and non-guarder males may thereby benefit through bet-hedging, at least in species with a heritable component to male mating phenotype.

Participation by non-guarder males can confer direct benefits to females through insurance of fertilisation. Infertility rates in nature can be high (Wedell *et al.*, 2002) and are often explained though sperm depletion following multiple ejaculations by males with high mating success (Nakatsuru & Kramer, 1982; Warner *et al.*, 1995). In *R. amarus*, females were shown to suffer a 20% reduction in the fertilisation rate of their eggs when spawning with a solitary guarder male as compared with a group of four non-guarders (Smith & Reichard, 2005). In *T. bifasciatum*, TP guarder males released an order of magnitude fewer spermatozoa than non-guarder IP males (Warner *et al.*, 1995). Further, guarder males with high mating success released fewer spermatozoa than guarders with a lower mating rate, implying sperm limitation in the most successful males.

Given the potential benefits of mating with non-guarder males, there are few studies that demonstrate choice of non-guarders by females. One example is in the rock-pool blenny (*Parablennius parvicornis*, Blenniidae). In this species, females were more responsive to male courtship in the nests of guarder males with associated non-guarders that also participated in mating (Oliveira *et al.*, 2002). Female coho salmon (*Oncorhynchus kisutch*, Salmonidae) performed more digging behaviour, which represents a measure of mate choice, when accompanied by mature male parr. Females also spent longer spawning, which correlates with the number of eggs laid, when mature parr took part in spawning (Watters, 2005). Similarly, female *L. macrochirus* laid three times more eggs when non-guarders males participated in spawning (Fu *et al.*, 2001). In an experimental study, female *R. amarus* preferred spawning close to non-guarder males and engaged in a specific behaviour that signalled that the female was about to oviposit (termed 'skimming' behaviour), which attracted additional non-guarder males (Smith *et al.*, 2007). Females also delayed oviposition when only a single guarder was present at a spawning, which had the effect of increasing the number of non-guarders that attempted to participate in matings (Smith & Reichard, 2005).

Female mating polymorphisms

Mating polymorphisms such as those observed in males are rarely reported in females. The absence of female mating polymorphisms in the literature may be because they are rare or, alternatively, because they do occur but are not recognised as such. Facultative brood parasitism, in which females that normally perform parental care place eggs or young stages under the care of other females (Chapter 9), is analogous to, but not directly comparable with, male mating polymorphisms. It could also be argued that ontogenetic sex change represents a form of mating polymorphism (Chapter 10), although, once again, the analogy with male mating polymorphisms is limited because the polymorphism is not confined to a single sex. However, there are examples of female mating polymorphisms that mirror those of males to a limited extent. Female *R. ocellatus* perform two distinct mating patterns, though without any evidence of morphological discontinuities among them. Female *R. ocellatus* can mate as one of a solitary pair. Here the female attends to the courtship of a guarder male and assesses the quality of the mussel spawning site (Kanoh, 2000). Female *R. ocellatus* assess male quality using olfactory cues that indicate the degree of genetic compatibility between male and female (Agbali *et al.*, 2010, 2012), and assess mussel quality based on the oxygen conditions inside the mussel, which indicates its suitability as a site of incubation for eggs and embryos (Smith *et al.* 2001b; Casalini *et al.* 2009). This discriminating pattern of mating may involve one or more non-guarding males, whose participation in mating the female appears to facilitate by performing soliciting behaviour, including 'skimming' behaviour (Smith *et al.* 2007), and by delaying oviposition (Smith & Reichard, 2005). Even with non-guarding males present, the female responds to courtship and spawns on the basis of mussel suitability as a site for oviposition and male quality. This pattern of mating contrasts with a non-discriminating female mating tactic, in which females spawn with groups of non-guarding males that do not perform courtship behaviour, and in the absence of female soliciting behaviour (Kanoh, 1996, 2000). These two mating phenotypes vary seasonally in natural populations, with non-discriminating behaviour declining in frequency over the course of the spawning season, while discriminating behaviour increases (Kanoh, 2000).

A comparable pattern of female mating is described in *T. bifasciatum* (Warner, 2001), which may similarly represent an example of female mating polymorphism.

Sexual conflict

When the genetic interests of males and females do not coincide, a sexual conflict arises (Parker, 1979; see also Chapter 2). Sexual conflict is recognised as an important component of sexual selection that drives male–female coevolution (Chapman *et al.*, 2003). Sexual conflict can occur through intra-locus conflicts, whereby a trait expressed in both sexes undergoes contrasting selection in each sex. An outcome is that genotypes that confer high female fitness tend to generate low male fitness, and vice versa. Intra-locus conflicts are probably responsible for generating sex-limited gene expression. Inter-locus conflict occurs when the outcome of male–female interaction has different optima for the sexes, with the result that selection for sexually antagonistic adaptations in one sex will result in counter-adaptation in the other. Sexual conflict represents a mechanism of sexual selection, exemplified by 'chase-away' selection (see the section 'Antagonistic genes'). Sexual conflict is considered an important aspect of sexual selection because it appears to be an inevitable outcome of sexual reproduction and underpins the rapid evolution of many genes involved in reproduction, which have significance for the evolution of genetic sex determination (Chapter 2). It can also potentially promote speciation (Arnqvist & Rowe, 2005; Hosken & Stockley, 2004).

Several examples of sexual conflicts in teleosts are described. Sexual conflict has been proposed as a mechanism for the elaboration of many male copulatory structures (Evans & Meisner, 2009), including the barbed spermatophores of male *O. setnai*. Female *O. setnai* possess protective 'genital pads' that appear to have evolved in females to protect them from injury from spermatophores (Grier, 1984; Fig. 8.14). Coercive matings and harassment of females are also features of many fish mating systems. In *P. reticulata*,

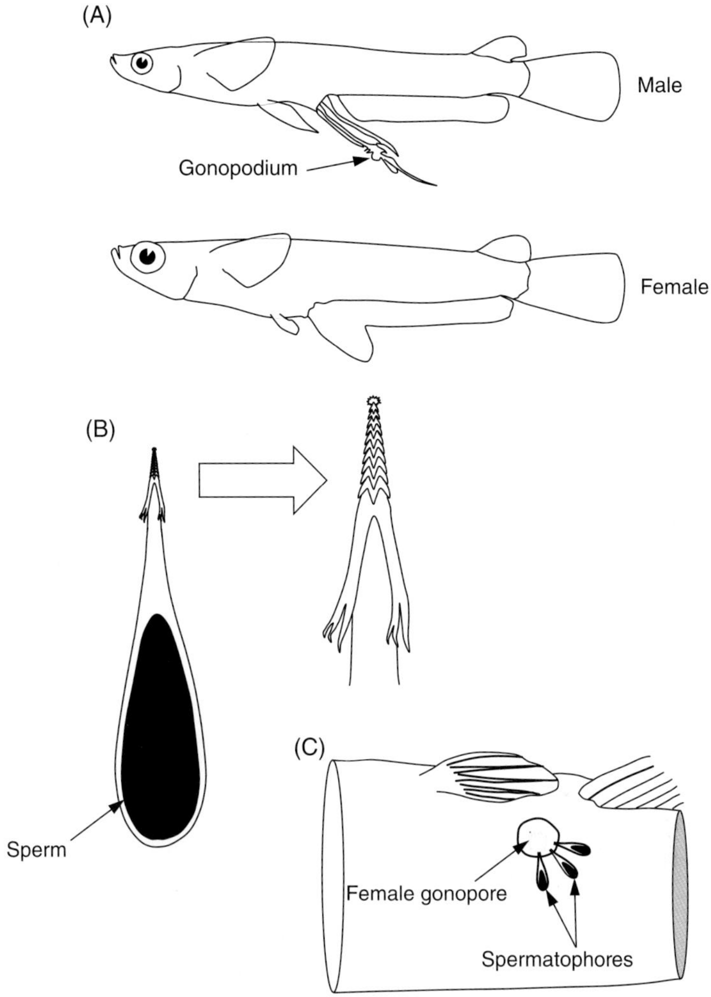

Figure 8.14 Sexual conflict in the Malabar ricefish (*Oryzias setnai*). (A) Male (top) with elaborate gonopodium. (B) Barbed spermatophore filled with sperm. (C) Spermatophores inserted into the thickened skin around the female genital opening (gonopore). (Source: Kulkarni 1940. Reproduced with permission of the Zoological Survey of India.)

males mate by performing courtship and obtaining a consensual mating or, in the event that females refuse mating, by quickly thrusting their gonopodia into the gonopore of the female and attempting to deposit sperm without female consent. Females can store the sperm of multiple males and are not sperm-limited, and they tend to resist male mating behaviour, avoiding males when possible (Magurran & Nowak, 1991; Magurran & Seghers, 1994).

In African killifishes (Nothobranchiidae), coercive male matings extend to heterospecific females. Male spotted killifish (*Nothobranchius orthonotus*, Nothobranchiidae) force matings with related female turquoise killifish (*Nothobranchius furzeri*, Nothobranchiidae), whereas matings within the respective species are more usually consenting (Polačik & Reichard, 2011; Fig. 8.15). Hybrid offspring have low viability, and the failure of male

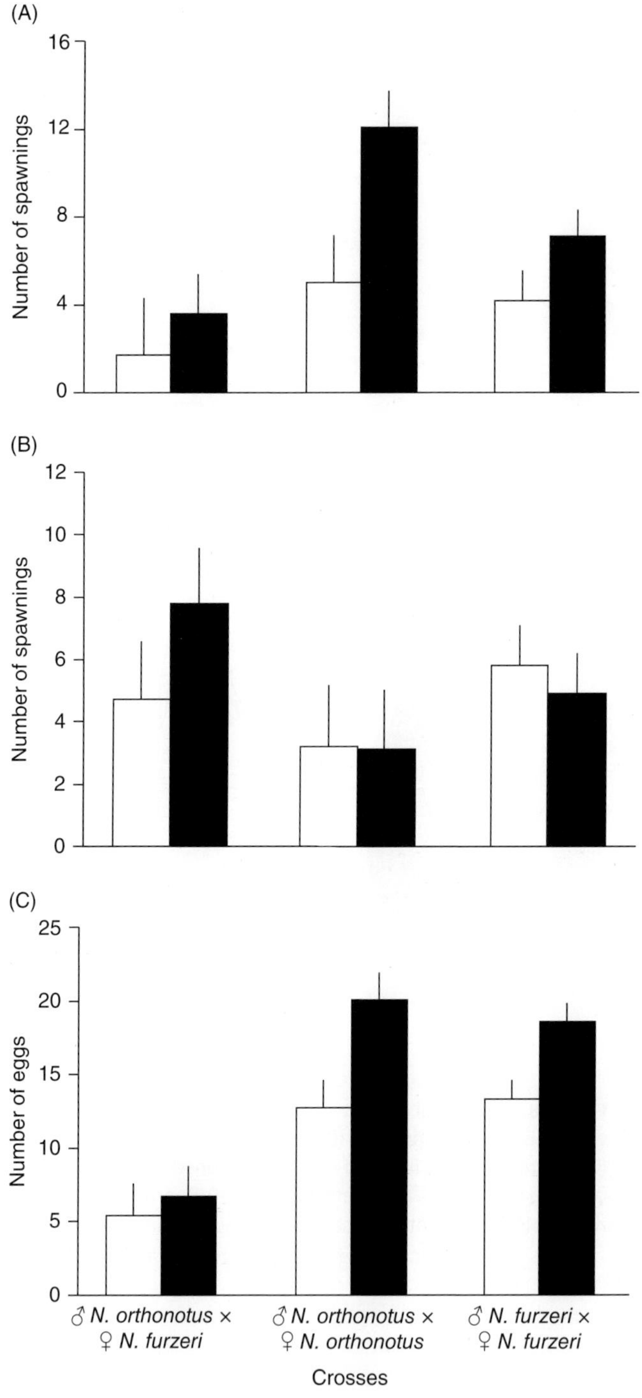

Figure 8.15 Heterospecific and conspecific matings between spotted killifish (*Nothobranchius orthonotus*) and turquoise killifish (*Nothobranchius furzeri*). Male *N. orthonotus* often force matings with female *N. furzeri*, thereby reducing female fitness. (A) Number of consentual spawnings; (B) number of forced spawnings; (C) number of eggs produced. Results after 24 h, white bars; after 72 h, black bars. (Source: Adapted from Polačik, M. & Reichard, M. 2011. Creative Commons Attribution License 4.0. http://creativecommons.org/licenses/by/4.0/.)

N. orthonotus to be more selective in mating was proposed to be because these fishes occupy temporary pools and the evolution of better discrimination of mates by males may carry a cost through lost matings associated with mistaken avoidance of conspecific matings (Polačik & Reichard, 2011).

Female *R. ocellatus* prefer mating with MHC-dissimilar males. This mating preference generates offspring with significantly greater survival and growth rates than matings with MHC-similar males (Agbali *et al.*, 2010). If females are allowed to mix freely with a group of males, the largest and most dominant males fertilise the majority of their eggs. However, if males are constrained and females are allowed to choose among them, females often fail to select the dominant males as spawning partners, suggesting a sexual conflict between female mate choice decisions and male dominance (Casalini *et al.*, 2009).

Mate choice in the Chinook salmon (*Oncorhynchus tshawytscha*, Salmonidae) also appears to be for MHC-dissimilar males (Garner *et al.*, 2010). In this species, females direct aggression at MHC-similar males, with the effect that females showed a greater tendency to mate with MHC-dissimilar males. However, male *O. tshawytscha* also harassed females that were unreceptive to mating, which, like *R. ocellatus*, resulted in sexual coercion by males coming into conflict with mate choice by females (Garner *et al.*, 2010).

Our understanding of sexual conflict has come primarily from studies on insect mating systems (Arnqvist & Rowe, 2005). Because teleosts display such a variety of reproductive modes, they offer exceptional material for understanding the significance of sexual conflict for vertebrate mating system evolution.

Phylogenetic history and mating system evolution

Within taxa, fishes show great variation in their mating systems. Despite this sometimes remarkable variety in modes of reproduction, phylogenetic history can place constraints on mating system evolution. For example, species with internal fertilisation as their ancestral state rarely, if ever, appear to evolve external fertilisation.

Comparative analyses that accommodate phylogeny can be extremely informative. Stockley *et al.* (1997) used a phylogenetic approach to examine traits associated with sperm competition in fishes, and established that relative testis size and sperm number tended to increase with greater sperm competition intensity. Mank *et al.* (2005) used a similar approach to test the predictions of several conflicting hypotheses for the evolution of parental care in fishes, showing that different forms of parental care have evolved independently on several occasions and that paternal care was associated with nest construction and sexual dichromatism. While this approach is potentially powerful, serious limitations to previous studies have been the rather poor resolution of phylogenetic data and the limited understanding of fish reproductive characters. Many attempts to use a phylogenetic approach to understand the evolution of mating systems rely on Breder and Rosen's (1966) dated review of fish modes of reproduction. More wide-ranging analyses, using newly available phylogenetic data, and a more comprehensive review of reproductive characters have the potential to provide greater insights into the evolution of fish mating systems.

Parental care is an additional important component of teleost mating systems. Parental care can influence the strength of sexual selection, and thereby plays a significant role in mating system evolution. Parental care in teleosts is addressed in the next chapter.

Chapter 9
Parental care

Reproductive Biology of Teleost Fishes, First Edition. Robert J. Wootton and Carl Smith.

Introduction

A successful outcome of courtship and mating is the production of offspring, with one or both parents, and occasionally siblings, potentially contributing to their care. How fish provide care, for how long, which parent cares, and the costs of caring represent key elements in parental care systems. Because of the great variability in their modes of parental care, teleost fishes offer unusually favourable material for understanding the evolution and maintenance of parental care.

Definition of care

Parental care can be defined as those processes, behavioural or physiological, following fertilisation that increase offspring fitness at the cost of the parent investing in other offspring (Trivers, 1972; Smith & Wootton, 1995a). This definition could be widened to include pre-fertilisation processes, but are excluded here because there may be some ambiguity as to whether the main function of such activities is to benefit offspring. Behaviour such as nest building prior to spawning does ultimately serve to increase offspring fitness, but may also have a role in courtship (Chapter 8). Fertilisation represents a precise moment at which parental care can be said to begin, particularly if fertilisation is external, and it is logical that parental care should only begin after the formation of a zygote and not before. Under this definition, all offspring-directed behaviours by parents after mating that carry a fitness cost represent parental care, as do all forms of bearing offspring, either internally or in brood pouches (Chapter 10). Fundamental to this definition is an assumption that parental care carries a cost to the parent (see the section 'Costs of parental care').

The function of parental care is to promote the survival and development of young stages. Most forms of parental care involve some form of protection against predation, including cannibalism. More sophisticated care can involve ventilation of offspring and, in some species, provision of nutrients to the young, either directly or through a maternal–foetal connection during internal brooding. Depending on the form of care provided, parental care in fishes can be described as either divisible or non-divisible. Divisible care can be shared by all offspring equally, e.g. protection of offspring against predators. This contrasts with non-divisible care, whereby a unit of care can only be devoted to a single, or a limited number of, offspring. Non-divisible care can take the form of feeding, with food provided to one offspring not available to another. The divisibility of care has implications for the evolution of parental care.

Distribution of parental care

A comprehensive taxonomic survey of parental care in teleosts has not been attempted since the monograph of Breder & Rosen (1966). Of the currently recognised 453 families of actinopterygian fishes recognised by Nelson (2006), some form of parental care is described in approximately 99 (22%) of them. Figures of between 20% and 30% of families in which parental care occurs are routinely cited (Blumer, 1979, 1982; Sargent & Gross, 1993; Mank *et al.*, 2005; Kolm, 2009), although these all ultimately rely on Breder & Rosen (1966) as their source. On the basis of current information, about one-quarter of all fish families include examples of parental care.

These figures may not reflect the distribution of care at the level of species. Some teleost families are monotypic (contain only a single species), while others contain several hundred species. In some families, parental care may be infrequent, while in others it may be performed by all species in the family. Parental care type can also vary widely within a family. Until a comprehensive review of the taxonomic distribution of parental care in fishes is available, the true distribution of parental care and the relationship between the distribution of care at the family level and species level will remain uncertain.

Modes of parental care

Protection from predators and disturbance

Parental care in fishes takes a variety of forms (Blumer, 1979; Keenleyside, 1979). Perhaps the most widespread is simply protection of young stages through aggressive defence. Aggressive defence is frequently linked with territoriality and is also associated with preparation of a spawning site or nest building (Chapter 7).

In some species, parental protection appears to be almost incidental to territory defence, with aggressive territoriality by males serving to protect eggs deposited in their territory. In the Pecos pupfish (*Cyprinodon pecosensis*, Cyprinodontidae), males often cannibalise eggs deposited in their territories. Females combat this egg eating by males by depositing their eggs in inaccessible crevices. Male egg cannibalism may be an important selective pressure on female choice of sites for oviposition (Kodric-Brown, 1983). Males do not provide any direct care of the offspring.

Female Pacific salmon (*Oncorhynchus* spp., Salmonidae), which are typically semelparous, guard their spawning nests ('redds') after completing spawning until they die. Nest guarding prevents other females from digging their own redds and thereby dislodging already buried eggs, which can be killed by gravel movement or eaten by egg predators. At high adult densities, egg mortalities arising from the reuse of redds can represent an important source of mortality (Hayes, 1987). Female guarding may last several days and is an important determinant of female reproductive success, with larger females, which are more aggressive, better able to defend nest sites (McPhee & Quinn, 1998).

In the case of iteroparous salmonids, such as Atlantic salmon (*Salmo salar*, Salmonidae), nests may be sited in two or more redds. After spawning, the females may leave the redd before resuming nest digging and spawning (Esteve, 2005).

Protection from hypoxia and desiccation

Parents may ventilate developing eggs and young stages. Ventilation of eggs can take several forms, but often involves the parent 'fanning' the offspring. Fanning entails driving a current of water over the eggs, typically using the pectoral fins (Fig. 9.1). To prevent themselves being displaced during fanning, parents simultaneously drive themselves forward using their tail, an activity analogous to 'running on the spot' and which is energetically expensive (see the section 'Costs of parental care'). Fanning has the effect of removing metabolic waste and delivering oxygenated water to the young, which are at the stages of development when they cannot adequately perform these functions. Fanning also has the effect of preventing the build-up of sediment that can stifle eggs and embryos by preventing adequate gas exchange. In cavity or tunnel nesting species, such as gobies and blennies, fanning is accomplished by the parent undulating the entire body rather than just the pectoral fins, to generate a stream of water that circulates the nest.

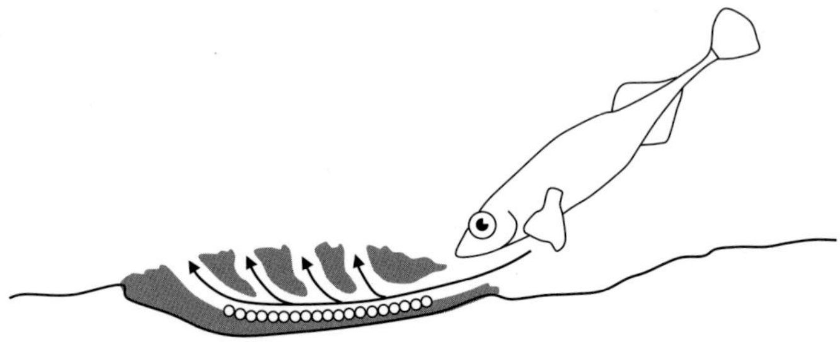

Figure 9.1 A male three-spined stickleback (*Gasterosteus aculeatus*) ventilating its eggs by fanning with the pectoral fins. Arrows indicate direction of water currents. Nest material is shown in grey. (Source: Adapted from Wootton 1976. Reproduced with permission of Elsevier.)

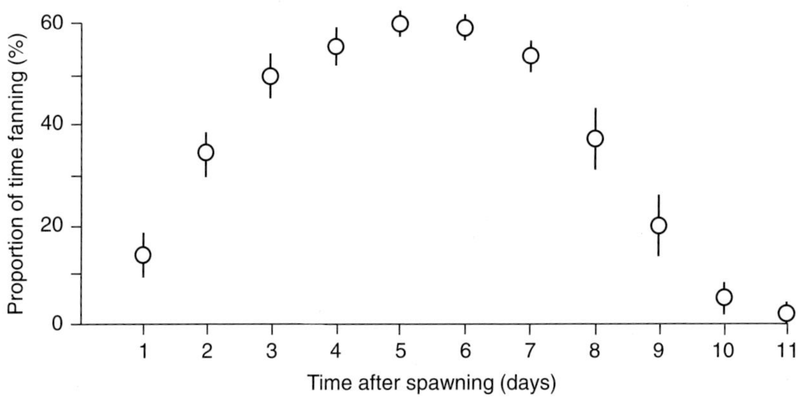

Figure 9.2 Mean (±1 SE) proportion of time spent by male three-spined sticklebacks (*Gasterosteus aculeatus*) fanning their nests over the parental care period at 20 °C. (Source: Adapted from Smith & Wootton 1995b. Reproduced with permission of Springer Science + Business Media.)

Fanning varies in response to specific environmental cues. In the three-spined stickleback (*Gasterosteus aculeatus*, Gasterosteidae), fanning by a parental male follows a distinct pattern once eggs are present in the nest. The frequency of fanning activity increases in duration over a period of 4–5 days and declines thereafter, ceasing after approximately 11–12 days (Fig. 9.2). Male sticklebacks are also responsive during parental care to water quality, and increase the amount of fanning behaviour if the concentration of dissolved oxygen is low and that of carbon dioxide is elevated (van Iersel, 1958). An increase in fanning has the effect of increasing the rate that oxygen is delivered to the eggs.

The adequate oxygenation of young stages in some species of cichlid is achieved by hanging the offspring from vegetation above the substrate. The embryonic stages of many species of substrate-brooding cichlids produce mucus threads from head glands that prevent the young from dispersing from the nest after hatching. In the rainbow cichlid (*Archocentrus multispinosus*, Cichlidae), parents spit embryos into submerged plants under low oxygen conditions, where they remain suspended from their mucus threads (Courtenay & Keenleyside, 1983). This behaviour appears to be a parental response to hypoxic conditions and may enhance offspring survival by lowering the density of offspring and placing them in better oxygenated water near the surface. Photosynthetic activity by the plants to which the embryos are attached might also oxygenate the water in their immediate vicinity, at least during daylight, further limiting the risk of hypoxia.

A comparable, if more radical, response to reduced oxygen availability is parental cannibalism (often termed 'filial cannibalism'; also see later in the chapter). Many species of fish engage in filial cannibalism (Smith & Reay, 1991), though the adaptive significance of the behaviour has long been unclear. Payne *et al.* (2002) used the beaugregory damselfish (*Stegastes leucostictus*, Pomacentridae), a coral reef species with adhesive eggs that are laid on a flat surface, to show that removal of small numbers of developing eggs from among egg clutches could speed up development. The surviving eggs not only developed faster after cannibalism, but had higher survival rates, with the overall production of offspring higher than if no cannibalism took place (Payne *et al.*, 2004). The effect of cannibalism appeared to come through increased circulation of oxygenated water to developing eggs and embryos and a reduction in local depletion of oxygen. Male *S. leucostictus* do not fan eggs in their nest, and cannibalism may be the only mechanism by which they can enhance ventilation of the developing eggs and embryos. Thus, filial cannibalism, at least in this species, represents a form of post-spawning clutch size adjustment in response to prevailing (or changing) environmental conditions.

The eggs of those atypical fish species that spawn on land, or whose eggs are exposed to air during development (Chapter 7), which include the capelin (*Mallotus villosus*, Osmeridae), Atlantic silverside (*Menidia menidia*, Atherinopsidae), grunion (*Leuresthes* spp., Atherinopsidae) and species of pupfish (Cyprinodontidae), galaxiid (Galaxiidae), sculpin (Cottidae) and pufferfish (Tetraodontidae), are usually adequately ventilated as oxygen is much more abundant in air compared with its concentration dissolved in water. Instead, the eggs and young stages risk desiccation, although in all cases of terrestrial spawning and development, eggs and embryos show adaptations that limit desiccation. In only one species with terrestrial spawning do the adults perform parental care, the splash tetra (*Copella arnoldi*, Lebiasinidae). After spawning on the underside of leaves hanging above the water surface (see Chapter 7 for description), the male prevents the eggs from becoming desiccated by flicking water at them using his tail. The parental male is sensitive to environmental conditions, moderating the rate of egg splashing during and immediately after rain when the eggs are at less risk of drying (Krekorian, 1976). In marine intertidal fishes, the parents may protect their eggs from desiccation by coiling their body around the egg mass, thereby retaining a film of water around the eggs and limiting their rate of drying (Potts, 1984; Coleman, 1992, 1999).

Protection from pathogens

In some species, parents frequently rub their bodies over the surface of the eggs they are guarding. This behaviour may be associated with the application of broad-range antimicrobial compounds that prevent fungal and bacterial infections in the eggs. The fringed darter (*Etheostoma crossopterum*, Percidae) has antimicrobial compounds present in the epidermal mucus of egg-guarding males (Knouft *et al.*, 2003). Parental males (which are the care-giving sex in this species) produced noticeably greater quantities of mucus than non-parental males. This mucus is secreted from a mucus cell-rich patch of tissue behind the head. In the peacock blenny (*Salaria pavo*, Blenniidae), antimicrobial compounds are produced by a pair of anal glands (Fig. 9.3). In this species,

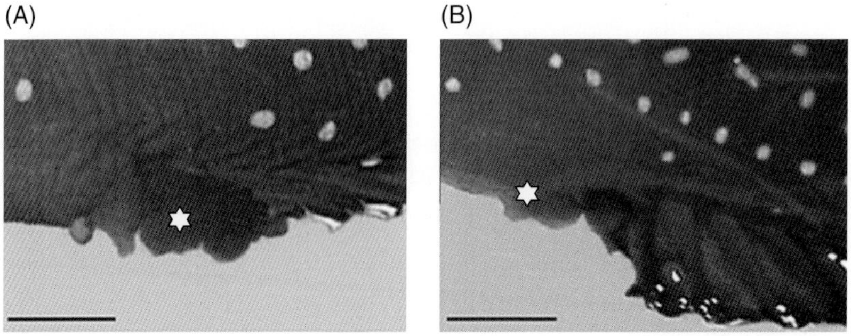

Figure 9.3 The anal glands of male peacock blennies (*Salaria pavo*), which have an antimicrobial function in protecting eggs and which may function in female mate choice. (A) Large anal glands; (B) small anal glands, in each case indicated by a star. (Source: Pizzolon *et al*. 2010. Reproduced with permission of John Wiley & Sons.)

females show a mating preference for males with larger anal glands, which may be able to provide greater protection from bacterial infection of the eggs (Pizzolon *et al*., 2010). The production of antimicrobial compounds for egg protection, and the morphological and behavioural traits associated with it represent a form of parental care that is now recognised in several species.

Nutrition

Feeding of offspring by parents is relatively rare in fishes, although there are a few examples. Probably the best known example is in the discus fishes (*Symphysodon* spp., Cichlidae), a group of specialised Neotropical cichlids. In *Symphysodon* spp. the parents produce epidermal mucus that is rich in amino acids, which the young nibble from their parent's body (Buckley *et al*., 2010). Feeding on parental mucus occurs over the first 30 days, peaking after approximately 15 days (Fig. 9.4). Thereafter, the young begin exogenous feeding and reduce their reliance on mucus. Young discus fish deprived of parental mucus following hatching have extremely low survival rates (Hildemann, 1959).

In the Midas cichlid (*Amphilophus citrinellus*, Cichlidae), young stages similarly consume parental epidermal mucus. In this species, the hormones prolactin, growth hormone and thyroxine occur in parental mucus (Schütz & Barlow, 1997). These hormones may play a role in promoting and regulating the development and growth of young stages, in addition to the nutritional benefits obtained from the proteins and mucopolysaccharides that are also present. Feeding on epidermal mucus occurs in a number of cichlid species, and possibly also in some species of bagrid catfishes (Sundara, 1962; Noakes, 1979).

While direct feeding of offspring by parents is rare, assistance with food finding is more frequent. Parental convict cichlids (*Amatitlania nigrofasciata*, Cichlidae), which show biparental care, exhibit two parental behaviours that increase food availability to their young. In 'fin digging', the parent lies with the ventral surface on the substrate and beats the pectoral fins rapidly, stirring up small food items. In response, the offspring dash to the site and feed on small particles of food dislodged by the parent (Keenleyside, 1991). In the same species, parents perform 'leaf lifting' in which they grasp a leaf or other submerged vegetation in their mouth and either drag or push it across the substrate, or lift it into the water column and release it. Like fin digging, the function of the behaviour appears to be to expose hidden food items to the young, which gather at

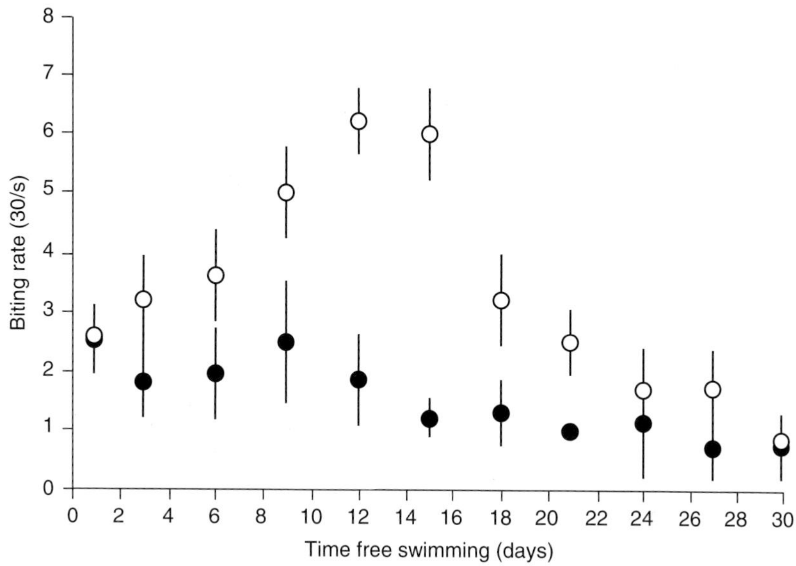

Figure 9.4 Mean (±1 SE) biting rate of larval discus fish (*Symphysodon* spp.) on the epidermal mucus of parents over the first 30 days after becoming free-swimming. White circles, without supplemental feeding; black circles, with supplemental feeding (*Artemia salina* 1 hour before recording). (Source: Adapted from Chong *et al.* 2005. Reproduced with permission of Elsevier.)

the site where the parent is active (Wisenden *et al.*, 1995). Non-breeding *A. nigrofasciata* perform fin digging and leaf lifting while foraging, but at a considerably lower rate than those caring for offspring. Parental female *A. nigrofasciata* perform both these behaviours at a significantly higher rate than males. A number of Neotropical cichlids also perform these parental food provisioning behaviours (Keenleyside, 1991).

External bearing

In some species, the eggs and young stages may be carried on, or attached to, the bodies of their parents or housed in special structures. The most rudimentary form of bearing is seen in some small oviparous species in which the eggs remain attached to the female by means of sticky filaments for a short interval after spawning, until they are brushed off against aquatic vegetation, to which they adhere. An example is the medaka (*Oryzias latipes*, Adrianichthyidae). The function of short-term transport of the eggs is uncertain.

The nurseryfish (*Kurtus gulliveri*, Kurtidae) is unique in that parental males possess a bony hook projecting from their foreheads that is used to carry a cluster of fertilised eggs (Berra & Neira, 2003; Fig. 9.5A). The hook is an extension of the skeleton and is displayed only in males (Fig. 9.5B). These fish inhabit turbid coastal rivers in northern Australia and southern New Guinea and the males probably carry the eggs until they hatch. Forehead brooding in this species probably protects the egg and embryo stages from smothering by silt and from being swept away by strong tidal water movements. The skin and tissue around the hook are highly vascularised, raising the possibility that there may be gas exchange and/or provision of nutrients to the developing embryos from the male (Berra & Humphrey, 2002). Engorgement of the tissues in this area with

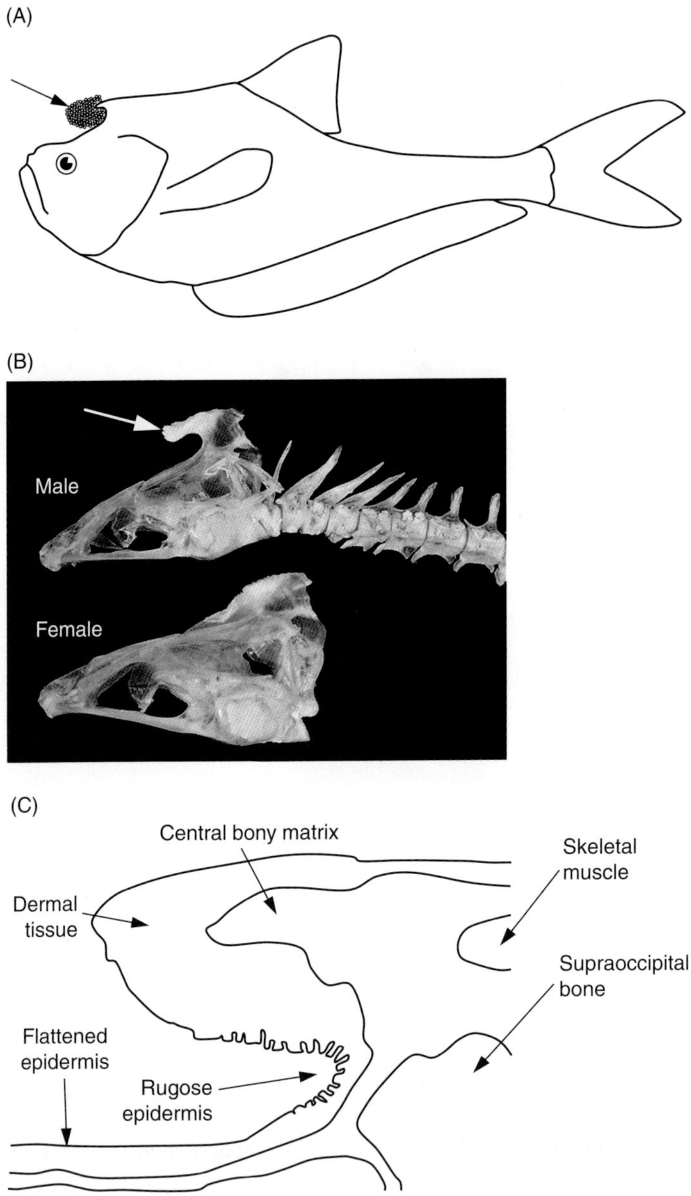

Figure 9.5 Adaptations for forehead brooding by male nurseryfish (*Kurtus gulliveri*). (A) A clutch of eggs attached to the supraoccipital hook, indicated by an arrow. (B) Skull of male and female nurseryfish illustrating the osteological foundation of the hook. (C) Diagram of a transverse section of the supraoccipital hook. (Source: Adapted from Berra & Humphrey 2002. Reproduced with permission of Springer Science + Business Media.)

blood may also assist in holding the egg mass in place (Fig. 9.5C). How the egg mass comes to be attached to the hook is unclear. The eggs may be fertilised after attachment, as males are occasionally caught with unfertilised eggs. It was proposed by Berra *et al.* (2004) that after spawning the male picks up the eggs using the hook, before releasing a cloud of sperm that it subsequently swims through to fertilise them. Spawning behaviour by captive nurseryfish has not been recorded. Alternatively, the eggs may be fertilised prior to attachment to the male's hook (Berra *et al.*, 2007). An inconclusive paternity analysis hinted that there may be cases of multiple paternity of egg masses carried by males (Sommer *et al.*, 2011).

Oral incubation of eggs and embryos in fish, termed 'mouthbrooding', is shown by a number of phylogenetically distinct families, including the climbing gouramies (Anabantidae), sea catfishes (Ariidae), cardinalfishes (Apogonidae), bagrid catfishes (Bagridae), cichlids (Cichlidae), jawfishes (Opisthognathidae), gouramies (Osphronemidae) and bony tongues (Osteoglossidae). In all cases, the eggs are usually collected in the mouth of one or both parents before or shortly after fertilisation. In many species of African mouthbrooding cichlid, after picking up the eggs, the female nibbles the anal fin or genital region of the male, who subsequently releases sperm, with the result that the eggs are fertilised in the mouth of the female. In some species of cichlid, mouthbrooding does not immediately succeed spawning. In the banded acara (*Bujurquina vittata*, Cichlidae), adhesive eggs are laid on leaf litter and parents aerate the eggs by fanning them with their pectoral fins until they hatch, after approximately 32 hours. Recently hatched young are subsequently orally brooded in the typical manner by both parents (Timms & Keenleyside, 1975). A variety of parental care patterns involving mouthbrooding, either immediately after spawning or with a delay until after hatching, are shown by the cichlids (Keenleyside, 1991).

The duration of mouthbrooding and the stages at which the young are released are variable, even within a family. Mouthbrooding may last over 40 days, with the young finally released as well-developed juveniles. In opisthognathids, the young are released as soon as they hatch, although the development rate of the eggs is slow in these fish and it may take up to 36 days to reach this stage in some species (Hess, 1993). After releasing independently swimming young, mouthbrooding may still continue. The young of some cichlid species remain with the parent after release and often forage as a shoal in proximity to the parent. If a potential predator approaches, the juveniles return to the parent, which allows them back into its mouth. The return of offspring to the parent can involve signals by the parent that are associated with specific fin movements and body posture (Fryer & Isles, 1972). Surprisingly, little research appears to have been conducted on this intriguing system of parent–offspring communication. Some predatory cichlids are reputed to specialise in preying on young stages and actively wrest eggs and embryos from the buccal cavities of mouthbrooding parents, so-called 'paedophagous' species. In the case of the Lake Malawi predatory cichlid *Champsochromis spilorhynchus* (Cichlidae), hunting adults appear to prompt parental mouthbrooders to recall their offspring and then intercept and eat the returning young before the parent can collect them (Ribbink & Ribbink, 1997).

While the young are held in the mouth of a parent, they are aerated and cleaned. An intermittent behaviour termed 'churning' is frequently seen in mouthbrooding fishes, in which the parent actively agitates the eggs and young stages. These movements appear to be crucial to normal development. In opisthognathids, in which care is exclusively performed by the male, it sporadically spits out the ball of developing eggs before picking them up again, which may serve the same function as churning. In some mouthbrooding cichlids, the parents are able to feed while mouthbrooding, and in some cases these feeding activities may provide food items to the developing young in the parent's mouth (Yanagisawa & Sato, 1990).

The buccal cavity of mouthbrooders is larger than that of related species without mouthbrooding. When only one sex performs mouthbrooding behaviour, the buccal cavity of the mouthbrooding sex is typically larger than that of the non-caring sex. In the paternal mouthbrooding Doederlein's cardinalfish (*Apogon doederleini*, Apogonidae), males have a larger buccal cavity than females during the breeding season, though not at other times, suggesting adaptive flexibility in buccal morphology (Okuda *et al.*, 2002).

A (possibly) unique mode of parental care comparable to mouthbrooding is that of brooding embryos in the gill chamber. This form of care has been proposed for amblyopsid cave fishes from North America, but wholly convincing evidence for this mode of care is so far lacking (Adams & Johnson, 2001).

Some fishes bear their young attached to their body surface, but with varying degrees of morphological adaptation for this mode of parental care. Skin brooding has been recorded in several families, including the pipefishes (Syngnathidae), armoured catfishes (Loricariidae) and frogfishes (Antennariidae and Tetrabrachiidae). Female banjo catfish (Aspredinidae) bear the eggs attached to their ventral surface in some species. Attachment of the eggs and embryos can be either directly to the body, e.g. in *Pterobunocephalus* spp. In other genera (*Platystacus*, *Aspredo* and *Aspredinichthys*), the eggs are suspended from stalks, called cotylephores, which are outgrowths from the body of the female. Cotylephores only develop during the breeding season. They comprise a pedicle or stalk with a cuplike structure (calyx) at its terminal end (Fig. 9.6). An individual fertilised egg is attached to a single calyx, which is highly vascularised. While possibly serving simply as a means of transporting the eggs and embryos, the cotylephores probably provide a site for gas exchange between mother and offspring, and possibly for the exchange of nutrients, but this has yet to be unambiguously demonstrated (Wetzel *et al.*, 1997).

An analogous adaptation is shown by the ghost pipefishes (*Solenostomus* spp., Solenostomidae). The eggs of these fishes are similarly brooded by the female, within a space formed by the fused pectoral fins, and attached to cotylephores, which appear to be structurally simpler than those of aspredinids (Wetzel & Wourms, 1995). These adaptations are an example of convergent evolution, as these two groups of fishes are phylogenetically distinct.

In Commerson's frogfish (*Antennarius commerson*, Antennariidae), a species of cryptic anglerfish that uses a modified fin ray as a lure to attract prey, the male bears the eggs on his lateral side close to the dorsal fin. The eggs of this species are visually highly conspicuous, especially as they are situated on the upper side of the body. The eggs of this and some related frogfish species may perform the function of lures that attract prey fish intent on feeding on them (Pietsch & Grobecker, 1980). If this is the case, this is the only known example of a species that has evolved to use its own offspring as bait to attract prey.

The pipefishes, seahorses and seadragons (Syngnathidae) show a gradation of adaptations for bearing eggs and young stages on their body surface in specialised brooding structures. Parental care is exclusively performed by the male. In some species, the eggs are simply attached to specialised sites along the abdomen of the male, but in others the eggs are partially enclosed by two pouch flaps that meet along the midline of the abdomen. The eggs of the seahorses (*Hippocampus* spp., Syngnathidae) are completely enclosed within a brood pouch or marsupium (Chapter 10). Parental care in all syngnathids lasts from fertilisation to the liberation of free-swimming juveniles. In addition to protection, parental care may involve gas exchange, osmoregulation and the transfer of nutrients and hormones to the

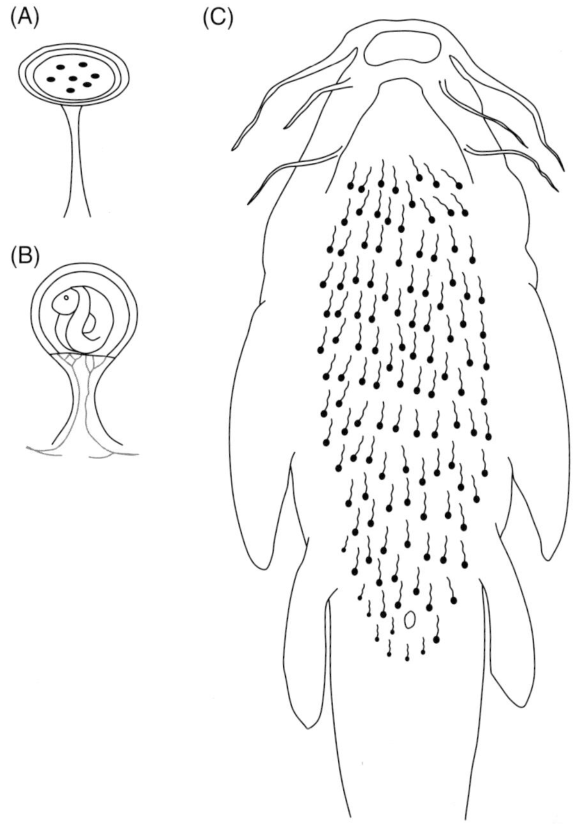

Figure 9.6 Adaptations for brooding young in the banjo catfishes (Aspredinidae). Females of the genera *Platystacus*, *Aspredo* and *Aspredinichthys* possess specialised structures, termed cotylephores, on their underside. (A) Cotylephores are typically about 1.5 mm in length and comprise a pedicle and calyx. (B) An egg attached to the calyx of a cotylephore, with the expanded pedicle showing vascularisation (in grey). (C) Cotylephores distributed over the ventral surface of an adult female. (Source: Adapted from Wyman 1859.)

developing embryos. The extent to which these processes occur in different species is not entirely clear, although there are clear interspecific differences (Carcupino *et al.*, 2002; Ripley & Foran, 2009). There is also evidence of a two-way exchange between father and offspring. Brooding males can absorb nutrients from embryos, at least in some species (Sagebakken *et al.*, 2010). How the male benefits from appropriating nutrients from its own offspring, or the impact of this process on developing embryos, is not yet clear.

Internal bearing

The retention of eggs in the reproductive tract of the female accompanied by internal fertilisation, termed viviparity, is common in chondrichthyian fishes (sharks and rays), but relatively rare in the teleosts, with just 14 families showing some form of internal bearing. Coelacanths (*Latimeria* spp.), which are placed in a separate class to the chondrichthyian and ray-finned fishes (the Sarcopterygii), also show internal bearing. (See Chapter 10 for a full discussion of teleost viviparity.)

Costs of parental care

The contribution of care by either parent can carry fitness costs, which will have a profound effect on which sex performs parental care and on the investment made in care. Two key questions present themselves. How are costs measured? And if a cost of parental care exists, how are costs managed to maximise lifetime reproductive success? To understand the costs of parental care, it is necessary to distinguish between 'parental expenditure' and 'parental costs' (Clutton-Brock, 1991). If parents expend time or energy on care, they may incur a fitness cost, but equally they may not. After spawning, female Pacific salmon (*Oncorhynchus* spp.) guard their nest site, thereby preventing the eggs from being dug up by other spawning females. The time and energy spent on this parental behaviour are measurable as expenditures, but carry no fitness cost to females because they have ceased spawning and will not spawn again until they die. A cost, then, is measured in terms of its effect on future reproduction, in accordance with Trivers' (1972) concept of parental investment. Translating parental expenditure on care into a parental cost is often not straightforward, although in many studies the unsupported assumption is made that expenditure on care necessarily carries a cost. How a parent manages the trade-off between current and future reproduction is summarised by Williams' principle, as formulated by Williams (1966), but which is an extension of earlier ideas proposed by Moreau (1944) and Lack (1947). Williams' principle holds that if adult survival is reliably high, parents will tend to reproduce over an extended period, with the result that current reproduction comprises a relatively small proportion of total lifetime reproductive success. Williams (1966) proposed the concept of the residual reproductive value (RRV), the capacity of an individual for future reproduction. Investment in current reproduction is assumed to erode RRV. An outcome of this hypothesis is that longer-lived adults should invest more heavily in future rather than in current offspring (see Chapter 11 for a fuller discussion). The converse is also the case: short-lived adults should invest heavily in current reproduction and relatively little in future reproductive events that are unlikely to occur. In the context of parental care, a prediction that can be made on the basis of Williams' principle is that longer-lived species should incur lower parental costs than short-lived species (Sargent, 1997). Predictions can also be extended to within-species comparisons. Individuals in poorer condition, with a lower probability of future survival, are predicted to invest a relatively greater proportion of resources in current reproduction than those in better condition. Investment in reproduction provides diminishing returns, i.e. a small investment in current reproduction generates a relatively high fitness benefit at low levels of investment, but at greater levels of investment, returns lower benefits. Because investment in reproduction generates diminishing fitness returns, the trade-off between current and future reproduction means that for a given species there is a predictable (in theory) optimal level of investment in current reproduction (Fig. 9.7).

Costs to parental care can come from a reduction in parental survival, or a reduction in the numbers of future offspring produced. This latter cost comprises two components; the numbers of offspring produced at each reproductive attempt; termed the 'batch fecundity', and the number of reproductive events.

Cost to parental survival

Parental care can inflict a survival cost if a parent is more susceptible to predators, a form of ecological cost (Magnhagen, 1992; Smith & Wootton, 1995a). Greater susceptibility to predators may come about because the demands of parental care

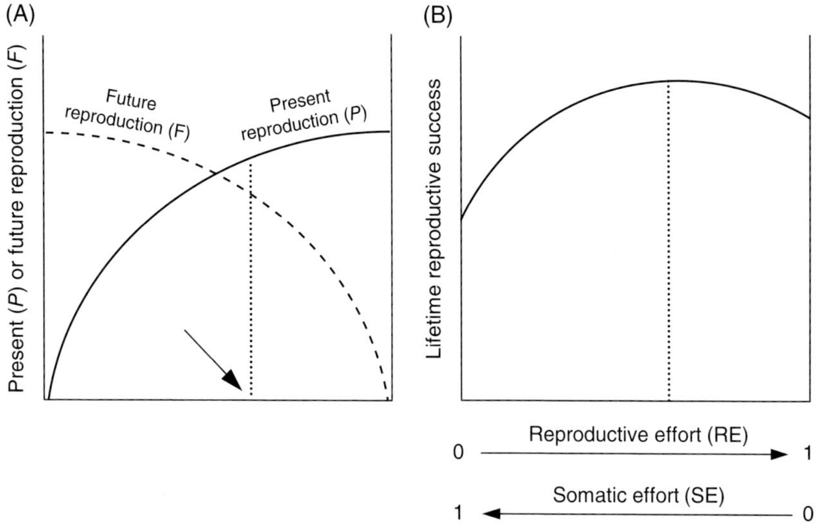

Figure 9.7 Graphical representation of Williams' principle showing the hypothetical relationship between present (*P*) and future reproductive success (*F*). (A) Current reproduction is assumed to increase with diminishing returns with greater reproductive effort (RE), while future reproduction is assumed to increase with diminishing returns with increasing somatic effort (SE). Resources allocated to reproduction cannot be used for maintenance, thus RE = 1 – SE. (B) Selection favours an optimal combination of investment in RE and SE (indicated by an arrow in panel A) at the point at which lifetime reproductive success is maximised. (Source: Sargent & Gross 1993. Reproduced with permission of Springer Science + Business Media.)

lead to alterations in appearance and behaviour. Guarding offspring against predators that can potentially eat both the parent and offspring presents the parent with particularly stark choices in terms of current and future reproduction. Pressley (1981) presented nest-guarding male *G. aculeatus* with dummy prickly sculpins (*Cottus asper*, Cottidae), a potential predator of both the eggs and the parent. Males attacked the model predator, thereby putting themselves at risk of predation. The propensity of males to engage in risky behaviour by attacking the predator was greater the more eggs they were guarding and the later the stage of egg development. This plasticity in male response suggests they are sensitive to the relative value of a group of offspring and corresponds with the predictions of Williams (1966). *G. aculeatus* with a low probability of future reproduction also appear to increase risk-taking by adjusting courtship activity and increasing investment in red nuptial colouration that may make them more visible to predators (Candolin, 2000). In the broad-nosed pipefish (*Syngnathus typhle*, Syngnathidae), parental males were 11 times more likely to be preyed upon while caring for young than when not reproducing (Svensson, 1988). The male carries its young in a ventral brood pouch, and the embryos, which are bright yellow, appear to disrupt his cryptic coloration. Parental male sand gobies (*Pomatoschistus minutus*, Gobiidae) are selectively targeted by bird predators, which may use the male's nests to identify their presence (Lindström & Ranta, 1992).

A survival cost may also arise if energy intake is compromised or the energy expended on care means that the risk of mortality is elevated, which represents a physiological cost of care. Bullhead (*Cottus gobio*, Cottidae) males, which are the parental sex, show an almost 10-fold increase in death rate during the breeding season that is matched by a deterioration in body condition (Marconato & Bisazza, 1988). Increased mortality among male *G. aculeatus* following breeding was attributed to parental males having depleted energy reserves, possibly as a result of providing parental care (Chellappa *et al.*, 1989). A substantially higher mortality rate of breeding male sticklebacks compared with non-breeders was also recorded by Dufresne *et al.* (1990). Interestingly, parental male *G. aculeatus* that were allowed to see and smell a female had significantly higher mortalities (approximately 20%) than isolated parental males (DeFraipont *et al.*, 1992). The body condition (and hence, it was assumed, future survival) of parental male *G. aculeatus* was also significantly reduced after allowing them to see potential predators of their offspring while caring, as compared with isolated parental males (Smith & Wootton, 1995b). Both these studies suggest that parental costs can be extremely sensitive to subtle variations in the parent's environment.

Cost to number of offspring produced

If the energy expended on parental care is non-trivial or parental care restricts feeding rate, the effect can be one of limiting the number or quality of future offspring. This cost can come about because fewer offspring are produced during future reproductive episodes, or because there are fewer future reproductive episodes. Greater energy expenditure or reduced energy intake can either directly limit energy available for further reproduction or can constrain somatic growth. The effect of reduced growth is a particularly costly one for females, as there is typically a strong relationship between size and fecundity in females, although there can also be an effect of size on male reproductive success (Clutton-Brock, 1991; Wootton, 1998; see Chapter 8).

In species with oral brooding of eggs and young stages, parental care often necessitates an interruption to feeding. This interruption can be for extended periods. In jawfishes (Opisthognathidae), males do not attempt to feed while they have eggs in their mouths, and instead place the eggs on the substrate inside their burrow during feeding bouts. However, males are still severely compromised in their feeding, with food intake 86% lower than that of non-brooding males (Hess, 1993). Whether this constraint to feeding translates into a fitness cost for males has not been demonstrated. The parent in mouthbrooding cichlids is typically unable to feed, or at least is severely constrained. Females of the Lake Victoria maternal mouthbrooder *Haplochromis argens* (Cichlidae) that performed parental care suffered a 61% reduction in body weight in comparison with control females, which had their eggs removed immediately after spawning and were able to feed. Parental females took 33% longer to respawn than control females, but did not produce smaller clutches (Smith & Wootton, 1994). They showed a cost to spawning frequency but not to batch fecundity as a result of restricted feeding. Thus, clutch size was maintained, but at a cost to breeding frequency. A possible explanation for this trade-off between the form of future reproductive costs is that there may be an optimal clutch size for mouthbrooding. By contrast, in the biparental mouthbrooding mango tilapia (*Sarotherodon galilaeus*, Cichlidae), females incurred a cost of parental care in terms of both reduced breeding frequency and clutch size (Balshine-Earn, 1995).

Despite the costs associated with restrictions placed on feeding, mouthbrooding is an otherwise efficient mode of care, as it requires little energy expenditure to perform.

There was no significant additional energy expenditure by parental females in the dwarf Victoria mouthbrooder (*Pseudocrenilabrus multicolor*, Cichlidae) compared with non-parental females (Mrowka & Schierwater, 1988).

Certain forms of parental behaviour can be energetically expensive and may carry fitness costs by reducing the capacity to produce further offspring. Female *A. nigrofasciata* that perform egg fanning behaviour suffer a reduced spawning frequency as a result of parental care in comparison with non-parental females (Smith & Wootton, 1995c), and the larger the brood that females care for, the smaller their investment in their subsequent brood (Lavery & Keenleyside, 1990). Fanning behaviour by male *G. aculeatus* is also energetically expensive, requiring approximately 12.2 J/g of parental tissue per hour, compared with 5.3 J/g per hour for a non-parental male (Smith & Wootton, 1999). Males that perform fanning experience restricted growth and reduced body condition as a result of care. If male *G. aculeatus* are exposed to a potential predator of its offspring, the effect on growth and condition is elevated (Smith & Wootton, 1995b). These energy demands may limit the ability of male *G. aculeatus* to attract further matings.

Parental care carries a cost for male parents if their breeding frequency is reduced as a result of providing care, either because parental care is incompatible with finding a mate and courtship, or indirectly because of the energetic demands of care. Examples include the spotnape cardinalfish (*Apogon notatus*, Apogonidae) (Kuwamura, 1983), Guyana leaffish (*Polycentrus schomburgkii*, Polycentridae) (Barlow, 1967), pearl-spot chromis (*Chromis notata*, Pomacentridae) (Ochi, 1985) and three-spot gourami (*Trichopodus trichopterus*, Osphronemidae) (Kramer & Liley, 1971). However, in many species, males suffer a limited cost of parental care, at least in terms of its impact on the prospect of future matings, and it is more typical that male parents continue to mate while already engaged in the care of previous broods (Gross & Sargent, 1985). The limited costs of care for males has implications for the evolution of parental care in fishes (see the section 'Evolution of parental care').

Sex role reversal and parental care

The availability of females that are ready to mate is usually more limiting than the availability of males. This situation leads, through the process of sexual selection, to well-characterised differences in sex roles; males court females and are more conspicuous than females, while females are typically choosy over mates and resources (Darwin, 1871; Fisher, 1930; Williams, 1966; see Chapter 8). In some species, constraints are placed on male reproduction to the extent that they become limiting to the reproduction of females. Under these conditions females come under stronger selection than males to obtain matings, with the evolutionary outcome that the sex roles can be reversed (Parker & Simmons, 1996) (Chapter 8).

Sex role reversal in fishes is associated with male parental care and is best recognised in pipefishes (*Syngnathus* spp., Syngnathidae) (Berglund & Rosenqvist, 2003). The reason for males being limited in their reproduction is linked to constraints on the rates at which they are able to 'process' young, both in terms of the number they can care for simultaneously and the time taken for young to achieve independence. In *Syngnathus* spp., in which the males carry the eggs in a brood pouch (Chapter 10), there is an upper limit on the number of eggs that can be carried. If a population of females is able to produce more eggs than males can accommodate, competition among females for males is a predicted outcome. In two species, *S. typhle* and the straight-nosed pipefish (*Nerophis ophidion*, Syngnathidae), females can produce eggs faster than males can care for them, with the result that the potential reproductive rate of females is higher than that of males. Although the rate of egg development speeds up at higher temperatures, the contraction of the parental period as a consequence is still insufficient to alter sex role reversal (Ahnesjö, 1995). As predicted by sexual selection theory, males of both these species are choosy about which females they mate with, while females are competitive in mating with males and show conspicuous colouration that is believed to function in female courtship (Berglund & Rosenqvist, 2003; see Chapter 8).

Phenotypic plasticity in care

Levels of parental investment are usually not fixed and parents can show flexibility in the amount of care they provide in response to changes in their environment, their condition and the status of the offspring. Thus, parents may increase or decrease the care they provide, or even abandon their offspring altogether in response to changes in the relative costs and benefits of current versus future reproduction (Carlisle, 1982; Clutton-Brock, 1991).

Number of offspring

Parents are sensitive to the number of offspring under their care. Parental care is often divisible, so one unit of care can be shared by all offspring equally. A consequence is that care of a large group of offspring is relatively more efficient than care of a small group. To a parent, then, not only is a large brood of offspring of greater absolute value, because they are more numerous, but the *per capita* cost of caring is also lower. One prediction is that parents will defend larger broods more vigorously than smaller broods, because they are more valuable. A second prediction is that small broods, or at least broods reduced to low levels, should be abandoned or cannibalised to enable a parent to restart care with a new, hopefully larger, brood (Sargent & Gross, 1993). There is good evidence for both these predictions. Female blue-point flag cichlids (*Aequidens coeruleopunctatus*, Cichlidae) guarded larger broods more tenaciously than small broods (Carlisle, 1985). Male *G. aculeatus* showed the same response (Pressley, 1981), as did male common gobies (*Pomatoschistus microps*, Gobiidae) (Magnhagen & Vestergaard, 1993). In the bluegill sunfish (*Lepomis macrochirus*, Centrarchidae), parental males fanned large egg clutches more frequently than small clutches (Coleman & Fischer, 1991) as did male rock bass (*Ambloplites rupestris*, Centrarchidae) (Sabat, 1994). There are numerous other examples to confirm that parents respond to large clutches with greater parental investment.

Conversely, the abandonment or cannibalism of small broods by the parent is also well documented (Dominey & Blumer, 1984; Smith & Reay, 1991; Smith, 1992), and can be adaptive because it releases parental time and energy otherwise required for offspring care to be used to prepare for subsequent broods. Cannibalism of an entire clutch of young appears more common than just abandonment of the eggs, presumably because consumption of the eggs or young stages provides an additional nutritional benefit that can be invested in future reproduction. In some cases, parental care by male parents may not be elicited if the number of eggs in their nest is too low (van den Assem, 1967; Kramer, 1973). Experimental reduction in the number of offspring below a threshold size can induce cannibalism in parental fish (Schwanck, 1986; Mrowka, 1987; Petersen & Marchetti, 1989). This threshold level may vary with respect to parental age-specific fecundity and life expectancy (Mrowka, 1987) and parental size, or the condition and age of offspring (Lavery & Keenleyside, 1990), although this has never been systematically investigated.

Offspring age

Parental investment is also responsive to the age of offspring. A general observation is that parental investment tends to increase with offspring age, but then to decline as the offspring near independence (Sargent & Gross, 1993). These changes in the pattern of care may simply reflect the care requirements of the offspring. However, a functional explanation for this pattern of care is provided by Sargent & Gross (1993) using a model of parental trade-offs between current and future reproduction. This model predicts that recently deposited eggs have relatively low value to a parent because their probability of survival is limited, though they are often most dependent on the parent for care at this stage. Thus, care is provided by the parent, but not at a maximum rate. As offspring age, their value increases as their probability of achieving independence increases, and consequently parents provide a greater level of care. Offspring that are older still have greater value to the parent, but the parent's ability to effectively provide care to offspring that are increasingly independent is limited, and care is curtailed as a consequence. This simple dynamic model of parental care is robust and mirrors empirical data extremely well.

Offspring quality

Offspring quality is rarely considered in the context of parental care, but may play a key role. A component of offspring quality with relevance to paternal care is the probability of relatedness of the male to the offspring for which it provides care. If the male's certainty of paternity is reduced, a prediction is that it will reduce its investment in current reproduction. Given that sneaky mating tactics are widespread and, in some species at least, common (Chapter 8), an appropriate response to a reduction in paternity might be expected. In some species, males do seem to modulate their parental investment in relation to the probability that they have been cuckolded. In *L. macrochirus*, parental males that were exposed to potential sneaked fertilisation during spawning reduced the level of care of the offspring, at least until the eggs hatched (Neff, 2003). After the eggs hatched, males were able to assess their genetic relatedness to offspring more directly, probably using olfactory cues, and adjusted their level of care appropriately in response. A similar pattern was demonstrated in *G. aculeatus* (Mehlis *et al.*, 2010). In this species, males were also able to discriminate eggs they had fertilised using olfactory cues, but they did not reduce their level of care in response to low paternity. However, males with a high proportion of eggs in their nests fertilised by another male showed a greater propensity to cannibalise these eggs, thereby terminating parental care. Thus, in two species in which sneaking by non-guarder males is relatively common, it appears that males are sensitive to reduced levels of paternity and respond by either limiting the care they provide or ceasing care altogether. These responses are adaptive if the cuckolded male is accurate in his ability to recognise that sneaking has taken place. Where an effective response to sneaking by the care-giving male has evolved, the effectiveness of sneaking as a reproductive tactic will tend to be eroded.

Food availability

Given that parental care is often energetically expensive for the parent, variation in food availability may impinge on parental care. If parents are able to track food availability, or their energy reserves, and adjust their level of care in relation to available resources, this provides them with a means of matching their energy expenditure on care with their energy intake while making provision for future reproduction. Some fish appear able to respond to food availability in this way. A positive relationship between food availability and the duration of egg fanning was shown by male *G. aculeatus*, with both the frequency and duration of fanning bouts correlated with ration (Stanley, 1983). The same relationship was shown for female *A. nigrofasciata* (Townshend & Wootton, 1985a). Male smallmouth bass (*Micropterus dolomieu*, Centrarchidae) reduced their level of parental care in tandem with a decline in their body condition over a breeding season (Mackereth *et al.*, 1999). In this species, larger males possessed greater energy reserves than smaller individuals, and this additional energy buffered the effects of energy expended on care, enabling large males to maintain a higher level of care. Intriguingly, male *M. dolomieu* that received additional food during care showed an increase in either the level of parental care they provided or subsequent survival, but not both simultaneously, suggesting constraints in the way energy can be allocated between current and future reproduction (Ridgway & Shuter, 1994).

Parental age

A parent that has reduced prospects of future reproduction should increase its relative investment in current reproduction (Williams, 1966). Consequently, older parents are predicted to invest more heavily in a current brood than would be the case for a younger individual. In cases where there is no prospect of future reproduction, such as in semelparous species, current reproduction should consume all resources, effectively resulting in the death of the parent. This pattern is observed, for example, in *Oncorhynchus* spp. A predicted increase in parental investment with age in iteroparous species, sometimes called 'terminal investment', has some empirical support (Roff, 2002), though the results of many studies are equivocal. There are some studies that show a tendency for older individuals to invest a greater proportion of resources in care than do young individuals. In the mosquitofish (*Gambusia affinis*, Poeciliidae), a livebearer, older females were shown to invest substantially more resources in offspring than were young females. Older females produced an almost 50% heavier clutch than young females after adjusting for differences in size (Fig. 9.8), in accordance with terminal investment theory (Belk & Tuckfield, 2010).

Availability of alternative mates

For males, parental care often places restrictions on their ability to mate with additional females, which represents a potential fitness cost of care. This cost is often less severe for females, which are principally constrained by the rate at which they produce eggs, rather than the rate at which they mate (Chapter 8). If males are able to abandon offspring, they may be able to increase their reproductive fitness, but only if the offspring they abandon have a prospect of surviving without their care (albeit at a lower rate than if the male were present) and if additional females are available with which to mate. These two conditions are sometimes satisfied in fish species with biparental care. In some species, the female is often able to raise young to independence without

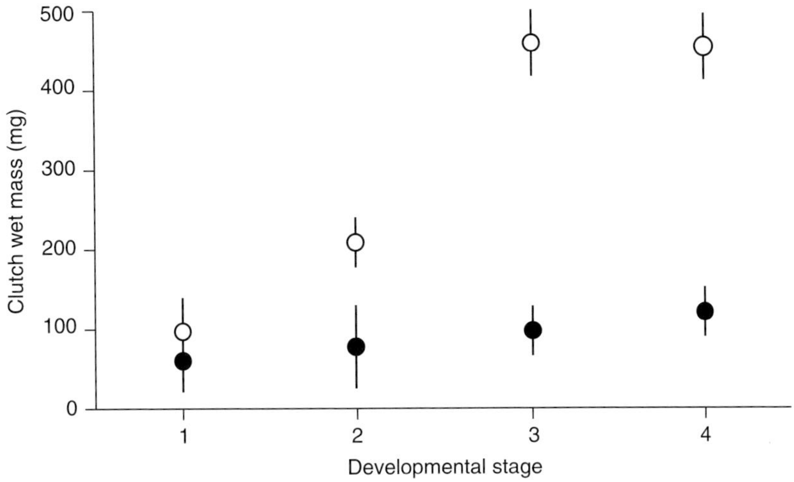

Figure 9.8 Least-squares mean (±1 SE) clutch wet mass (mg) as a function of female age class and stage of embryonic development in the mosquitofish (*Gambusia affinis*). White circles, old females; black circles, young females. (Source: Adapted from Belk & Tuckfield 2010. Reproduced with permission of John Wiley & Sons.)

assistance from the male, especially if the risk of predation of the young is limited. Male Panama cichlids (*Cryptoheros panamensis*, Cichlidae) sometimes desert their partners after spawning, and the probability of desertion is significantly higher if the risk of predation on the offspring is low (Townshend & Wootton, 1985b), suggesting that males are sensitive to the survival prospects of their offspring from predation. Older offspring have a higher chance of survival than young stages, and male biparental cichlids are more likely to desert later in a parental cycle than early on (Keenleyside *et al.*, 1990; Keenleyside & Mackereth, 1992).

The frequency of male mate desertion is also sensitive to their probability of remating. A female-biased sex ratio promotes male desertion in both *A. multispinosus* and *A. nigrofasciata* (Keenleyside 1983, 1985). Food availability can also influence whether a male deserts its partner. The reproductive rate of females is strongly dependent on food availability (Townshend & Wootton, 1985a; Wootton, 1998), so low food availability will tend to increase the interval before a female is able to spawn again after completing parental care. In biparental species, this may also encourage the male to desert and find another mate rather than wait for the same female to ovulate again (Townshend & Wootton, 1985b).

Helper care systems

Helping dominant individuals to raise offspring is a feature of the mating system of several mammal and bird species, but also of several species of lamprologine cichlids, such as *Neolamprologus pulcher* (Cichlidae), from Lake Tanganyika (Taborsky, 1984; Hatchwell & Komdeur, 2000; Heg *et al.*, 2005), although it is otherwise rare. In cichlid helper care systems, parental pairs obtain assistance from subordinate individuals. Some subordinates, usually the smallest, are the offspring from previous broods, but parentage analysis has shown that this is not always the case, and older unrelated subordinates may also function as helpers (Heg *et al.*, 2009). Helpers perform the same behaviours as the parents, fanning eggs and free embryos and protecting the young stages from predators, which represent a form of alloparental care (see the section 'Alloparental care'). Pairs with the assistance of helpers have higher reproductive success than those without assistance, although accepting the assistance of helpers may put the offspring at risk of cannibalism (Taborsky, 1985). Some helpers may themselves be reproductively active, with female subordinates occasionally able to produce their own clutches, fertilised by the dominant male, and male subordinates sometimes fertilising the eggs of dominant females. A consequence is that the dominant pair may obtain additional reproductive opportunities, in addition to assistance with parental care, but the reproductive success of the dominant male may also be eroded by the sneaking behaviour of subordinate male helpers (Heg *et al.*, 2009; Mitchell *et al.*, 2009). The benefit to the helpers appears to be through group membership, which gives them access to resources and protection, with a consequence that helpers have a greater probability of survival than non-helping subordinates, and thereby a greater chance of obtaining a dominant breeding position in the future. Helpers may also have the opportunity of limited mating opportunities as subordinates. For helpers that are related to the dominant pair, there is an additional fitness benefit through kin selection, as these individuals are assisting their parents in producing siblings to whom they are genetically closely related. The evolutionary and behavioural dynamics of cichlid helper systems raise intriguing questions and are the subjects of ongoing research.

Alloparental care

It is not uncommon to find that the recipients of parental care are not related to the care-givers, and in some cases are not even the same species. Care of non-descendant young is termed alloparental care and appears to contradict the conventional predictions for the evolution of parental care. However, there are a variety of explanations for the widespread occurrence of alloparental care in teleosts (Wisenden, 1999).

Sneaky mating and egg dumping

If a male parent is successfully cuckolded, a likely outcome is that he will provide care to unrelated offspring. Male parents, in some cases, may recognise that a sneaky mating has occurred, either because a sneaker was present during mating or because they are able to recognise the relatedness of offspring using olfactory cues. The evolutionary significance of sneaky mating is well recognised (discussed in Chapter 8). Although there are clear fitness costs to a male parent from sneaking, in terms of both lost paternity and wasted parental care invested in unrelated offspring, marginal benefits have also been proposed, in the form of a dilution of predation. Thus, if sneaking is relatively limited, so that the male parent sires at least some offspring, there may be a fitness benefit from caring for unrelated young, as there is a reduction in the probability that an attack by a predator will result in one of the parent's offspring being eaten. However, it is difficult to see how the fitness cost of lost paternity can be offset by even a substantial benefit of reduced probability of predation of related offspring, and dilution clearly could not be selected as an adaptive parental tactic.

The female equivalent of sneaking is termed 'egg dumping', and although common in birds (Yom-Tov, 1980), it has rarely been recorded in fish. One example is in the glass knifefish (*Eigenmannia virescens*, Sternopygidae) in which females attempt to deposit their eggs in the nests of territorial females (Hagedorn & Heiligenberg, 1985), though with no subsequent parental care by either males or females.

Egg stealing, kidnapping and nest takeovers

A series of extraordinary behaviours that involve usurpation of a parental male from his nest site, or the outright theft of eggs or young, can result in a male parent forcibly adopting young fathered by other males. The function of these behaviours appears to be in acquiring offspring to the advantage of the stepfather, possibly through a dilution effect, or by making the male more attractive to females. Egg stealing is a well recognised, but incompletely understood, behaviour performed by *G. aculeatus* (van den Assem, 1967; Wootton, 1971). Egg-stealing male *G. aculeatus* stealthily approach the nests of other males, before quickly entering the nest, grabbing a cluster of eggs in their mouths then departing, often pursued by the resident male. Egg-stealing males usually possess a nest and place the stolen eggs inside it. If the male does not have a nest, the eggs are invariably eaten (Wootton, 1976). The fitness advantage to a male of stealing eggs appears to be in increasing his attractiveness with regard to mate choice, as female *G. aculeatus* seem to have a preference for spawning in nests that already contain eggs (Rohwer, 1978; Ridley & Rechten, 1981); however, some studies have failed to show this effect (Jamieson & Colgan, 1989). A female preference for spawning in nests that already contain eggs has been shown in other fish species (Petersen & Marchetti, 1989; Sikkel, 1989; Forsgren *et al.*, 1996; Lindström & Sargent, 1997), and this may be a widespread response by females, although it is unclear whether the preference is for the male or the nest (Jamieson, 1995). In either case, the presence of eggs in a male's nest, whether obtained from a spawning female or by theft, can have the effect of increasing a male's subsequent success in attracting mates. Alternatively, egg stealing may benefit the male through dilution. There are some examples of males kidnapping free-swimming young stages in cichlids (McKaye & McKaye, 1977; McKaye, 1991), but this behaviour appears to be rare.

Males may also evict rival males from territories and continue care of the young belonging to the displaced male. This behaviour has been recorded in a range of teleost fishes, including the tessellated darter (*Etheostoma olmstedi*, Percidae) (DeWoody *et al.*, 2000), the fathead minnow (*Pimephales promelas*, Cyprinidae) (Unger & Sargent, 1988), and *P. minutus* (Lindström & Pampoulie, 2005). The benefit to the male again appears to be through either increased attractiveness to females or dilution.

An unusual form of nest takeover is 'piracy'. Here a male drives off a resident male, spawns with a female in the departed male's nest, then abandons the nest to the original male, which then resumes care of the nest, including the eggs fertilised by the pirate male (van den Berghe, 1988). As such, this behaviour represents an overt form of sneaking (Chapter 8), but results in a parental male caring for offspring that are not genetically related. Examples include shell-nesting Lake Tanganyika cichlids (Mboko & Kohda, 1999).

Brood parasitism

Brood parasitism involves a parent placing its eggs or young stages in the care of an unrelated individual, thereby avoiding the costs associated with parental care. The behaviour has been most comprehensively studied in birds, notably in cuckoos (Cuculidae) and cowbirds (Icteridae), but is also a feature of the reproduction of fishes and warrants greater attention by behavioural and evolutionary biologists. Brood parasitism can have a severe impact on host fitness, resulting in the destruction of all the host's offspring. However, the offspring of the host and parasite may be reared together, with limited effects on host reproductive fitness (Rothstein, 1990).

Brood parasitism can occur among members of the same species (intraspecific brood parasitism) or between different species (interspecific brood parasitism). In cichlid fishes, intraspecific brood parasitism has been called 'farming out' (Yanagisawa, 1985), and involves the transfer and subsequent abandonment of young close to conspecifics with young, which subsequently incorporate the abandoned young into their own brood. The behaviour is often associated with species with biparental care in which one parent has been deserted, though not exclusively, and intact pairs also engage in brood parasitism (Ochi & Yanagisawa, 1996). Farming out of young may be more common for single parents because the parental costs of attempting to rear a brood alone may be higher than for a pair, and the success rate of single parents is often lower than for a pair. Brood parasitism, then, is an alternative parental tactic to total brood abandonment or cannibalism of offspring when parental costs are increased. In the scale-eating Lake Tanganyika cichlid *Perissodus microlepis* (Cichlidae), a parent farming out its young will first gather a group of them in its mouth and approach a shoal of young belonging to another pair, before spitting its young into the shoal. This process is repeated until the entire brood has been disposed of. The new foster parents may attack the introduced young if they are a different body size to their own, though the parent that attempts to farm out its offspring often searches for shoals of young of the same body size to its own (Yanagisawa, 1985; Ochi & Yanagisawa, 1996).

Adopting parents may not suffer a substantially greater cost of caring, as the additional costs of caring for extra young may be limited, and acceptance of unrelated young may benefit them through a dilution effect. Parents of Central American cichlids species will not adopt young that are larger than their own offspring (Fraser, 1996). Smaller young have poorer anti-predator escape responses and consequently should suffer higher rates of predation. The adoption of more susceptible offspring is termed the 'selfish shepherd effect' and increases parental fitness through diverting predation to non-kin young (Wisenden, 1999). Wisenden & Keenleyside (1994) tested the selfish shepherd effect in wild *A. nigrofasciata* and showed that adopted young were more likely to be eaten by predators in accordance with predictions. Adopted young also grew more slowly than genetically related young, suggesting limited food competition by the adopted young with the parent's own offspring.

Interspecific brood parasitism, like intraspecific brood parasitism, can involve eggs as well as older stages. Some North American pebble-nesting cyprinids, which do not perform parental care, deposit their eggs in the nests of parental species and thereby obtain care for their young. For example, river chub (*Nocomis micropogon*, Cyprinidae) construct a substantial and complex nest to which they attract females to spawn and defend against other males (Maurakis *et al.*, 1991) (Chapter 7). Species such as the common shiner (*Luxilus cornutus*, Cyprinidae) use the nests of *N. micropogon* for spawning and thereby benefit from the parental activities of this species. It is not unusual for every *N. micropogon* in a population to have the eggs of other species in their nests (Vives, 1990). The cues used by brood parasites in locating and identifying a host include visual cues associated with the movements of a parent over the substrate, but also possibly chemical cues (Hunter & Hasler, 1965). Odour cues associated with the release of sperm and eggs by hosts may also trigger spawning in parasites (Hunter & Hasler, 1965), demonstrating coevolution between parasite and host.

Other cases of interspecific brood parasitism in fishes involve even more complex coevolution between host and parasite, comparable to that of avian brood parasites (Lotem & Nakamura, 1998). One of the best-known examples is that of the cuckoo catfish (*Synodontis multipunctatus*, Mochokidae) from Lake Tanganyika. This catfish exploits mouthbrooding cichlids by simultaneously spawning with them. Female mouthbrooders quickly pick up their own eggs after spawning, but also pick up the fertilised eggs of *S. multipunctatus*. The catfish eggs hatch more quickly than the host cichlid and the young *S. multipunctatus* eat the cichlid embryos inside the mouth of the parental cichlid. The cichlid subsequently broods a mouthful of young *S. multipunctatus* until they emerge as independent juveniles. Spawning *S. multipunctatus* do not appear to specialise in specific cichlid species, though whether they have species preferences remains to be formally tested. *S. multipunctatus* is an obligate parasite of mouthbrooding cichlids, but a related species, *Synodontis petricola* (Mochokidae), may be a facultative brood parasite.

Other brood parasites include the Japanese minnow (*Pungtungia herzi*, Cyprinidae), which exploits male parental care in the Japanese aucha perch (*Siniperca kawamebari*, Percichthyidae). *P. herzi* targets nests with recently deposited eggs, which ensures the minnow eggs hatch before the host eggs and thereby receive protection throughout development. *P. herzi* seeks *S. kawamebari* nests with large numbers of eggs, possibly because parental *S. kawamebari* will sometimes abandon nests with relatively few eggs (Baba, 1994). Male *S. kawamebari* (the parental sex) show no responses to minnow eggs in their nests, although spawning females appear to avoid this parasitism. Female *S. kawamebari* were less likely to spawn in nests that contained large numbers of eggs, which are the nests preferred by *P. herzi*. Females also produced smaller clutches during the period when parasitism by *P. herzi* was most likely. These female traits may have evolved as evolutionary responses to parasitism by *P. herzi* (Baba & Karino, 1998). *P. herzi* exploit other nesting species, including the dark sleeper (*Odontobutis obscura*, Odontobutidae). As with *S. kawamebari*, *P. herzi* are selective about the use of *O. obscura* nests with respect to the stage of parental care of the host and the numbers of eggs already in the host's nest (Hyoi & Nagata, 2000).

Cooperative care

Alloparental care may also evolve through cooperative breeding. Here, several parents may jointly engage in parental care of a mixed brood of young and jointly defend them from predators. This behaviour is not common, probably because parents are usually intolerant of other adults during parental care, although there are instances of cooperative parental care in cichlids. The benefit of cooperative care is presumably because a group of adults is more vigilant and effective in driving away predators than a solitary parent or pair (Dugatkin, 2002). This behaviour may also evolve where nursery or nesting areas are limiting and reproducing adults are forced into aggregations. Examples of cooperative care include the redbreast tilapia (*Tilapia rendalli*, Cichlidae) (Ribbink *et al.*, 1981), red empress cichlid (*Protomelas taeniolatus*, Cichlidae) (Ribbink, 1990) and green chromide (*Etroplus suratensis*, Cichlidae) (Ward & Wyman, 1977). In each case, it is unclear whether cooperative breeding is obligate or facultative.

Cooperative parental care is not confined to single species, and there are several cases of interspecific cooperative care. Cooperative defence of mixed broods of young cichlids and kampoyo, a bagrid catfish (*Bagrus meridionalis*, Bagridae), was maintained by parental cichlids and catfish in Lake Malawi. In several cases, *B. meridionalis* alone were found to care for mixed broods, suggesting that this association may alternate between interspecific cooperation and brood parasitism by the cichlids (McKaye, 1985). A feature of this association was that two species of young cichlid, *Copadichromis pleurostigmoides* (Cichlidae) and *Ctenopharynx pictus* (Cichlidae), were only ever seen in association with *B. meridionalis*, suggesting a mutualistic or parasitic relationship with them. An alloparental association also occurs between the Nicaragua cichlid (*Hypsophrys nicaraguensis*, Cichlidae) and the predatory guapote (*Parachromis dovii*, Cichlidae). Groups of *H. nicaraguensis*, in conjunction with pairs of adult *P. dovii* (presumably the parents), defended young *P. dovii* from predation by other cichlids, principally *A. nigrofasciata*, Midas cichlids (*Amphilophus citrinellus*, Cichlidae) and neet cichlids (*Hypsophrys nematopus*, Cichlidae), as well as juvenile bigmouth sleepers (*Gobiomorus dormitor*, Eleotridae) (McKaye, 1977). With the added protection of *H. nicaraguensis*, the survival of young *P. dovii* was significantly enhanced.

Misdirected care

Perhaps the most parsimonious explanation for many cases of alloparental care is that it simply represents misdirected care, with no fitness value to the alloparent (Coyne & Sohn, 1978). Parental fish are highly motivated to protect young stages and in many cases may be unable to distinguish their own offspring from unrelated young. Mobile young stages often stray from their parents or are displaced when predators attack a nest or brood. The inadvertent adoption of such young by unrelated parents is probably not uncommon. The costs of policing care by parents, including expelling unrelated offspring, is potentially expensive if the error rate in discriminating genetically related and unrelated young is significant. Further, if the fitness cost to a parent of caring for small numbers of genetically unrelated offspring is trivial, the strength of selection to evolve mechanisms to discriminate and expel unrelated young may be limited, and a low level of alloparental care may be a normal feature of parental care in many teleosts.

Filial cannibalism

A surprisingly productive area of research has accompanied the observation that parents occasionally eat some of their own developing offspring while caring for them. This seemingly non-adaptive behaviour demands explanation and has attracted the attention of evolutionary ecologists. To understand the function of filial cannibalism, a distinction needs to be made between what has been termed total or whole-clutch cannibalism and partial filial cannibalism. Whole-clutch cannibalism is associated with small broods and can be performed by a male or female parent or both. Whole-clutch cannibalism is comparable to abandonment of young and occurs most frequently when a brood is small in number, or in biparental species when one parent deserts its mate. Complete cannibalism of broods that fall below a threshold number is found in a number of species, including *G. aculeatus* (van den Assem, 1967), spotted tilapia (*Tilapia mariae*, Cichlidae) (Schwanck, 1986) and Cortez damselfish (*Stegastes rectifraenum*, Pomacentridae) (Petersen & Marchetti, 1989). Parents often also cannibalise the last eggs remaining in their nests after the rest have hatched (Rohwer, 1978; Smith, 1992). The adaptive value of whole-clutch cannibalism comes from the parents ceasing parental care and recovering energy invested in the eggs themselves (in the case of a female) or their care (both sexes). As such, whole-clutch cannibalism represents a comprehensive switch from current to future parental investment. The reason this switch is efficient in fish is because parental care is shareable, which means it is relatively less profitable for a parent to rear a small brood than a large one. Whole-clutch cannibalism is not only sensitive to brood size; desertion by a mate can also trigger cannibalism in biparental species, as solitary parents have greater parental costs than those in pairs, which increases the relative cost of parental care. Cannibalism by deserted parents is also more likely if desertion occurs early in the parental cycle (Lavery & Keenleyside, 1990; Keenleyside & Mackereth, 1992). For a deserted parent, the benefit–cost ratio of raising offspring from early in the parental cycle may be too low to be adaptive. Older young, which require less care and have a greater chance of reaching independence, have a higher benefit–cost ratio and it may pay to continue care in those circumstances. Thus, whole-clutch cannibalism is readily explained as a trade-off between current and future reproduction, and is contingent on the relative costs of parental care and the current status of a brood of young in comparison to that of a prospective future brood.

By contrast, partial-clutch filial cannibalism by male parents has proved more difficult to explain. Here, the male parent consumes a small proportion of the eggs under its care over the course of their development. Rohwer (1978) proposed that this behaviour had an energetic basis, with hungry males eating offspring to provide the energy necessary to complete care of all the offspring. These ideas were later elaborated and stated more formally by Sargent (1992). Under this hypothesis, males effectively parasitise female reproductive investment to maximise their own fitness, an example of an intersexual conflict. A simple prediction from this hypothesis is that poorly provisioned males should be more likely than well-fed males to engage in partial-clutch cannibalism. Several studies support this prediction. Male river blennies (*Salaria fluviatilis*, Blenniidae) that ate eggs in their nests were shown to maintain their body condition as food availability declined (Vinyoles *et al.*, 1999). Similarly, parental males that had their diets supplemented tended to reduce filial cannibalism in *S. rectifraenum* (Hoelzer, 1992) and *P. microps* (Kvarnemo *et al.*, 1998). However, there are also studies that show no relationship between food availability and filial cannibalism, including in *G. aculeatus* (Belles-Isles & Fitzgerald, 1991) and fantail darter (*Etheostoma flabellare*, Percidae) (Lindström & Sargent, 1997). Indeed, one study showed that Florida flagfish (*Jordanella floridae*, Cyprinodontidae) that received a reduced ration cannibalised fewer eggs than well-fed males, directly contravening a key prediction of Rohwer's (1978) hypothesis (Klug & St. Mary, 2005). The energetic benefits of filial cannibalism relative to the energetic costs of parental care may also be trivial in some species (Smith, 1992). Thus, although intuitively appealing, the assumption that filial cannibalism is necessarily linked directly to nutrition appears unsatisfactory, at least in some cases. Simply demonstrating that cannibalistic males obtain a nutritional benefit through cannibalism is not evidence that the behaviour has evolved for this function. Parents that engage in whole-clutch cannibalism are also likely to obtain a nutritional advantage as a result of eating their offspring, but an energetic benefit to whole-clutch cannibalism need not be invoked to explain its occurrence.

An alternative explanation for partial clutch filial cannibalism was proposed by Payne *et al.* (2002), who argued that selectively cannibalising small numbers of offspring could have the effect of increasing the overall rate of development and survival of a brood of young through better oxygenation of the remaining offspring. Artificially reducing embryo densities in *S. leucostictus* significantly increased overall embryo survival rates under low oxygen conditions (Payne *et al.*, 2002). A striking feature of filial cannibalism was that the spatial pattern of egg removal from a clutch of eggs by the male was random. By contrast, when eggs were eaten by predators, including unrelated

S. leucostictus males, they were removed in clumps. A random removal of eggs had the effect of increasing the rate of embryo development throughout the clutch, whereas removing eggs from a single region in the way that predators did had only a localised effect. Payne *et al.* (2004) used a model to illustrate that a parent that tracked environmental conditions and removed eggs in accordance with prevailing oxygen conditions could greatly improve its reproductive success. Cannibalising a proportion of a brood in a variable environment could also be beneficial if developing eggs and embryos faced a risk of suffocation. An advantage of filial cannibalism as a parental tactic to ventilate eggs and embryos is that it carries no cost through energy expenditure in the way that egg fanning does (Smith & Wootton, 1995a). The hypothesis of Payne *et al.* (2002, 2004) regarding partial-clutch cannibalism has received support from subsequent studies. Notably, Klug *et al.* (2006) subsequently extended Payne *et al.*'s (2002) ideas to propose a more general hypothesis, arguing that filial cannibalism is adaptive through limiting density-dependent mortality of eggs, which potentially arises both through oxygen limitation and the toxic effects of elevated levels of metabolic waste.

Evolution of parental care

The enormous variation in the modes of parental care in teleost fishes, and its wide phylogenetic distribution (Fig. 9.9) show that the behaviour has evolved on multiple occasions in this group. A consequence is that phylogenetic analyses of parental care in teleosts are potentially informative about the conditions under which different modes of care have evolved.

The most likely ancestral state in bony fishes with respect to parental care is of no care (Baylis, 1981; Mank *et al.*, 2005). How paternal, maternal and biparental care have subsequently evolved from a presumed ancestral state of no care in fishes has been the subject of speculation. There are two principal models of how the different forms of care might have evolved from a condition with no parental care. The stepping-stone model proposes that male parental care tends to evolve from no care, with biparental care evolving from paternal care and maternal care from biparental care (Barlow, 1974; Gittleman, 1981; Perrone & Zaret, 1979). An alternative model of care evolution is for the independent origin of the different forms of care (Mank *et al.*, 2005). The best evidence to date is equivocal, although there is marginal support for the latter model over the former. A large-scale phylogenetic analysis demonstrated that paternal care has evolved independently from a state of no care on at least 22 occasions in external fertilising

Figure 9.9 Supertree of actinopterygian fishes with the mode of parental care plotted. The number of surveyed families in each clade is shown. For polyphyletic orders, monophyletic clades are designated with an upper-case letter and comprise the following lower-level taxa: Perciformes A, Blennioidei; Perciformes B, Sphyraenidae, Polynemidae, Menidae; Perciformes C, Carangidae; Perciformes D, Gobioidei; Perciformes E, Emmelichthyidae, Lutjanidae; Perciformes F, Zoarcoidei; Perciformes G, Labroidei, Acanthuroidei; Perciformes H, Notothenoidei, Percidae; Perciformes I, Scombroidei, Stromateidae; Perciformes J, Gobioidei; Gasterosteiformes A, Syngnathoidei; Gasterosteiformes B, Gasterosteioidei; Scorpaeniformes A, Cottoidei; Scorpaeniformes B, Scorpaenoidei; Scorpaeniformes C, Dactylopteroidei; Osmeriformes A, Osmeriformes B, Alepocephalidae, Platytrocidae. (Source: Adapted from Mank *et al.* 2005. Reproduced with permission of John Wiley & Sons.)

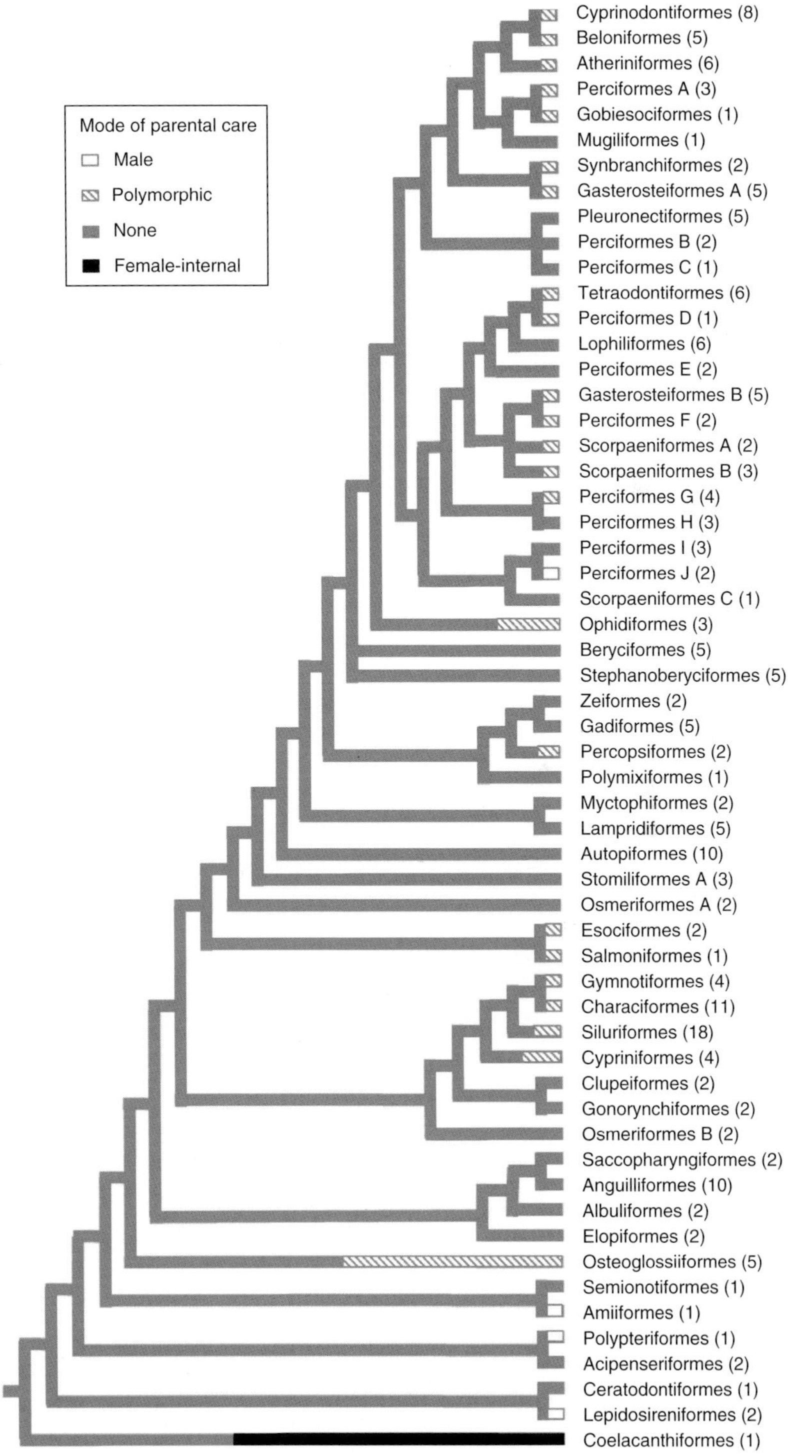
Mode of parental care
Male
Polymorphic
None
Female-internal
Cyprinodontiformes (8)
Beloniformes (5)
Atheriniformes (6)
Perciformes A (3)
Gobiesociformes (1)
Mugiliformes (1)
Synbranchiformes (2)
Gasterosteiformes A (5)
Pleuronectiformes (5)
Perciformes B (2)
Perciformes C (1)
Tetraodontiformes (6)
Perciformes D (1)
Lophiliformes (6)
Perciformes E (2)
Gasterosteiformes B (5)
Perciformes F (2)
Scorpaeniformes A (2)
Scorpaeniformes B (3)
Perciformes G (4)
Perciformes H (3)
Perciformes I (3)
Perciformes J (2)
Scorpaeniformes C (1)
Ophidiformes (3)
Beryciformes (5)
Stephanoberyciformes (5)
Zeiformes (2)
Gadiformes (5)
Percopsiformes (2)
Polymixiformes (1)
Myctophiformes (2)
Lampridiformes (5)
Autopiformes (10)
Stomiliformes A (3)
Osmeriformes A (2)
Esociformes (2)
Salmoniformes (1)
Gymnotiformes (4)
Characiformes (11)
Siluriformes (18)
Cypriniformes (4)
Clupeiformes (2)
Gonorynchiformes (2)
Osmeriformes B (2)
Saccopharyngiformes (2)
Anguilliformes (10)
Albuliformes (2)
Elopiformes (2)
Osteoglossiiformes (5)
Semionotiformes (1)
Amiiformes (1)
Polypteriformes (1)
Acipenseriformes (2)
Ceratodontiformes (1)
Lepidosireniformes (2)
Coelacanthiformes (1)

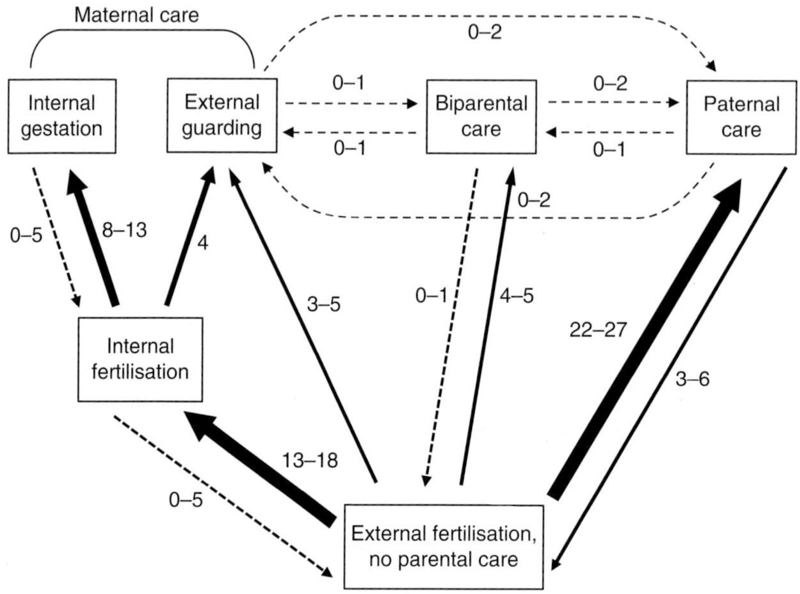

Figure 9.10 Maximum parsimony inferences for independent evolutionary transitions among modes of parental care in actinopterygian fishes. Arrow size reflects relative number of transitions, and numbers adjacent to arrows represent the minimum and maximum numbers of transitions. Dashed arrows indicate transitions that may not have occurred. (Source: Mank *et al.* 2005. Reproduced with permission of John Wiley & Sons.)

species, while maternal care has evolved at least eight times in internal fertilisers (Fig. 9.10; Mank *et al.*, 2005; see also Chapter 10). While it is comprehensive, this analysis is incomplete, as a little less than half of all teleost families were included. Biparental care and maternal care in externally fertilising species are also relatively uncommon, and it is premature to conclude that these two forms of care can only arise independently.

Despite ambiguities about the evolution of different forms of care, there are some broad patterns that appear robust. The first is that in species with external fertilisation, paternal care is the most common form of care. Paternal care is also strongly associated with territoriality and nest building (Gittleman, 1981; Smith & Wootton, 1995a; Mank *et al.*, 2005; Mank & Avise, 2006). Nest building, while not strictly a parental behaviour, tends to enhance the effectiveness of care (Chapter 7). Similarly, territoriality, which primarily serves to exclude rival males from sites of reproduction, would appear to be a prerequisite for nest building and care. Neither activity is compatible with the presence of competitors or potential predators of young stages. Paternal care is associated with both territoriality and pair spawning (Ah-King *et al.*, 2005), further indicating that paternal care will only evolve when males are able to exclude rivals and thereby assure paternity of offspring.

The other major pattern of parental care evolution is the association between internal fertilisation and maternal care. In this case, low paternity assurance may constrain the evolution of paternal care. In addition, the evolution of relatively complex anatomical and physiological traits by females to facilitate viviparity may so limit a male's ability to contribute to parental care that a paternal contribution to care is effectively obsolete. The evolution of maternal viviparity may be an irreversible process with no clear cases of transitions from this state back to oviparity in teleosts (Mank *et al.*, 2005; see Chapter 10 for a fuller discussion).

Parental care in fishes also tends to be more associated with freshwater than marine species (Baylis, 1981). The reason for the lower occurrence of care in marine fish may be linked to the higher frequency of pelagic eggs in marine systems, which parents cannot readily guard and which are usually well oxygenated by water movement (Gross & Sargent, 1985). Freshwater environments may more frequently experience hypoxic conditions, which demands specialised adaptations by eggs and embryos or active ventilation of the young stages by a parent. Other reproductive traits suggested to have coevolved with parental care include egg size and clutch size. In species with parental care the proposal is that both are greater; however, in neither case is the evidence wholly convincing (Kolm, 2009).

Parental care and the traits associated with it contribute to the overall life-history strategy of an individual. Ultimately, an understanding of how the various reproductive traits influence reproductive success can only be understood in the context of life-history theory (Chapter 11). However, teleosts also display a variety of unusual modes of reproduction, including viviparity, unisexuality and hermaphroditism. These modes of reproduction are addressed in Chapter 10.

Chapter 10
Unusual reproductive modes

Reproductive Biology of Teleost Fishes, First Edition. Robert J. Wootton and Carl Smith.

Introduction

The majority of teleost species are gonochoristic (Chapter 3) and oviparous (Chapter 4), yet the diversity of reproduction in teleosts is illustrated by species with unusual modes of reproduction (Blüm, 1986; Jameson, 1988). One of these modes, viviparity, has evolved several times in the phylogeny of the vertebrates and, excluding the monotremes, is universal in the Mammalia, common in the Chondrichthyes, occurs sporadically in the Amphibia and Reptilia, but is absent in the Aves. It is rare in the teleosts. Parthenogenesis, in which only the female genome passes from generation to generation, does not occur in the Chondrichthyes, Aves and Mammalia, but does occur rarely in the teleosts, Amphibia and Reptilia. Hermaphroditism is rare in the vertebrates, essentially absent in the Chondrichthyes, Aves and Mammalia, but occurs in a number of teleost families. The range of reproductive modes found in the teleosts reflects the lability of their sex determination mechanisms (Chapter 3). This diversity of reproductive modes in teleosts provides an opportunity for analysing the adaptive significance of these modes, and this partly accounts for the disproportionate attention that has been paid to teleosts with unusual modes of reproduction.

Viviparity in teleosts

Viviparity is usually defined as the situation in which eggs are fertilised internally, then retained in the maternal reproductive system, where they undergo development. Hatching precedes or coincides with parturition and the result is the birth of free-living neonates (Wourms, 1981; Wourms *et al.*, 1988). For teleosts, the definition has to be modified, replacing 'maternal reproductive system' by 'parental reproductive system', because in one taxon, the Syngnathidae (pipefishes and seahorses), the males carry the developing embryos to parturition in a brood pouch (Stölting & Wilson, 2007). Male viviparity, as shown by the Syngnathidae, is rare in the animal kingdom and this raises questions about the nature of sexual selection in species in which the roles of males and females have been, at least partly, reversed (see Chapter 9). Here, male viviparity is treated separately from female viviparity.

Even in female teleosts, the nature of the viviparity differs from that of other jawed vertebrates, because in teleosts the gonoducts connecting the ovaries to the external environment are not a derivative of the embryonic Müllerian ducts, but represent new structures that develop from the ovary and the peritoneum (Chapter 4). These structures, for reasons that are unclear, do not seem to have the potential to act as uteri, in which the embryos can develop, the situation most clearly seen in the Chondrichthyes and Mammalia (Wourms *et al.*, 1988). Uniquely, in viviparous teleosts, the embryos develop within the ovary – intra-ovarian gestation.

Two basic types of ovarian gestation occur (Wourms, 1981; Wourms *et al.*, 1988). With few exceptions, the egg is fertilised within the ovarian follicle. In some species, the fertilised egg is then ovulated, released from the follicle into the ovarian lumen, in which the embryo continues its development to hatching and parturition. This is intraluminal gestation. In other species, the embryonic development takes place within the follicle, termed intrafollicular gestation. In some species, the developing egg is released from the follicle to complete its development in the ovarian lumen, but in other species the development continues within the follicle until the embryo has developed close to term (Blüm, 1986; Wourms, 1981; Wourms *et al.*, 1988). Thus, in viviparous teleosts, the ovary acts as both the source of the eggs and as the site in which the eggs develop to term and parturition.

Systematic distribution of viviparity in teleosts

Of the approximately 30 000 species of teleosts, only just over 500 (i.e. < 2%) are viviparous. These are distributed across 14 families (Mank *et al.*, 2005), although this number may increase as more species are studied (Wourms *et al.*, 1988; Pandian, 2011). All the viviparous species so far described come from evolutionary advanced teleosts (Paracanthopterygii and Acanthopterygii; see Chapter 1). The sporadic taxonomic distribution of viviparity, the diversity of the habitats in which viviparous species occur and phylogenetic analysis demonstrate that it has evolved independently on several occasions in the teleosts (Mank *et al.*, 2005). Phylogenetic analysis has suggested that viviparity has evolved independently at least four times in the Atherinomorpha, once in the Beloniformes and three times in the Cyprinodontiformes (Mank & Avise, 2006), with the goodeids (Goodeidae), four-eyed fish (Anablepidae) and livebearers (Poeciliidae) consisting almost entirely of viviparous forms (Table 10.1).

Internal fertilisation

A prerequisite for the evolution of viviparity is the evolution of internal fertilisation, such that the male deposits spermatozoa at the entrance of or within the female gonoduct. Internal fertilisation in teleosts has evolved on more occasions than viviparity. It has been identified in at least 21 families from about nine orders (Mank *et al.*, 2005; Evans & Meisner, 2009). In species with internal fertilisation but not viviparity, the eggs are fertilised within the female reproductive tract, but then spawned at an early stage of embryogenesis (Abe & Munehara, 2009; Evans & Meisner, 2009).

In the sculpins (superfamily Cottoidea), about 21 genera have internal fertilisation, but viviparity occurs only in the Baikal oilfishes (Comephoridae), a family of cottids endemic to Lake Baikal in eastern Russia (Abe & Munehara, 2009). The male cottid has a flexible, tube-like genital papilla that is inserted into the gonoduct of the female to deposit the spermatozoa. Fertilisation only occurs after the egg has ovulated into the ovarian lumen and is only completed after the eggs are released into the external

Table 10.1 Distribution of viviparity in teleosts. (Sources: Data from Wourms *et al.* 1988; Mank *et al.* 2005; Mank & Avise 2006.)

Order	Family	Number of species	Proportion (%) viviparous
Ophidiiformes	Bythitidae	107	100%
	Aphyonidae	22	100%
	Parabrotulidae	3	100%
Atheriniformes	Hemiramphidae	78	3%
Cyprinodontiformes	Goodeidae	38	90%
	Anablepidae	6	86%
	Poeciliidae	137	99%
Scorpaeniformes	Scorpaenidae	330	33%
	Comephoridae	2	100%
Perciformes	Embiotocidae	23	100%
	Clinidae	75	80%
	Labrisomidae	75	21%
	Zoarcidae	65	3%

environment. Internal fertilisation in cottids is typically found in species in which the females spawn several clutches over a breeding season and is frequently associated with parental care (Chapter 9).

In viviparous kelp blennies (Clinidae), the sperm is transferred to the female by a protruding, intromittent papilla, whose form and size show interspecific variation (Fishelson *et al.*, 2006). Internally, the paired gonoducts of the male open into a muscular ampulla, which also receives a duct from the urinary bladder. The ampulla then opens into the external intromittent papilla. Spermatozoa are transferred to the female gonoduct in spermatozeugmata (Chapter 4).

The priapiumfishes (Phallostethidae) area family of minute teleosts found in South Asia, all of which have internal fertilisation, but not viviparity. They have an unusual, bizarre intromittent organ, the priapium (Evans & Meisner, 2009). This consists of a modification of the pelvic fin (ctenactinium) or some other structure (toxactinium), positioned under the head at the base of the throat. The genital papilla, testes and urinogenital opening are all displaced anteriorly, lying in front of or below the pectoral fin (Fig. 10.1). The mating biology of these unusual and poorly studied fishes, and their mechanism of fertilisation are not well understood. Males, at least of some species, produce spermatozeugmata (Chapter 5). The male clasps the female with the aid of the priapium, and deposits free spermatozoa or spermatozeugmata over or near the female genital opening from where the spermatozoa migrate into the oviduct (Mok & Munro, 1997). The females do not give birth to live young, but instead lay fertilised eggs (Grier & Parenti, 1994).

The anal fin is often modified for internal fertilisation (Evans & Meisner, 2009). Intromittent morphology in fishes is best described and studied in the cyprinodont fishes, which possess gonopodia. In the four families of cyprinodonts with internal fertilisation (Rivulidae, Goodeidae, Anablepidae, Poeciliidae), the gonopodium comprises modified anal fin rays, but the degree of modification of this structure varies across the order. In the Poeciliidae, the gonopodium is used to deposit spermatozeugmata at, or in the entrance of, the female gonoduct (Constantz, 1989; Evans & Meisner, 2009). In the male, of the nine anal fin rays, rays 3, 4 and 5 elongate and thicken, with ray 3 being the

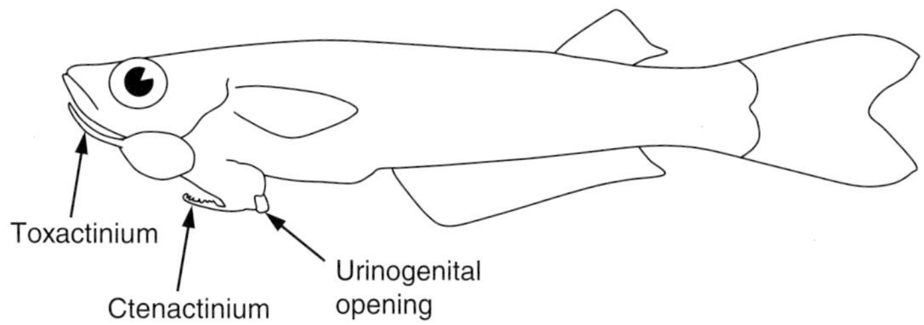

Figure 10.1 A male priapiumfish (*Phallostethus cuulong*) showing the position of the priapium, comprising the ctenactinium and toxactinium, which are used in mating to clasp a female and deposit spermatozoa or spermatozeugmata over or near the female genital opening. (Source: Adapted from Shibukawa *et al.* 2012. Creative Commons Attribution 3.0 Unported License, https://creativecommons.org/licenses/by/3.0/.)

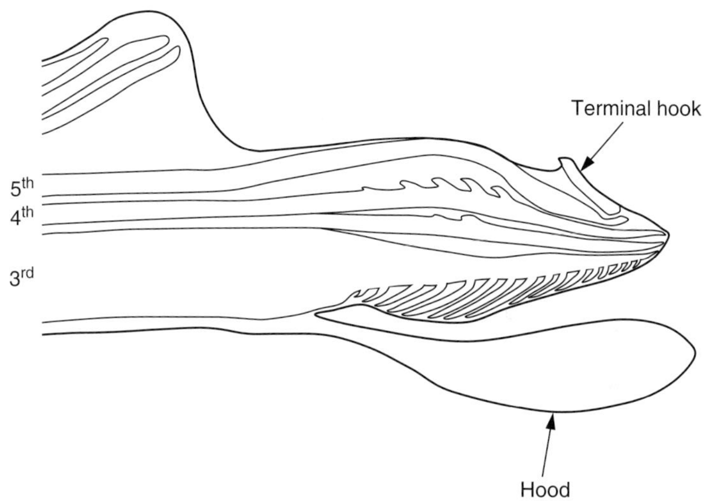

Figure 10.2 Lateral view of the distal tip of the male gonopodium of the guppy (*Poecilia reticulata*). Numerals indicate anal fin ray numbers. (Source: Adapted from Gasparini *et al.* 2011. Creative Commons Attribution License 4.0, http://creativecommons.org/licenses/by/4.0/.)

thickest and the main supporting element of the gonopodium (Fig. 10.2). The rays consist of bilaterally paired, opposing non-fused bone segments, between which are blood vessels, nerves and connective tissue. The gonopodia of poeciliids are usually held in a resting position against the body of the male, but are rotated through 180° for mating, when the tip is inserted into the female gonopore. Specialised muscle and associated ligaments allow the male to swing the gonopodium forward and aim it at the opening of the female gonoduct. When the gonopodium is swung forward, the rotation of the anal fin rays forms a partially enclosed tube. There is interspecific variation in the dorsal tip of the gonopodium and the form of the gonopodial hood, which is a vascularised tissue over the anterior region of the fin ray. The gonopodia can possess hooks, barbs, spines and serrations, which are used as characters for species identification (Rosen & Bailey, 1963; Greven, 2011), but which may also be functional.

In the Malabar ricefish (*Oryzias setnai*, Adrianichthyidae), males produce pointed, barbed spermatophores and possess an intricately shaped anal fin that functions as a gonopodium. The male uses his gonopodium to insert the spiked spermatophores into the tissue around the female genital pore. The spermatophore remains fixed in place because of the whorl of barbs at its tip (Fig. 8.14). After a delay, the spermatophore ruptures and spermatozoa released from it migrate into the female genital opening. Females sometimes have multiple spermatophores attached to them, and the area around their genital pore is covered by 'genital pads', thickened and hardened skin that serves as a defence against the effects of stabbing with spermatophores (Grier, 1984). The evolution of genital pads appears to be an example of 'chase-away' selection for female resistance to injury during mating (Arnqvist & Rowe, 2005; see also Chapter 8).

Members of the atherinomorph family, the Hemiramphidae that live in fresh or brackish water have internal fertilisation. In these species, the anal fin of the male is modified into an andropodium, which may be used to orientate an elongated genital papilla so that it is near the female genital opening (Evans & Meisner, 2009).

A key point is that internal fertilisation, with the required adaptations in the males, has to evolve before viviparity can evolve, but does not necessarily lead to viviparity. It is a necessary but not a sufficient trait. Internal fertilisation may increase the confidence of the male in his paternity, unless the female has the ability to store sperm. For the female, internal fertilisation may increase the proportion of her eggs that are successfully fertilised.

Other Orders of teleost that show internal fertilisation include characiform, siluriform, osmeriform, ophidiiform, scorpaeniform and perciform fishes (Evans & Meisner, 2009).

Lecithotrophy and matrotrophy

A key feature of viviparity in teleosts is the degree to which the female supplies the developing embryo with nutrients in addition to those supplied in the form of yolk. In lecithotrophy, the developing embryo depends entirely on the yolk deposited in the oocyte during vitellogenesis prior to fertilisation (Chapter 4). In matrotrophy, the embryo is also nourished by nutrient transfer from the female to the developing embryo after fertilisation (Constantz, 1989; Wourms, 1981; Wourms *et al.*, 1988; Marsh-Matthews, 2011). There can be a continuum from full lecithotrophy to full matrotrophy, in which the female supplies little or no yolk prior to fertilisation (Pollux *et al.*, 2009; Marsh-Matthews, 2011). The distinction between lecithotrophy and matrotrophy can be inferred from the weight changes of embryos during development. While lecithotrophic species tend to lose approximately 35–40% of their dry mass during development, matrotrophic species either maintain or gain weight. In some species, weight gains can be 20–30 times higher than the dry weight of the egg at fertilisation (Reznick *et al.*, 1996b, 2002b). Variation in provisioning has evolved among closely related species, e.g. among members of the genus *Poeciliopsis* (Poeciliidae), which show marked variation in maternal contribution. The embryo can be almost wholly reliant on matrotrophic transfer of nutrients, e.g. in the least killifish (*Heterandria formosa*, Poeciliidae) (Scrimshaw, 1944). Matrotrophy in other species may be facultative, e.g. in the sailfin molly (*Poecilia latipinna*, Poeciliidae) (Trexler, 1997).

The position of a species along the lecithotrophy–matrotrophy continuum is usually quantified as the matrophy index (MI), defined as the dry mass of the neonate (DW_{neo}) divided by the dry mass of the newly fertilised egg (DW_{egg}; Pollux *et al.*, 2009). A better index is the logarithm of MI, defined as $\log(DW_{neo}/DW_{egg})$. The advantage of log(MI) is that it has better statistical properties than the simple ratio, MI, it takes negative values when the dry mass declines between fertilisation and hatching, and it is symmetrical around a value of MI = 0. In oviparous species, in which the developing embryo is entirely dependent on yolk, the MI is typically about 55–75% [$\log_{10}$(MI) = –0.260 to –0.125]. These values indicate a loss in mass as the developing embryo uses a portion of the yolk to meet its metabolic expenditures. Lecithotrophic, viviparous species have similar values. Values that approach MI = 1, i.e. $\log_{10}$(MI) = 0, indicate that some nutrients are being transferred from the female to the embryo. When MI = 1, the transfer has been just sufficient to meet the metabolic costs of the developing embryo. Values of MI greater than 1, i.e. log(MI) > 0, indicate that the transfer from the female has been sufficient to support a growth in dry mass of the embryo during the gestation. Table 10.2 shows examples of the matrophy index for a range of teleost families.

In the Poeciliidae, species range from the fully lecithotrophic, such as the Amazon molly (*Poecilia formosa*, Poeciliidae), *P. reticulata* and green swordtail (*Xiphophorus helleri*, Poeciliidae), to species that are specialised matrotrophs such as *H. formosa* and elongate toothcarp (*Poeciliopsis elongata*, Poeciliidae) (Pollux *et al.*, 2009). Whereas the lecithotrohic species have large, yolky oocytes about 2 mm in diameter, the specialised matrotrophs have small oocytes about 0.4 mm (*H. formosa*) to 0.8 mm in diameter (*Poeciliopsis*) (Constantz, 1989). During oogenesis, such matrotrophs show only limited vitellogenesis. The mature oocyte of *H. formosa* contains a large oil droplet, but only a few yolk globules (Uribe & Grier, 2011).

The long-lived, marine rockfishes (Scorpaenidae) include oviparous species, species with internal fertilisation and viviparous species (Haldorson & Love, 1991). The viviparous species have intraluminal gestation and are mainly lecithotrophic but some, such as the black rockfish (*Sebastes melanops*, Scorpaenidae), show some degree of matrotrophy (Table 10.2).

The extent of matrotrophy tends to correlate with the degree of development of specialised maternal and embryonic tissues that form the interface between mother and offspring. Thus, lecithotrophy can be considered as the most primitive form of viviparity, with relatively unspecialised structures, while selection for greater maternal transfer of resources to developing embryos has driven the evolution of a suite of derived features in a series of independent evolutionary events (Wourms, 1981).

Ovarian modifications in viviparity

Even in lecithotrophic, viviparous species, in which the female makes no nutrient contribution to the developing embryo, the ovary must still be able to meet the respiratory demands, the needs for osmotic regulation and the processing of metabolic wastes of the developing embryos (Wourms, 1981; Wourms *et al.*, 1988). In one sense, these demands have already been made by the developing oocytes, but those of the developing embryos are much more intense. A further problem, not posed by the developing oocyte, is that the embryo has the potential of inducing an immunological response by the female. At least in some species, the zona pellucida may protect the embryo from any immunological defence mounted by the female (Hogarth, 1973). For matrotrophic species, the ovary must have adaptations that allow it to supply the embryos with their nutritional requirements.

In teleosts such as the Atherinomorpha and surfperches (Embiotocidae), the paired ovaries have fused to form a single ovary with a single oviduct opening to the external environment at a fleshy papilla just posterior to the anus (Wourms, 1981; Wourms *et al.*, 1988; Aranzábal *et al.*, 2009). In the viviparous rockfishes, the right and left ovaries fuse posteriorly and lead to a single gonoduct. Curiously, in the viviparous genera, *Sebastes*

Table 10.2 Matrotrophic investment (MI) in viviparous species. (Source: Adapted from Wourms *et al*. 1988. Reproduced with permission of Elsevier.)

Order	Family	Species	Egg mass (mg)	Neonate mass (mg)	Change in mass (mg)	MI	Log_{10}(MI)
Cyprinodontiformes	Goodeidae	*Goodea atripinnis*	0.245	3.15	2.905	12.86	+1.11
		Chapalichthys encaustus	0.12	3.38	3.26	28.17	+1.45
		Ameca splendens	0.21	31.7	31.49	150.9	+2.17
	Anablepidae	*Anableps anableps*	0.049	149	148.95	3041	+3.48
		A. dowei	0.108	910	909.89	8426	+3.93
		Jenynsia lineata	0.024	5.8	5.78	241.7	+2.38
	Poeciliidae	*Belonesox belizanus*	9.9	6.9	−3.0	0.70	−0.16
		Poeciliopsis monacha	2	1.26	−0.74	0.63	−0.20
		Poeciliopsis turneri	0.18	3.39	3.21	18.83	+1.27
		Heterandria formosa	0.017	0.68	0.66	40	+1.60
Scorpaeniformes	Scorpaenidae	*Sebastes norvegicus*	3	2	−1	0.67	−0.18
		S. melanops	0.071	0.067	−0.004	0.94	−0.03
Perciformes	Embiotocidae	*Embiotoca lateralis*	4.42	910	905.6	205.9	+2.31
	Clinidae	*Clinus superciliosus*	0.047	16.2	16.15	344.7	+2.54
	Zoarcidae	*Zoarces viviparus*	20	240	220	12	+1.08

and *Sebasticus*, the basic structure of the ovary is similar to that in many oviparous species (Chapter 4), whereas in *Helicolenus*, there is a highly modified ovary. *Helicolenus* has internal fertilisation, but spawns the fertilised eggs embedded in a gelatinous mass, and the structure of the modified ovary allows for that embedding (Koya & Muñoz, 2007).

The ovaries of viviparous species are usually characterised by increased vascularisation and an increased area of the internal epithelium. These increases ensure an adequate supply of oxygen and removal of metabolic wastes for the embryos. The viviparous rockfishes have intraluminal gestation and are characterised by high fecundities, often numbering in the hundreds of thousands (Haldorson & Love, 1991). The ovary has a dual arterial system, which forms an arterial loop by the confluence of branches of the anterior and posterior ovarian arteries, to meet the respiratory demands of the numerous embryos (Wourms, 1981; Koya & Muñoz, 2007).

The poeciliids have intrafollicular gestation. In the lecithotrophic species, the wall of the follicle shows little modification. In the matrotrophic species, the follicles become highly modified, with the follicular walls becoming highly vascularised and developing elongated villi, which are covered with a layer of secretory cells (Wourms, 1981; Wourms *et al.*, 1988; Constantz, 1989). The ovarian epithelium of the onesided livebearer (*Jenynsia lineata*, Anablepidae) develops highly vascularised, villous extensions of the inner ovarian epithelium. These extensions enter the gill chambers of the developing embryos.

In the Goodeidae and Embiotocidae, both of which have intraluminal gestation, the inner ovarian epithelium lining the ovary is thrown into a series of lamellae, called ovigerous folds (Wourms, 1981). In the Embiotocidae, the height of the epithelial cells increases during early gestation, reaching a maximum during late gestation. These cells secrete the fluid that fills the ovarian cavity. There is an increase in the vascularisation of the ovary, with large numbers of branches ramifying beneath the epithelium. The ovigerous folds form compartments around the developing embiotocid embryo (Wiebe, 1968; Wourms, 1981).

Testis modification in viviparity

Although the Atherinomorpha are unusual in having restricted lobular testes (Chapter 4), it is not obvious why this should be an adaptation for viviparity. The main male adaptation is related to the need to fertilise the female internally. The spermatozoa are usually deposited in a bundle, such as the spermatozeugmata of the poeciliids or the spermatophores of ophidioid species.

In the shiner perch (*Cymatogaster aggregata*, Embiotocidae), there is a curious temporal pattern to spermatogenesis and oogenesis, with the cycles of the males and females about 6 months out of phase (Wiebe, 1968).

Gestation

Timing of events

Gestation is the period from the fertilisation of the eggs until the neonates are born (Wourms, 1981; Wourms *et al.*, 1988). The simplest, although not the most common, sequence of events occurs in the viviparous eelpouts (Zoarcidae). The unfertilised oocyte is ovulated into the ovarian lumen and is fertilised there. All the embryonic development, including hatching, takes place intraluminally. In most viviparous teleosts, fertilisation takes place within the ovarian follicle. In some species, the temporal sequence is fertilisation, hatching within the follicles, ovulation and eventually parturition. An alternative sequence is fertilisation, ovulation, hatching within the ovarian lumen and parturition.

Within families, the sequence may vary among species (Wourms, 1981). In the brotulas (Bythitidae), *Ogilbia* has almost simultaneous ovulation and hatching, whereas in *Cataetyx*, hatching occurs some time after ovulation. In the Goodeidae, fertilisation and ovulation can be almost simultaneous, as in the twoline skiffia (*Skiffia bilineata*, Goodeidae). The Anablepidae show two patterns. *Jenynsia* embryos initially develop intrafollicularly, but then are released from the follicle into the ovarian lumen to complete their embryonic development. By contrast, the development of the embryos of *Anableps* takes place within the follicle until close to term. This is also the pattern in the Poeciliidae. In the Embiotocidae, such as *Cymatogaster*, fertilisation and ovulation are closely associated, so that essentially all the embryonic development takes place in the ovarian lumen. The fertilisation of the oocytes and embryonic development in the highly fecund viviparous rockfishes both take place in the ovarian lumen. A similar pattern occurs in the viviparous Zoarcidae. In the eelpout (*Zoarces viviparus*, Zoarcidae), the eggs in the lumen hatch after about 2 months and parturition occurs after another two months.

Embryonic nutrition

In matrotrophic species, there must be mechanisms by which the female can transfer nutrients to the developing embryo. There are three main paths by which nutrients may be transferred. All or a part of the body surface, such as the fins, of the embryo may be able to absorb nutrients from the fluid secreted by the ovarian epithelium. Secondly, part of the alimentary canal of the embryo may become specialised for nutrient uptake. In some species, outgrowths of the hindgut called trophotaenia project into the ovarian lumen. Trophotaenia have evolved independently in several lineages of viviparous teleosts. Thirdly, some form of placenta may develop. Here, placenta is used as a descriptive term and does not imply any homology between the various forms of placenta found in viviparous vertebrates. A placenta is defined as a close apposition or fusion of embryonic organs to maternal tissues (Wourms *et al.*, 1988). Within a teleostean family, there may be considerable variation in the adaptations for embryonic nutrition.

Within the Ophidiiformes, the false brotulas (Parabrotulidae) have trophotaenia that are probably derived from the alimentary canal (Wourms, 1981). The apical surface of the cells of the trophotaenia are covered with microvilli, increasing their surface area. Similar structures occur in *Microbrotula* and *Oligopus*, both members of the Bythitidae. In the key brotula (*Ogilbia cayorum*, Bythitidae), late-stage embryos have ovigerous bulbs ('ovarian nipples') in their mouths. These bulbs, derived from the ovarian epithelium, have fluid-filled vesicles.

A range of adaptations occur in the Cyprinodontiformes (Wourms, 1981; Wourms *et al.*, 1988). The Goodeidae show a range from lecithotrophy to extreme matrotrophy. The least specialised matrotrophy occurs in *Ataeniobius*, in which the yolk reserves are adequate for the early stages of embryonic development, but subsequently, ovarian fluid is absorbed in the alimentary canal. More characteristic of the Goodeidae is the development of trophotaenia by the embryo, which is the main site of nutrient absorption. These take one of two forms – rosette or ribbon-like – which extend from the perianal region into the ovarian lumen. *Goodea* has rosette trophotaenia, which are short, blunt processes united at their base. Species such as *Allephorus* and *Ameca* have long, ribbon-like processes. The epithelium of the trophotaenia is nearly identical to the absorptive embryonic cells.

The Anablepidae exhibit high matrotrophic indices (Table 10.2). The family has two distinct patterns of matrotrophy. In *Jenynsia*, where the latter part of the development takes place in the ovarian lumen, processes of the inner ovarian epithelium invade the opercular opening and fill the mouth and pharyngeal cavity of the embryo. In *Anableps* spp., there is a prolonged gestation within the follicle and a complex follicular placenta develops. This initially consists of an extra-embryonic development of the embryonic pericardial sac, a follicular space and the follicle wall. The pericardial sac expands ventrally to form a 'belly sac', but is gradually obliterated by a coelomic component of the 'belly sac'. This placental mechanism is augmented by an enlargement of the alimentary canal, which is lined with cells with vascularised villi.

Phylogenetic analysis showed that placentas have evolved several times within the Poeciliidae (Reznick *et al.*, 2002b; Meredith *et al.*, 2011), including three independent origins within the genus *Poeciliopsis*. These analyses suggested that placentotrophy could evolve in as little as 750 000 years within a lineage.

In the viviparous Poeciliidae, an expanded pericardial sac forms the embryonic component of the placenta. In *Heterandria* and *Xiphophorus*, the pericardial sac expands over the anterior portion of the embryo, enveloping it in a double-layered, highly vascularised sac. In *Gambusia* and *Poecilia*, the pericardial sac is ruptured by the enlarging head, forming a 'neck strap'. In *Poeciliopsis*, the pericardial sac spreads ventrally and posteriorly to form a portion of a 'belly sac', while the coelom expands to form a portion of the posterior part of the belly sac. The follicular placenta is formed from the follicular wall, with its villi, the follicular space and the embryonic pericardial tissue. In some species, the vascularised yolk sac may also form part of the follicular placenta (Wourms, 1981; Wourms *et al.*, 1988; Constantz, 1989).

In the Perciformes, the Embiotocidae show an adaptation whereby the unpaired fins hypertrophy and are highly vascularised. Later in development, the embryonic hindgut becomes hypertrophied. The epithelium of the ovigerous folds becomes closely apposed to embryonic tissue. In the Clinidae, the embryonic body surface and fins are the sites of nutritional transfer.

Superfetation

The usual situation in viviparous species is for the female to carry a single brood of embryos through to parturition. However, in some taxa, most noticeably in the Poeciliidae, younger eggs are fertilised before the birth of older clutches, resulting in superfetation (sometimes termed 'superembryonation'; Wourms, 1981; Constantz, 1989; Pires *et al.* 2011). There are multiple broods at different embryonic stages developing simultaneously in the ovary. In some species, there may be only two broods; for example, in the labrisomid blennies (Labriosomidae), there are typically only two clutches, one in the follicles and one in the ovarian lumen. In extreme cases, the number can range from five, as in the blackstripe livebearer (*Poeciliopsis prolifica*, Poeciliidae), to nine, as in *H. formosa*. In *Clinus superciliosis* (Clinidae), as many as 12 simultaneous broods can occur. One consequence of superfetation is that the interval between successive births is reduced. In poeciliids with single broods, the interval between successive births ranges from about 23 to 75 days, whereas in species with superfetation, the inter-brood interval ranges from 1 to 10 days.

An unusual form of superfetation occurs in *Hemirhamphodon pogonognathus* (Hemiramphidae). The ovaries are long and narrow and the ova and embryos are arranged in a single file. The ova and embryos in the early stages of development are in the anterior portion of the ovary and the embryos in the later stages of development are in the posterior portion of the ovary. Typically, there are only one or two embryos in the same stage of development. This arrangement has been graphically described as a linear assembly line for embryos (Reznick *et al.*, 2007).

In matrotrophic species, the female has the problem of how to allocate resources between the different broods. In the Gila topminnow (*Poeciliopsis occidentalis*, Poeciliidae), as vitellogenesis supplies the younger clutch with yolk, little is allocated to the older clutch. After eggs in the younger clutch have been fertilised, nutrients are again transferred to the older brood (Constantz, 1989). Superfetation is abandoned when food is in short supply.

Within the Poeciliidae, both lecithotrophic and matrotrophic species can show superfetation. In some lineages, placentotrophy evolved before superfetation, whereas in other lineages, superfetation evolved before placentotrophy (Meredith *et al.*, 2011). However, there was a correlation, which implies that the presence of the one trait facilitated the evolution of the other.

Endocrine control of viviparity

Studies on poeciliids and viviparous species of the Scorpaenidae and Zoarcidae have suggested that the endocrine control of vitellogenesis and oocyte maturation is similar to that occurring in oviparous species (Chapter 6). The role of endocrines in the maintenance of pregnancy is less clear and may differ between species with intrafollicular gestation and species with intraluminal gestation. In the poeciliids, the levels of progestogens falls after oocyte maturation, but in the scorpaenids the level of 17α,20β dihydroxy-4-pregnen-3-one (17α,20β-P), a maturation-inducing hormone, remains high during pregnancy (Koya, 2008). The endocrinological control of superfetation, in which several stages of oocyte development and embryogenesis coexist, is still poorly understood. Parturition in the poeciliids is initiated by prostaglandins, which are implicated in ovulation in oviparous species (Chapter 4). The capacity of ovarian follicles to synthesise prostaglandins increases from the middle to late pregnancy, but declines after the neonates have been born (Koya, 2008).

Evolution of viviparity and matrotrophy

The rarity of viviparity in the teleosts suggests either that the conditions that favour the evolution of viviparity are rarely encountered or that the lack of oviducts derived from the embryonic Müllerian ducts (Chapter 4) have made it difficult for viviparity to evolve in the teleosts. There is a striking contrast between the teleosts and the Chondrichthyes, both of which have spent their entire evolutionary history in water. The cartilaginous fishes do have oviducts derived from the Müllerian ducts and a high proportion of the taxa has some form of viviparity (Blüm, 1986; Wourms *et al.*, 1988).

A clear potential advantage of viviparity is that it protects the young during the developmental stages when they are most vulnerable to predation or adverse environmental conditions (Wourms *et al.*, 1988; Clutton-Brock, 1991). It permits optimum conditions for growth and development of the embryos and provides the opportunity for direct provisioning of embryos (Wourms & Lombardi, 1992). Because of their lack of mobility and small size, the eggs of oviparous teleosts are particularly vulnerable to predation and unfavourable environmental conditions (Wootton, 1998) (Chapter 9). For the parent, viviparity may extend the time over which the parent is responsible for its progeny in comparison to a comparable oviparous species. Viviparity may also have effects on the mobility of the parent and could impose a constraint on the size of the clutch (Clutton-Brock, 1991).

The conditions under which viviparity is favoured have been formalised (Blackburn & Evans, 1986; Clutton-Brock, 1991). It is favoured if the survival of the eggs and embryos is higher when retained by the parent than if eggs are released into the environment. However, if the survival of the parent from the beginning of one reproductive attempt to the next is lower for the viviparous than for the oviparous form, viviparity is inhibited. It is also inhibited if the clutch size in the viviparous form is lower than in the oviparous form. Viviparity will also be favoured where small offspring are at a disadvantage to larger ones (Trexler & DeAngelis, 2003; Trexler *et al.*, 2011).

The teleost families in which viviparity occurs fall roughly in four groups, but if there are common features between the groups that could account for the evolution of viviparity, they are not obvious.

The first group consists of the Atherinomorpha, which has the highest proportion of viviparous species (Reznick *et al.*, 2007). Viviparous species in the halfbeaks (Hemiramphidae) are small-bodied and occur in fresh or brackish waters in South-east Asia. The marine species in the family are oviparous. The livebearers (Poeciliidae), goodeids (Goodeidae) and four-eyed fishes (Anablepidae) are all small to medium-sized fish found in fresh or brackish water in sub-tropical and tropical America, although the Poeciliidae also includes the oviparous Aplocheilichthyinae from Africa and Madagascar. In the Atherinomorpha, there is a strong association between the invasion of freshwater habitats and the evolution of viviparity.

In the Percomorpha, the viviparous forms are marine species. The surfperches (Embiotocidae) are small to medium species living in inshore habitats, including kelp beds, tidal pools and rocky reefs, found predominantly in the north-east Pacific (one species is found in freshwater) (Helfman *et al.*, 2009).

The rockfishes (Scorpaenidae) include both oviparous and viviparous species. They are benthic, marine species. As the common name of the family suggests, they are often associated with rocky reefs, but they occur over a wide range of depths and are not confined to inshore areas.

The fourth group of families that includes viviparous forms is taxonomically diverse. It includes the ophidiiform families the false brotulas (Parabrotulidae), viviparous brotulas (Bythitidae) and aphyonids (Aphyonidae), and the perciform families, the kelp blennies (Clinidae) and labrisomid blennies (Labrisomidae). The common feature here is body shape. All are elongated, shallow-bodied fish (blenny-like), which usually indicates a benthic mode of life. They are typically marine and some occur in the abyssal depths of oceans.

The stages in the evolution of viviparity include the initial evolution of internal fertilisation, then the retention of the fertilised egg in the follicle or the ovarian lumen, with the egg dependent on the yolk provided by the female, the lecithotrophic condition. The next step in some lineages is the evolution of matrotrophy, with the female supplying some or virtually all of the nutrients required by the developing embryo after fertilisation.

Modelling studies suggest that matrotrophy evolves if the environment provides a high and stable level of food resources (Trexler & DeAngelis, 2003; Trexler *et al.*, 2011), which ensure that the parent can maintain a supply of nutrients to the embryos until parturition. This allows the production of offspring at a high rate.

What advantages accrue to the parent with the evolution of matrotrophy? A comparison of viviparous Hemiramphidae and Poeciliidae failed to find any consistent relationship between matrotrophy and either offspring size or offspring number (Reznick *et al.*, 2007). In the poeciliid genus *Poeciliopsis*, the matrotrophic species in the 'northern clade' tended to have more and smaller offspring than lecithotrophic species. By contrast, in the 'southern clade', the matrotrophic species tended to have fewer and larger offspring (Reznick *et al.*, 2007). However, the matrotrophic poeciliids generally produce broods at shorter intervals than the lecithotrophic species, given adequate food availability.

It has been proposed that in matrotrophic species in the Poeciliidae and Zenarchopteridae, the female devotes a smaller proportion of her body mass to reproduction than in lecithotrophic species (Reznick *et al.*, 2007; Pollux *et al.*, 2009). As a consequence, the body of the female is less distended and the swimming performance of the matrotrophic female is superior to that of the lecithotrophic female, with consequent advantages in avoiding predation and escaping unfavourable conditions. This hypothesis compares two forms of viviparity and needs to be distinguished from the hypothesis that viviparity exerts a cost in terms of reduced locomotory competence compared with oviparous species that shed the eggs soon after ovulation.

In some taxa, matrotrophy is associated with the evolution of some form of placenta. The obvious explanation for the evolution of the placenta and its elaboration is that it is driven by selection imposed by the environment, which favours the efficient transfer of nutrients from the parent to the embryos. An alternative hypothesis has suggested that the evolution of the placenta reflects a genomic conflict between the parent and the developing embryo caused by the prolonged contact between the embryo and the parent (Reznick *et al.*, 2007; Pollux *et al.*, 2009). As yet, there is little evidence for this hypothesis in viviparous teleosts.

A study of the northern clade of *Poecilipsis* suggested that the evolution of the placenta probably facilitated early maturation. Life-history theory (see Chapter 11) predicts that high mortality rates of adults caused by environmental factors such as predation favour the evolution of an early age at maturation (Pires *et al.*, 2011).

In both the Hemiramphidae and Poeciliidae, there is an association between the evolution of a placenta and superfetation. This is not an absolute correlation. In both taxa, there are species with lecithotrophy and superfetation and species with matrotrophy but not superfetation (Reznick *et al.*, 2007). In the northern clade of *Poeciliopsis*, there was high positive correlation between superfetation and the MI, but a high negative correlation with age at first births, offspring size at birth and the interval between successive births (Pires *et al.*, 2011).

In the Poeciliidae, one species, *Tomeurus gracilis* (Poeciliidae), displays zygoparity (embryoparity), in which fertilisation is internal but with females laying fertilised eggs or retaining them internally until shortly before hatching. Like viviparous poeciliids, males produce spermatozeugmata and possess restricted lobular testes. Thus, *T. gracilis* shares several apomorphic reproductive traits with viviparous poeciliids, including spermatozeugmata, testis morphology and a gonopodium, but also displays plesiomorphic traits characteristic of oviparous taxa, such as a thick zona pellucida with filaments (Parenti *et al.*, 2010). Surprisingly, this species is not basal in the Poeciliidae, but displays primitive traits that might be expected in a basal species in the family.

The most plausible explanation for the evolution of viviparity and the subsequent evolution of matrotrophy in teleosts is the selective environment imposed by predation intensities both on sexually mature adults and on eggs and free-living young stages. High predation rates on eggs will favour viviparity, whereas high predation rates on adults will favour the evolution of placentas and superfetation. However, this question is far from resolved. The long-lived rockfishes, for example, appear not to fit the general pattern.

Male viviparity in the Syngnathidae

The majority of teleostean species do not show parental care (Chapter 9). However, amongst those that do show such care, male parental care (paternal care) is the commonest mode (see Chapter 9 for discussion). Typically, this paternal care consists of the defence of the eggs and young stages that have been spawned on the substratum or in a nest (see Chapter 7). The most extreme form of paternal care in the teleosts is shown by the family Syngnathidae, the pipefishes and seahorses. In this family, the male is viviparous (Clutton-Brock, 1991; Stölting & Wilson, 2007). The gravid female transfers her eggs to the male to be fertilised. Male viviparity is rare in the animal kingdom and its occurrence in the Syngnathidae is another example of the exceptional reproductive diversity of the teleosts.

The Syngnathidae comprises about 300 species and has marine, brackish and freshwater representatives (Wilson & Orr, 2011). The basic morphological form is that of an elongated cylinder and the main propulsion is provided by the dorsal fin without involving the caudal fin. This mode of locomotion frees up the ventral surface of the fish and this surface is the site of the adaptations for the male viviparity. Within the Syngnathidae, there are two phylogenetic lineages, the Gastrophori and the Urophori, based on the position of the brood pouch (Wilson *et al.*, 2003; Stölting & Wilson, 2007). In the Gastrophori, the female attaches the eggs to the abdomen of the male, while in the Urophori the brood area is on the tail (Fig. 10.3). Within each of these lineages, there is an increase in the complexity of the brood pouches of the males, with the complexity reaching its highest level in the seahorses of the Urophori.

There are five levels of complexity of the brooding structures (Wilson *et al.*, 2003). The simplest, as seen in the straight-nosed pipefish (*Nerophis ophidion*, Gastrophori), is an unprotected area on the ventral surface of the fish (Fig. 10.4). The eggs are attached to the surface of the male with mucus. The eggs of *N. ophidion* are round and only a small area of the egg surface contacts the male body. In a second level of complexity, the eggs are enclosed in individual membranous compartments, as seen in *Dunckerocampus* (Gastrophori) and *Phyllopteryx* (Urophori) (Fig. 10.4). The next level of complexity, as in *Doryrhamphus* (Gastrophori), sees the protection of the eggs in a brood pouch with pouch plates (Fig. 10.4). This is the greatest level of complexity of development of brood pouch structure in the Gastrophori. In the Urophori, two further levels of complexity are reached. In the pipefishes (*Syngnathus* spp.), folds of skin develop ventral-laterally and grow towards the midline. In some species, each fold grows back towards

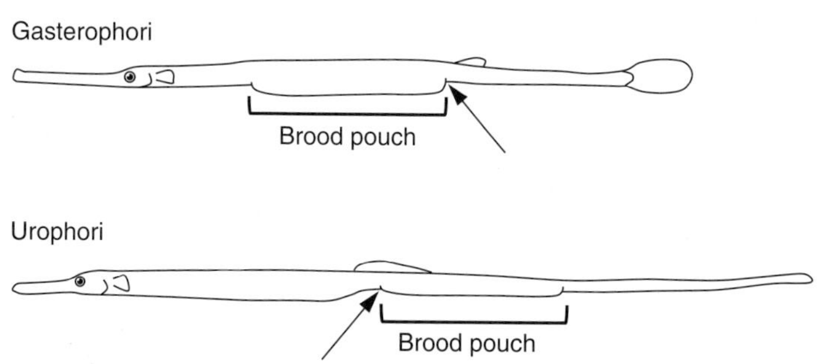

Figure 10.3 Distinction between the Urophori and Gastrophori, monophyletic clades within the Syngnathidae, defined by the position of their brood pouch, either under the abdomen (Gastrophori) or under the tail (Urophori). Arrows indicate the position of the anus.

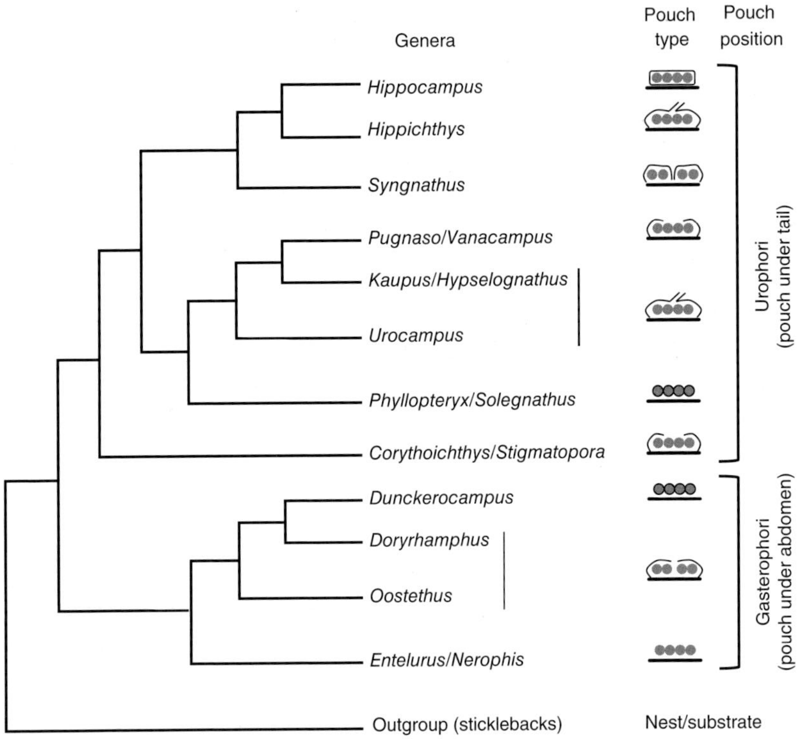

Figure 10.4 Simplified consensus phylogenetic tree for syngnathid species. Brood pouch types show increasing complexity within the Urophori and Gastrophori. Gasterophori: *Entelurus* and *Nerophis* unprotected area on the ventral surface of the male's abdomen; *Dunckerocampus*, eggs enclosed in individual membranous compartments; *Oostethus* and *Doryrhamphus*, brood pouch with pouch plates. Urophori: *Phyllopteryx/Solegnathus*, eggs enclosed in individual membranous compartments; *Corythoichthys/Stigmatopora* and *Pugnaso/Vanacampus*, semi-inverted bilateral pouch folds; *Kaupus/Hypselognathus*, *Urocampus* and *Hippichthys*, everted bilateral pouch folds; *Syngnathus*, inverted bilateral pouch folds; *Hippocampus*, enclosed pouch. (Source: Adapted from Wilson *et al.* 2003. Reproduced with permission of John Wiley & Sons.)

the surface of the male, so that two brooding compartments are formed. In other species, the folds meet in the midline forming a single compartment (Fig. 10.4). After the female has deposited the eggs in the brooding compartment, the compartments are essentially isolated from the external environment by the folds. The seahorses (*Hippocampus* spp.) show the highest level of complexity. The brood pouch is completely enclosed, communicating with the external environment only through a small, anterior pore. As the complexity of the brood pouch increases, the developing embryo is more and more isolated from the external environment and consequently more dependent on the male for its needs.

Protection of the developing embryos is provided by all types of brooding structure. With increasing isolation from the external environment, the demands of the embryos for oxygen, the removal of metabolic waste products, and the osmo- and ionic-regulation of the brood pouch fluid fall more and more on the parental male. Although the female lays yolky eggs, any shortfall in the nutrients supplied in the eggs must be met by the male.

In *N. ophidion*, the epithelium of the brood area is similar to the epithelium of the rest of the body, although the dermis is better vascularised in the region of the brood surface. The contact between the male and the egg is tenuous and the male seems to act only as a substratum on which the female can deposit eggs (Carcupino *et al.*, 2002). There is both histological and physiological evidence that the osmotic and ionic characteristics of the fluid bathing the developing embryos in the brood pouch are regulated in *Syngnathus* spp. (Carcupino *et al.*, 2002; Ripley, 2009). The dermis is more vascularised than in *N. ophidion* and the thinner, folded epithelium increases the surface area and so may facilitate the supply of oxygen to, and the removal of metabolites from, the embryos. In *Hippocampus* spp., the eggs are pear-shaped and the narrow end of the egg is embedded in skin folds in the brood pouch. The epidermis and dermis interdigitate and the dermis is highly vascularised. There is some evidence in *Syngnathus* and *Hippocampus* of the transfer of nutrients from the male to the developing embryo, supplementing the nutrients supplied in the yolk by the female (Kvarnemo *et al.*, 2011). There is also the possibility that the developing embryos may obtain some nutrients released from eggs supplied by the female, which have failed to develop and break down.

The brood structure in syngnathids also relates to mating patterns (Carcupino *et al.*, 2002; Wilson *et al.*, 2003). A receptive male *N. ophidian* receives an entire clutch from a single female. The female can produce another clutch before the male has completed the gestation of the brood and will mate with another male. The reproductive success of the female depends on the number of males she obtains. *Syngnathus* spp. are polygamous. The structure of the brood pouch, with its folds, allows the male to mate several times without losing eggs previously deposited in the pouch. In many species, males can carry the eggs of several females, while the females can deposit the eggs of a single clutch among several males. The male can adjust the position of embryos in the pouch and partially open the pouch to release neonates. *Hippocampus* spp., with their enclosed pouch, are monogamous, with males and females forming stable pairs. The time taken by a male to incubate the eggs in its brood pouch determines when it can re-mate, and hence its reproductive rate. If the rate at which the females can produce new clutches is high, then the gravid females may have to compete for access to a limited number of receptive males, leading to sex role reversal (see Chapter 9). It is noteworthy that the monogamous seahorses do not show sex role reversal, because the reproductive rates of the partners are essentially the same.

Technical difficulties have hampered the study of the neural and hormonal control of pregnancy in male syngnathids (Sobell & MacKenzie, 2011). Histological studies on *Hippocampus* found that the Leydig cells, responsible for androgen synthesis (Chapter 5), show seasonal cycles, indicating that androgen synthesis is maximal during spermatogenesis and the development of the brood pouch, but is suppressed during gestation. Androgens do not appear to be necessary to maintain pregnancy. Indeed, in the broadnosed pipefish (*Syngnathus typhle*, Syngnathidae), plasma levels of testosterone (T), 11-ketotestosterone (11-KT) and hydroxytestosterone (OHT) were lower in pregnant males than in non-pregnant males in breeding condition. Histological evidence from *Hippocampus* suggested that pituitary cells producing prolactin and corticotrophic hormones were active during pregnancy. In other teleosts, prolactin has a role in osmoregulation and so prolactin may be implicated in the regulation of the brood pouch fluid. Neurohormones such as isotocin and argentine vasotocin (AVT), by analogy with other vertebrates, may also be involved in the regulation of gestation and parturition. It is not clear whether the endocrine control of male viviparity shares common features with the endocrine control of paternal behaviour in other teleosts. There is no evidence that the levels of T, 11-KT or oestradiol-17β (E_2) differ between males from species with conventional sex roles and those with reversed sex roles.

Male pregnancy in the animal kingdom is so rare that it is not clear under what circumstances it evolves. In the Syngnathidae, the more complex brood structures seem to have evolved from the condition in which the male simply acts as a substratum for the eggs (Wilson *et al.*, 2003). The Syngnathidae typically live in structurally complex habitats, such as reefs or eel grass beds, which offer many potential surfaces on which the female could lay eggs to be defended by the male. It is possible that high levels of predation favour the use of the male as a substratum for the eggs rather than an immobile surface. An advantage for the male is that the intimate contact between the female and male during the transfer of eggs ensures that the male has a high confidence in its paternity of the developing embryos, though this is at the cost of the commitment the male has to make to the offspring.

Unisexual teleosts

Sexual reproduction is the characteristic mode of reproduction in vertebrates. It comes as something of a surprise that there are a few vertebrate taxa, about 70, in which reproduction is a form of parthenogenesis, with only the female genome being transmitted from generation to generation (Neaves & Baumann, 2011). This mode of reproduction occurs in the teleosts, amphibians and reptiles. In the teleosts, there are seven or eight taxa with parthenogenetic reproduction producing unisexual, all-female populations (Pandian, 2011). In all vertebrates, the unisexual taxa are the result of hybridisation between sexually reproducing, closely related species. Hybridisation is not uncommon between species of teleosts, so the rarity of unisexual biotypes suggests that either the unisexuals are at a major disadvantage when in competition with sexual species or the conditions under which unisexual biotypes evolve are extremely rare (Stock *et al.*, 2010). The hybrid origin of unisexual vertebrates means that they have high levels of heterozygosity, which is maintained because of the mode of reproduction. The parthenogenetic reproduction of teleosts and amphibians is unusual, because the eggs of the unisexual taxa only develop if they are activated by spermatozoa from a closely related sexual species. The unisexual taxa are 'sperm parasites' on sexual taxa. A consequence is that the distribution of the unisexual form must overlap with the sexual species that can supply the sperm. In unisexual reptiles, this sperm parasitism is not required for the development of the parthenogenetic eggs and so reproduction is an unmodified from of parthenogenesis (Neaves & Baumann, 2011).

Asexual teleosts show two modified forms of parthenogenesis: gynogenesis and hybridogenesis (Moore, 1984; Schlupp, 2005; Lampert & Schartl, 2008; Neaves & Baumann, 2011).

In gynogenesis, the spermatozoa activates the development of the egg (Chapter 4), but the male chromosomes in the spermatozoa break down within the cytoplasm of the egg and usually make no contribution to the genome of the developing embryo. Inheritance is clonal. The usual mechanism is for the normal meiosis seen during oogenesis in sexual species (Chapter 4) to be replaced by a modified meiosis. In the pre-meiotic oocyte, the number of chromosomes doubles, a process known as endomitosis, with the subsequent reduction division restoring the diploid number of chromosomes. There is no recombination, so the maternal genome is passed on to the next generation unchanged (Pandian, 2011; Fig. 10.5). In hybridogenesis, oogenesis produces haploid eggs (Chapter 4), which are then fertilised by spermatozoa from a male of a closely related sexual species, producing a diploid zygote, with the somatic cells of the developing embryo containing the genomes of both the female and the male.

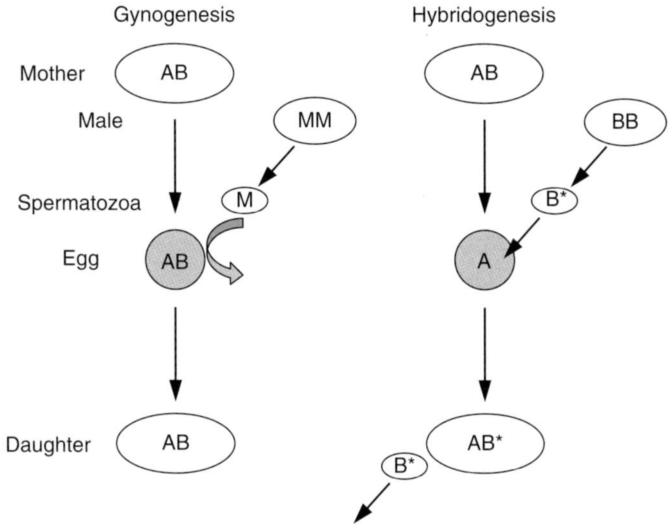

Figure 10.5 Summary of two modes of reproductive parthenogenesis. Maternal diploid genome (AB), paternal diploid genome (MM). In gynogenesis, spermatozoa of a closely related species activate development of the egg, but the male chromosome breaks down within the egg cytoplasm and makes no contribution to the embryo, with inheritance wholly clonal. In hybridogenesis, a haploid egg is produced without meiosis and fertilised by the spermatozoa of a closely related species to produce a diploid zygote. During subsequent oogenesis, the male chromosomes are excluded and inheritance is hemiclonal. In both gynogenesis and hybridogenesis, only the female genome is inherited. (Source: Adapted from Lampert & Schartl 2010.)

However, during oogenesis (Chapter 4), the male (paternal) chromosomes are excluded and only the female (maternal) chromosome is incorporated into the oocytes and subsequent eggs (Fig. 10.5). Inheritance is hemiclonal. In both modes, only the female genome is passed on to the next generation.

The biological concept of the species does not apply to parthenogenetic organisms and this complicates the naming of the unisexual taxa of teleosts. The usual solution is to name the taxa as if they were sexually reproducing forms and to call the taxa biotypes.

Unisexual vertebrates are rare, and unisexual teleosts represent only about 0.03% of all teleost species. However, they have attracted attention because their study helps to address two profound biological questions: the adaptive significance of sex and the mechanisms by which sexual and asexual organisms can coexist (Schlupp, 2005). The unisexual teleosts provide examples where unisexual and sexual species that are closely related live in close proximity.

Systematic distribution of unisexual biotypes in teleosts

Because the unisexual biotypes usually closely resemble the parental sexual species, it is possible that more unisexual biotypes will be described in the future. Eight unisexual biotypes have been identified (Pandian, 2011). All come from three orders, the Cyprinodontiformes and Atheriniformes (both Atherinomorpha) and Cypriniformes. The cyprinodontiform asexuals occur in two genera of the viviparous Poeciliidae, *Poecilia* and *Poeciliopsis*, both from Central America, while the atheriniform unisexuals are found in the genus *Menidia* (Atherinidae) off the coast of the Atlantic and Gulf of Mexico. In the Cypriniformes, unisexual biotypes occur in the Asian genus *Carassius* (Cyprinidae), the Iberian genus *Squalius* (Cyprinidae) and the North American genus *Phoxinus* (Cyprinidae), the European genus *Cobitis* (Cobitidae) and the Japanese genus *Misgurnus* (Cobitidae). The two poeciliid unisexual biotypes have received the most attention of the teleost unisexuals.

Cyprinodontiformes

Poeciliid unisexuals

Poeciliid unisexuals face a particular problem. Like their sexual congeners, they are viviparous and this mode of reproduction requires internal fertilisation. The male and female copulate, with the male transferring sperm to the female's genital tract using a gonopodium. Consequently, the unisexuals potentially have to compete with the sexual females to obtain copulations. The males gain no genetic advantage from copulating with females of the unisexual biotype and so might be expected to avoid such copulations.

Poeciliid unisexuals: *Poecilia formosa*

The unisexual Amazon molly (*Poecilia formosa*, Poeciliidae) lives in rivers in north-eastern Mexico and Texas and was the first unisexual vertebrate to be recognised. It is typically diploid, reproducing by gynogenesis. The meiotic division that normally occurs in oogenesis is suppressed (Balsano *et al.*, 1989; Chapter 4). *P. formosa* is a result of hybridisation between two sexual poeciliids, the shortfin molly (*Poecilia mexicana*) and the sailfin molly (*Poecilia latipinna*). *P. mexicana* was the female parent of *P. formosa* and *P. latipinna* the male parent (Lampert & Schartl, 2008). The males of the parental species provide the spermatozoa that activate the eggs of the unisexual species. Nuclear and mitochondrial DNA suggest that the hybridisation derived from a single event that took place approximately 280 000 years ago, which is approximately 840 000 generations. Although *P. formosa* probably represents a monophyletic clade (Stock *et al.*, 2010), there are clones that differ genetically, presumably because of the accumulation of mutations in each clone since the original hybridisation. The frequency of unisexual *P. formosa* relative to the sexual species varies widely. In the Rio Purificacion, the frequency varies from about 9% to 50%, while in the Soto la Marina drainage, frequencies can reach close to 100% (Moore, 1984).

In two river systems, triploid *P. formosa* occur alongside the sexual species and the diploid unisexuals. The triploid ($3n$) is formed when the spermatozoa of the male is retained by the diploid ($2n$) egg. The triploid biotype can be designated as *P. 2mexicana-latipinna*, indicating the chromosome dosage from the female and the male. The triploids reproduce gynogenetically and are probably of relatively recent origin (Moore, 1984; Lampert & Schartl, 2008). The appearance of triploids, both in nature and in captivity, implies that the mechanism of spermatozoa exclusion may sometimes fail (Pandian, 2011).

Poeciliid unisexuals: *Poeciliopsis*

In contrast to *P. formosa*, there is evidence of several hybridisation events that generated unisexual *Poeciliopsis*. These unisexual biotypes are distributed in rivers of north-western Mexico and include both diploid and triploid forms. The sexual species providing the maternal genome is the headwater livebearer (*Poeciliopsis monacha*, Poeciliidae), while

Table 10.3 Unisexual forms of *Poeciliopsis*, showing ploidy level, sexual hosts and modes of reproduction. (Source: Vrijenhoek 1984. Reproduced with permission of Springer Science + Business Media.)

Unisexual biotype	Ploidy	Sexual host	Reproductive mode
P. monacha-lucida	2*n*	*P. lucida*	Hybridogenetic
P. monacha-occidentalis	2*n*	*P. occidentalis*	Hybridogenetic
P. monacha-latidens	2*n*	*P. latidens*	Hybridogenetic
P. 2monacha-lucida	3*n*	*P. monacha*	Gynogenetic
P. monacha-2lucida	3*n*	*P. lucida*	Gynogenetic
P. monacha-lucida-viriosa	3*n*	*P. viriosa*	Gynogenetic

species providing spermatozoa are the Gila topminnow (*Poeciliopsis occidentalis*), clearfin livebearer (*Poeciliopsis lucida*), lowland livebearer (*Poeciliopsis latidens*) and the chubby livebearer (*Poeciliopsis viriosa*) (Vrijenhoek, 1984; Schultz, 1989).

The diploid biotypes are *P. monacha-lucida*, *P. monacha-occidentalis* and *P. monacha-latidens*. These diploid biotypes reproduce by hybridogenesis, i.e. hemiclonal inheritance. The female produces haploid (1*n*) eggs that contain only genes originally from *P. monacha*. The eggs are fertilised by spermatozoa from a male *P. lucida*, *P. occidentalis* or *P. latidens*, depending on which sexual species are sympatric with the unisexual biotype (Table 10.3). During oogenesis, the *monacha* chromosomes at meiosis align, unpaired, with the paternal chromosomes on the metaphase plate, where they attach to a unipolar spindle. They are then drawn to a single pole and become the sole genetic contribution to the egg nucleus. There is no recombination, so the maternal genes are transmitted as a single linkage unit (Schulz, 1989). The triploid biotypes reproduce by gynogenesis, so the spermatozoa only activates the egg and contributes no genetic material to the zygote and inheritance is completely clonal (Table 10.3). Again, the frequency of unisexual biotypes varies widely, with mean frequencies in a series of river systems running from south to north ranging from 84% to 5.2% (Balsano *et al.*, 1989).

Poecilia formosa and the *Poeciliopsis* unisexuals both show high levels of clonal diversity and this is of importance for some of the hypotheses that seek to explain how the sexual and unisexual biotypes can coexist.

Atheriniformes

The Atheriniformes contains a coastal unisexual biotype, distributed on the coast of the Gulf of Mexico, the Texas silverside (*Menidia clarkhubbsi*, Atherinopsidae). The parental sexual species are the inland silverside (*Menidia beryllina*, Atherinopsidae) and the tidewater silverside (*Menidia peninsulae*, Atherinopsidae). There is evidence of multiple hybridisation events leading to the formation of unisexual clones. *M. clarkhubbsi* reproduces by gynogenesis and triploids are rarely encountered. Samples contained a mean of only 2.8% of unisexuals (Pandian, 2011).

Cypriniformes

The Cypriniformes include five identified unisexual biotypes, from a disparate range of geographical locations. They belong to two families, the Cyprinidae (carps and minnows) and the Cobitidae (loaches).

Phoxinus eos-neogaeus

In North America, in a swathe of territory encompassing northern states of the USA and the border between Canada and the USA, there is a unisexual cyprinid *Phoxinus eos-neogaeus*, which is the product of hybridisations between females of the finescale dace (*Phoxinus neogaeus*, Cyprinidae) and males of the northern redbelly dace (*Phoxinus eos*, Cyprinidae) (Elder Jr & Schlosser, 1995; Angers & Schlosser, 2007; Pandian, 2011). *Phoxinus eos-neogaeus* unisexuals are diploid and reproduce gynogenetically, with the eggs activated by spermatozoa from one of the sexual species. The picture is complicated, because a relatively high proportion of the diploid eggs are fertilised by the spermatozoa forming triploids, or mosaics, which are individuals made up of cells that are either diploid (2*n*) or triploid (3*n*). The triploids and some mosaics reproduce non-clonally. In contrast to *Poecilia* and *Poeciliopsis*, within populations of *P. eos-neogaeus* there is little or no within-clonal diversity. Analyses of mitochondrial and nuclear DNA from individuals from a wide range of geographical locations suggest that there have been several independent hybridisation events leading to clonal lineages. The same analyses indicated that the hybridisations may have occurred in a glacial refuge during the Pleistocene Wisconsin glaciation, with subsequent dispersal as the ice retreated northwards during the last post-glacial period (Angers & Schlosser, 2007). In many localities, the unisexual *P. eos-neogaeus* form a high proportion of the total population of the two sexual species and unisexual biotype, with the proportions changing with the type of habitat (Elder Jr & Schlosser, 1995).

Tropidophoxinellus alburnoides

In the Iberian Peninsula, a complex set of biotypes occurs in rivers of Portugal and western Spain. The complex is *Tropidophoxinellus* (*Leuciscus* or *Rutilus* or *Squalius*) *alburnoides*, the multiple generic names reflecting the difficulty of classifying this cyprinid (Alves *et al.*, 2001). The paternal sexual species is *Squalius carolitertii* (Cyprinidae) in the northern drainages and *Squalius pyrenaicus* (Cyprinidae) elsewhere. The maternal species has not been identified. The mode of reproduction seems to depend on the genome and the degree of ploidy. Diploid (2*n*), triploid (3*n*) and tetraploid (4*n*) individuals occur naturally. Diploid females may reproduce hybridogenetically, discarding the paternal genome during oogenesis. Diploid eggs may also develop gynogenetically but frequently the diploid egg is fertilised, generating a triploid, as occurs in *P. eos-neogaeus*. Triploids with *S. carolitertii* as the paternal sexual species reproduce by hybridogenesis, but triploids with *S. pyrenaicus* as the paternal sexual species reproduce by a mode called 'meiotic hybridisation', in which both diploid and haploid eggs are produced. In the *T. alburnoides* complex, the situation is dynamic, with a continual shifting between genomic composition. Like *P. eos-neogaeus*, *T. alburnoides* is a biotype that probably originated in a glacial refuge in the last Ice Age and has subsequently dispersed through river systems.

Carassius gibelio

The Prussian carp (*Carassius gibelio*, Cyprinidae), which is of Asian origin, also displays a complex pattern of sexual and unisexual forms (Zhou & Gui, 2002). Unisexual biotypes reproducing by gynogenesis occur in north China and Russia, and are identified as *C. gibelio*, the gibel carp. In Japan, the unisexual form is identified as the ginbuna carp, *Carassius langsdorfii* (Cyprinidae), while the unisexual Sugu carp (*C. auratus* var. Sugu) occurs in south China. The eggs of the unisexual *C. gibelio* are activated by spermatozoa of closely related species, such as the common carp (*Cyprinus carpio*, Cyprinidae) and crucian carp (*Carassius carassius*, Cyprinidae). The triploid gynogen is believed to be derived from an ancient hybridisation event, with *C. auratus* the maternal ancestor and *C. carpio* the paternal ancestor. Hybridisation is relatively common in the Cyprinidae (Maitland & Campbell, 1992), but so far these are the only unisexual biotypes described.

Cobitis

Unisexual biotypes occur in the cypriniform spined loaches in the family Cobitidae. In central Europe, members of the genus *Cobitis*, primarily two sexual species, the spined loach (*Cobitis taenia*) and *Cobitis elongatoides*, form a complex of sexual, unisexual, triploid and tetraploid biotypes through reciprocal hybridisation (Janko *et al.*, 2007).

The unisexual female clones are gynogenetic and are dependent on males of the sexual species to provide spermatozoa. Diploid clones (*C. elongatoides-taenia*) occur in a hybrid zone between the two sexual species. Evidence suggests that spermatozoa from the sexual species fuse with unreduced eggs from diploid clonal females in approximately one-third of cases, resulting in triploid progeny, whereas about two-thirds of progeny remain diploid (Janko *et al.*, 2007). The triploid biotypes can be designated as either *C. 2elongatoides-taenia* or *C. elongatoides-2taenia*, and arise by the incorporation of spermatozoa from male *C. elongatoides* and *C. taenia*, respectively. These triploid biotypes represent the dominant forms across the range of the parental species. In an ingenious study, Choleva *et al.* (2012) reconstructed the evolutionary path from sexuality to clonality and polyploidy using experimental crosses of *C. taenia* and *C. elongatoides*, thereby providing support for a hybrid origin for clonal diploid and triploid *Cobitis*. Significantly, the study demonstrated that the emergence of polyploidy is a by-product of clonality, not its cause.

A related cobitid, the pond loach (*Misgurnus anguillicaudatus*, Cobitidae), also comprises a complex of sexual and unisexual forms (Itano *et al.*, 2007). However, the clonal unisexual biotypes have a restricted geographical distribution, with some evidence that their numerical abundance is low (Morishima *et al.*, 2008).

Problems posed by unisexual teleosts

Developmental

In unisexual teleosts, there is some disruption of the normal meiosis in oogenesis and in gynogenetic species of fertilisation, compared with sexually reproducing teleosts (Chapter 4). These impediments pose the question of how these changes are controlled and regulated. There seems to be no evidence on whether the neuroendocrine control of oogenesis is altered in unisexual biotypes. One problem that needs to be explained is the modification of the process of meiosis that prevents the reduction division and maintains the diploid nature of the oocyte that is characteristic of gynogenetic biotypes. For gynogenetic biotypes, what prevents the spermatozoa nucleus from fertilising the egg? For biotypes with hybridogenesis, what controls the failure of the paternal genome to be transmitted in the oocytes? In gynogenetic biotypes, there can be some 'leakage' of the paternal genome into the clonal lineage (Schlupp, 2005; Lampert & Schartl, 2008). This leakage can take two forms. The first is a failure of the paternal chromosomes to fully break down and, as a consequence, small fragments of the paternal chromosomes are incorporated in the genome of the clonal lineage. These fragments are called microchromosomes. The second mechanism is when the spermatozoa successfully fertilise the egg and a polyploid is formed. It is unclear how important such leakage is for the evolution of the clones of unisexual biotypes.

Evolutionary implications

Unisexual biotypes seem to have a major evolutionary advantage over closely related sexual species. Other things being equal, a unisexual biotype will produce two female offspring for every female produced by the sexual female. This is because the sexual female also produces males. This is called the 'cost of meiosis' (Maynard Smith, 1978). Consequently, the rate of increase in population abundance of the unisexual biotype is twice the rate for the sexual species, and, in the long run, the unisexual biotype should replace the sexual species. If the unisexual biotype is a 'sperm' parasite of the sexual species, then the unisexual biotype would also become extinct. The ability of asexual biotypes to avoid the cost of meiosis poses the question of why sexual species are so common and have not been out-competed by asexual biotypes.

However, there is a major disadvantage for asexual biotypes. Once a clone is established – in teleosts this is a consequence of a hybridisation event – the genome of the clone is fixed and can only change because of mutation. Most mutations are disadvantageous and so the clone accumulates deleterious mutations and the fitness of the clone declines with extinction as the end point. This process of the accumulation of deleterious mutations is termed Müller's ratchet (Maynard Smith, 1978). These deleterious mutations cannot be purged by the recombination that characterises sexual reproduction, nor can beneficial mutations be brought together by recombination in a single genome.

With unisexual biotypes living in close association with closely related sexual species, the teleosts seem to provide a situation in which the advantages and disadvantages of the two modes of reproduction can be studied. Most studies have focused on the poeciliids, *Poeciliopsis* and *Poecilia*. The evidence that the unisexual *P. formosa* may have originated approximately 280 000 years ago suggests that the clones have not suffered from Müller's ratchet, although this may not be true for other unisexual lineages.

Most discussion of the unisexual poeciliids has focused on developing models that predict the long-term coexistence of closely related sexual and unisexual forms. These models fall into three categories: ecological factors, behavioural regulation, life-history regulation and meta-population dynamics.

The most influential ecological model developed to explain the coexistence of sexual and unisexual *Poeciliopsis* in water bodies in north-western Mexico is the frozen niche-variation (FNV) model (Vrijenhoek, 1989; Wetherington *et al.*, 1989). It assumes that distinct clones had independent origins from sexual ancestors. Each clone represents a fixed sample of the genotypes that existed in the sexual gene pool. This means that each clone is adapted to a limited niche, and because each genome is fixed, a clone is restricted to this niche. By contrast, the sexual species have genetic variability, which is continually being refreshed by recombination. Consequently, the sexual species are adapted to a range of niches, whereas the clones are adapted to narrow niches. There is some evidence that clones do differ in aspects of their ecology such as diet, but whether these are sufficient to allow coexistence is unclear. There is also evidence for *Poeciliopsis* that clones have been produced independently many times, so natural selection could select clones from a wide selection of genomes. This is not true for the unisexual *P. formosa*, which seems to be monophyletic.

The FNV model cannot account for the coexistence of the unisexual *P. eos-neogaeus* with its sexual ancestors because it is characterised by low clonal diversity within populations. Indeed, the range of habitats that *P. eos-neogaeus* occupies has led to the suggestion that it represents a general-purpose genotype that can succeed in such a range (Angers & Schlosser, 2007), in direct contradiction of the predictions of the FNV model. While the FNV has a number of deficiencies, there is currently no coherent alternative model to account for the occurrence of unisexual biotypes.

Mate choice is an important aspect of the reproductive behaviour of poeciliids (see Chapter 8). The unisexual poeciliids have a particular problem, in that they require copulation with males of closely related sexual species. The males are essentially wasting sperm, because the unisexual biotypes exclude the male genome, whether by gynogenesis or hybridogenesis. Coexistence requires that the unisexuals can mate successfully with the males (Schlupp, 2005; Schlupp & Riesch, 2011). There are several possibilities. The males may mistake the unisexual females for sexual females, or, if mating is not costly for the male, make the 'best of a bad job', essentially mating with any females that become available. The females of unisexuals and sexuals are viviparous and have only a short period of receptivity to courting males (Chapter 8), so receptive females are usually in short supply. It is also likely that pheromones produced by females signalling their sexual receptivity (Chapter 6) may be identical or similar between the sexual and unisexual biotypes, so males will attempt to copulate with both biotypes. Coexistence may also be favoured if mate choice is frequency-dependent, such that males tend to mate with the biotype that is at the lowest frequency in the population. The outcome of such frequency-dependent mate choice can be an equilibrium mixture of biotypes. A further behavioural mechanism favouring coexistence is mate copying (Chapter 8), in which individuals tend to mate with individuals that have mated with other individuals. These mechanisms would not operate in teleost taxa in which mate choice is not an important component of reproductive behaviour.

A further mechanism that may lead to coexistence of sexual and unisexual biotypes is life-history regulation. A unisexual biotype could coexist with sexual species if the lifetime reproductive success of the unisexual biotype was about 50% of that of the sexual species because of the cost of meiosis. A theoretical study based on *P. formosa* suggested that unisexual clones and sexual species could coexist as a metapopulation, in which a balance exists between extinctions and recolonisations.

It is likely that there is no general explanation for the coexistence of sexual and unisexual biotypes in teleost fishes, but the explanation varies from taxon to taxon and with geographical location. Some of the hybridisations that have led to unisexual species may be a consequence of related species being driven into sympatry by the southward movement of ice masses during the last Ice Age. There is a need for studies of the fecundity and survival schedules in populations that consist of mixtures of sexual species and their closely related unisexual biotypes so that preliminary estimates of the fitness of the various biotypes can be made.

Although hybridisation, at least in some families, is relatively common, unisexual biotypes in the teleosts are rare. There are two possible explanations. The first is that unisexual forms go extinct at a much faster rate than related sexual forms. A second is that the conditions required for the evolution of unisexual biotypes, especially the modifications that are required for meiosis in oogenesis and the mechanisms of egg activation or fertilisation, are difficult to acquire (Stock *et al.*, 2010).

Hermaphroditism in teleosts

Perhaps no other feature sets the reproductive biology of the teleosts apart from that of other jawed vertebrates than the incidence of hermaphroditism. Hermaphroditism is the occurrence, in the same individual, of functional male and female gametes during its lifetime and is not uncommon in the animal kingdom (Avise, 2011). However, the vast majority, perhaps 99%, of vertebrates are gonochoristic (Avise, 2011) and even within the teleosts the majority are also gonochoristic (Chapter 2). Hermaphroditism occurs in about 2% of teleost species (Sadovy & Liu, 2008). The emphasis here is on the functional nature of the gametes. Even in gonochoristic vertebrate species, individuals occur in which the gonads may, because of developmental abnormality, contain both male and female germ cells (Chapter 6). Such individuals should be identified as intersexes, to distinguish them from functional hermaphrodites (Sadovy & Liu, 2008).

Two modes of functional hermaphroditism occur in the teleosts (Sadovy & Liu, 2008; Avise, 2011; Pandian, 2011). In the less common mode, simultaneous or synchronous hermaphroditism, the gonad contains functional male and female gametes at the same time. This situation raises the possibility of self-fertilisation, in which the sperm of an individual fertilises the eggs of the same individual. In teleosts, self-fertilisation has only been well described for two closely related species. The usual pattern is for cross-fertilisation, with an exchange of gametes between individuals.

The second and commoner mode of hermaphroditism in teleosts is sequential or serial hermaphroditism. In this mode, an individual, after the onset of sexual maturity, functions initially as one sex and then at some point in its life changes sex. Typically, this sex change is irreversible.

Teleosts show three patterns of sequential hermaphroditism. In protandry, the individual starts its reproductive life as a functional male and then changes to a functional female. In protogyny, the most common pattern in teleosts, the functional female phase precedes the functional male phase. In monandric, protogynous species, all the males are secondary males, being derived from females. In diandric species, there are two types of male. Primary males attain sexual maturity as males, whereas secondary males are derived from sex-changed functional females. The equivalent states in protandrous species are monogyny, in which all females arise from sex-changed males, whereas in digyny there are both primary and secondary females.

A parallel classification of males is based on their secondary sexual traits. Initial phase (IP) males are small and resemble females in their colouration. Terminal phase (TP) males are larger and, at least in some species, have a distinctive colouration. TP males may be secondary males, derived from sex-changed females, or primary males that have grown sufficiently large to compete for matings with other TP males.

A third, rarer mode of sequential hermaphroditism in teleosts is bidirectional sex change (Munday *et al.*, 2010). In these species, the sex change can occur in both directions, female to male (protogyny) and male to female (protandry). In some species, the change can be reversible.

Phylogenetic distribution of hermaphroditism in teleosts

Although it is probable that the numbers will change as the reproductive biology of more species is described, there are about 500 hermaphroditic teleost species representing as many as 48 families from 10 orders (Sadovy & Liu, 2008; Avise & Mank, 2009; Avise 2011; Pandian, 2011). Most are sequential hermaphrodites, with at least 31 families containing protogynous species, whereas about 10 families contain protandrous species (Lawson, 2011). Simultaneous and bidirectional hermaphrodites also occur.

A remarkable feature of hermaphroditism in teleosts is its scattered distribution across the phylogenetic tree of the group. This pattern implies that hermaphroditism has evolved independently on several occasions. It occurs in phylogenetically basal clades, such as the Anguilliformes, but is most prevalent in the evolutionarily advanced Perciformes (Table 10.4). Even within an evolutionary clade such as the Gobiidae, there

Table 10.4 Distribution of hermaphroditism across teleostean orders. Confirmed examples are in Roman type, and provisional examples are in italics. In all orders listed, gonochorism also occurs. (Sources: Data from Sadovy & Liu 2008; Avise & Mank 2009; Avise 2011.)

Order	Family	Mode of hermaphroditism
Anguilliformes	Muraenidae	Simultaneous; *protogyny*
Clupeiformes	*Clupeidae*	*Protandry*
Cypriniformes	*Cyprinidae*	*Protogyny*
	Cobitidae	*Protogyny; protandry*
Stomiiformes	Gonostomatidae	Protandry
Aulopiformes	Chlorophthalmidae	Simultaneous
	Ipnopidae	Simultaneous
	Scopelarchidae	Simultaneous
	Alepisauridae	Simultaneous
	Bathysauridae	Simultaneous
	Notosudidae	*Simultaneous*
	Evermannellidae	*Simultaneous*
	Paralepididae	*Simultaneous*
Cyprinodontiformes	Rivulidae	Simultaneous
	Poeciliidae	*Protogyny*
Synbranchiformes	Synbranchidae	Protogyny
Scorpaeniformes	Caracanthidae	Protogyny
	Platycephalidae	Protogyny
Perciformes	Centropomidae	Protandry
	Latidae	Protandry
	Serranidae	Simultaneous; protogyny; serial
	Pseudochromidae	Protogyny; serial
	Nemipteridae	Protogyny
	Lethrinidae	Protogyny
	Sparidae	Simultaneous; protogyny; protandry; serial
	Centracanthidae	Protogyny
	Pomacanthidae	Protogyny; serial
	Cirrhitidae	Protogyny; serial
	Pomacentridae	Protogyny; protandry
	Labridae	Protogyny
	Scaridae	Protogyny
	Pinguipedidae	Protogyny
	Trichonotidae	Protogyny
	Gobiidae	Protogyny; serial
	Moronidae	*Protandry*
	Grammatidae	*Protogyny*
	Malacanthidae	*Protogyny*
	Polynemidae	*Simultaneous*
	Chaetodontidae	*Protogyny*
	Terapontidae	*Protogyny*
	Cepolidae	*Protogyny*
	Cichlidae	*Protogyny*
	Odacidae	*Protogyny*
	Eleginopsidae	*Protandry*
	Percophidae	*Protandry*
	Blenniidae	*Protogyny*
Tetraodontiformes	*Balistidae*	*Protogyny*
	Ostraciidae	*Protogyny*

is phylogenetic evidence of the independent evolution of hermaphroditism on several occasions (Cole, 2010a).

The scattered phylogenetic distribution of hermaphroditism partly reflects the habitat distribution of hermaphroditic species. They occur almost exclusively in marine habitats, particularly the deep seas and coral reefs. Hermaphroditic species are absent, or almost completely absent, from freshwater habitats, estuaries, the epipelagic marine zone and the continental shelf. Only two hermaphroditic species have been provisionally identified in the most speciose freshwater order, the Cypriniformes (Avise, 2011); by contrast, about 260 species of the speciose marine order Perciformes are hermaphroditic (Avise, 2011).

Gonadal organisation in hermaphroditic fishes

Hermaphroditic teleosts display a diversity of solutions to the problem of the organisation of the gonads, even within a single clade such as the Gobiidae. The problem is that the gonad has to accommodate both functional female and male gametes. This problem is more acute for simultaneous than for sequential hermaphrodites.

In most simultaneous hermaphrodites, e.g. the perciform serranines such as *Hypoplectrus*, the gonoducts for the sperm and the eggs are separate (Sadovy & Liu, 2008). This arrangement ensures that self-fertilisation cannot occur within the same gonad. The exceptions to this are the cyprinodontiform mangrove rivulus, *Kryptolebias* (formerly *Rivulus*) *marmoratus* (Rivulidae) and *Kryptolebias hermaphroditus* (formerly known as *K. ocellatus*), which are found in coastal mangrove habitats from Florida to Brazil (Earley *et al.*, 2012; Tatarenkov *et al.*, 2012). In these species, free spermatozoa occur in the posterior part of the ovarian lumen, allowing self-fertilisation to take place. These species are the only hermaphroditic vertebrates for which there is good evidence of self-fertilisation commonly occurring (see the section 'Simultaneous hermaphrodites').

The family Gobiidae provides a good example of the range of gonadal organisations that can occur in sequential hermaphrodites, even within a single evolutionary clade (Cole, 2010a,b).

The least complex organisation is found in the protogynous *Coryphopterus* genus. Initially, the gonad is an ovary, with no visible testis tissue, and after maturation it is a functional ovary. During the sex change, seminiferous lobules develop, which completely obliterate the ovarian tissue. During the ovarian phase, somatic cells form precursive accessory gonadal structures (pAGS). As the seminiferous lobules form, the cells of the pAGS proliferate and develop numerous epithelial-lined lobules, forming the accessory gonadal structures (AGS). These AGS may represent a novel feature associated with some hermaphroditic gobies. These structures recall the sperm duct glands (sensu Miller, 1984) found in many gonochoristic gobies.

In the protogynous goby genus *Eviota*, the gonad contains both ovarian and testicular tissue. With the sex change from functional female to functional male, there is a proliferation of spermatocytes, and with the associated somatic cells these become organised into seminiferous lobules. With the sex change, there is a reduction in the proportion of ovarian tissue, while the pAGS differentiate into the AGS. In the functional male phase, the gonad retains a central lumen with healthy, early-stage oocytes scattered throughout the interstices of the seminiferous tubules.

A novel gonadal organisation occurs in the protogynous genus *Lythrypnus*. The gonad contains both ovarian and testicular tissue, but there is a localised concentration of male spermatogenic tissue rather than the intermingled pattern of ovarian and testicular tissue as in *Eviota*.

In gobiid genera, such as *Trimma* and *Bryaminops*, from the onset of gonadal organisation the gonad is partitioned into ovarian and testicular components, separated by a boundary layer of connective tissue. In the sex change from female to male, the ovarian component becomes reduced and the ventral region expands, becoming organised into seminiferous lobules, and the pAGS develops into the AGS.

A curiously complex gonadal organisation characterises the Indo-Pacific genus *Gobiodon*. In the functional female, the gonad has no visible spermatogenic tissue, but with the sex change, the gonadal organisation becomes more complex. There are three contiguous, intergraded zones. Dorsally, there is a gametogenic region consisting of seminiferous lobules and primary oocytes. A medial stromal region separates the gametogenic region from a ventrolateral lobular region. This complexity recalls that seen in some gonochoristic gobies (Chapter 5). Directly associated with the gonoduct is a pair of lobular structures, which may be homologous with the sperm duct glands (sensu Miller, 1984) of gonochoristic gobies (Chapter 5).

In those protogynous, hermaphroditic teleosts that are diandric, such as the wrasses (Labridae) and parrotfishes (Scaridae), the structure of the gonad can, in some but not all cases, distinguish between primary and secondary males (Sadovy & Liu, 2008). In primary males, the testes were typical of those of gonochoristic species (Chapter 5). In secondary males, after the sex change, the testes retain characteristics of the previous female phase, including an ovarian-like lumen, while newly formed sperm sinuses arise within the walls of the original ovarian lumen. However, it is not always possible to distinguish between primary and secondary males on the basis of the organisation of the gonad in the male phase. In the protogynous seabasses and groupers (Serranidae), the pattern of testicular tissue developing within the ovary varies from complete separation between the testicular and ovarian components, as in black seabass (*Centropristis striatus*, Serranidae), to complete integration, as in *Epinephelus* (Serranidae).

Endocrinology of reproduction in hermaphroditic species

In teleosts, as in other vertebrates, gametogenesis is regulated by the brain–pituitary–gonad (BPG) axis (Chapter 5). The main messengers in this axis, the gonadotrophin-releasing hormone (GnRH) of the brain and the two gonadotrophins produced by the pituitary, follicle-stimulating hormone (FSH) and the luteinising hormone (LH), do not have sex-specific forms. They regulate both oogenesis and spermatogenesis, and the two processes show many parallels (Chapter 5). In a sense, the axis is inherently hermaphroditic. Indeed, the endocrine control of reproduction in hermaphroditic teleosts shares many of the characteristics of the control in gonochoristic species. The main differences are often in the timing of events rather than the nature of the regulatory processes.

In both simultaneous and sequential hermaphroditic species, the female phase is characterised by high plasma levels of oestrogens, especially E_2, whereas the male phase is characterised by high plasma levels of androgens. The belted sandbass (*Serranus subligarius*, Serranidae) is a simultaneous hermaphrodite in which there is a tendency for the spawning role to switch from female to male with increased body size. This switch is accompanied by increasing concentration of the androgen 11-KT and the progestogen, 17,20β,21-trihydroxy-4-pregnen-3-one (20β-S; Goodwin, 2010a,b; Lawson, 2011). A key characteristic of species with sequential hermaphroditism is that the sex change is typically evoked by a change in the social environment. This is a form of environmental sex determination (Chapters 2 and 6), which indicates that the social environment affects the endocrine control of reproduction.

In the protandrous fire clownfish (*Amphiprion melanopus*, Pomacentridae), mature males have higher levels of 11-KT, but lower levels of T, E_2 and androstenedione (AD) than do females. Levels of 11-KT drop only when the sex change to female is completed, while E_2 stays low until the change is completed. AD shows a similar pattern. For T, the level is intermediate in the male, drops until midway through the sex change and then increases to high levels in the female (Goodwin, 2010a,b; Lawson, 2011).

Samples taken from a natural population of the protogynous saddleback wrasse (*Thalassoma duperrey*, Labridae) showed that levels of 11-KT were relatively low in females but high in TP males. Levels of T were similar between females and TP males and did not vary across the sex change from female to male. The IP males were almost indistinguishable from females and they had lower levels of 11-KT and lower synthetic capacity than TP males. *In vitro* studies demonstrated that females have a higher capacity for synthesis of E_2 and TP males have a greater capacity for the synthesis of 11-KT (Goodwin, 2010a). An increase in serum levels of 11-KT was correlated with a shift in the colouration of the anal fin that is typical of TP males in the wrasse *Pseudolabrus sieboldi* (Labridae) (Ohta *et al.*, 2008a).

Similar patterns in the levels of E_2 and 11-KT also occur in protogynous hermaphroditic gobies. Even outside the breeding season, in the blackeye goby (*Rhinogobiops nicholsii*, Gobiidae), whole-body levels of 11-KT were higher in males and E_2 levels were higher in females (Goodwin, 2010b).

Thus, the endocrine status of sequential hermaphrodites shows similarities with that of gonochoristic species, with E_2 prominent in mature females and 11-KT prominent in mature males (Chapter 6). An apparent exception to this pattern is seen in the IP males of some diandric species such as *T. duperrey* and the stoplight parrotfish (*Sparisoma viride*, Scaridae) (Lawson, 2011). These males are too small to defend a spawning site against large TP males, but may engage in group spawning and sneaking fertilisations by interfering with pair spawnings between TP males and females (Chapter 8). In colouration, IP males resemble females and the endocrine status of IP males more resembles that of females than that of TP males, with low levels of 11-KT and moderate levels of T and E_2. Yet, in relation to body size, their testes are large and proficient at producing sperm. This endocrine state recalls that of the gonochoristic species, the three-spined stickleback (*Gasterosteus aculeatus*, Gasterosteidae) (Chapter 6). In this species, during the spawning season, the male has vivid nuptial colours, aggressively defends a territory and has high levels of 11-KT. After the breeding season, 11-KT drops to low levels and spermatogenesis takes place. At least in some teleosts, high levels of 11-KT in the male seem to be more associated with well-developed secondary sexual traits and behavioural traits related to reproduction rather than spermatogenesis, which may require only low levels of androgens to be initiated.

Species with serial sex change maintain both ovarian and testicular tissue, but with only one type active at one time. In the rippled coral goby (*Gobiodon erythrospilus*, Gobiidae), higher levels of E_2 were observed when an individual was functioning as a female, but levels of 11-KT did not differ between the female and male phases (Goodwin, 2010b).

Although the gonadotrophins FSH and LH are the same in males and females, a study of *P. sieboldi* identified differences in the temporal pattern of the expression of mRNAs for FSH and LH between spawning males and females (Ohta, 2008b). In a laboratory study, the fish spawned between 06:00 and 09:00 hours. For the females, expression of the mRNA for the subunit LH-β (Chapter 6) increased from 12:00 and peaked at 18:00 hours, when the final stage of maturation began. Expression for FSH-β peaked at 12:00 hours. In males, expression of LH-β increased from 21:00 and peaked at 03:00 hours, and was associated with spermiation. FSH-β peaked at 0:00. These changes in functional females and males have parallels with the changes seen in males and females of gonochoristic species (Chapter 6).

The gonadotrophin (GTH) receptors in the gonads also show changes associated with a sex change. The Okinawa rubble goby (*Trimma okinawae*, Gobiidae) is a serial sex changer, in which the gonad contains both ovarian and testicular tissue, but only one tissue is active at any one time. Expression of the mRNAs for the receptors of FSH and LH was strongly biased to the component of the gonad that was producing functional gametes. However, as the social environment changed, there were shifts in the expression of the receptor mRNAs within 24 hours of the initiation of sex change (Goodwin, 2010a,b). The mechanism driving the change in receptor mRNAs is still to be clarified.

The number of neurons in the pre-optic area (POA) of the brain that are immunoreactive to GnRH (GnRH-ir) is higher in TP males of the bluehead wrasse (*Thalassoma bifasciatum*, Labridae) than in females. A similar difference occurs between males and females of the protogynous ballan wrasse (*Labrus bergylta*, Labridae). The functional significance of this difference is unclear.

The neuroendocrine AVT also shows differences between functional males and females in some hermaphroditic species. TP males of *T. bifasciatum* have higher levels of the mRNA for AVT than females or IP males. The serial sex-changing goby, *T. okinawae*, shows rapid and reversible changes in the number of AVT-producing cells with serial sex changes. AVT may play a role in territorial and dominance behaviour of males and hence in defining the social environment (Lawson, 2011).

The noradrenergic, serotonergic and dopaminergic neuroendocrine systems in the brain have also been implicated in sex change in the protogynous *T. duperrey*, with the noradrenergic system stimulating the initiation and completion of a sex change from female to male (Lawson, 2011). The serotonergic and dopaminergic systems may inhibit sex change.

A key factor in sex change in sequential hermaphroditic species is likely to be the activity of aromatase genes, the gonadal form *cyp19a1* and the brain form *cyp19b1*. Aromatase is the enzyme that catalyses the synthesis of the oestrogen, E_2, from the androgen, T, in both the gonad and the brain (Chapter 5). It is an important step in the feminisation of an individual. Thus, up-regulation of the aromatase genes will tend to drive the sex change from male to female. Down-regulation and the consequent decline in the activity of aromatase causes a switch to the synthesis of 11-KT from T, which thus tends to drive the masculinisation of an individual. In gonochoristic species, this is an either/or event: an individual matures either as a male or as a female. If the up- or down-regulation of the activity of the aromatase genes takes place after the acquisition of sexual maturity, then a change of sex is likely.

In the protogynous *T. duperrey*, aromatase activity in the ovary declined during the sex change from female to male. By contrast, the expression of the enzyme 11β-hydroxylase, required for the synthesis of 11-KT, increased with the development of the testicular tissue. The bluebanded goby (*Lythrypnus dalli*, Gobiidae) is a serial sex changer. During the socially induced sex change from female to male, both the brain and gonadal aromatases show declines in activity, but the decline in the activity of the brain aromatase is more rapid than that of the gonadal aromatase. The decline in brain aromatase activity correlated with behavioural changes as the female became a male (Goodwin, 2010a,b). It has to be determined how the up- or down-regulation of the aromatase genes is regulated and this may depend on the nature of the response elements in the promoter region of each gene. The promoter region of the gonadal aromatase gene in *T. okinawae* contains several oestrogen response elements (Chapter 5).

The overall picture that emerges is that the endocrine states of the male and female phases in hermaphroditic species are comparable to the states in male and female gonochoristic species. Whereas in gonochoristic species, the sex of the individual is determined before the onset of sexual maturity, in hermaphroditic species the sex is determined after the onset of sexual maturation, and in the case of simultaneous hermaphrodites, the sex change occurs over short time intervals.

Simultaneous hermaphrodites

Self-fertilisation

Given the diversity of the modes of reproduction in the teleosts, it is noteworthy that there are only two well documented examples of self-fertilising species (Harrington Jr, 1975; Earley *et al.*, 2012; Tatarenkov *et al.*, 2012). These species, *K. marmoratus* and its close relative, *K. hermaphroditus*, may be the only hermaphroditic species in the Cyprinodontiformes, although the order contains species with unusual modes of reproduction for teleosts, including viviparity and a form of parthenogenesis. *K. marmoratus* and *K. hermaphroditus* inhabit mangrove stands along the western Atlantic coast from Florida to northern South America, including the Caribbean islands. These species are amphibious and capable of surviving emersion for up to 66 days. They aggregate in large numbers in mangrove litter and holes in decaying, emergent mangrove logs (Fig. 10.6; Taylor *et al.*, 2008).

Figure 10.6 The self-fertilising hermaphrodite, the mangrove rivulus (*Kryptolebias marmoratus*). (A) Its typical mangrove habitat in Belize; (B) galleried mangrove logs in which the fish is sometimes found; (C) emersed specimens, indicated by arrows, on mangrove wood; (D) an adult hermaphrodite specimen (TL 40 mm). (Source: D. Scott Taylor, Brevard County Environmentally Endangered Lands Program, 91 East Drive Melbourne, FL 32904, USA. Reproduced with permission of D. Scott Taylor.)

In some populations, e.g. in Florida, the rates of self-fertilisation are such that the fish belong to highly inbred lineages with extremely low levels of heterozygosity, essentially clones. This situation parallels the low levels of within-clonal diversity found in parthenogenetic cyprinodont species (see the section 'Unisexual teleosts'). In addition to hermaphroditic individuals, pure males do occur at varying frequencies across populations and these are responsible for occasional outcrossing events, which may help to maintain the genetic variability of populations in addition to mutation (Avise, 2011). Pure males are produced when a hermaphroditic individual spawns unfertilised eggs into the external environment and they are fertilised by spermatozoa from a male. Laboratory studies have demonstrated that there is an element of environmental sex determination (ESD; Chapter 2) in the reproductive biology of *K. marmoratus*. Pure males are produced at temperatures from 19°C or lower (Harrington, 1975). Notably, the geographic range of *K. marmoratus* is more extensive than any other inshore-dwelling coastal fish species in the Americas (Taylor *et al.*, 2008).

Cross-fertilisation

For simultaneous hermaphrodites, outbreeding, with cross-fertilisation between individuals, is the normal mode of reproduction (Avise & Mank, 2009; Avise, 2011). The total number of species is about 40 and they are largely found in two distinct habitats, the deep sea and reefs, especially coral reefs (Sadovy & Liu, 2008).

Simultaneous hermaphroditic species from deep sea habitats come from phylogenetically relatively basal teleosts. The moray eels (Muraenidae, Anguilliformes) have two or three species of simultaneous hermaphrodites. The Aulopiformes contains a similar, small number of simultaneous hermaphrodites (Sadovy & Liu, 2008; Avise & Mank, 2009; Avise, 2011). Their habitat makes the study of the reproductive biology of these deep sea fishes a technical challenge.

Because of their relative accessibility, the simultaneous hermaphroditic species of coral reefs are much better studied, especially members of the Serranidae (Perciformes). The subfamily Serraninae contains four genera (*Diplectrum*, *Hypoplectrus*, *Serraniculus* and *Serranus*), about 14 species in total (Avise, 2011). These serranids show a variety of mating behaviours that ensure cross-fertilisation and hence out-breeding (Mank & Avise, 2009; Avise, 2011; Pandian, 2011).

The black hamlet (*Hypoplectrus nigricans*, Serranidae) illustrates a mating behaviour called egg trading. Individuals are usually solitary, but at spawning they form pairs. In a pair, each individual acts alternately as a female and a male. Each individual parcels up the eggs to be spawned into batches. The fish that behaves first as the female initiates the courtship and releases a parcel of eggs to be fertilised by the partner acting as the male. The individuals then swap roles for the next spawning. This alternation of roles continues as up to nine parcels of eggs are spawned by each individual. The partners tend to stay together to complete the spawning sequence; however, they may change partners on subsequent days. In the harlequin bass (*Serranus tigrinus*, Serranidae), the pattern of mating is similar, but the eggs are not divided into parcels. This species may have permanent monogamy, so the problem of desertion by the individual playing the male role is much reduced. A variation is seen in the barred serrano (*Serranus fasciatus*, Serranidae), in which some larger individuals become males, losing their female function. Such males spawn in a pair with a hermaphroditic female from a harem of females they defend.

Sequential hermaphrodites

Sequential hermaphroditism is by far the commonest mode of hermaphroditism in the teleosts, with between 200 and 300 species identified (Sadovy & Liu, 2008; Avise & Mank, 2009; Avise, 2011). The change within the lifetime of an individual from one functional sex to the other raises an obvious question: what are the proximate factors that evoke the change? The ultimate evolutionary question of when in the lifetime the change should be made is considered in the following sections. In those sequential hermaphrodites that have been well studied, the social environment has proved to be the crucial proximate factor (Shapiro, 1984; Warner, 1988; Goodwin, 2010a,b). This effect of social environment on sex determination represents a form of ESD (Chapter 2). In protogynous species, in which a large dominant male defends a harem of females or a favourable spawning site, the loss of the male results in a female changing sex. This is usually, but not invariably, the largest female. In species that live in groups, the ratio of females to males is the crucial factor. If a male is lost, a female changes sex to maintain the ratio. In protandrous species, it is the loss of a reproductively active female that stimulates a sex change by an accompanying male. In both protogynous and protandrous species, the behavioural changes that are associated with the sex change occur before the morphological changes in the gonad are initiated.

Protogyny

Protogyny is particularly associated with reproductive modes in which large TP males can dominate other individuals and so sequester a harem of females or a favourable spawning site (Robertson & Warner, 1978; Warner & Robertson, 1978). On coral reefs, where the majority of protogynous species occur, such spawning sites may be limited, further facilitating domination by large males.

A well-studied example is the wrasse, *T. bifasciatum* (Warner & Robertson, 1978; Charnov, 1982; Avise, 2011). Large TP males establish territories at suitable sites, usually down-current edges of a coral reef, which the females visit to spawn. The species is dichromatic, with a major difference between the large TP males and the smaller females and IP males. A TP male has a bright blue head, black and white stripes on its mid-section, and the posterior body is green. Females and IP males are yellowish. This distinctiveness of TP males suggests the operation of strong sexual selection on the TP (Chapter 8). IP males are not totally excluded from reproduction, but their success depends, at least partially, on the population abundance. At low abundances, TP males can effectively defend their spawning territories and IP males can only spawn by sneaking into pair spawnings between the TP males and a female (Chapter 8). At high population abundances, defence of a territory becomes much more difficult and IP males engage in group spawning, in which several males spawn with a single female. In one study, 98% of the spawning of IP males took place in group spawning, but only 5% of spawnings by TP males involved more than one male. A feature of IP males is that their testes are much larger relative to body size than those of TP males. This feature is also seen in non-guarder or sneaker males of gonochoristic species with alternative mating tactics (Chapter 8). The difference in testes size between IP and TP males reflects the intensity of sperm competition in the group spawning of IP males as opposed to the pair spawning of TP males. A similar pattern of testes size is also seen in other species with TP and sexually active IP males.

The Indo-Pacific bluestreak cleaner wrasse (*Labroides dimidiatus*, Labridae) has a strictly haremic breeding system, with a single TP male controlling a group of females. This species is monochromatic, with no difference in colour between sexually mature females and males. The lack of obvious sexual dimorphism in appearance suggests there is no strong sexual selection on the males (Robertson & Warner, 1978), although an alternative explanation is that natural selection on distinctive colour patterns, which signal cleaning behaviour to clients, is stronger (Cheney *et al.*, 2009).

Another protogynous pattern is seen in the Indo-Pacific sea goldie (*Pseudanthias* (*Anthias*) *squamipinnus*, Serranidae). Individuals live in groups that can vary considerably in size. Individual males spawn with several females within a group. Usually it is the largest female that changes sex if a male is lost (Shapiro, 1984).

These examples show that within the protogynous mode of reproduction in coral reef fishes, there can be both intra- and interspecific variation in the details (Pandian, 2011). The one freshwater family that includes protogynous species is the eel-like synbranchids, with examples such as the Asian swamp eel (*Monopterus albus*, Synbranchidae) and the Neotropical marbled swamp eel (*Synbranchus marmoratus*, Synbranchidae), which also show interpopulation variation.

Protandry

Protandry is much less common than protogyny, occurring in about 21 genera spread over six families (Pandian, 2011). In the Perciformes, it occurs most notably in the anemonefishes (Pomacentridae), but also in the seabreams (Sparidae) and snooks (Centropomidae). The relative rarity of protandry may reflect that in the absence of large, dominant males, the size advantage accruing to females from an increase in size through the associated increase in fecundity (Chapter 11) is not conspicuously different from any increase for males, especially in species with group spawning (Avise, 2011). Sperm production by relatively small males is sufficient to fertilise all the eggs produced by even a relatively large female.

The best-studied protandrous species are the anemonefishes, *Amphiprion* spp. (Sadovy & Liu, 2008; Avise, 2011). Their unique characteristic is the intimate relationship between fish and sea anemone and this has probably been the major factor of the evolution of their protandry (Chapter 7). The anemonefish can find refuge within the tentacles of the sea anemone, without being stung by the nematocysts on the tentacles of the sea anemone. A reproductive group comprises a dominant female, a smaller mature male and a few unrelated juveniles. If the female is lost, the male changes sex and is replaced either by a juvenile becoming mature as a male or by another male moving to the anemone. The reproductive output of the social group is determined by the fecundity of the female, which is a function of female size.

The black porgy (*Acanthopagrus schlegeli*, Sparidae) is a protandrous species, which is used in aquaculture in Japan. An individual matures as a male at about 2 years old, with the gonad dominated by testicular tissue, but with ovarian tissue in nests of cells. After spawning, the testicular tissue regresses and ovarian tissue proliferates, but does not progress beyond the primary oocyte stage. Testicular tissue then proliferates again as the ovarian tissue regresses and the fish spawns as a functional male. This cycle of regression and proliferation is repeated, leading to a third spawning as a male. The testicular tissue then regresses, persisting only as a remnant. After this, primary oocytes initiate vitellogenesis and the individual becomes a functional female.

Serial change

Nothing illustrates better the lability of the teleost gonad than the existence of sequential hermaphrodites that display serial sex change (Munday *et al.*, 2010; Pandian, 2011). In these species, change of sex in both directions can occur: female to male and male to female. In some sex-changing species, an individual can change sex more than once in its lifetime. Serial sex change is commonest in the Gobiidae (Perciformes), with about 11 species displaying this pattern of hermaphrodites. It has also been identified, at least in the laboratory, in other Perciformes: the Serranidae (seabass and groupers), the

Pseudochromidae (dottybacks), the Pomacanthidae (angelfishes), the Cirrhitidae (hawkfishes) and the Labridae (wrasses). About 25 species with serial sex change have been identified so far.

The gobies with serial sex change are typically cryptic, small-bodied fishes, living in close association with coral (Munday *et al.*, 2010). *T. okinawae* is a polygynous goby that lives in crevices and on the underside of corals. A typical breeding group consists of a single male and several females. If the male is lost, one of the females changes sex. The change from male to female occurs if a single male enters a group that already contains a male. Other gobies, such as the redhead goby (*Paragobiodon echinocephalus*, Gobiidae), are monogamous. The commonest arrangement is a breeding pair, and any other individuals present are non-breeding juveniles. The protogynous female to male change occurs either when a single adult female changes sex and forms a breeding pair with a juvenile that has matured as a male, or when two females find themselves occupying a coral and one changes sex. The protandrous male to female change is rarer, but if a male moves and forms a pair with another male, one changes sex. When mature, *L. dalli* has a gonad that resembles that of a simultaneous hermaphrodite. The gonad has functional ovarian and testicular tissue, but the individual fish functions only as a male or as a female in successive breeding events. This means that the physiological and behavioural aspects of simultaneous hermaphroditism are partially decoupled and the outcome is a mode of sequential hermaphroditism (Avise, 2011). The situation also complicates the identification of serial sex change, distinguishing it from simultaneous hermaphroditism.

Adaptive significance of hermaphroditism

If two gonochoristic adolescents meet at random, the probability that they represent the two sexes is 0.5. If two hermaphroditic adolescents meet, the probability that they represent the two sexes is 1.0. This divergent outcome suggests that hermaphroditism will be favoured over gonochorism when population densities are low and encounters with other individuals are rare (Sadovy & Liu, 2008; Avise, 2011). In the teleosts, hermaphroditism occurs in deep sea fishes, where population densities are low and the problem of finding a mate is acute. However, most hermaphroditic teleosts, both simultaneous and sequential, are associated with coral reefs. A characteristic of many coral reef teleosts is that the early life-history stages are pelagic and the subsequent recruitment of juveniles to a reef depends on the vagaries of oceanic currents. Post-settlement dispersal is also often limited. This means that the densities of settlers may, under some conditions, be low and the chances of meeting a member of the opposite sex correspondingly limited.

The problem of meeting an appropriate sexual partner is completely solved by species with self-fertilisation (Avise, 2011). However, it is unclear why self-fertilisation, which is otherwise common in plants and invertebrates, is so rare in vertebrates. The rarity of self-fertilisation in the teleosts, described in detail only for *K. marmoratus*, suggests that either there is intense selection against self-fertilisation because of inbreeding depression, or the modifications to the reproductive system required for self-fertilisation are difficult to overcome in vertebrates.

With hermaphrodites, there is the fundamental question of how resources should be allocated between male and female functions to maximise fitness, the problem of sex allocation (Charnov, 1982). For simultaneous hermaphrodites, the problem is what allocation of resources to the male and female functions should be made in each breeding season? For sequential hermaphrodites, the sex allocation problem includes two components. The first has to do with the order in which the sexes should be manifest – i.e. should it be the male function or the female function that is entered at sexual maturity? The second component relates to when in its lifespan the individual should respond to a change in its environment by changing sex. If it changes sex, an individual must increase its lifetime reproductive success by more than it could by not changing sex. This can be expressed in terms of life-history theory (Warner, 1988). If the probability of surviving to age *x* is l_x and the fecundity at age *x* is m_x, then if the product of $l_x \times m_x$ increases faster for one sex than the other, then an individual that changes sex will have a higher lifetime expected reproductive success than one that does not. This will depend on the population abundance and the regimes of growth, fecundity and mortality encountered in the local population and the genetic propensity to change sex. Thus the timing of sex change will vary with local environmental conditions, rather than being genetically fixed for a species.

The best-supported hypothesis that explains the timing of sex change is the size-advantage hypothesis (Ghiselin, 1969; Warner, 1988; Avise, 2011). The model argues that if the size-specific reproductive success is higher at smaller body sizes for one sex, but higher at larger body sizes for the other sex, then sex change is favoured. Because size and age are positively related, the argument can also be made in terms of age as well as size. If males are at an advantage at smaller body sizes and females are at an advantage at larger body sizes, then protandry is favoured. If females have the advantage at smaller body sizes and males at larger body sizes, then protogyny is favoured (Fig. 10.7).

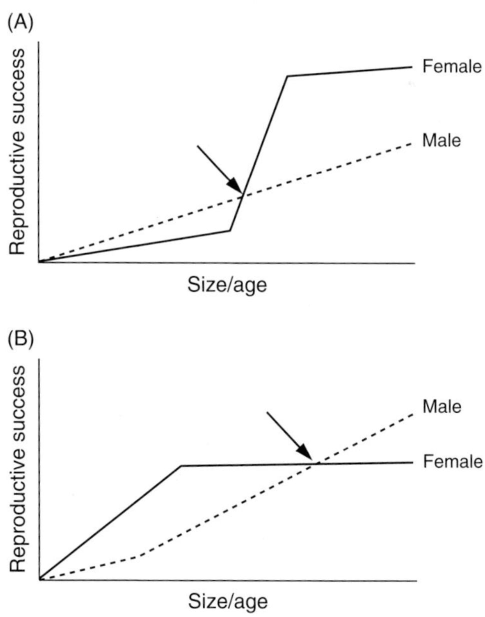

Figure 10.7 Conceptual model for the evolution of sequential hermaphroditism. Arrows indicate the size or age at which sex change is selected. (A) Protandry is favoured if female fecundity is strongly correlated with body size, while males experience a limited size advantage during competition for fertilisations. (B) Protogyny is favoured when large males are able to monopolise matings with females and exclude smaller males from matings, while female fecundity shows a weaker relationship with body size.

Considerable variation in the size and age at sex change might be expected as populations respond to different abiotic and biotic environmental conditions. However, an analysis of 61 populations representing 52 species with sequential hermaphroditism demonstrated some striking regularities (Allsop & West, 2003). If L_{50} is defined as the body length at which 50% of the fish are the second sex (male for protogynous species, female for protandrous species) and L_{∞} is defined as the maximum length reached, the mean value for the 61 populations of the ratio of L_{50}/L_{∞} was 0.79. In other words, the fish tended to change sex when their length was about 80% of their maximum length. If τ is defined as the age at sex change and α as the age at maturity, then the mean value of τ was 2.5α. The relationship between L_{50} and L_{∞} was not significantly affected by mating system or by whether the species was protandrous or protogynous. The relationship between τ and α was slightly, but significantly, affected by whether the species was protandrous or protogynous and, if the latter, by whether the species was monandric or diandric. The regularities suggest that the relationship between male fertility and body size is approximately the same across species that change sex in the same direction. Even more surprisingly, the regularities suggest that the effect favours a change of sex at the same relative size for both protandrous and protogynous species.

A survey of the wrasses (Labridae), a family that shows an array of reproductive modes, provided evidence for the size-advantage hypothesis. There was a significant correlation between sequential hermaphroditism and the degree of size-related reproductive skew, such as is seen if a small number of large males monopolise matings (Kazencioglu & Alonzo, 2010). The presence of mate sequestration behaviour as seen in haremic mating systems or with breeding territories was correlated with protogyny. In the Labridae, there was evidence that the rate of evolution from gonochorism to protogyny within the clade was about three times faster than the evolution of gonochorism from protogyny. This latter evolutionary transition was significantly less likely to happen when the size advantage was stronger.

For monogamous, serial sex-changing gobies, the size-advantage hypothesis does not seem to be valid because there is no difference in the rate at which the reproductive success of males and females increases with size (Munday *et al.*, 2010). A plausible explanation is that the risk of predation is extremely high for these small gobies if they move from the shelter of their coral. The ability to change sex in either direction would enable an individual to mate with another individual it encounters if its original partner were lost.

The evolution of sequential hermaphroditism from gonochorism may be inhibited by reproductive costs associated with the sex change. One of these may be the cost of reorganisation of the gonads, with resources being expended on the change that could otherwise be directly invested in gametes. A second cost is the loss of reproductive output while the change is taking place. This time cost has two components. The first is the time taken to complete the change from one functional sex to the other. The second component is the time taken for the individual that has changed sex to become reproductively successful in its new role. In species that are seasonal breeders, the time cost can be minimised if the sex change takes place outside the breeding season. For example, the protogynous spotty wrasse (*Notolabrus celidotus*, Labridae) in the waters off New Zealand breeds from late July to November, but sex change occurs between November and May (Jones, 1980).

The time taken for the sex change varies between species. In the Gobiidae, *Trimma* can change from one functional sex to the other in less than a week. In *Gobiodon* and *Paragobiodon*, the change can take more than 2 weeks. The rate of sex change in *L. dalli* is temperature-dependent, with the change being more rapid in warmer waters

(Munday *et al.*, 2010). The protogynous wrasse, *T. bifasciatum*, can complete a sex change in about 8–20 days (Goodwin, 2010b; Lawson, 2011), whereas *T. duperrey* takes about 60 days to complete the change. Both species show the behavioural changes associated with the sex change before the gonadal changes.

After the change is completed, it may take time for the individual to become reproductively competent. In a study of *T. bifasciatum*, TP males spent a minimum of 120 days before achieving appreciable mating success. However, they did grow faster than sexually active females, which increased the chances of the males reaching a size at which they could successfully compete for matings and enjoy high reproductive success (Warner, 1988). A faster response was achieved by the bucktooth parrotfish (*Sparisoma radians*, Scaridae) (Muñoz & Warner, 2003; Pandian, 2011). After the removal of the dominant male from a harem, the individual that changed sex started spawning about 10 days later and its fertilisation success increased, reaching about 80% at 18 days after the removal.

At a population level, the cost of change can be quantified by the transitional load, which is the percentage of the population that is sexually inactive because of sex change (Pandian, 2011). For 16 protogynous species, representing seven families, the mean transitional load was 10.4%. This load may be a function of the ratio of males in the local population. However, given that in many coral reef fishes, recruitment to a reef is decoupled from reproduction on the reef because of the planktonic phase in their early life history, it is not clear that the transitional load has any significance at the local population level.

Why are hermaphrodites found in teleosts?

There are numerous examples in non-teleostean jawed vertebrates of large, dominant males sequestering females either in harems or on breeding territories, but only in the teleosts is this associated with protogynous hermaphroditism. An evolutionary puzzle concerns why hermaphroditism is likely to evolve within teleostean clades. The most plausible explanation is that it is a consequence of the embryological origin of the gonads (Sadovy & Liu, 2008; Avise & Mank, 2009). In most jawed vertebrates, the somatic cells of the gonads are derived from two distinct embryological tissues, which form the medulla and cortex of the gonad. By contrast, the somatic cells of the gonads of the teleost develop from a single precursor tissue, the peritoneal epithelium (Chapter 3). In the teleost gonad, both the primordial germ cells and the gonadal somatic cells are bipotential and, depending on the genetic and environmental conditions, in gonochoristic species the gonads will develop as either ovaries or testes. A common pattern in gonochoristic teleost species, such as the zebrafish (*Danio rerio*, Cyprinidae), is undifferentiated gonadal development (Chapter 3), in which the gonads initially contain ovarian tissue, which in a portion of individuals gets invaded and replaced by testicular tissue. Even in these gonochoristic species, before the onset of sexual maturity, the gonad goes through a phase when both ovarian and testicular tissues coexist. At this stage, the individual is a non-functional hermaphrodite. For functional hermaphroditism to evolve, either the germ cells have to retain their potentiality after the onset of sexual maturity, or both ovarian and testicular tissue can coexist and function in the gonads (Sadovy & Liu, 2008). The evolution of hermaphroditism in teleosts may only require a change in the timing of the onset of sexual maturation.

The atypical mode of development of the gonads with its associated lability and bipotentiality means that all teleosts seem to be teetering on the edge of hermaphroditism, and in some environmental circumstances this leads to functional hermaphroditism. With this feature of gonadal differentiation, it is possible that the teleosts had initially evolved as proto-hermaphrodites, with the potential thereafter to evolve various modes of functional hermaphroditism (Sadovy & Liu, 2008) and in a variety of clades within the teleosts (Avise & Mank, 2009).

Deep-sea anglerfish dwarf males: quasi-hermaphroditism

Finding a mate in the deep sea environment poses major problems because of the low population densities and the lack of light, except for bioluminescence (Helfman *et al.*, 2009). The extraordinary diversity of modes of reproduction in the teleosts is further illustrated by some deep-sea anglerfishes of the order Lophiiformes, which have solved the problem of finding a mate through the evolution of parasitic dwarf males. These males become permanently attached to the female and so form a quasi-hermaphroditic unit (Pietsch, 1976, 2005; Helfman *et al.*, 2009). The deep-sea anglerfishes belong to the sub-order Ceratioidea, which are found in the bathypelagic zone below 300 metres. They have evolved from less specialised bottom-dwelling Lophiiformes. Phylogenetic analysis suggests that parasitic dwarf males have evolved independently several times in the ceratioids.

All these deep-sea anglerfish show an extreme sexual dimorphism, with dwarf males only about 10% the size of the females (Fig. 10.8A). At lengths of 6.2–7 mm, the smallest

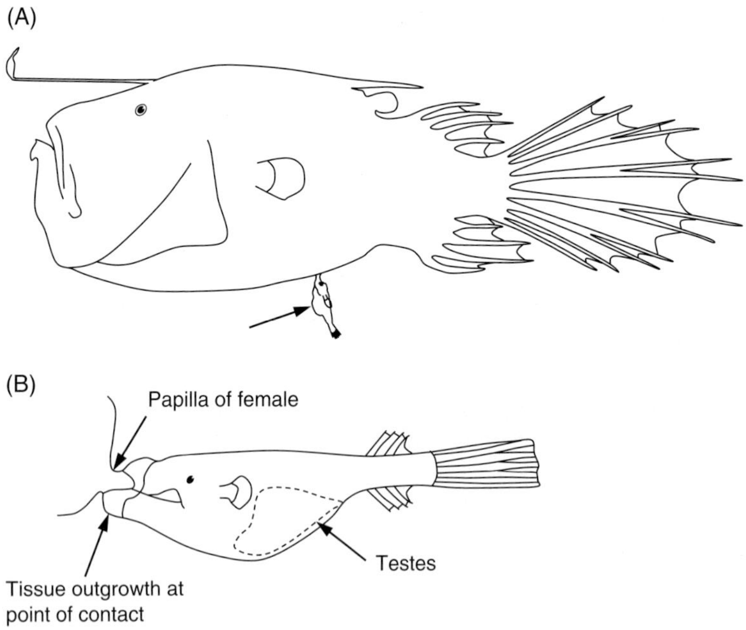

Figure 10.8 Sexual parasitism in deep-sea anglerfish. (A) A female triplewart seadevil (*Cryptopsaras couesii*) with a parasitic male attached, indicated by an arrow. (B) A parasitic male of Kroyer's deep-sea angler fish (*Ceratias holboelli*) showing the outgrowth of tissue from the snout and lower jaw of the male at the point of contact with the female. The (internal) testes are indicated by a dotted line, which can occupy more than half the volume of the coelomic cavity. (Source: Adapted from Pietsch 2005. Reproduced with permission of Springer Science + Business Media.)

of these dwarf males with fully developed testes are amongst the smallest known sexually mature vertebrates. These dwarf males also lack the lure apparatus that is characteristic of anglerfishes. A second feature of dwarf males is that, at metamorphosis, the normal jaw teeth are lost, but are replaced by forceps-like denticles at the anterior tips of the jaws. These denticles are used by the male to grasp a female (Fig. 10.8B). Males tend to have well-developed eyes or olfactory organs, or both. It is assumed that olfaction can be used to detect pheromones released by a female and help the male to find the female. Once close, the male can use visual cues provided by the bioluminescence of the female.

Amongst deep-sea anglerfish species, the degree of association between females and dwarf males ranges from a fully parasitic relationship, with the dwarf male permanently attached to a female, to a brief association between the sexes.

In some families, the parasitic mode is obligatory. In these taxa, males do not reach sexual maturity unless attached to a female, and females do not become gravid unless they have an attached male. Sexual maturity seems to be determined by the parasitic association rather than the age or size of the fish. The attachment of the male to the female is achieved by outgrowths from the snout and tip of the lower jaw of the male, which fuse with the skin of the female. In some species, the heads of the males become fused to the skin of the female from the tip of the lower jaw to the rear of the skull. In other species, the male is carried at the tip of an elongated stalk of female tissue. The circulatory systems of the male and the female become associated so the female can transfer nutrients to the male, whose alimentary canal and associated organs tend to degenerate, while the testes grow. This arrangement recalls the close association between maternal and embryonic tissues in some matrotrophic species (see the section 'Viviparity in teleosts'). Another parallel is that in some anglerfish species, a papilla of female flesh fills the buccal cavity of the dwarf male, a feature also seen in the female–embryo relationship of some viviparous fishes. It is not clear how the dwarf male avoids any immunological rejection by the female. It is assumed that once a male is attached to a female it survives until the female dies.

Typically only one male attaches to a female, usually on the ventral surface of the female with the male upside down. This orientation probably allows a close association between the genital openings of the male and female. In some species, more than one male may attach over a range of positions on the female. For other families of the ceratioids, the association between male and female is temporary and so no parasitic association develops. In such species, the sexes can achieve sexual maturity independently of each other. Some ceratioids may have facultative dwarf male parasitism, with the male attached for shorter or longer periods, providing another example of the lability of reproduction in the teleosts.

Chapter 11

Reproduction and life-history evolution

Reproductive Biology of Teleost Fishes, First Edition. Robert J. Wootton and Carl Smith.

Introduction

A life-history strategy is the suite of traits expressed by an individual that determine how it spreads its reproduction over its lifetime. Life-history theory predicts that selection will favour a pattern of allocation of resources to reproduction over the lifetime of an individual that will maximise its contribution of offspring to the next generation (Roff, 1992, 2002; Stearns, 1992).

Teleosts display an extraordinary range of life histories. The killifish *Nothobranchius kadleci* (Nothobranchiidae) can achieve sexual maturity at a length of 31 mm after 17 days of development from hatching, the most rapid development of any vertebrate. Some populations of this species have a maximum lifespan of 11 weeks and a generation time of just 32 days (Reichard, 2010; Blažek *et al.*, 2013). It produces multiple clutches that comprise a high proportion of a female's total energy budget. Males are brightly coloured and intensely aggressive in competing for matings and suffer high mortality rates during intrasexual contests. Ageing, or senescence, which represents intrinsic mortality rate with age, is unusually rapid in these fishes. This life-history strategy contrasts with some salmonid populations in which females do not achieve maturity until after 10 years, but subsequently die after spawning (Groot & Margolis, 1991). Some marine species, especially if not subjected to exploitation by commercial fisheries, become sexually mature only after several years, but then continue to spawn at yearly intervals for several decades. Still other life histories, such as those of some cichlids, may comprise a comparatively long lifespan, with the regular production of modest clutches of large eggs, in which a monogamous pair of adults invest substantially in parental care. This multiplicity of life histories is partly a reflection of the extensive range of aquatic habitats occupied by teleosts. However, within a given habitat, teleosts have evolved numerous life histories to exploit that habitat. Even within a species, some individuals display strikingly divergent life-history traits. Life-history theory seeks to predict the direction of evolutionary responses of age-specific reproductive traits to environmental change, within the bounds imposed by physiological, ecological, genetic and phylogenetic constraints. Thus, life-history theory provides a means of understanding how the multitude of reproductive strategies displayed by fishes has evolved.

Three main theoretical approaches have been used (Roff, 2002; Stearns, 1992). Demographical models use life-table data to predict the values that life-history variables such as age at maturity, age-specific fecundity and egg size should take to maximise fitness. Templet models aim to identify the life histories that are particularly characteristic of given habitats. A third approach seeks to identify non-dimensional, invariant traits that characterise teleost life histories. The raw data for all these approaches are the life-history traits that make up a reproductive strategy.

Life-history variables

A number of reproductive variables contribute to structuring an organism's life-history strategy, although many do not vary independently of each other. The interdependence of life-history variables makes identification of those traits on which selection acts notoriously difficult. Ultimately, all those variables that influence fitness and show heritable variance have the capacity to shape life-history evolution.

Size and age at maturity

The age, and hence the size, at which sexual maturation is achieved is a fundamental life-history variable. It represents the age at which resources are no longer channelled solely to growth, maintenance and storage, but also to reproduction. The transition from the juvenile to the adult stage is regarded as irreversible (Chapter 5). Maturation, in a life-history context, is defined as the age or size at which spawning, in the case of oviparous species, or parturition for viviparous species, first occurs. Age at maturity differs from the age at which an individual is functionally capable of mating but does not engage in reproduction. In the case of semelparous species, i.e. species that reproduce only once in a lifetime, age at maturation corresponds approximately to an individual's lifespan.

Age and size at maturity are associated because it takes time to grow to a given size, and in natural populations it is difficult to disentangle the two. Both variables can have a powerful influence on fitness and are predicted to be sensitive to selection (Stearns, 1992). In teleosts, body size and fecundity are often correlated, mortality is frequently size-dependent and growth is indeterminate (Wootton, 1998). Consequently, size rather than age at maturation may be the more critical trait for understanding life-history evolution.

Age and size at maturity tend to have opposite effects on fitness. Other things being equal, a reduction in the age of maturity is a potent method of increasing fitness. This is because it reduces the generation length, which can be approximately defined as the average age at which a mother first reproduces and the average age at which her daughter reproduces. A reduction in the age at maturity reduces the probability of the individual dying before reproducing. However, an early age of maturity has the disadvantage that it occurs at a small body size. In many teleosts, fecundity and body size are positively correlated, so early maturation at a small body size means females have a low fecundity. Small size of males may also be associated with a limited capacity to compete for a territory or nest site, attract males and exclude rivals during mating (Chapters 8 and 9).

The range of ages of maturation in teleosts is extremely wide. At one extreme, males of the viviparous dwarf surfperch (*Micrometrus minimus*, Embiotocidae) are mature at birth, and in the first weeks after birth they inseminate newborn females (Schultz, 1993a,b). These females store the sperm and do not reach sexual maturity until several months after birth. By contrast, marine flatfish of the genera *Hippoglossus* and *Hippoglossoides* (Pleuronectidae) may not become sexually mature until about 15 years old.

Shifts in variables such as juvenile or adult mortality rates are expected to have predictable impacts on age and size at maturity (Stearns & Koella, 1986; Roff, 2002). A common, but not invariable, pattern is for the age and size at maturation to decrease as conditions for growth improve.

Fecundity and clutch size

Because of the diversity of life-history strategies in the teleosts, this life-history trait has to be carefully defined (Wootton, 1998). For semelparous species, such as *Anguilla* (Anguillidae) and many *Oncorhynchus* (Salmonidae) species, there is only a single breeding season, with no fish surviving to a second breeding season (Chapter 1). Fecundity is simply the number of eggs produced in that first and only breeding season. For iteroparous species, in which individuals can survive to breed more than once, lifetime fecundity (F) is the total number of eggs produced by a female over her lifetime, essentially lifetime reproductive success. Breeding season fecundity (F_b) is the total number of eggs produced over a breeding season. For a total spawner, such as Atlantic salmon (*Salmo salar*, Salmonidae) and northern pike (*Esox lucius*, Esocidae), this is the number of eggs spawned during the brief spawning period. However, for species that are batch spawners, the definition of fecundity is more complex. A simple definition is the product of the mean number of eggs per batch or clutch (F_{batch}) and the number of spawning episodes over a breeding season. For batch spawners with determinate fecundity (Chapter 4), the potential number of eggs is determined by the start of the breeding season, but may be down-regulated by pre-ovulatory atresia (Chapters 4 and 6) to generate the realised breeding season fecundity. For batch spawners with indeterminate fecundity, the breeding season fecundity depends on the environmental conditions, particularly the feeding conditions, which determine the number of spawning episodes and the interval between successive spawning episodes (Townsend & Wootton, 1984; Smith & Wootton, 1995a,b). For some species, such as tropical killifish, *Nothobranchius* spp., the interval is only a day, but in other species it may be several days, or spawning may follow a semi-lunar or lunar cycle (Chapter 6). Batch fecundity is the number of eggs produced at each spawning episode and can be approximately equated to the number of eggs produced by a total spawner.

Batch or total fecundity (F) is an allometric function of body size (Fig. 11.1; Wootton, 1998). If body size is measured as length (L), the relationship takes the form:

$$F = aL^b$$

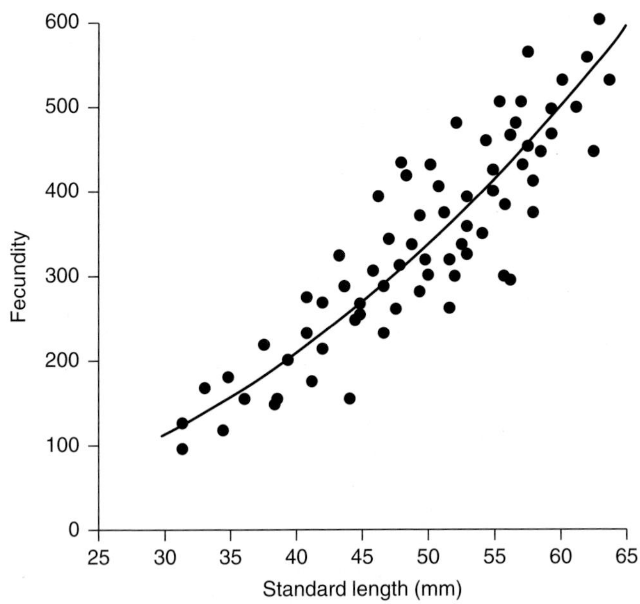

Figure 11.1 Allometric relationship between fecundity and body size for European bitterling (*Rhodeus amarus*) based on dissecting the ovaries of 77 females prior to spawning. The relationship between size and fecundity takes the form $F = aL^b$, where F is fecundity and L is standard length (measured from the tip of the snout to the base of the caudal fin). The value fitted for parameter a in this model is 0.07 and for parameter b is 2.165. (Source: Data from Smith *et al.* 2000.)

which can also be expressed as the linear relationship:

$$\log F = \log a + b \log L$$

The equivalent relationship when size is measured as body mass (*W*) is:

$$F = cW^d$$

or

$$\log F = \log c + d \log W$$

On geometric grounds, because the eggs occupy a volume of the body cavity before spawning, the value of the exponents would be expected to be: $b = 3.0$ and $d = 1.0$ respectively.

However, there is both inter- and intraspecific variation in the allometric parameters, *b* and *d* (Wootton, 1979; Kamler, 1992). In a survey of 62 species, the value for *b* varied from about 1.0 to 5.0, but the adaptive significance of this variation remains to be explored.

The other determinant of batch (total) fecundity is the allocation of resources to each egg, thus determining egg size. If total egg volume (TV) is defined as the product of batch fecundity and mean egg volume, the allometric relation for 238 species took the form:

$$\text{TV} = aL^{3.09}$$

which implies that total egg volume tends to increase isometrically with body length, as suggested by geometric considerations (Wootton, 1992). Batch fecundity is the result of total resource allocation to the ovaries and the subsequent partitioning of that allocation to individual progeny (bestowal or reproductive allotment).

Many of the mechanical constraints that shape clutch size in terrestrial organisms, such as climbing, crevice living and foraging, do not apply to teleosts because of the support provided by water. However, there are potentially negative effects on swimming performance because of the distension of the abdomen in a gravid female. In the viviparous mosquitofish (*Gambusia affinis*, Poeciliidae), there were no significant effects of pregnancy on hydrodynamics resulting from a change in body shape, although pregnancy did impose aerobic constraints on swimming speed (Plaut, 2002). By contrast, in the guppy (*Poecilia reticulata*, Poeciliidae), another livebearer, pregnancy had the effect on females of limiting maximum velocity and the distance travelled in fast-start escape responses, which function in the evasion of predators (Ghalambor *et al.*, 2004). In populations of *P. reticulata* in which adults experienced a high predation risk, selection on reproductive effort favoured the production of large clutch sizes, with the outcome that the impact of pregnancy intensified predation risk on females. In this species, high predation selects for increased clutch size, thereby eroding the escape response of females to predators.

Egg and offspring size

Teleosts as a group have adopted a reproductive strategy of high fecundity and small egg size (Winemiller & Rose, 1993). This is in contrast to that other important ectothermic vertebrate component of marine waters, the Chondrichthyes, the cartilaginous fishes, which have a reproductive strategy of low fecundity but large propagules (Freedman & Noakes, 2002; Helfman *et al.*, 2009). A frequency distribution of egg sizes of teleosts is skewed towards the smaller egg sizes, suggesting selection to minimise egg size (Fig. 11.2; Wootton, 1979). The smallest teleostean egg has a diameter of about

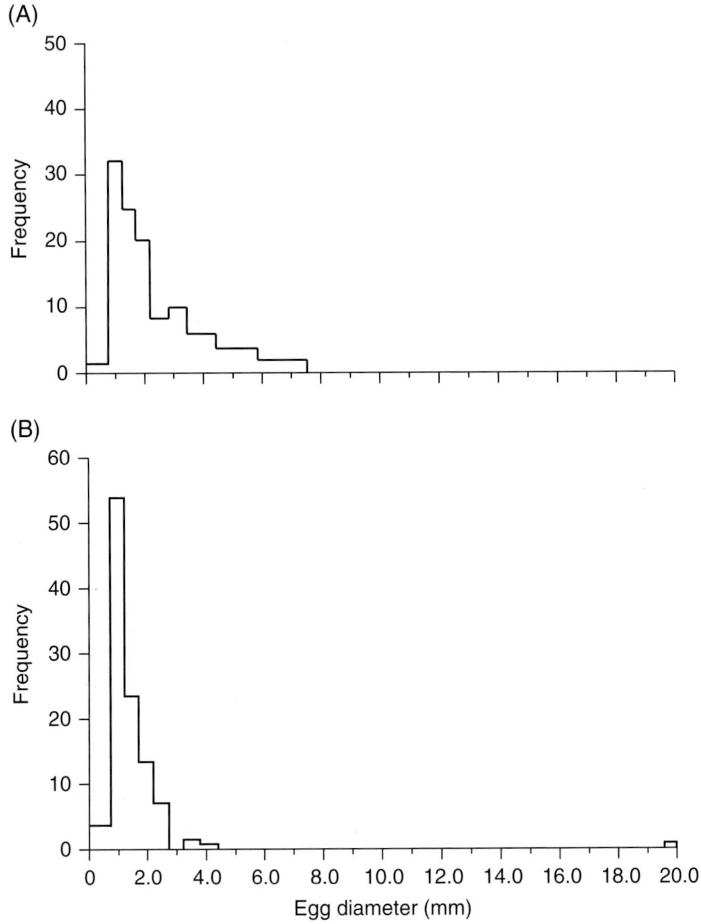

Figure 11.2 Frequency histogram of egg sizes of: (A) freshwater; (B) marine fishes. (Source: Adapted from Wootton 1979. Reproduced with permission of Elsevier.)

0.25 mm and the largest a diameter of about 20 mm. These largest eggs are those of mouthbrooding marine catfish (Ariidae), but these are outliers in the distribution of teleostean egg sizes. The modal egg diameter for marine species is about 1.0–2.2 mm, but slightly larger for freshwater species (Wootton, 1979; Kamler, 1992). Within marine species, there is a tendency for demersal eggs to have a bigger diameter than pelagic eggs (Thresher, 1988). Within demersal spawners, egg size tends to increase with latitude, but the same trend is not shown strongly by marine pelagic spawners. As a consequence, at low latitudes, the mean egg size of demersal and pelagic spawners is similar, but the difference increases with latitude.

Egg diameter is frequently used as a measure of egg size. However, care has to be taken. The important variable is the size of the yolk mass, which supplies the resources for the developing embryo. In some species, the yolk does largely fill the egg, but in other species, the yolk occupies a smaller volume and the perivitelline space between the egg plasma membrane and the chorion is large (Chapter 4). An important correlation is that between egg size and the size of the hatchlings. In general, larger eggs produce bigger hatchlings (Kamler, 1992).

The strategy of high fecundity and small eggs was probably characteristic of the ancestors of the teleosts. Phylogenetically basal actinopterygians, such as the Acipenseriformes (sturgeons and paddlefish), the Lepisosteiformes (gars) and Amiiformes (bowfin), all have relatively high fecundities and small egg sizes (Helfman *et al.*, 2009).

The adaptive significance of the high fecundity–small egg size strategy has still to be clarified. One modelling study suggested that the strategy is an adaptation to the situation when resources are patchily distributed over a relatively large spatial scale (Winemiller & Rose, 1993). The role of food availability for the evolution of the strategy was suggested by a second modelling study, which suggested that the value of the ratio between predation mortality and specific assimilated intake of food is critical for the strategy to be adaptive (Andersen *et al.*, 2008).

Egg size is a reflection of female investment in each oocyte and correlates with the quantity of yolk allotted to each offspring. A change in egg size faces two potential trade-offs. The first is a trade-off between egg size and clutch size. Within the limits of a given reproductive allotment, this means that selection for greater egg or embryo size will necessarily result in a decline in clutch size. Brown trout (*Salmo trutta*, Salmonidae) from Norwegian rivers showed this trade-off between egg mass and fecundity (Jonsson & Jonsson, 1999). A second trade-off is between egg size and mortality. Larger eggs contain more yolk, have greater hatching success, produce larger larvae with a reduced susceptibility to starvation, lower risk of predation and greater food capture abilities (Blaxter & Hempel, 1963; Bagenal, 1969; Miller *et al.*, 1988; Pepin, 1991; Trippel, 1998; Kamler, 2008). Consequently, the fitness benefit of selection for increased egg size must outweigh the fitness loss of reduced fecundity.

Egg size can also correlate negatively with survival. This effect may arise because larger eggs tend to take longer to hatch and reach independence, making them susceptible to predators for longer (Steele, 1977).

Whether selection acts on egg size or clutch size is not always clear. Different environmental conditions may favour one trait over the other under different conditions, and both may be selected in different selection regimes. At one extreme is the possibility that egg size is selected to maximise offspring fitness. Given the general advantages that accrue to large offspring, this would mean selection for large eggs, but at the cost of a reduction in the fecundity of the female. The other alternative is that egg size is selected to maximise parental fitness. For a female, this is the product of fecundity and the survival of the progeny to independence, which is when maternal effects no longer influence the survival of the progeny. There is then an optimal egg size that maximises parental fitness. The optimum size depends on the rate at which progeny survival increases with egg size and the rate at which fecundity declines with egg size (Smith & Fretwell, 1974). This model implies that optimum egg size will vary between environments, depending on how the environment affects maternal fecundity and the relationship between progeny size and progeny fitness.

A study on *S. salar* showed that egg size, measured as mass, had evolved as a response to selection on maternal fitness rather than offspring fitness (Einum & Fleming, 2000). This study controlled for the effects of other maternal or genetic traits that potentially correlated with egg size. Problems of identifying the relationship between progeny size and fitness and of calculating the fecundity of females, especially in species with indeterminate fecundity, can make identification of the optimum egg size in a population difficult (Rollinson & Hutchings, 2013). However, on a broad taxonomic scale, empirical data for marine teleosts suggest that propagule size may be the key variable, with optimal egg size a function of size-dependent mortality and mean water temperature (Wootton, 1994b).

The Smith & Fretwell (1974) model predicts that egg size will differ among environments, and environments can vary both temporally and spatially. Substantial seasonal decreases in egg size are well described in teleosts. For example, in marine teleosts in the North and Irish seas, both planktonic and demersal eggs show consistent decreases in size over the course of the spawning season (Fig. 11.3; Bagenal, 1971). Notably, there is little geographic variation in egg size in these fishes. By contrast, a study of Coho salmon (*Oncorhynchus kisutch*, Salmonidae) revealed 255% variation in egg size among females within populations. No significant component of this variance could be attributed to seasonal patterns, but the breeding season of this semelparous total spawner is relatively short (Fleming & Gross, 1990). In this case, most of the variation in egg size was attributable to variation in female size. There were highly significant differences in egg mass between populations, with mean egg mass in a population tending to decline as latitude increased, but with mean fecundity per female tending to increase.

Differences in egg size, particularly seasonal patterns, need not necessarily reflect adaptive change in response to selection. The decline in female condition over the

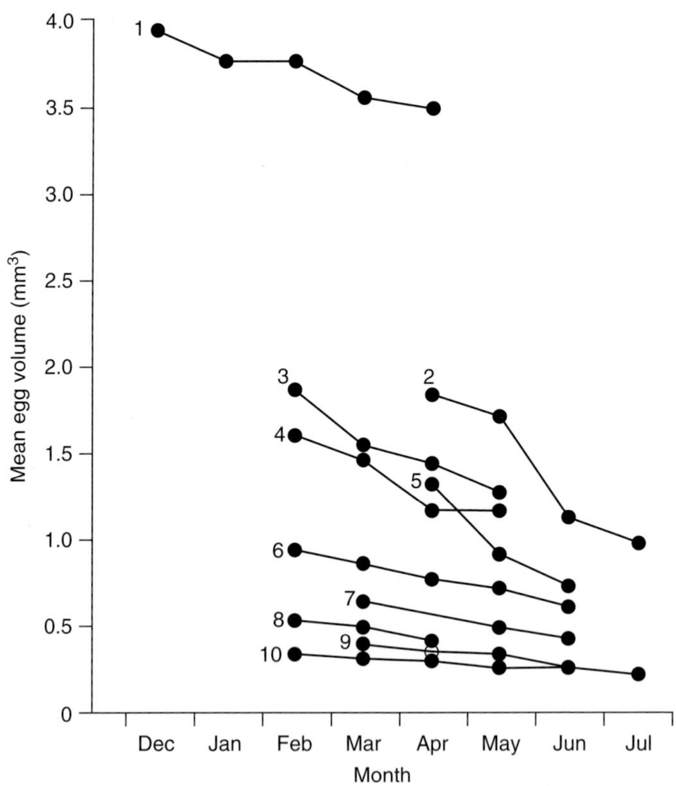

Figure 11.3 Seasonal decline in mean egg volume (mm^3) in North Sea fishes; 1, plaice (*Pleuronectes platessa*); 2, grey gurnard (*Eutrigla gurnadus*); 3, haddock (*Melanogrammus aeglefinus*); 4, Atlantic cod (*Gadus morhua*); 5, Dover sole (*Solea solea*); 6, whiting (*Merlangius merlangus*); 7, sprat (*Sprattus sprattus*); 8, flounder (*Platichthys flesus*); 9, fourbeard rockling (*Enchelyopus cimbrius*); 10, dab (*Limanda limanda*). (Source: Bagenal 1971. Reproduced with permission of John Wiley & Sons.)

course of a spawning season, as well as the impact of environmental variables such as temperature and food availability, can result in changes in egg size and total reproductive allotment (Wootton, 1977; Pepin & Myers, 1991; Trippel, 1998). Caution must be exercised in interpreting data on variation in egg size in teleosts, as it appears to show plasticity that may have no adaptive basis.

Variation in egg size is associated with parental care in fishes, with care-giving species tending to produce larger eggs than non-carers. Within care-giving families, the extent of care also appears to correlate positively with egg size (Sargent *et al.*, 1987). Several hypotheses have been proposed to account for this pattern (Shine, 1978, 1989). The core argument is that larger embryos require more care, primarily because they take longer to develop and require more oxygen. Whether selection acts on extending the length of embryo and larval stages, e.g. if mortality rates of juveniles after hatching are relatively high, or because selection favours larger egg size for some other reason, is unclear. Another mechanism by which large egg size and extended parental care might be selected is through assortative mating (Shine, 1989). If larger males provide more extensive care and larger females, which typically produce larger eggs, prefer larger males, a positive association between length of care and egg size might evolve. The weakness of this hypothesis is that it confounds intra- and interspecific patterns. Empirical data to support the association between egg size and parental care are sparse and contradictory (Kolm & Ahnesjö, 2005) and the association warrants further research, particularly the application of phylogenetic comparative methods.

In some taxa, additional trade-offs may play a role in the evolution of egg size. In the bitterling fishes (Subfamily Acheilognathinae), which lay their eggs in the gills of living freshwater mussels, the evolution of egg size and shape may be constrained by selection to ensure the eggs fit between the gill lamellae of host mussels (Liu *et al.*, 2006; Reichard *et al.*, 2010; Kitamura *et al.*, 2012; Fig. 11.4). The lazy gudgeons (*Sarcocheilichthys* spp., Cyprinidae) similarly use living freshwater bivalves for oviposition. Unlike bitterling, which place their eggs in the mussel gill cavity, *Sarcocheilichthys* spp. deposit their eggs in the mussel mantle cavity. Their eggs are unusually large, exceeding 5 mm in diameter in some species, which is largely because of the development of an outsized perivitelline space after fertilisation (Balon, 1999). The remarkable size of their eggs is an adaptation to prevent them being lost from the mantle cavity through the inhalant siphon of the host mussel.

Incubation time

Other things being equal, the longer the period an embryo spends in development, the larger it will be as a hatchling and the longer the interval before exogenous feeding begins. Longer incubation correlates positively with egg size and will tend to buffer young stages against starvation, and may also limit the range of predators to which they are exposed. The cost of a long incubation is a greater risk of predation that comes with prolonged development (Sargent *et al.*, 1987). Consequently, long incubation periods are often associated with parental care.

For freshwater species in the temperate zone of the Northern Hemisphere, the time to hatching and the time from hatching to exogenous feeding are highly, but inversely, correlated with water temperature (Telechea & Fontaine, 2010). Families such as the salmonids have large eggs, which are incubated at low temperatures and for a long period, with the larvae hatching at a large size. In salmonids, the eggs are protected from predation by being buried in gravel in a nest dug by the female. Many cyprinid

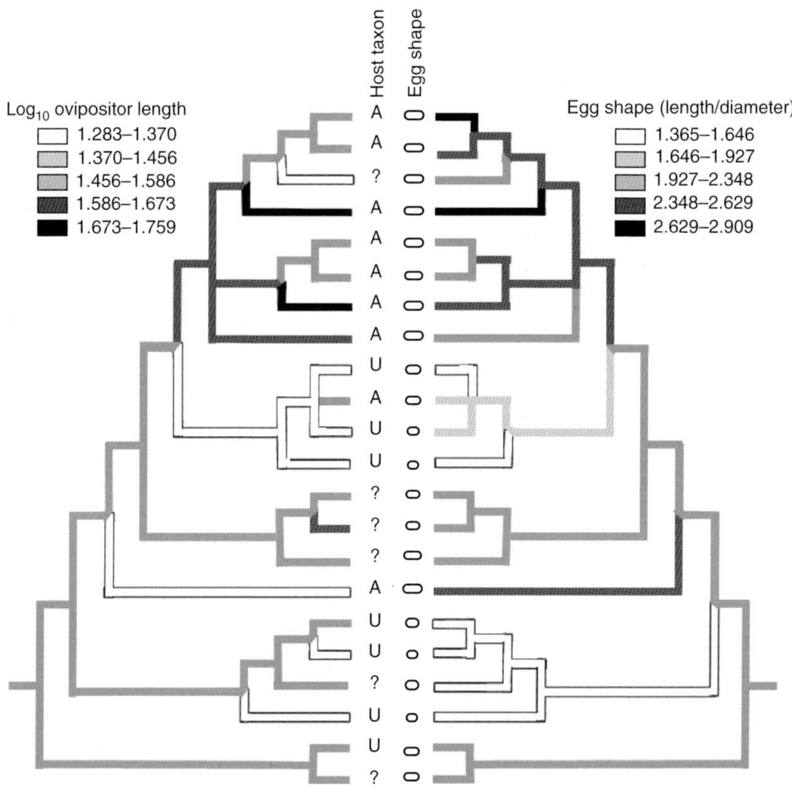

Figure 11.4 Reconstructed ancestral states for ovipositor length and egg shape (length/diameter ratio) on the phylogenetic tree of the tabira bitterling (*Acheilognathus tabira*) from 22 sites in Japan. The different shading of branches denotes ancestral states. For each bitterling population, host mussel subfamily (A, Anodontinae; U, Unioninae; ?, unknown) and average egg shape are shown. The figure illustrates that the evolution of elongated ovipositors and elongated eggs has occurred repeatedly in different tabira bitterling lineages in discrete geographic regions. (Source: Adapted from Kitamura *et al.* 2012. Reproduced with permission of John Wiley & Sons.)

species produce small eggs that develop quickly at high temperatures and hatch as relatively small larvae. Egg and larval traits of these temperate teleost species, especially size and the temperature requirements for development, are closely linked (Telechea & Fontaine, 2010).

Exceptions to the general rule that long incubation times are correlated with large eggs and early stages are species that produce diapausing eggs. Diapause is an interruption to development and is rare in vertebrates, although it is displayed by two groups of fishes, the cyprinodontiform killifish and cypriniform bitterling. In annual killifish, taxa that occupy temporary water bodies produce eggs with a thickened chorion, which are buried in mud or organic litter. During dry periods, when the water body in which the eggs are spawned has dried up, the egg enters an arrested state of development, often lasting months. Development is resumed once the water body again fills with the onset of rain (Wourms, 1972). Some of the periods spent in diapause are obligate, but others may be facultative, occurring only if the conditions require diapause. In bitterling fishes, which spawn in the gills of bivalve molluscs, developmental arrest is seen in autumn-spawning species and corresponds with the lowest-temperature winter months (Kawamura & Uehara, 2005).

Reproductive lifespan

In iteroparous species, the period over which reproduction occurs is the reproductive lifespan and is a highly variable life-history trait. High risk of mortality selects for rapid maturation and short reproductive lifespan, accompanied by rapid senescence. Thus, reproductive lifespan represents a trade-off between selection that increases the number of reproductive events after maturation and age-specific intrinsic mortality rate. Senescence is a negative outcome of selection for increased reproduction, and has the effect of increasing mortality rate and reducing fecundity. An outcome of this trade-off is that selection that constrains investment in offspring is predicted to limit senescence. Early senescence is displayed by annual killifish (Nothobranchiidae), which display unusually rapid growth and early maturation followed by high reproductive investment until (an early) death.

In semelparous species, reproductive lifespan is restricted to a single breeding season, even if age at maturation does not occur until after several years of development, such as in Pacific salmon species (*Oncorhynchus* spp.). Lifespan and reproductive lifespan are discrete traits. Preventing reproduction in semelparous species can greatly extend lifespan, which implies that mortality is a direct result of physiological changes associated with reproduction (Chapters 4 and 6). In sockeye salmon (*Oncorhynchus nerka*, Salmonidae), lifespan can be approximately doubled from 4 to 8 years through castration (Robertson, 1961).

Parental care

The provision of parental care influences the mortality rates of offspring, and the extent of care varies substantially among teleosts, although the majority of species do not show parental care (Chapter 9). Parental guarding limits predation of eggs and young stages, but parental care can also include ventilation of the eggs and keeping them free from infection by pathogens. In some viviparous and oviparous species, the parent provisions the young (Chapter 9). Care carries a cost to the caregiver (Smith & Wootton, 1995a), with implications for future reproduction. The extent of care appears, at least ostensibly, to correlate positively with egg size, as well as with other life-history traits. Parental care is also associated with extended periods of development. Parental care shows substantial flexibility, with adjustment in the level of care, including abandonment of young and filial cannibalism, in response to environmental variation (Chapter 9).

Other life-history traits

The major life-history traits listed in the previous sections are not the only ones to shape fish life-history evolution. For some taxa, other traits may experience strong selection. Thus, in poeciliids, the degree of superfetation and extent of lecithotrophy in relation to matrotrophy are significant (Johnson & Bagley, 2011; see Chapter 10). In other taxa, such as bitterling fishes, male investment in sperm production may be a key life-history variable (Smith *et al.*, 2009, 2013). In many cases, the understanding of these 'secondary traits' in life-history evolution is limited.

Basic concepts in life-history theory

Trade-offs

Given the array of reproductive traits, the key evolutionary problem is to identify the combination of traits that, in a given environment, maximise the chances that an individual will pass its genes on to the next generation, in other words, maximise its fitness.

The two fundamental quantities related to fitness are age-specific survival (l_x) and age-specific fecundity (m_x; Stearns, 1992; Roff, 1992, 2002). In teleosts, both l_x and m_x are related to size (Wootton, 1998), so a third relevant variable is age-specific growth rate (g_x). An ideal organism, a 'Darwinian demon', would start reproducing at an early age, have high age-specific fecundities and low age-specific mortality rates, resulting in an extremely long life. No teleosts have this combination of life-history traits, which implies that the maximisation of fitness is constrained by trade-offs (Roff, 2002; Table 11.1).

These trade-offs essentially arise because of the phenomenon of resource allocation. The rate of intake of resources, including energy, by an individual through its feeding activities will be finite. Consequently, allocation of resources by the individual to an activity such as reproduction will mean that fewer resources are available for activities related to survival, including growth (Fig. 11.5). Selection must operate against a background of temporal variation in population size and environmental stochasticity, with unpredictable outcomes for the success of a given set of trade-offs.

Costs of reproduction

Closely related to the concept of trade-off is the concept of the cost of reproduction. Cost, in this context, is measured in units of fitness. A central assumption of life-history theory is that reproduction carries a cost, measurable in terms of future reproductive output. Individuals that invest heavily in current reproduction are predicted to show elevated mortality and a limitation in their capacity to invest in future reproduction (Williams, 1966; Charlesworth, 1980; Stearns, 1992; Roff, 2002). If reproduction was not costly, constraints on reproduction would solely be limited by physiology, allometry and phylogeny (Wootton, 1994a).

Table 11.1 Pairwise tradeoffs between life-history traits. Traits are as follows: n, number of female offspring per female per breeding attempt; s_1, survivorship from birth until first breeding; s_2, survivorship of adults between successive breeding attempts; t_1, age at first breeding; t_2, interval between successive breeding attempts. (Source: Adapted from Wootton 1998. Reproduced with permission of Elsevier.)

Trade-off	Interpretation
n vs. s_1	Parental provisioning is critical to offspring survival
n vs. s_2	Elevated risk of death associated with reproduction
s_1 vs. s_2	Parent(s) protect offspring from predators
t_1 vs. s_1	Feeding for juveniles is risky
t_1 vs. s_2	Parent(s) feed offspring
t_2 vs. s_2	Feeding for parents is risky
n vs. t_1	Large offspring limit the number of offspring produced
n vs. t_2	More offspring limits adult breeding frequency
t_1 vs. t_2	Greater investment by parents accelerates offspring development
s_1 vs. t_2	Greater investment by parents enhances offspring survival

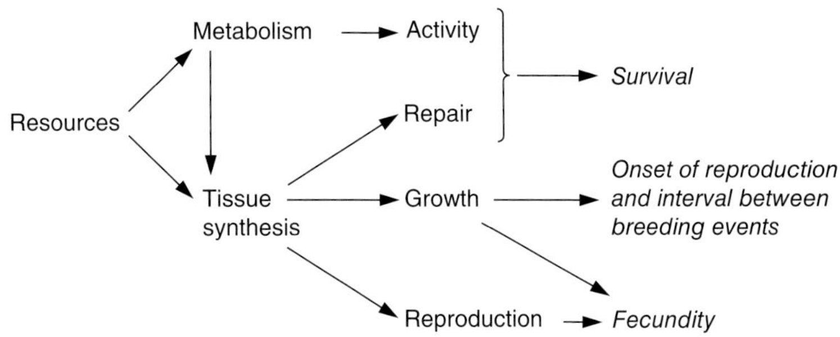

Figure 11.5 Resource allocation among metabolic demands. Effect on components of fitness are shown in italic. (Source: Adapted from Calow 1985. Reproduced with permission of Springer Science + Business Media.)

Costs can be classed as physiological or ecological. Physiological costs of reproduction are those that constrain the capacity of an individual to maintain their long-term condition, and include adverse effects on processes such as metabolic performance, energy and nutrient storage, growth, immunocompetence, recovery from injury, DNA repair and resistance to oxidative stress. Ecological costs are those that place an individual at greater risk of predation, injury or infection by pathogens. In teleosts, an adverse effect on growth constitutes a cost because fecundity is typically positively correlated with size. A reduction in growth, therefore, constrains future reproductive output.

Measuring the costs of reproduction

Costs of reproduction can be measured empirically through identifying phenotypic correlations between traits, by undertaking experimental manipulations, from quantitative estimates of genetic correlations between traits and from correlated responses to selection (Reznick, 1985; Roff, 2002).

Negative phenotypic correlations between life-history traits can illustrate how investment in current reproduction can have negative impacts on parental survival, growth or future allocation to reproduction. In three populations of brook trout (*Salvelinus fontinalis*, Salmonidae) in Newfoundland, reproduction imposed an over-winter survival cost on individuals, which was greater for older individuals (Hutchings, 1994). Similarly, in *S. trutta*, survival rates of spawning individuals reflected somatic declines in body lipid content (Berg *et al.*, 1998). A complication of using phenotypic correlations to identify costs of reproduction is that among-individual variation in condition will confound effects. Individuals in poor condition may have low current investment in reproduction, as well as low future survival. Conversely, those individuals in good condition will have the capacity to invest heavily in both current and future reproduction (Wootton, 1998). Comparisons among populations can suffer from similar shortcomings, with variation in habitat quality generating spurious correlations. Life-history trade-offs can reflect negative genetic correlations between traits through pleiotropic effects or linkage disequilibrium (Roff, 2002). Another limitation to interpreting negative phenotypic correlations is that they may not reflect underlying genetic effects, and it is possible to measure positive phenotypic correlations between traits that have negative genetic correlations. Caution must especially be exercised in assigning causality to negative phenotypic correlations (Reznick, 1985).

Undertaking experimental manipulation of some aspect of reproductive investment and then measuring the outcome in terms of growth, survival and fecundity has the advantage of focusing on the costs of specific components of reproduction, such as nest building or parental care. Studies of this type are relatively readily accomplished and can provide valuable insights. Removal of offspring from parental fish, or exposure of offspring to predators to heighten levels of parental care, have demonstrated costs associated with parental care of reduced growth, reduced frequency of reproduction and greater risk of predation (Pressley, 1981; Smith & Wootton, 1994; Balshine-Earn, 1995; Okuda, 2001; Steinhart *et al.*, 2005). Experimental manipulation of ration has demonstrated that reproductive costs are typically greater when resources are withheld from parents (Hirschfield, 1980; Stanley & Wootton, 1986; Smith & Wootton, 1995a,b). In some cases, adults may show adaptive responses to elevated costs by reducing parental investment (Townshend & Wootton, 1984), while in others, parental investment (and therefore parental costs) are maintained irrespective of energy budget (Hirschfield, 1980).

Like phenotypic correlations, experimental manipulations ultimately assume a genetic correlation between traits, which is rarely demonstrated. Another drawback is that experimental manipulations may lack biological realism, e.g. if experimental manipulations exceed the range of conditions to which the study species is naturally exposed. In this case, an outcome might be that experimental subjects fail to show adaptive responses simply because they do not possess the capacity to respond to the manipulation.

A further approach to measuring costs is to use a full sib/half sib breeding experiment to identify genetic correlations between life-history traits. Using the appropriate design, it is possible to quantify genotype × environment interaction effects on life-history traits (Falconer & Mackay, 1996). An extension of this approach is to perform selection experiments whereby some aspect of an organism's life history is manipulated in an experimental group and the life-history outcome is measured in both the manipulated group and a corresponding set of controls, with a subsequent test of genetic correlation among the traits. Genetic correlations combined with selection provide a measure of direct causal relationships among life-history traits, and thereby a qualitative measure of response to selection. In this respect, they are superior to purely phenotypic correlations. The limitations of genetic correlation studies are that the outcome may be specific to the environment in which they are conducted. In changed conditions, e.g. under food deprivation, correlations and especially interactions may alter. Costs of reproduction may exist in natural conditions that are not demonstrated in a laboratory setting (Reznick, 1985). Additionally, the degree of inbreeding in the study population under test may constrain the outcome of genetic correlation studies (Rose, 1984).

Evidence of costs of reproduction

Mortality is a key cost of reproduction and plays a major role in shaping life-history evolution in teleosts and other taxa (Roff, 1992). A survival cost of reproduction can be caused by the additional energetic burden that arises from energetic investment in gametes (Chapter 4), secondary sexual traits or activities associated with reproduction, such as migration (Chapter 7), courtship, competition for mates (Chapter 8) and parental care (Chapter 9). Increased mortality can also be associated with reproductive behaviours themselves, which make an individual more susceptible to predation, injury or disease (Bell, 1980; Roff, 1992).

An experimental study using the medaka (*Oryzias latipes*, Adrianichthyidae), a species with daily spawning, demonstrated a significant mortality cost of reproduction

(Hirschfield, 1980). The number of sick and dead fish in the study correlated positively with the ratio of energetic investment in eggs to energy acquired in food intake. Costs of reproduction also included a reduction in body mass and tissue energy content.

Within a population of the live-bearing guppy (*P. reticulata*), females were constrained in redirecting energy from reproduction to somatic growth when prevented from reproducing, suggesting a limited cost of reproduction (Reznick, 1983). However, among populations, females that invested most in reproduction had limited somatic growth. The total energy budget of females was comparable in all populations. Notably, it was in populations with greater risk of predation that females devoted a higher proportion of their energy to reproduction, which corresponds with the predictions of life-history theory (see the section 'Predictions of demographic models').

In a study of migratory and resident populations of the three-spined stickleback (*Gasterosteus aculeatus*, Gasterosteidae), genetic correlations among life-history traits were extensive and varied among populations, suggesting that there are contrasting selection pressures on life history in different environments (Snyder, 1991a). Heritabilities of life-history traits were low and with limited additive genetic variance but strong non-additive effects, which implies that these traits were under strong selection.

Several studies have demonstrated potential reproductive costs of parental care (Smith & Wootton, 1995a). Costs include reduced female batch fecundity, breeding frequency and parental survival (see also Table 11.1). An effect of care on batch fecundity comes from reduced growth rates of females due to elevated energy expenditure during care, such as from egg fanning, or constraints on feeding, such as from oral brooding of eggs (Chapter 9). In males, breeding frequency can be limited when parental care and courtship and mating behaviours are incompatible. However, this is relatively rare and in most cases males continue to attract mates while engaged in the care of offspring from previous matings. For females, the energetic cost of care and constraints on feeding imposed by care not only result in reduced fecundity through an indirect effect on growth, but also limit the rate at which they produce eggs. Females that cannibalise their offspring, lose them to predators or have them removed experimentally are able to respawn much more quickly than those that complete a full parental cycle. In two species of mouthbrooding cichlid, respawning occurred 33% and 38% sooner when a brood was removed from a female compared with those that completed care (Smith & Haley, 1987; Smith & Wootton, 1994). In a substrate-brooding cichlid species, the interval was 45% sooner (Smith & Wootton, 1995b). In biparental species, which are frequently monogamous over several spawning cycles, constraints on female respawning rate also represent a cost to the male. This constraint on males can be overcome by abandoning the female or 'bigamous' behaviour, whereby they pair with more than one female and provide care to two broods of young simultaneously (Smith & Wootton, 1995a).

Parental care can also impose a survival cost on a parent. This cost may come from an elevated risk of predation associated with care. In *G. aculeatus*, the bright colouration of males may put them at greater risk of detection by predators (Semler, 1971). Parental males are also belligerent in defence of their nests and place themselves at risk of predation (Pressley, 1981). Parental care can also increase risk of death through greater energy expenditure and injury during fights with rivals. In the bullhead (*Cottus gobio*, Cottidae), parental males suffered a 10-fold increase in mortality rate during the breeding season, accompanied by a substantial decline in body condition (Marconato & Bisazza, 1988). Male *G. aculeatus* that successfully reproduced suffered a 35–40% higher mortality rate in comparison with males that failed to breed in a natural population (DeFraipont *et al.*, 1992).

The spawning costs of migration in salmonids can be substantial (Chapter 7); in some cases fish traverse several thousand kilometres from feeding to spawning grounds, which is energetically expensive and greatly elevates their risk of predation (Hvidsten & Lund, 1988; Jonsson, 1991; Jonsson *et al.*, 1997; Bohlin *et al.*, 2001).

Reproductive effort

Reproductive effort is a second concept that has been important in the development of life-history theory. Intuitively, reproductive effort is the proportion of available resources that are allocated to reproduction (Fig. 11.6). If the reproductive effort at age x, E_x, is zero,

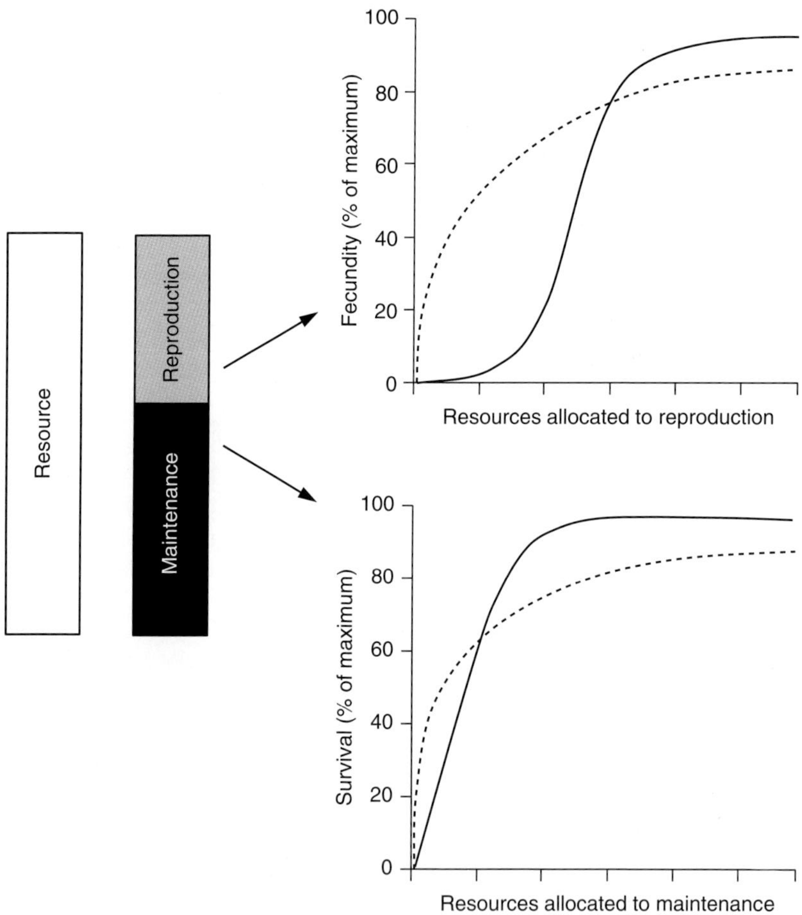

Figure 11.6 Intra-individual trade-off in reproductive effort, survival and fecundity. For a given level of resources (white area), an individual allocates a proportion to its reproduction, termed its reproductive effort. Increasing reproductive effort increases the resources put into reproduction (grey area), while limiting resources available for the individual's somatic maintenance (black area), potentially affecting its survival. The trade-off between fecundity and survival is dependent on the allocation of resource to reproductive effort and maintenance costs, and these may show a variety of non-linear responses. (Source: Brommer 2000. Reproduced with permission of John Wiley & Sons.)

then no reproduction takes place. At the other extreme, if all the available resources are allocated to reproduction, then there can be no post-reproductive survival. Thus, age-specific fecundity, m_x, will vary from zero to a maximum, as E_x increases up to a maximum. If the concept of trade-offs is combined with that of reproductive effort, then trade-off curves between age-specific fecundity and age-specific survival, l_x, can be described. Analysis has shown that the shape of such trade-off curves is important in defining the structure of life histories. An empirical definition of reproductive effort is the proportion of an individual's energy budget that is allocated to reproduction over a meaningful, biological timescale (Hutchings, 2002). For example, age-specific reproductive effort, E_x, is the proportion of energy allocated to reproduction at age *x*.

Reproductive effort, as a proportion of the total energy budget of an individual, may vary among individuals depending on age, sex or condition. However, accurately measuring reproductive effort is not straightforward. One approach has been to use the gonadosomatic index (GSI) as a surrogate for reproductive effort. The GSI is the relative size of the gonads as a proportion of either total body mass, $GSI = M_G/M_T$, or body mass excluding the gonads, $GSI = M_G/M_S$, where M_G is the mass of the gonads, M_S is the somatic weight (i.e. the body mass excluding the gonads) and M_T is the total body mass. An example of the use of GSI as a surrogate for reproductive effort came in a survey of 20 northern temperate marine species (Gunderson & Dygert, 1988). There was a high, positive correlation between GSI, calculated for wet mass, and instantaneous natural mortality, *M*, suggesting a trade-off between allocation to reproduction and post-breeding survival.

The use of GSI is overly simplistic because it ignores energy turnover, as do other physiological measures of reproductive effort (Hirschfield & Tinkle, 1975; Stearns, 1992). Energy expenditure on reproduction is the most meaningful measure of reproductive effort, and there are a number of techniques that can be applied to quantify energy expended on reproductive activities (Pianka & Parker, 1975; Calow, 1979). For example, the energy expended on parental behaviour by male *G. aculeatus* was measured by respirometry and ration manipulation (Smith & Wootton, 1999). Respirometry was used to measure the rate of oxygen consumption while males performed parental care behaviour, with oxygen consumption converted to a rate of energy expenditure using an oxycalorific equivalent. Manipulation of an experimental ration was used to obtain an estimate of the amount of food required by a male to support parental behaviour when fish weight was constant. The energetic value of the food fed to males was known, which meant the rate of energy expenditure on parental care could be estimated. However, reproductive effort on its own does not provide a measure of reproductive costs, which are quantified in terms of an individual's current and future reproductive success. Reproductive effort does, though, provide insights into how constraints on life-history evolution arise (Stearns, 1992). Thus, while difficult to measure, reproductive effort is a key variable that reflects how selection has shaped the way resources are allocated among the components of an individual's energy budget.

Demographic models of life-history evolution

Life histories can be seen as processes by which individuals in a population vary their age-specific expenditure of reproductive effort, E_x, in response to changes in age-specific survival, l_x, and age-specific fecundity, m_x, induced by physiological, environmental, ecological and genetic changes (Hutchings, 2002).

Demographic models seek to predict life-history patterns that evolve in environments that favour particular age-specific schedules of mortality, growth and fecundity.

The 'environment' comprises all biotic and abiotic conditions or challenges that impinge on these demographic variables. An assumption of demographic models of life-history evolution is that fitness is maximised at some optimal schedule of reproduction with age, with less optimal life histories assumed to be less successful and natural selection eroding genetic variance for life-history traits that do not generate the optimum life-history strategy in a given environment.

Measures of fitness

The concept of fitness is central to evolutionary biology, but a definition that is suitable for quantifying fitness empirically, an operational definition, has proved more difficult (Brommer, 2000; Roff, 2008).

A central concept in demography is the intrinsic rate of natural increase, r_m, which is defined by the Euler–Lotka equation:

$$1 = \sum_{x=\alpha}^{\infty} \exp(-r_m x) l_x m_x$$

where α is age at maturity, l_x is the probability of surviving from birth to the beginning of age class x, and m_x is the expected number of offspring for a female in age class x. For practical purposes, the variables l_x and m_x are estimated from population studies by measuring how many offspring are produced in each age class, x, and their subsequent survival from one age class to the next. The intrinsic rate of increase, r_m, is the per capita rate of increase, dN/Ndt, where N is population density and the rate of increase is not limited by the population's own density. Intrinsic rate of natural increase is a commonly used measure of fitness.

If the population is numerically stable, it is at equilibrium and $r_m = 0$, then the Euler–Lotka equation reduces to the net reproductive rate, R_o, defined as $\Sigma l_x m_x$, which is also used as a measure of fitness for populations at, or close to, equilibrium. These two measures of fitness are related because, for an equilibrium population, $R_o = 1$ and $r_m = 0$, so $R_o - 1 = r_m$.

Reproductive value, V_x, is a third demographic parameter that has been important in the development of demographic models of life-history evolution. It essentially measures future population growth that is a consequence of reproduction at age x compared with that of a newborn individual, V_0, where $V_0 = 1$. At a given age x, the reproductive value V_x can be partitioned into two components, so $V_x = m_x + \text{RRV}$, where m_x is fecundity at age x. The residual reproductive value (RRV) is the expected reproductive output of an individual aged x if it survives its current reproductive attempt, defined as $p_x V_{x+1}/V_0$, where p_x is the probability of surviving to age x. The intrinsic rate of natural increase is maximised if, at each age x, the quantity $m_x + \text{RRV}$ is maximised.

Both r_m and R_0 provide an index of fitness, but they may predict contradictory outcomes for the same life-history model. This complication arises when density dependence impinges on population growth, when populations are polymorphic with more than one life-history strategy, and when populations experience environmental variation. A key question yet to be addressed is to understand under what realistic biological conditions r_m and R_o fail as measures of fitness. A more general measure of fitness, but more difficult to use in developing models, is the invasion exponent (Roff, 2008). This derives from the concept of the evolutionarily stable strategy (ESS) and describes the ability of a mutant adopting a novel life-history strategy to invade an established population.

A limitation of demographic models based on the Euler–Lotka equation is that it assumes that the population has a stable age distribution and is very large. The assumption of population stability will rarely be met; many fish populations are characterised by strong variation in recruitment among years (Wootton, 1998). Despite this assumption, the model is not wholly unrealistic, as the Euler–Lotka equation can also describe the spread of an allele in a population, rather than population growth, and modest deviations from stability will not qualitatively alter predictions (Stearns, 1992).

Another feature of demographic models of life-history evolution is an assumption that the parameters of the model, specifically survival and fecundity, are age-specific. While age may play a role, size rather than age is the chief determinant of survival and fecundity in fishes. Survival, growth and fecundity are also likely to be sensitive to variation in physical and physiological state. Attempts have been made to characterise individuals by body condition rather than age in demographic models of life-history evolution (Caswell, 1989), and models of this type can contradict conventional life-history predictions (McNamara *et al.*, 2009). The challenge in this context is to design empirical studies to examine the role of condition rather than age in life-history evolution, as well as the mechanistic basis to life-history trade-offs.

Predictions of demographic models

Demographic models that assume a constant environment predict that when juvenile mortality is high compared with adult mortality, selection favours the characteristics of an iteroparous life history. The traits are delayed maturity at a large body size, a long life that contains several breeding episodes and a moderate reproductive allocation at each breeding episode. If adult mortality is high compared with juvenile mortality, selection favours the characteristics of a semelparous life history or an abbreviated iteroparous life history. This implies no (semelparity) or limited (abbreviated iteroparity) post-breeding survival. At each breeding episode, reproductive allocation is high.

Demographic models developed assuming a constant environment need to be modified when the environment varies unpredictably over time in ways that affect the age-specific rates of survival, fecundity and growth. In such a stochastic environment, the intrinsic rate of increase is related to the deterministic rate of increase by the relationship: $r_s = r_d - 1/2(\sigma_r^2)$, where r_s is the intrinsic rate of increase in a stochastic environment, r_d is the rate of increase in the constant environment, and σ_r^2 is the variance of *r* over time. Stochastic estimates of *r* are always less than the deterministic estimates (Hutchings, 2002; Roff, 2002). In a stochastic environment, selection should act to reduce the variance in individual fitness over generations, even if this means the individual fitness within a generation is smaller than it would be in a stable environment. The strategy is to spread the reproductive risk across time through traits, such as several spawning events in a breeding season and several reproductive seasons over a lifetime. Such a strategy minimises the risk that no progeny will be produced because of adverse conditions at critical times and is often called a 'bet-hedging strategy' (Hutchings, 2002; Roff, 2002).

A study of the Tasmanian pygmy perch (*Nannoperca australis*, Percichthyidae), a freshwater teleost from south-eastern Australia, provides a potential example of bet hedging (Morrongiello *et al.*, 2012). The species occupies a range of habitats, which differ in their exposure to drought conditions, with the streams more susceptible to drought in more variable and unpredictable environments. Female *N. australis* living in unpredictable environments produced eggs of more variable size than females living in more stable environments. This was interpreted as a bet-hedging strategy by females in the less predictable habitats in which there is uncertainty about the optimal egg size. Such females diversify the range of egg sizes they produce. An alternative, non-adaptive hypothesis is that in the less predictable environments, females have poorer control over the bestowal of yolk that each oocyte receives.

Demographic models predict that the effect of stochastic environmental variability on the evolution of life-history traits depends on the life-history stage most vulnerable to the environmental variation. When the variability causes wide fluctuations in the survival of juveniles, selection favours long life, late maturity and many breeding episodes in a lifetime. When variability causes wide variations in adult survival, selection favours early reproduction, high fecundity and only a few reproductive episodes in a lifetime. At the extreme, this leads to semelparity, with only a single breeding episode.

Thus, demographic models of life-history evolution predict that if juvenile mortality is high or variable compared with adult mortality, traits characteristic of iteroparity are favoured. When adult mortality is high or variable compared with juvenile mortality, traits characteristic of semelparity are favoured.

Empirical studies of life-history evolution

Early, but influential, models have demonstrated the strength of using demographic principles that incorporate, either explicitly or implicitly, a cost of reproduction.

An evolutionary problem is why, amongst the salmonids, most *Oncorhynchus* spp. of the northern Pacific Ocean are semelparous, while the *Salmo* spp. of the northern Atlantic are iteroparous. A model based on salmonid life history asserted that at each age class, x, the value m_x+RRV was maximised, thus maximising r_m (Schaffer, 1974a,b, 1979). The model assumed that there were trade-offs between current reproduction, m_x and survival from the age x to x_{+1}, p_x, and growth between x and x_{+1}, g_x. In the relationship between m_x and p_x, the trade-off was assumed to take one of two forms, convex upwards or concave (Fig. 11.7). If the trade-off curve was convex, then intermediate values of m_x and p_x were optimal, maximising m_x+RRV, defining an iteroparous life history. At a given age x, the concave curve predicted that m_x+RRV was maximised if either m_x was zero, indicating no reproduction, but maximising p_x, or m_x was maximised and there was no post-breeding survival, i.e. $p_x = 0$. The latter is the semelparous solution. If the reproductive biology of *Oncorhynchus* spp. means that the chances of post-breeding survival are extremely low, then the best solution is to maximise fecundity and use no body reserves to support post-breeding survival, i.e. a semelparous life history.

There is also the possibility that the shape of the trade-off curve between m_x and p_x is more complex. Then it is possible to have both iteroparous and semelparous solutions. This illustrates the importance of determining the shape of the trade-off curve in real populations, a goal that has proved difficult to achieve.

The development of demographic models for populations in which density-dependent effects are small led to clear predictions for the age at maturity and reproductive allocation when age-specific mortality rates change, e.g. from the effects of predation. These predictions were tested in a brilliant series of field and experimental studies on populations of the viviparous *P. reticulata* on the Caribbean island of Trinidad. The focus was on the response to selection imposed by size-specific predators. In some sites, the chief predator is the millet (*Crenicichla alta*, Cichlidae), a predatory pike cichlid that feeds on sexually mature *P. reticulata*. At other sites, predation comes mainly from Hart's rivulus (*Anablepsoides hartii*, Rivulidae), a toothcarp that is a predator of juvenile *P. reticulata*. High predation of adult *P. reticulata* at *C. alta* sites results in sexual maturation at a smaller size, with increased reproductive allotment producing a greater

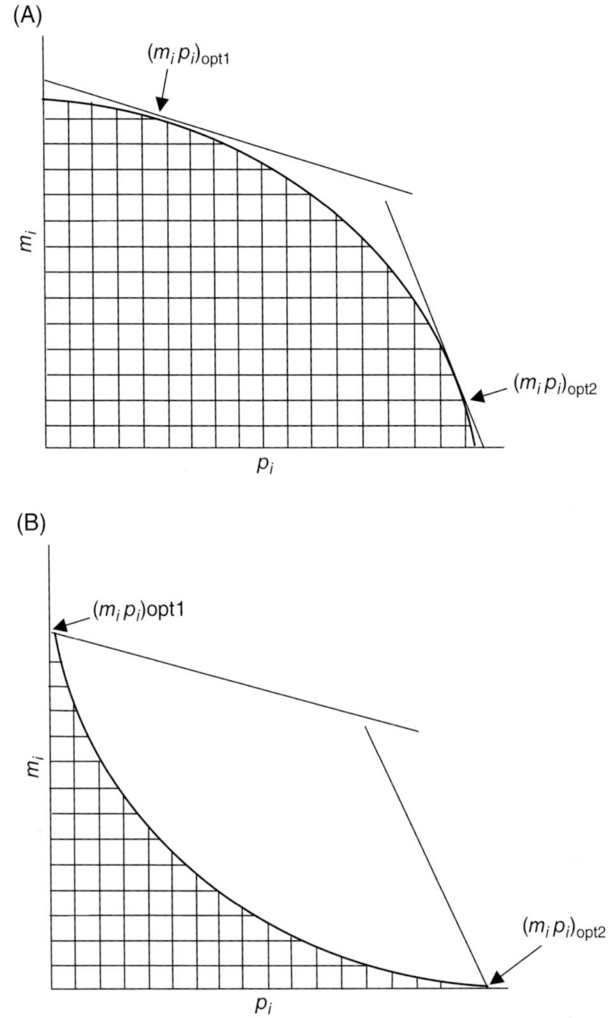

Figure 11.7 Trade-off curves between fecundity at age i, m_i, and the probability of surviving from age i to $i+1$, p_i. (A) Convex relationship, for which optimal combination of fecundity and survival are intermediate values of m_i and P_i (iteroparity). Two examples are illustrated, one with high fecundity and poor post-breeding survival $[(m_i p_i)_{opt1}]$ and one with low fecundity and high post-breeding survival $[(m_i p_i)_{opt2}]$. (B) Concave trade-off curve for which optimal combinations are either no reproduction at age i, so $m_i = 0$, or no post-breeding survival, p_i (semelparity). Straight lines define combinations of m_i and p_i, which yield the same intrinsic rate of increase. The hatched areas are the combinations of m_i and p_i that are physiologically possible. (Source: Wootton 1998. Reproduced with permission of Elsevier.)

number of smaller offspring. In accordance with the predictions of life-history theory, the opposite response to predation of young stages was observed at *A. hartii* sites (Reznick & Endler, 1982; Reznick *et al.*, 1990). When *P. reticulata* from *C. alta* sites were experimentally transferred to *A. hartii* sites, where adults were exposed to limited predation, the transferred populations showed shifts in life-history traits that were characteristic of *A. hartii* populations, such as an increase in size at maturation and offspring size. Controlled experimental laboratory studies over two generations showed that these changes in life-history traits were heritable.

Although these studies provided support for the life-history theory, there was one anomalous result. When mortality rates in natural populations were estimated, it was confirmed that the mortality rates were higher at the *C. alta* sites, but the expected differences in size-related mortality were not found (Reznick *et al.*, 1996b). The mortality rates were higher at the *C. alta* sites, but the difference was similar for all size classes of *P. reticulata*. This suggests other factors are playing a role in the evolutionary response of *P. reticulata* to a change in the predation regime.

A potentially relevant factor is density dependence. The demographic models of life-history evolution that assume selection tends to maximise r_m also assume that there are no density-dependent effects acting on the population. A change in predation regime is likely to change the density of the prey population. An increase in predation will tend to reduce the prey population, but this will make more resources, such as food, available to the survivors, supporting faster growth rates. Subsequent studies confirmed that at least some *P. reticulata* populations do experience density dependence, with high density resulting in reduced offspring size, increased adult mortality and reduced fat storage by females (Reznick *et al.*, 2012). Predation, in addition to any direct selective effects it has, can have indirect effects on life-history traits, because of its effects on resource levels mediated by density.

Interestingly, these experimental introductions of *P. reticulata* also imposed selection on *A. hartii*, because adult guppies feed on juveniles of this species, with *P. reticulata* predation expected to select for evolution of delayed maturity and reduced reproductive allotment. However, *P. reticulata* predation also had the indirect effect of substantially reducing *A. hartii* densities by 42%, so the *A. hartii* experienced enhanced growth rates through reduced competition (Walsh & Reznick, 2010; Walsh *et al.*, 2011). The outcome for *A. hartii* life-history evolution was earlier maturation, greater reproductive allotment, higher fecundity and significantly smaller eggs than fish from control sites within a 25-year period. If predation alone were acting on *A. hartii* life-history evolution, the prediction was for delayed maturation and reduced reproductive investment. This alternative life-history response demonstrates how the interaction of two ecological variables, in this case predation and density, with concomitant effects on growth rate, can substantially alter predicted outcomes of life-history evolution.

In *Nothobranchius* spp. killifish, which are adapted to living in seasonal water bodies in dry and semi-arid African environments, all adults die when their habitat desiccates, with partially developed embryos surviving in a developmental diapause until the next rainy season (Wourms, 1972). Age at maturity corresponds with degree of aridity; at sites with a high degree of aridity, sexual maturity is achieved more rapidly than at sites where the habitat is less ephemeral. These patterns are found both within and among species (Reichard *et al.*, 2009b; Polačik *et al.*, 2011; Terzibasi Tozzini *et al.*, 2013).

Habitat invasion

The invasion of a new habitat can result in the rapid evolution of life-history traits. The freshwater vendace (*Coregonus albula*, Salmonidae) invaded the sub-arctic Pasvik watercourse during the 1980s (Bøhn & Amundsen, 2001). This species is an efficient zooplanktivore and excluded a native European whitefish (*Coregonus lavaretus*, Salmonidae) from the watercourse. During the early phase of its invasion, while still at low population density, the prediction was that *C. albula* would exhibit high reproductive allotment, early maturation, and small but abundant eggs. In the later stages of the invasion, when the population density was higher, reduced reproductive allotment, later maturation and fewer larger eggs were predicted. *C. albula* established in the upper reaches of the Pasvik system first and later spread to the lower reaches. A 10-year study examined life-history traits of *C. albula* in the two sections and showed a distinct pattern of size at maturation, fecundity and egg size, which matched predictions (Fig. 11.8). These changes in *C. albula* life-history variables were rapid and, while they

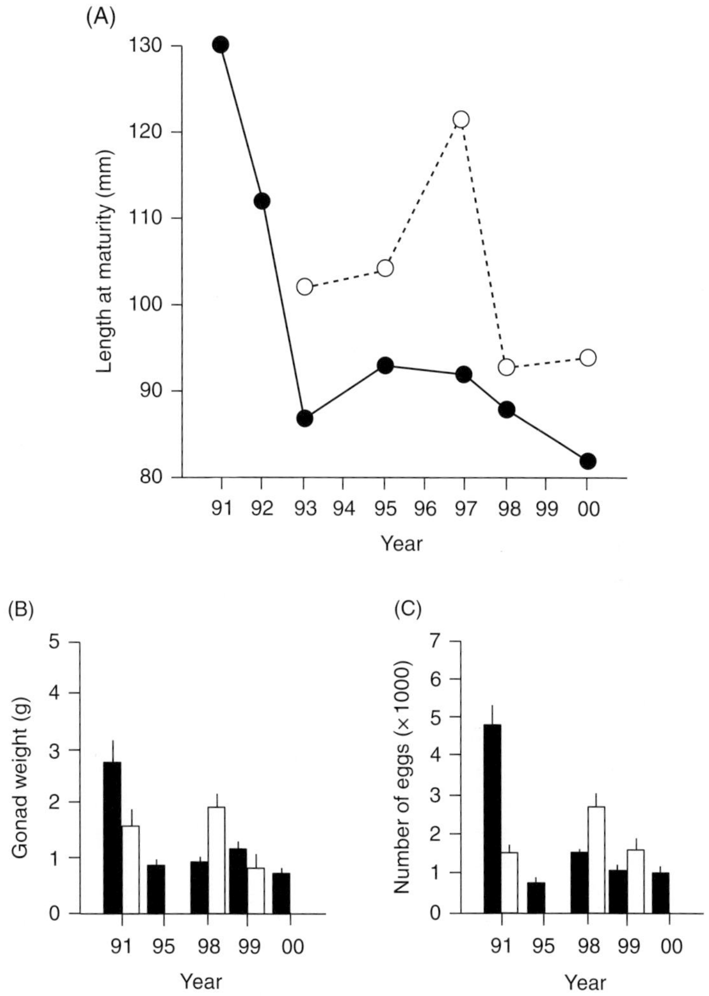

Figure 11.8 Length at first reproduction of invasive vendace (*Coregonus albula*) in the Pasvik watercourse. In the early stages of invasion, vendace showed high reproductive allotment. In the later stages of invasion, reproductive allotment decreased. (A) Length at maturation (solid line with black circles, late stages of invasion, dashed line and open circles, early stages of invasion). (B) Mean (+SE) female gonad weight (black bars, late stages of invasion; white bars, early stages of invasion). (C) Mean (+SE) fecundity (black bars, late stages of invasion; white bars, early stages of invasion). (Source: Adapted from Bøhn *et al.*, 2004. Reproduced with permission of John Wiley & Sons.)

may have arisen through selection, the observed changes might also have been the result of phenotypic plasticity.

Environmental gradients

Local adaptations to environmental gradients provide further empirical evidence for life-history responses to selection. In populations of *S. salar* that display alternative mating strategies, the anadromous phenotype appears at a lower frequency in the upper reaches of rivers, with the frequency of the resident phenotype correlating with altitude (Baum *et al.*, 2004). A comparable relationship is obtained with distance upstream (Aubin-Horth *et al.*, 2006). The prevalence of resident phenotypes is associated with headwater spawning sites in Chinook salmon (*Oncorhynchus tshawytscha*, Salmonidae) (Healy, 1991), Dolly Varden (*Salvelinus malma*) (Armstrong & Morrow, 1980), *S. fontinalis* (Curry, 2005) and rainbow trout (*Oncorhynchus mykiss*, Salmonidae) (Narum *et al.*, 2008). The decline in anadromous phenotypes with increasing altitude and distance upstream is associated with migration costs selecting against anadromy (Dodson *et al.*, 2013).

Latitudinal gradients are also associated with predictable patterns in the frequency of alternative male mating strategies in salmonids. Anadromy is associated with high latitudes, while residency is associated with lower latitude sites at high altitude. This pattern of distribution of male phenotypes appears to be driven by water temperature, with critically high water temperatures limiting anadromy and confining the resident phenotype to low-temperature refuges in river headwaters (Dodson *et al.*, 2013).

In the anadromous American shad (*Alosa sapidissima*, Clupeidae) on the Atlantic coast of the USA, the proportion of iteroparous repeat spawners increases with latitude. In the north of their range in New Brunswick, the proportion of repeat spawners is high, while in the south of their range in Florida the life history is characterised by semelparity (Leggett & Carscadden, 1978). This trend in life history is explained by the thermal environment of their natal rivers. In the north of their distribution, the environment is relatively harsh and unpredictable, and spawning success and recruitment are variable. In the south of their range, temperatures are more favourable for egg, larval and juvenile survival. Life-history theory predicts that in southern rivers selection will favour earlier age at maturity and greater reproductive allotment, which manifests itself as semelparity.

Temperature is also an important explanatory factor for the latitudinal patterns shown by two European freshwater fishes, the European perch (*Perca fluviatilis*, Percidae) and the roach (*Rutilus rutilus*, Cyprinidae) (Heibo *et al.*, 2005; Lappalainen *et al.*, 2008). These two species are widely distributed in Europe and both have a broad latitudinal range from the warm temperate to the sub-arctic zones. This latitudinal range means that populations experience different mean temperatures. This has effects on growth rates because teleosts are ectothermic.

In *P. fluviatilis*, growth rates decrease with latitude, as do both juvenile and adult mortality, so maximum age increases with latitude. Reproductive allocation as measured by GSI decreases with latitude, while age and length at maturity increase with latitude. Age at maturity is positively correlated with the ratio between juvenile and adult maturity. In *R. rutilus*, growth, fecundity and reproductive lifespan are related to latitude, with the decrease in fecundity with latitude being compensated by the increase in reproductive lifespan.

With the low or nonexistent gene flow between many freshwater populations, each population provides an independent picture of the effect of factors such as temperature on the evolution of life-history traits. Phylogenetically, *P. fluviatilis* and *R. rutilus* are not closely related, so the similarity of their responses to the latitudinal gradient indicates a common selective response to the environmental conditions. The responses are also compatible with the predictions of life-history theory.

Habitat templet model

Demographic models are primarily focused on predicting the effect of environmental changes on life-history traits in populations. A different level of analysis is the habitat templet and related models (Southwood, 1977). These built on the concept of *r*- and *K*-selection (MacArthur & Wilson, 1967). The habitat templet model assumes that spatial and temporal heterogeneity in the physical and biological environment permits habitats to be classed as distinct categories, which provide a templet for life histories. The archaic term 'templet', rather than 'template', infers flexibility in these habitat categories, with an organism occupying a particular habitat category capable of modifying its environment within broad limits (Southwood, 1988). The spatial heterogeneity of habitats is considered as patches of habitat that provide a favourable environment. Temporal heterogeneity arises from the rate and magnitude of change in conditions in a habitat over time.

Southwood (1988) recognised three habitat axes that permitted differentiation of life histories into distinct categories within the three-dimensional space these axes generate. The first templet axis is 'disturbance', which reflects the temporal stability of a habitat. Disturbance considers the rate at which habitat change proceeds in relation to the habitat requirements of the organism of interest, particularly its generation time. The second axis is 'adversity' and reflects the degree of (un)favourability of the environment to the organism attempting to occupy it. Productivity is an approximation of adversity, but other factors, such as salinity, the presence of toxins, dissolved oxygen concentration or temperature, may also contribute to adversity. The final dimension of the model is the 'biotic interactions' axis. This axis does not define the habitat in abiotic terms, but instead describes the impact of the biotic environment on an organism, such as predation, parasitism and intra- and interspecific competition and trophic complexity.

A modification of this habitat templet model for riverine systems emphasised two axes: spatial variability and temporal variability, reflecting the characteristics of rivers (Townsend & Hildrew, 1994). An attempt to apply this model to the teleost fauna of the upper River Rhône in France failed to find clear associations between species life-history traits and habitat utilisation (Persat *et al.*, 1994). It was not clear whether this lack of correspondence between species traits and habitats represented a failure of the model or the particular circumstances of the upper Rhône. In a completely different environment, the Scotian Shelf and Bay of Funday in the North-west Atlantic, two habitat axes were identified: natural disturbance, a characteristic of the sea floor, and scope for growth, an estimate of local energy available for growth and reproduction (Fisher *et al.*, 2011). As in the Rhône, no clear relationship was identified between the habitat axes and the life-history traits of the species living in the geographical area.

Although the verbal arguments used in templet models often seem persuasive, their empirical success when applied to teleosts has been limited. The challenge in using the templet model to understand life-history evolution is in demonstrating adaptive responses to habitat features conforming to axes that have been defined *a priori*.

An alternative but related approach uses a statistical method to cluster species that have similar life-history traits and, if clusters were identified, to relate such clusters to

Table 11.2 Winemiller's triangular demographic life-history model.

Life-history strategy	Typical life-history traits	Habitat characteristics
Opportunistic	Short generation length Low age-specific fecundity Low age-specific survival	Frequent or stochastic change on small temporal and spatial scales
Equilibrium	Long generation length Low age-specific fecundity High age-specific survival	Low variation in quality of the environment Strong direct and indirect biotic interactions
Periodic	Long generation length High age-specific fecundity Low age-specific survival	Large-scale cyclical or spatial variation

particular types of environment. An analysis of 216 North American marine and freshwater teleosts belonging to 57 families used principal components analysis to group species on the basis of 16 life-history traits (Winemiller & Rose, 1992). The study identified a primary association among large adult size, late maturation, greater longevity, high fecundity, small egg size, a short spawning season and few reproductive bouts. An additional association linked parental care with larger eggs, an extended spawning season and multiple reproductive bouts. When species were separated on the basis of both these sets of association, three distinct syndromes of life-history traits emerged. These were designated as 'periodic', 'opportunistic' and 'equilibrium' (Table 11.2). The periodic strategy is typical of many fishes and predicts a trend of delaying maturation to enable a larger size to be achieved. A large body size permits the production of large clutches, but also buffers the individual against periodic periods of suboptimal environmental conditions when food availability is restricted or physical conditions depart from an optimum, such as winter or dry seasons. The opportunistic life-history strategy is associated with early maturation, small body size, frequent reproduction and an extended reproductive season. These traits result in rapid population turnover and generate high rates of population increase. Examples include killifishes and other species that occupy constantly changing habitats in which adults experience high mortalities. The equilibrium life-history strategy is associated with adaptation to habitats that are resource-limited and where density-dependent competition may be important. This syndrome is characterised by late maturation, parental care and relatively small clutches, and is typified by many cichlids. While many species fall into these distinct syndromes, others occupy intermediate positions on the life-history adaptive surface (Fig. 11.9). For example, medium-sized fishes, such as sunfishes (Centrarchidae), show seasonal reproduction, moderately large clutches and parental care. Other studies have suggested that the salmonids form a separate cluster, implying that there have been phylogenetic constraints on the evolution of salmonid life histories (Wootton, 1984b; Teletchea *et al.*, 2009).

A study of West African tropical rivers suggested that Winemiller's model is relevant for habitats not included in the original analysis (Tedesco *et al.*, 2008). The species distributions reflected hydrological seasonality. For rivers that had drainages characterised by a short favourable period for reproduction, the teleost fauna was dominated by species that had a periodic strategy. Species with an equilibrium strategy were associated with drainages with more than 6 months of wet season. The

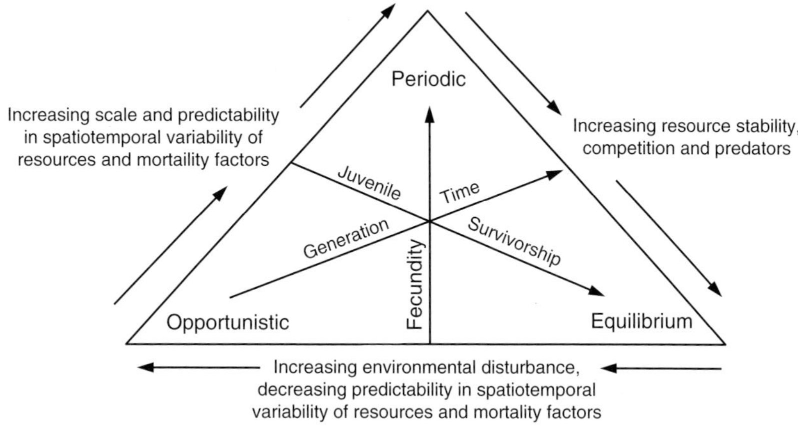

Figure 11.9 Winemiller's triangular demographic life-history model. Environmental gradients select for life-history strategies that are characterised by optimisation of the demographic parameters of generation time, age-specific survival and age-specific fecundity, thereby generating 'opportunistic', 'periodic' and 'equilibrium' life-history syndromes. (Source: Winemiller 2005. Reproduced with permission of Canadian Science Publishing.)

habitat acts as filter, selecting reproductive strategies, although most habitats do contain a mixture of strategies, but with different proportions of periodic, opportunistic and equilibrium species. In a sense, life histories act as strategies for sampling suitable habitats (Wootton, 1994b).

A simplified form of the model was used to relate life-history strategies to population dynamics in species living in the arctic Barents Sea (Bjørkvoll *et al.*, 2012). Species were classified along a 'slow–fast' continuum of variations in life history. 'Slow' species had a long life expectancy but low annual reproduction rates, while 'fast' species had an early age of maturity and a relatively short lifespan. In the Barents Sea, the fast species were pelagic, such as the capelin (*Mallotus villosus*, Osmeridae), while the slow species were deep-water species such as rockfish (*Sebastes* spp., Sebastidae). This classification clearly relates to the 'periodic' and 'opportunistic' strategies identified in the original Winemiller & Rose (1992) model.

A disadvantage of the templet approach is that, if other life-history traits are used in the clustering analysis, a different set of clusters is extracted from the data (Teletchea *et al.*, 2009). An analysis of the reproductive traits of 65 freshwater teleost species from the temperate Northern Hemisphere used a largely different set of variables from that used by Winemiller & Rose (1992). Both analyses identified the salmonid cluster, but otherwise the classification of the freshwater species differed from the Winemiller & Rose (1992) model. Ten reasonably homogeneous clusters of species were identified by the cluster analysis, largely on the basis of when in the year spawning took place and whether parental care was provided.

A benefit of templet and related models is that species can be assigned to specific life-history syndromes on the basis of readily measured traits, which can have useful application, e.g. in fisheries management (Winemiller, 2005). A drawback of this approach is that, like other empirical models, although the patterns of life-history traits may reflect fundamental biological processes, the model does not assist our understanding in why these distinct syndromes arise.

Dimensionless variables and life-history invariants

At an even higher level of generality are the attempts to identify assembly rules for life histories, which have broad application (Charnov *et al.*, 2013). These rules refer to relationships among mortality rate, growth rate, size and age at maturation, and reproductive allocation after maturation. It is assumed that natural selection favours some special relationships among these traits (Winemiller & Rose, 1992; Charnov, 1993). This approach has three elements: the choice of a suitable growth model, the identification of dimensionless numbers, and life-history invariants.

The growth model most frequently used for teleosts is the well-established von Bertalanffy growth model (VBGM; Wootton, 1998). This takes the following form:

$$L_t = L_\infty\left[1-\exp(-kt)\right] \text{ for growth in length}$$

$$W_t = W_\infty\left[1-\exp(-kt)\right]^3 \text{ for growth in mass}$$

where t is age, L_t and W_t are length and mass at age t, respectively, L_∞ and W_∞ are asymptotic length and mass, i.e. the hypothetical maximum size, and k is a growth coefficient. The two parameters that determine the shape of the growth curve are k and the asymptotic size.

Dimensionless ratios were first identified during a search for a method of estimating natural mortality in a population (M), which is usually difficult to measure directly, from other more easily estimated values (Beverton & Holt, 1959). An example of a dimensionless variable is the ratio of natural mortality (M) to the VBGM growth rate, k. Both M and k have the dimension of per unit time, so the ratio (k/M) is dimensionless. Quantitative studies suggest that this ratio is similar across teleost species, so M could be estimated from k. Ratios that take similar values across species are identified as being invariant. Other life-history invariants that have been recognised include L_d/L_∞ and aM, where a is the age at maturity. The existence of life-history invariants is assumed to reflect some fundamental and universal constraint on life-history evolution (Charnov, 1993, 1997).

The VBGM was used to predict age at maturity, a, for a range of teleost species, assuming that there was a mortality cost of reproduction and that net reproductive rate, R_0, was a suitable measure of fitness (Roff, 1984). There was a high correlation between the predicted and observed values of a for a sample of 32 species. Further development of this model suggested that predictions of a and reproductive allocation depended on the joint action of trade-offs between reproductive allocation and growth and between reproductive allocation and mortality (Roff *et al.*, 2006).

The use of dimensionless ratios and life-history invariants has been further elaborated to identify assembly rules for vertebrates (Charnov *et al.*, 2001, 2007, 2013). An analysis that used data from ectothermic lizards and endothermic mammals, but not teleosts, defined reproductive effort (RE) as the mass allocated to reproduction per unit time as a proportion of adult mass and lifetime reproductive effort (LRE) as the product RE$\times E$, where E is the average adult lifespan. This product is a dimensionless variable. The data from lizards and mammals suggest that a female can only produce a mass of offspring approximately equal to 1.4 times her own body mass in a lifetime. Whether this generalisation can be applied to teleosts has still to be determined. If the generalisation is valid and does apply to teleosts, it implies an underlying set of constraints on the reproduction of vertebrates.

Further development of models based on invariants suggested that much of the variation in life-history traits between teleost species can be related to two variables.

The first variable is the production rates of individuals (*A*) and the second is the extrinsic mortality rate around the age of maturity (M_a). Extrinsic mortality is that imposed by the external environment. The analysis suggested that species can only grow to a large size if external mortality rates are low and so the lifespan is long. Reproductive effort can be low. When external mortality is high, only a small body size can be achieved and a high reproductive effort is required. A third relevant axis is egg or progeny size, which is determined by how reproductive allocation is divided between eggs or progeny, but it was not included in this analysis. The analysis does provide an alternative method of classifying teleost life histories.

Life-history invariants are derived from empirical observations of life history, and often show strikingly high correlations, which is unusual in evolutionary and ecological studies. Consequently, life-history invariants have been assumed to reliably reflect some fundamental and universal constraints on life-history evolution (Charnov, 1993, 1997). Such invariants are typically identified by plotting the two traits against each other, with both on a logarithmic scale. If the relationship is invariant, the plot has a slope of 1 and an intercept of 0.

A difficulty with the concept of life-history invariants comes from the observation that while a plot of the logarithm of one life-history trait against another often generates a linear relationship, this relationship does not imply invariance and could result from relationships unrelated to a fixed invariant ratio (Nee *et al.*, 2005). Indeed, invariant correlations are anticipated even from randomly generated data drawn from a uniform distribution that are plotted on a log-log scale. The problem arises because one life-history trait is effectively being regressed on itself, and this effect is exaggerated when traits vary by orders of magnitude. For example, consider the life-history traits L_a (length at sexual maturity) and L_∞ (asymptotic length, the length a fish would achieve if it were to live for an infinite period) – a fish must achieve sexual maturity at a size smaller than its maximum possible length, and thus $L_a < L_\infty$. When this relationship is scaled up over several orders of magnitude, from the smallest goby species (which barely exceeds 10 mm) to the largest marlin (which may grow to over 5000 mm) and is plotted on a logarithmic scale, a strong positive relationship is inevitable. The implication is that dimensionless ratios may actually not be invariant (Nee *et al.*, 2005). Despite this, mean values for life-history traits can still vary among taxa in biologically interesting ways and might reflect differences in underlying life-history trade-offs.

Plasticity of life-history traits

The models of life-history evolution largely assume that the phenotypic changes predicted reflect underlying genetic changes. By contrast, phenotypic plasticity is the capacity of a single genotype to express a range of phenotypes in response to environmental variation (Fordyce, 2006). Variation in life-history traits among populations may partly or wholly reflect phenotypic plasticity rather than underlying genetic correlations (Reznick, 1985; Stearns & Koella, 1986; Stearns, 1989). As selection acts on the phenotypes of individuals, plasticity is significant in permitting rapid responses to temporal and spatial fluctuation in environmental conditions, and influences adaptive responses to short-term selection. Plasticity also facilitates dispersal and invasion of new environments, because adaptation to the new environment can occur within a generation and does not require genetic changes, which require inter-generational change.

Data from several studies suggest that phenotypic plasticity can explain substantial variation in teleost life-history traits. A small population of approximately 150 mosquitofish

(*Gambusia affinis*, Poeciliidae) was introduced to Hawaii in 1905. Approximately 150 generations later, Stearns (1983) investigated the life-history traits of populations that occupied reservoirs with either stable water levels or in which water levels rapidly fluctuated, resulting in elevated adult mortality rates. Reproductive allocation varied between reservoir types, with fish from fluctuating reservoirs exhibiting greater levels of reproductive allocation and larger body size. However, most of the variance in these variables was attributed to plastic responses to environmental change. Genetic differences among populations were attributed to founder effects and genetic drift.

A powerful approach to measuring the extent to which plastic response to changed environments can explain life-history variation is through the use of transplant and common-garden experiments. *C. gobio* has a wide distribution in western Europe; populations in northern England, occupying cold infertile upland streams, exhibited life-history traits that differed from populations in warmer and more productive chalk streams in southern England (Mann *et al.*, 1984). Females in northern populations spawned only once each year, showed delayed maturity and had a relatively long lifespan. Southern females spawned several times each year, matured earlier but had a shorter lifespan. Reciprocal transplants of *C. gobio* between northern and southern sites showed that fish adopted the reproductive characteristics of the local population almost entirely (Table 11.3), suggesting that much of the variance in life-history traits could be attributed to plastic responses to contrasting environmental conditions (Mann *et al.*, 1984).

In the livebearer *Brachyrhaphis rhabdophora* (Poeciliidae), populations in Costa Rica that co-occurred with predators achieved sexual maturity at a smaller size and produced a greater number of smaller offspring than populations that were predator-free (Johnson & Belk, 2001). This pattern was repeated among drainages. A common-garden experiment was used to examine whether these differences in life history were due to genetic differentiation among populations in response to selection from predators, or whether the pattern could be explained by phenotypic plasticity (Johnson, 2001). In this case, life-history traits persisted for at least two generations in common conditions in the laboratory, implying a significant genetic component to these reproductive traits.

Phenotypic plasticity can be described using reaction norms, which for each genotype comprise a plot of the value of a phenotypic trait in contrasting environments. If there is no phenotypic plasticity, the plot will show no trend, whereas phenotypic plasticity is indicated by a trend in the plot. An example would be the relationship between the age of maturity and temperature for a given genotype. While informative about phenotypic responses to environmental variation, reaction norms do not differentiate

Table 11.3 Impact of reciprocal transplants between northern and southern populations on the reproductive biology of the river bullhead (*Cottus gobio*). (Source: Data from Mann *et al.* 1984.)

Experimental site	**Northern**		**Southern**	
Source population	**Northern**	**Southern**	**Northern**	**Southern**
Spawning period	–	April/May	March/May	March/May
Female fecundity	–	91	712	1110
Egg batches per female	–	0.75	2.60	3.42
Eggs per batch	–	121	297	357
Eggs/g total weight	–	33	127	192
Mean egg dry weight (mg)	–	–	1.3	1.4

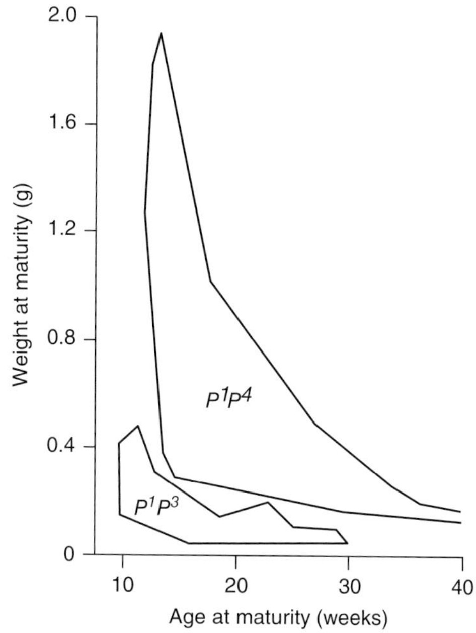

Figure 11.10 Bivariate plot of ages and weights at maturity for two genotypes (P^1P^3 and P^1P^4) of southern platyfish (*Xiphophorus maculatus*), reared over a range of ration levels. The range of values for each genotype is delineated. (Source: Adapted from McKenzie *et al.* 1983. Reproduced with permission of American Society of Ichthyologists and Herpetologists.)

the underlying mechanisms or consequences of phenotypic plasticity (Hutchings, 2011). Decision rules denote the mechanism of response to environmental variation. Decision rules and reaction norms are optimised when they yield the maximum fitness in a given environment.

Reaction norms can also be bivariate, as in the plot of age at maturity against size at maturity (Stearns & Koella, 1986). Such a bivariate plot can also reveal genetic differences in reaction norms. This is shown by the plot of age and size at maturity for two genotypes of the southern platyfish (*Xiphophorus maculatus*, Poeciliidae) (Fig. 11.10; McKenzie *et al.*, 1983). Such studies illustrate that reaction norms are genetically determined and so, in principle, are subject to selection (Roff, 2002; Hutchings, 2011)

Constraints on life-history evolution

The possible combinations of life-history traits is finite. No 'Darwinian demon', a fish that starts reproducing at an early age, with high age-specific fecundity and low age-specific mortality, resulting in an extended lifespan, exists. Yet such a life-history strategy would be favoured under selection, suggesting that there are constraints on life-history evolution. Gould & Lewontin (1979) drew attention to the problem of speculative adaptive accounts of optimality and this conclusion is strongly supported by life-history theory (Caswell 1989). The interaction of life-history traits will affect how selection acts, potentially producing responses to selection that are unexpected. Selection can only act on life-history traits within the constraints imposed by the genetic covariance among them, resulting in life-history trade-offs. Constraints can also arise from the incompatibilities of demographic and allometric traits. Thus, while an increase in clutch size might have a positive effect on fitness, the impact of increased clutch size on body shape may have a negative effect on female fitness by hampering locomotion.

Life-history evolution can also be constrained by phylogenetic history. Related taxa, e.g. the salmonids, tend to occupy similar ecological niches, presumably because their immediate ancestors occupied similar environments. Consequently related taxa tend to exhibit similar life histories, with the implication that similar selection pressures over the course of the evolutionary history of the group has limited the combinations of life-history traits that have evolved. Constraints might also arise because genetic variation for certain trait combinations does not exist within a particular taxonomic group. Comparative analysis of life history thus demands a phylogenetically based approach in order to control for the constraints imposed by evolutionary history (Harvey & Pagel, 1991; Ricklefs & Starck, 1996).

In the discussion of constraints, two meanings of constraint are used. The first is where constraint refers to a bias, in which evolution in a particular direction is impeded but not stopped. The second meaning refers to combinations of traits or evolutionary trajectories that cannot be achieved by the organisms (Roff & Fairburn, 2007). Care has to be taken to distinguish between these two meanings.

Alternative life-history strategies

A striking feature of some species is the expression of starkly different life histories in the same population. These mating polymorphisms represent alternative life-history strategies and have attracted great interest in behavioural and evolutionary biology (Taborsky, 2008) (Chapter 8). Alternative life-history strategies are typically associated with males, which may achieve sexual maturity early in life and at a smaller body size than the typical pattern. Males displaying an alternative mating strategy may express a distinct phenotype, with alternative mating phenotypes showing a discontinuous distribution in the population, and may show greater investment in testes and sperm production. In some cases, they fail to display species-typical reproductive behaviours, such as migration, nest construction and parental care, but may alternatively display strategy-specific mating behaviours, such as female mimicry or sneak fertilisations. In some cases, they appear to have reduced longevity.

The best-described examples of alternative mating strategies are in the salmonids (Dodson *et al.*, 2013). In many populations of *S. salar*, males exhibit two distinct life-history strategies. All male *S. salar* develop as juveniles in freshwater, and thereafter there are two alternative pathways for development. Some males undergo the process of smoltification, a complex sequence of morphological, physiological and behavioural changes that prepare the young fish for seaward migration. These individuals are anadromous and return to freshwater once they are sexually mature after one winter (anadromous grilse) or after multiple winters at sea (anadromous salmon). An alternative strategy is residency as a precocial parr, with males remaining in their natal river and achieving sexual maturity either without migration or after a brief migration. These males mature as parr at a size of only 70–150 mm, substantially smaller than grilse and salmon, which achieve a size of 500–1000 mm at maturity. During spawning, anadromous males actively defend ripe females from other males. Resident males associate with courting anadromous males and females and compete with each other to sneak fertilisations when spawning occurs. The resident and anadromous phenotypes are sympatric and in most populations are not genetically differentiated; thus both life histories derive from the same gene pool. Other salmonids display a striking range of alternative mating phenotypes, typically in males, but in some cases also in females (reviewed by Dodson *et al.*, 2013).

Body size and growth rates are often reliable predictors of male mating phenotype in salmonids, although their impact can operate in opposite directions. In populations of *S. trutta*, freshwater residency is sometimes associated with rapid early growth (Olsson *et al.*, 2006), and in others with migration (Forseth *et al.*, 1999). In many salmonids, there appears to be a critical threshold in body size above which an alternative life history is expressed, although this may vary with growth rate, and other threshold mechanisms may also play a role in determining life-history phenotype. Control of life history has a substantial genetic component, which means selection can act on life-history evolution. In *S. salar* there is substantial additive genetic variance for body size at migration, and selection appears to act on threshold body size. The gene expression profile of resident and migratory males varies, with significant differences associated with genes controlling metabolism, growth, behaviour, learning and memory (Aubin-Horth *et al.*, 2009; Chapter 8).

The mechanism by which alternative life-history strategies are maintained in the same population is generally considered to be through negative frequency-dependent selection (Dodson *et al.*, 2013). The prediction is that the reproductive success of each alterative strategy should vary inversely with the frequency of individuals that display the strategy. In the context of salmonid life histories, in a population dominated by anadromous males, resident males are predicted to enjoy higher reproductive success, with the result that selection will favour the resident phenotype, and vice versa. Thus the frequency of two alternative strategies should achieve equilibrium at a point where the average fitness of each strategy is equivalent. An alternative model is the status-dependent selection hypothesis of Gross (1996). Here, alternative strategies reflect variation in physical condition or social status, with one strategy linked to high condition or status and the alternative strategy linked to a lower condition or status. Under this model, the average fitness of individuals adopting the alternative strategy is expected to be lower, although there is a predicted switch point between mating strategies (in this context, termed 'tactics') when fitness will be equivalent. While the average fitness of the two tactics is not equivalent, the prediction is that an individual will achieve its maximum fitness given its condition or status. This model of negative frequency-dependent selection is appealing, because it appears applicable to a broad range of mating systems, and a capacity to switch between mating tactics is a feature of a range of species in response to condition and age (Brockmann & Taborsky, 2008). However, this model has been criticised on the basis that it assumes heritable variation for the switchpoint is either absent or minimal, thereby constraining its responsiveness to selection (Shuster & Wade, 2003; Shuster, 2010). While negative frequency-dependent selection appears to be a valid mechanism to maintain alternative life-history strategies in the same population, it has rarely been demonstrated, and the challenge is to show that this process operates in nature, while distinguishing whether alternative life-history strategies generate equal or unequal average fitness.

The adoption of alternative strategies may also be influenced through phenotypic plasticity, with the expression of alternative mating strategies being the outcome of an interaction between genetic and environmental variation. In *S. salar*, the critical threshold in body size, which underpins the migratory or resident strategy in males, is determined by juvenile growth rates; faster growth is typical of resident mature parr, while slow early growth is associated with anadromy. Growth rate in *S. salar* is a function of the environment, potentially implicating a significant environmental component governing the adoption of the two strategies. A study of a population of *S. salar* in Newfoundland showed that the fitness of resident and anadromous male strategies depended on competition with rivals and age at maturity (Hutchings & Myers, 1994). A simple threshold model of maturation assumed a growth rate threshold had to be exceeded at a specific critical time of each year, typically in autumn, for maturation to proceed the following spring (Thorpe, 1986). Growth rate thresholds were assumed to have a genetic basis and to vary among individuals, with the frequencies of alternative genotypes displaying different thresholds that were maintained through frequency-dependent selection. Thus, the interaction of growth rate due to environmental variability, mate competition and individual variation in growth rate thresholds generates an evolutionarily stable continuum of reproductive strategy frequencies. This model illustrates how alternative life-history strategies might be maintained through the response of threshold traits to environmental variability. An understanding of the regulation of the expression of alternative life-history strategies will come from an understanding of the neuroendocrine and endocrine control of reproduction in alternative mating phenotypes.

Conclusions

From the initial studies of Fisher (1930) and Cole (1954), life-history theory has developed as a major theme in evolutionary ecology. This theory makes explicit predictions about how life-history variables should respond under selection, and responses to selection on traits is a means of identifying life-history trade-offs (Reznick, 1983). A prediction of life-history theory is for adaptive responses to changes in selection pressure, e.g. through changes in the intensity of predation, or the life stages on which predation acts. There are numerous cases that demonstrate life-history responses in teleosts, with examples coming from the impacts of environmental change, environmental gradients, and the naturalisation of fishes outside their natural range, as well as experimental studies of artificial selection.

Life-history theory provides the basis for predicting the effect of environmental changes, such as changes caused by climatic change, changes in land use and other anthropogenic effects. Commercial fishing is also considered to be a potential mechanism for altering teleost life histories (Chapter 12). Commercially exploited stocks often show younger age and smaller size at maturation, and this is sometimes taken as evidence for evolved responses to selection imposed by fishing mortality. An alternative explanation is that the changes in life-history traits seen in exploited fish stocks arise from compensatory plastic responses to reduced population size. The relative significance of compensatory responses or evolved life-history responses that reflect genetic change is a subject of ongoing debate (Law, 2000; Marshall & Browman, 2007; Palkovacs, 2011; see Chapter 12 for a full discussion).

Changes in life-history traits, whether reflecting phenotypic plasticity or genetic changes, require changes in the underlying physiological processes that control reproduction and related traits. As yet, these links are poorly understood. For changes in traits such as age and size at maturity, fecundity and egg size, there must be changes in the neuroendocrine and endocrine regulatory processes that control reproduction (Chapter 5).

Chapter 12

Reproduction, fisheries and aquaculture

Reproductive Biology of Teleost Fishes, First Edition. Robert J. Wootton and Carl Smith.

Introduction

Reproduction has direct implications for the exploitation, management and cultivation of fishes. The success of fisheries management is contingent on an understanding of reproductive parameters, and the way in which these change in populations undergoing exploitation. The control of reproduction is a key element of aquaculture, and developments in genetic biotechnology offer powerful tools for controlling the volume and quality of juvenile fish for ongrowing, and potentially for expanding the range of species under cultivation.

Reproduction and fisheries management

The goal of fisheries management is to maximise the biomass of fish removed from a population, in the context of fisheries management generally termed a 'stock', while ensuring sufficient adult fish remain to maintain its long-term viability. Traditional fisheries models are based around estimates of fish growth rates and natural mortality, weighed against the impact of fishing mortality. Reproduction is integrated into fisheries models principally through the relationship between adult stock size and the number of new recruits added to the population (Pitcher & Hart, 1982). 'Recruitment' is an ambiguous term in fisheries biology, though it is generally used to refer either to the addition of individuals to the adult population, or that component of the population susceptible to fishing. Thus the age of recruitment is the age at which an individual either achieves sexual maturity or becomes at risk of capture; these ages may correspond, but not necessarily.

Stock–recruitment relationships relate the abundance of reproductive adults to the size of future cohorts of offspring. Fisheries models also incorporate key reproductive traits as measures of stock reproductive potential. There is growing recognition that species exhibiting specific modes of reproduction may have greater susceptibility to overexploitation, and also that overexploitation can shape fish life-history evolution, including reproductive traits.

Stock–recruitment relationships and reproduction

The nature of the relationship between adult stock size and the number of offspring that ultimately recruit to the stock, termed the stock–recruitment relationship, summarises a number of key processes relating to reproduction. There are several stock–recruitment models, the best known being the Beverton & Holt (1957), Ricker (1975) and Shepherd (1982) models (Fig. 12.1). Each differs in the assumptions

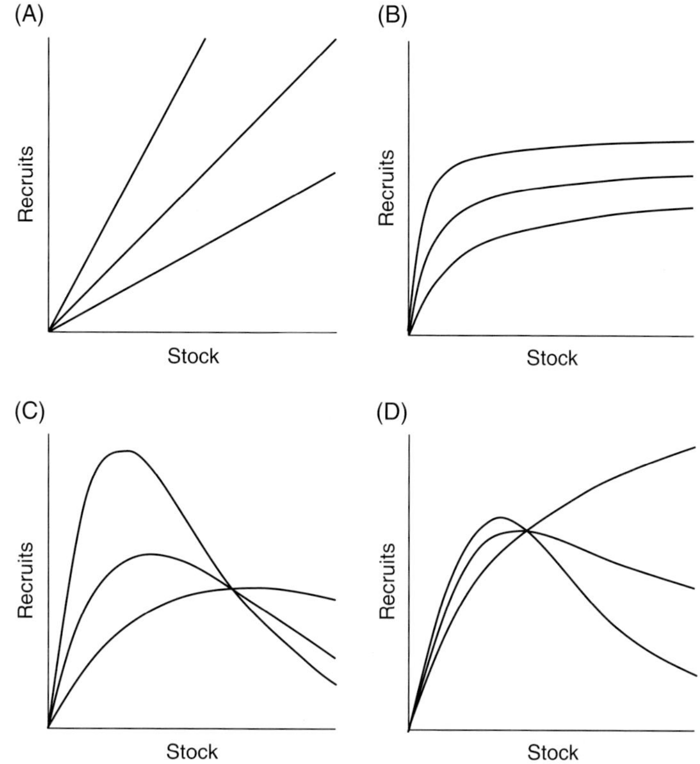

Figure 12.1 Stock–recruitment relationships. (A) Density-independent relationships; (B) Beverton–Holt curves; (C) Ricker curves; (D) Shepherd curves.

of how density dependence limits recruitment at high spawning stock sizes. The relationship between stock size and recruitment is a central question in fisheries biology, and fitting data and selecting among different models have taxed fisheries scientists for decades. It is paradoxical that most fisheries management problems relate to stocks reduced to low levels, where models tend to correspond in their predictions.

The Beverton–Holt model assumes the relationship between adult stock and recruitment approaches an asymptote at high stock size. The relationship takes the form:

$$R = \frac{S}{\alpha + \beta S}$$

where R is the number of recruits, however defined, and S is the number or biomass of spawners. The parameters α and β are parameters of the curve. The Beverton–Holt model includes a density-dependent component through the decrease in the number of recruits per spawner as a function of increasing spawning stock size. This model is believed to apply to species in which there is competition among juveniles that results in mortality, e.g. over limiting food or space. It could also apply to species in which competition for food among older and younger cohorts limits egg production by spawners. Alternatively it may apply to species in which predators can impose mortalities on a cohort of juveniles without a time lag in their impact (Wootton, 1998).

The Ricker model describes a dome-shaped relationship between spawners and recruits, with a peak in recruitment at some intermediate spawner abundance, described by the relationship:

$$R = \alpha S e^{-\beta S}$$

This stock–recruitment model can apply to species in which cannibalism by older fish on younger cohorts is a significant cause of mortality. It might also arise when adults damage one another's spawning sites. For example, when female salmonids dig spawning redds as a prelude to oviposition (Chapter 7), they can displace eggs previously laid by other females (termed redd superimposition), thereby killing them. Empirical data indicate that redd superimposition occurs at an accelerating rate with increases in female density, in accordance with predictions of the model (Fukushima *et al*., 1998). A Ricker curve might also be generated if disease transmission from adults imposes significant mortalities on juveniles.

The Shepherd model incorporates features of both these models and takes the form:

$$R = \frac{\alpha S}{1 + (S/K)^{\beta}}$$

Shepherd's model includes fewer constraints on how spawner density impinges on recruitment, but also includes less biological reality (Garrod & Horwood, 1984). The parameter K in the model represents a threshold stock size above which density-dependent effects increasingly impose constraints on recruitment.

While stock–recruitment models can be fitted to empirical data, the fit is often unconvincing, and the utility of these models is clearly quite limited (Wootton, 1998; Hilborn & Walters, 2001; Rose *et al*., 2001). Stock–recruitment relationships are typically composed of a time-series of data points that are not statistically independent. An outcome is that under conditions generating low recruitment, spawning stock size will decrease, suggesting a stronger stock–recruitment relationship than actually exists (Gilbert, 2002). One explanation for the weak predictive power of stock–recruitment models may not be because such a relationship does not exist, but rather that great variability clouds the relationship. Thus, high fecundities linked with unpredictable juvenile mortality rates result in a weak relationship between spawning stock and recruitment (Fig. 12.2). An alternative hypothesis is that recruitment is driven wholly, or at least largely, by random processes linked to environmental conditions (Gilbert, 1997), with the typically weak relationship between spawning stock size and recruitment simply an artefact of the data. The 'recruitment states' hypothesis proposes that recruitment switches between different states with different mean levels of recruitment without a direct relationship between stock size and recruitment, although because of the nature of time-series data, a spurious correlation is predicted in response to low levels of recruitment (Gilbert, 1997). A meta-analysis of marine teleosts supported the recruitment states hypothesis, while a comparable analysis for salmonid populations suggested a real relationship between spawning stock size and recruitment (Gilbert, 1997). These contrasting results may reflect differences in the spawning biology of these two groups. Despite a lack of consensus of the significance of stock–recruitment relationships in fisheries biology, at a broad scale a relationship undoubtedly does exist. Thus, recruitment will always tend to be greatest when stock size is high, while recruitment will tend to be constrained when stock sizes are low (Myers, 1997).

The ineffectiveness of simple stock–recruitment models has been attributed to their failure to incorporate sufficient biological and environmental data (Rothschild, 1986;

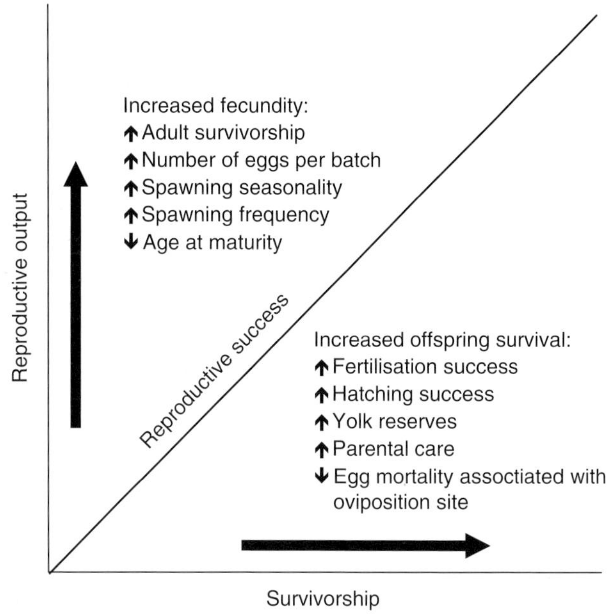

Figure 12.2 Reproductive success comprises two components: reproductive output and survival rate of the offspring. A suite of reproductive traits potentially influence these variables. Reproductive output can be enhanced through increased adult survival, greater batch fecundity, an extended spawning season, increased spawning frequency and a reduction in age at maturity. Offspring survival can be increased through enhanced fertilisation success, improved hatching rates, increased yolk reserves, reduced egg mortality through oviposition decisions and elevated levels of parental care. (Source: Lowerre-Barbieri 2009. Reproduced with permission of Taylor and Francis Group.)

Trippel *et al.*, 1997; Wootton, 1998). Numerous life-history variables are likely to contribute to reproductive output for a given stock, including the sex ratio, age structure, timing of maturation, fecundity, egg size, spawning frequency, generation time, oviposition choices, mating system and parental care system, among others. However, these data are rarely incorporated into fisheries models (Trippel *et al.*, 1997; Wootton, 1998; Lowerre-Barbieri, 2009).

An alternative to stock–recruitment relationships for fisheries management is the spawning potential ratio (SPR; Goodyear, 1989; Lowerre-Barbieri, 2009). This is the ratio of estimated fecundity under current levels of fishery exploitation to that in the absence of fishing, and is expressed as:

$$SPR = \frac{P_{\text{fished}}}{P_{\text{unfished}}}$$

where *P* is the potential fecundity. Potential fecundity is the average lifetime production of eggs per recruit in a fished and unfished population in the absence of density dependence (Goodyear, 1989). An estimate of SPR < 0.2 indicates a stock at high risk of collapse. The approach to fisheries management using SPR has the advantage that it assumes a relationship between stock and recruitment, but without explicitly having to define it (Goodyear, 1989; Lowerre-Barbieri, 2009).

Reproductive traits and population dynamics

Fisheries models tend to oversimplify fish reproductive strategies, possibly because the focus of fisheries science has traditionally been on fish yields, and the subtleties of reproduction do not always have an obvious impact on this parameter. A more mundane explanation is simply that adequate data on the reproductive biology of even the most important commercial species are not available to be incorporated into fisheries models. Integration of reproductive variables into fisheries models, especially for multispecies fisheries, also presents a substantial challenge in terms of the complexity of models. The two key reproductive variables incorporated into models are age at maturity and fecundity, although others will also play a role.

Age at maturity

The age at which a stock achieves sexual maturity is a significant variable, and is typically estimated for a population by fitting a logistic curve to data for age (or size as a proxy) and proportion of the population that are sexually mature. The curve provides an estimate of the probability of maturity at different ages or sizes. The age at which 50% of the age class are mature is taken as the age (or size) at maturity for the stock. The form of the maturation curve varies among species, and the size range may be within a narrow or extended age/length band (Fig. 12.3). The age at maturity for males and females may differ, and data for the sexes are often plotted separately (Fig. 12.4). The goal of fisheries management is to ensure that fish, and especially females, are not captured before they reach maturity. Harvesting sexually immature fish can result in 'recruitment overfishing', whereby too few fish achieve sexual maturity to sustain the stock through reproduction. This contrasts with 'growth overfishing', which describes a situation in which fish are harvested before they have reached their maximum growth potential, but without necessarily threatening recruitment to the stock.

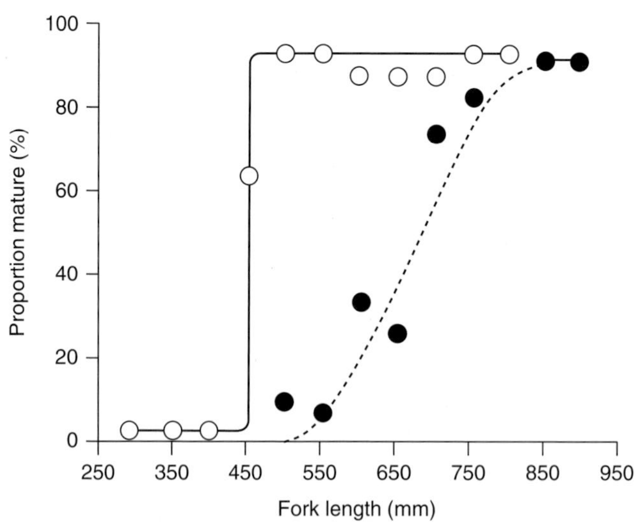

Figure 12.3 Proportion of sexually mature female green jobfish (*Aprion virescens*) (solid line, open circles) and deepwater longtail red snapper (*Etelis coruscans*) (dashed line, closed circles), two species of Hawaiian snapper, showing contrasting maturation ranges. (Source: Adapted from Everson *et al.* 1989.)

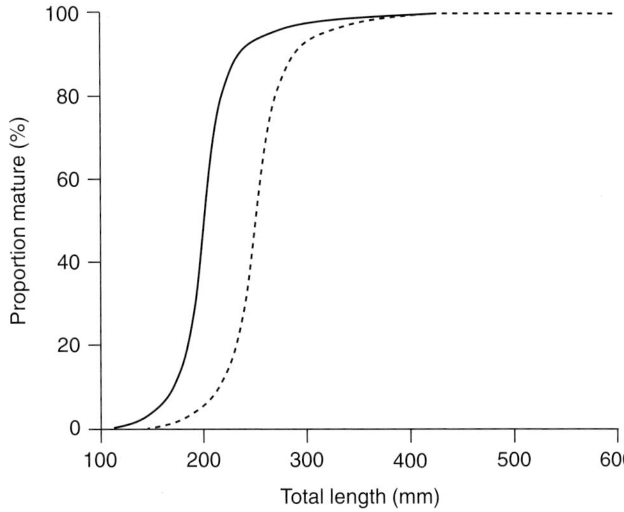

Figure 12.4 Cumulative distribution function for proportional maturity (%) for male (solid line) and female (dashed line) spotted weakfish (*Cynoscion nebulosus*), illustrating a different pattern of maturation for the sexes. (Source: Adapted from Lowerre-Barbieri 2009. Reproduced with permission of Taylor and Francis Group.)

Age at maturity in fish is a key life-history variable that responds to selection (Chapter 11). The physiological process of maturation is complex and controlled by the neuroendocrine and endocrine systems of the fish (Chapter 5), linked through the central nervous system to external stimuli (Chapter 6). High levels of fishing appear to induce shifts in the age at maturity of a stock, with heavily fished stocks tending to mature at a younger age. These changes in age at maturity in response to fishing pressure have been taken as evidence of fisheries-induced evolution (see the section 'Fisheries-induced evolution'). Changes in age at maturity can also arise through density-dependent compensation, with greater food availability in populations reduced to low levels by fishing enabling maturation at a younger age. In the case of density-dependent compensation, size at maturity is predicted to be unchanged, even though age at maturity is reduced. Where both show a decline in response to elevated fishing pressure, the implication is that the response is a fisheries-induced evolution effect.

Fecundity

The number of eggs a female produces is termed her fecundity. Fecundity will depend on spawning dynamics for a given species (Chapter 4) and can be measured over a spawning bout (termed 'batch fecundity', F_{batch}), a spawning season ('breeding season fecundity', F_b), or a female's entire lifetime ('lifetime fecundity', F) (see Chapter 11). In fisheries biology, breeding season fecundity is the most frequently used measure of female egg production, as this provides an estimate of potential recruitment in subsequent years. Batch fecundity typically correlates with female body size, with the relationship taking the form:

$$F_{batch} = aL^b$$

where L is female length, a is the intercept of the function, and b is the length exponent, which characteristically takes a value close to 3, though potentially ranging from 1 to 5 among species (Wootton, 1998; Chapter 11). However, even within species, this parameter shows substantial variation, with some females producing more eggs at a given size than others, while the same female may produce different batch sizes under different prevailing conditions. In a fisheries context, *b* is a critical variable, as it potentially underpins recruitment. Despite its significance, variation in this parameter, among and within species and individuals, is poorly understood. Variation in F_{batch} is hypothetically controlled by a variety of abiotic and biotic conditions. For example, female Atlantic cod (*Gadus morhua*, Gadidae) showed 80–90% variation in batch fecundity in response to natural increases in food availability and elevated environmental temperatures during the period of vitellogenesis (Kjesbu *et al.*, 1998). Pronounced latitudinal clines in size-specific fecundity are well described in a number of marine and freshwater fishes (Conover, 1992; Kokita, 2003; Heibo *et al.*, 2005). While productivity and temperature are frequently implicated in these latitudinal patterns, the mechanisms are often obscure.

In species that are total spawners, such as salmonids and many gadoids, the entire annual production of eggs is released in a single spawning event; thus batch and breeding season fecundity are equivalent ($F = F_{batch}$). In other species, breeding season fecundity will depend not only on the magnitude of batch fecundity, but also on the frequency with which spawning occurs over the spawning season. In these species, which include many commercial species, such as anchovies (Engraulidae) and herrings (Clupeidae), estimates of breeding season fecundity are reliant on histological methods to identify the number of spawning bouts typical of the population. A further complication is that batch spawners may show determinate or indeterminate fecundity. In the former case, breeding season fecundity is fixed at the beginning of the spawning season, with a gradual depletion of oocytes with each spawning event. Not all oocytes may be released, and some may be resorbed through the process of atresia (Chapters 4 and 6), although at least the maximum breeding season fecundity can be estimated prior to the onset of spawning. In the case of indeterminate fecundity, new oocytes develop over the course of the spawning season, with breeding season fecundity governed by the rate of oocyte development, spawning and atresia (Wootton, 1998). This mode of spawning presents challenges to the fisheries biologist in making reliable estimates of population fecundity. Determinate fecundity is associated with fish at high latitudes, while indeterminate fecundity is typically encountered in tropical species.

Mode of reproduction and susceptibility to overexploitation

Certain mating systems may increase susceptibility to overexploitation. Migratory species often congregate in large numbers to spawn, or follow predictable routes to spawning grounds, thereby making them more vulnerable to capture (Chapter 7); examples include salmonid and tuna (Scombridae) species. Benthic and pelagic spawners may show different resiliencies to fishing pressure, as may oviparous and viviparous species (Cowen & Sponaugle, 1997). Despite the clear implications of mode of reproduction and mating system for susceptibility to overexploitation, not much systematic research has examined this question. In the context of life-history evolution, there is a tendency to predict that small-bodied, short-lived, early-maturing, fast-growing species, such as engraulids, clupeids and some scombrids, are able to withstand high levels of fishing pressure (Dulvy *et al.*, 2004). The converse prediction is that large-bodied, long-lived, late-maturing, slow-growing species are more susceptible to fisheries exploitation. Several comprehensive analyses and case studies support this generalisation (e.g. Fromentin & Fonteneau, 2001; Halpern & Warner, 2002; Rochet *et al.*, 2011; Jørgensen & Holt, 2013), although others do not.

Russ & Alcala (1998) used life-history characteristics of 14 families of reef fishes at Sumilon and Apo Islands in the Philippines to predict their probable vulnerability to fishing. While large predators, such as groupers (Serranidae) and snappers (Lutjanidae), proved susceptible to fishing in accordance with predictions, the response of other families contradicted predictions based solely on life history (Table 12.1). To properly compare among exploited species or stocks to identify life-history traits that predispose them to overexploitation, it is necessary to use a phylogenetic comparative approach to ensure independence of the data. However, as a general principle, species with high rates of population increase, short lifespan and early maturity appear more resilient to fishing pressure, even after controlling for phylogeny. This life-history strategy corresponds with the 'opportunistic' life-history classification of Winemiller (2005) (Chapter 11).

In some pelagic-spawning planktivorous species, particularly engraulids and clupeids, cannibalism of eggs and larvae by spawning adults imposes extremely high mortalities, in some cases sufficient to limit population size (Hunter & Kimbrell, 1980; Hourston *et al.*, 1981; Valdes *et al.*, 1987; Folkvord, 1997). Cannibalism, more than other forms of predation, has the capacity to regulate population size in a density-dependent manner (Ricker, 1954; Smith & Reay, 1991; Folkvord, 1997). In some populations, persistent cannibalism by older age cohorts can generate population cycles by suppressing recruitment for several years through inter-cohort cannibalism, followed by a burst of recruitment after the eventual extinction of the older cohorts through senescence. The cycle is repeated as the age cohort produced in the year of strong recruitment imposes mortality on subsequent cohorts (Smith & Reay, 1991).

The impact of fisheries on reproductive traits

Fishing mortality can be substantial; in many heavily fished stocks, mortality from fishing greatly exceeds natural mortality (Jackson *et al.*, 2001). A consequence is that fisheries can shape fish population parameters, either through direct effects on fish populations, as a result of plastic responses by fish to elevated mortality, or through evolved responses to selection imposed by fishing mortality.

Age and size structure

Fishing is typically size-selective, as a consequence of either the mesh sizes of fishing nets and traps or the sizes of bait and hooks. Fishermen also often specifically target the largest individuals in a stock because of their relatively higher market value. Restrictions on fishing gear are usually in the form of increased mesh sizes to protect immature juveniles. In some fisheries, a minimum size limit for fish that can be landed is also imposed, again to protect juveniles. The effect of size selectivity is, predictably, to remove larger and older individuals from the population with the outcome that mean age and body size decline as fishing pressure increases. Alteration of the age and size structure of exploited populations is one of the most profound direct impacts of fishing on a stock (Trippel *et al.*, 1997). The most immediate impact of removing larger individuals from the population is that smaller and younger fish make a greater contribution to the spawning stock, with first-time spawners potentially making a substantial contribution. In a stock of Arcto-Nowegian cod (*Gadus morhua*, Gadidae), the contribution of fish over 11 years old to spawning stock biomass exceeded 60% at the start of the 20th century. This figure had declined to less than 5% by the first decade of the 21st century, with an increasing reliance on first-time spawners to sustain reproduction and the majority of spawners comprising fish aged 5–7 years (Fig. 12.5).

Table 12.1 Life-history characteristics of coral reef fishes at Apo (A) and Sumilon (S) Islands, their predicted vulnerability to fishing pressure, the observed impact of fishing, and the correspondence between predicted and observed vulnerabilities to fishing.

Taxon	Body size	Potential longevity	Natural mortality	Growth rate	Recruitment rate	Predicted vulnerability to fishing?	Observed effect of fishing	Correspondence between observed and predicted effects
Acanthuridae	Medium	Long-lived	Medium/low	Medium	Medium	Intermediate	Weak (A, S)	No
Anthiinae	Small	Short-lived	High	High	High	No	Weak (A, S)	Yes
Caesionidae	Medium	Medium	High	High	High	No	Strong (A, S)	No
Carangidae	Large	Long-lived	Low	Low	Low	Yes	Strong (A, S)	Yes
Chaetodontidae	Medium	Medium	Medium	Medium	Medium	Intermediate	Weak (A), Strong (S)	No
Labridae	Medium	Medium	Medium	Medium	Medium	Intermediate	Weak (A, S)	No
Lethrinidae & Lutjanidae	Large	Long-lived	Low	Low	Low	Yes	Strong (A, S)	Yes
Mullidae	Medium	Medium	Medium	Medium	Medium	Intermediate	Weak (A, S)	No
Pomacentridae	Small	Medium	High	High	High	No	Weak (A, S)	Yes
Scaridae	Medium	Medium	Medium	Medium	Medium	Intermediate	Weak (A)	No
Serranidae	Large	Long-lived	Low	Low	Low	Yes	Strong (A, S)	Yes

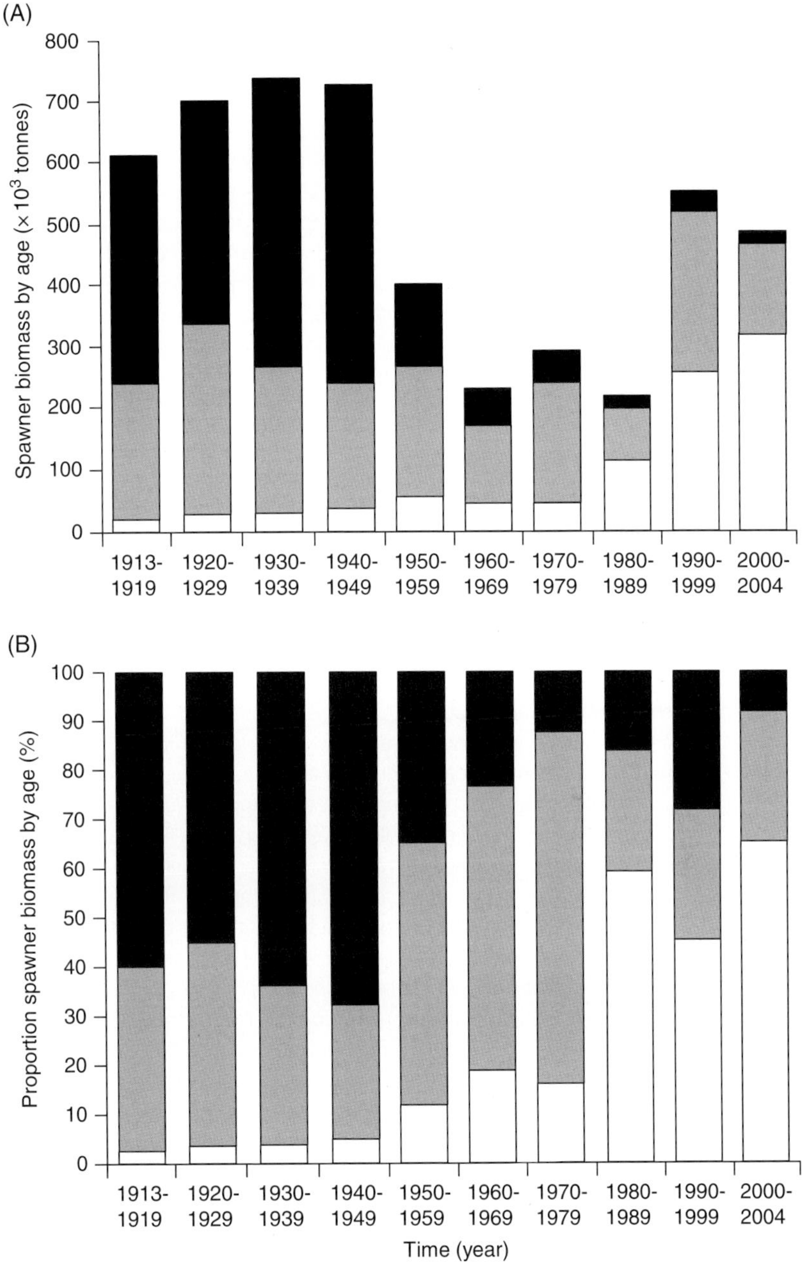

Figure 12.5 The effects of overexploitation on population age structure in Arcto-Norwegian cod (*Gadus morhua*) from 1913 to 2004 for three age groups, 4- to 7-year-old spawners (white section), 8- to 10-year-old spawners (grey section) and 11- to 13-year-old spawners (black section). (A) Spawning stock biomass (SSB); (B) proportion of spawner biomass. Over a 90-year period the stock has become increasingly reliant on first-time spawners to maintain reproduction. (Source: Adapted from Ottersen 2008. Reproduced with permission of Canadian Science Publishing.)

Conventional stock–recruitment models assume that all mature fish in a population make an equivalent contribution to spawning stock biomass. However, this assumption may be flawed, with research suggesting that first-time spawners may be less reproductively viable than older individuals (Trippel, 1998; Green, 2008). Females in poor condition at the onset of spawning produce fewer eggs per unit weight than those in better condition (Tyler & Dunn, 1976; Lambert & Dutil, 2000). Younger females may be less resilient than larger and older females in years in which food supply is limited or environmental conditions are unfavourable for reproduction, with the outcome that first-time spawners will have lower fecundities than predicted for their size and they may also suffer higher rates of post-spawning mortality (Trippel *et al.*, 1997; Lambert & Dutil, 2000). The duration of the spawning period is longer for larger than it is for smaller females, especially for batch spawners, and the onset of spawning is also typically earlier (Trippel *et al.*, 1997). Recruitment success in marine species has been linked to plankton production, with larvae hatching either too early or too late and risking starvation (Cushing, 1972; Leggett & Deblois 1994). A more protracted spawning season has a greater probability that young stages will correspond with the optimum period for feeding and survival, thereby favouring reproduction by older females. The early onset of reproduction will also ensure a longer growing season for larval stages, which can be critical for overwinter survival in temperate species (Sogard, 1997).

Gamete quality can also be a function of body size and age in fishes. Across a wide range of teleost species, egg size correlates positively with female body size (Wootton, 1998). Larger eggs have greater hatching success and produce larger larvae (Pepin, 1991; Trippel, 1998). Larger larvae tend to have reduced susceptibility to starvation, lower risk of predation and greater food capture abilities (Blaxter & Hempel, 1963; Miller *et al.*, 1988; Trippel, 1998; Tamada & Iwata, 2005), and consequently better growth rates (Einum & Fleming, 1999; Kamler, 2008). Spermatozoa viability may also correlate with male size in some commercially exploited species, but evidence is mixed (Trippel & Neilson, 1992; Rakitin *et al.*, 1999).

Size selectivity can potentially also have an impact on fish mating systems. In species with dominance hierarchies or territoriality, older and larger males tend to dominate mating (Chapter 8). Removal of these individuals through fishing will have the effect of increasing the opportunities for smaller, subordinate males to mate, while potentially limiting female mate choice, mating opportunities and fertilisation success (Rowe & Hutchings, 2003). In the great northern tilefish (*Lopholatilus chamaeleonticeps*, Malacanthidae), the mating system features construction of a burrow by large males, which dominate mating (Grimes *et al.*, 1988). Long-line fishing is selective for these large dominant males, thereby enabling smaller males to participate more frequently in mating. An outcome is that all the eggs produced by females may not be fertilised by the subordinate males, which have smaller testes in this species (Grimes & Turner, 1999). A simplifying assumption of fisheries models, and specifically stock–recruitment relationships, is that spawning stock biomass is directly proportional to reproductive potential. However, the overall impact of shifting the age and size structure of fish populations downwards through fishing will be to depress population reproductive output below that predicted by conventional fisheries models, which may explain why many fish stocks have failed to recover from overfishing in the way that these models predict (Trippel *et al.*, 1997).

Population sex ratio

Fishing can alter fish population sex ratios, either as a consequence of size selectivity in species with pronounced size dimorphism, or because one sex is more susceptible to capture for other reasons, such as greater levels of activity, greater boldness, or because the sexes occupy different habitats (Rowe & Hutchings, 2003). In tropical fisheries, many species are hermaphroditic (Chapter 10). Size selectivity in these species will have the effect of skewing the sex ratio, in the case of protogynous hermaphrodites generating a female bias in the population, and in protandrous hermaphrodites a male bias. For example, the gag (*Mycteroperca microlepis*, Serranidae) is a slow-growing protogynous grouper that occurs on reefs along the Atlantic coast of North and South America. Males, which are larger and more aggressive than females, are selectively removed by the fishery for this species. During 1976–1982, the proportion of males in the population comprised 19.6%. This figure had fallen to just 5.5% by 1995 in a fishery along the south-east coast of the USA, while in the Gulf of Mexico, males comprised only 1.9% of the stock (McGovern *et al.*, 1998).

In some species, sex change is socially induced, e.g. in response to changes in density, sex ratio or size distribution, in which case removal of one sex will theoretically result in a compensatory response through an increased rate of sex change, albeit with a lag of weeks to months while sex change takes place (Chapter 10). In other hermaphroditic species, initiation of sex change is size- or age-specific and a bias in the sex ratio will be maintained while sex selectivity persists. In the Venus tusk fish (*Choerodon venustus*, Labridae), a protogynous hermaphrodite, sex change is socially induced. This species forms a significant component of a line fishery that is size-selective and thereby imposes greater fishing pressure on males. This sex-specific mortality resulted in a compensatory response, with a constant sex ratio maintained through a shift in the age and size at which sex change occurred (Platten *et al.*, 2002; Fig. 12.6).

Alonzo & Mangal (2005) explored the responses to fishing for populations displaying different patterns of protogynous sex change. Four patterns of sex change were investigated, though sex change at a fixed size proved most sensitive to fishing pressure when the fishery was size-selective, with populations ultimately crashing when fishing selectivity targeted fish at a mean size below that at which sex change occurred. Each pattern of sex change showed a different response to fisheries exploitation, illustrating the importance of an understanding of reproductive life history for fisheries models (Alonzo & Mangal, 2005). These conclusions are supported by empirical studies (Chopelet *et al.*, 2009; Lorance, 2011; Mariani *et al.*, 2013). Changes to population sex ratio through size-selective fishing are not confined to hermaphroditic species, and a skew in the sex ratio can also be imposed on gonochorists if the growth rates or age/size at maturity of the sexes differ (Fenberg & Roy, 2008).

In species with parental care, the care-giving sex may be more susceptible to capture, e.g. through aggressive defence of a nest or territory. In fishes, it is males that are more commonly the care-giving sex (Chapter 9) and a skew in the population sex ratio towards females in a parental species might have the effect of limiting female reproduction because of a failure to find a male able to undertake care of her eggs. Similarly, in lek-breeding species (Chapter 8), in which males defend small territories where they display to females and engage in aggressive or ritualised contests, males may have greater susceptibility to capture. In Lake Turkana, Kenya male Nile tilapia (*Oreochromis niloticus*, Cichlidae) gather in a shallow bay called Ferguson's Gulf in which they form an extensive lek and spawning takes place. Breeding males in Ferguson's Gulf are a target of fishing activity (Stewart, 1988; Kolding, 1993), although the impact on the population sex ratio or population reproductive output is unknown. Thus, despite the profound impacts on the sex ratio of some fish stocks as a result of fishing, the effects at the population level are poorly understood. Circumstantial evidence that sex-selective fishing may have a significant impact comes from studies that demonstrate changes to the onset of sex change in hermaphroditic species (Platten *et al.*, 2002; Hamilton *et al.*, 2007; Caselle *et al.*, 2011).

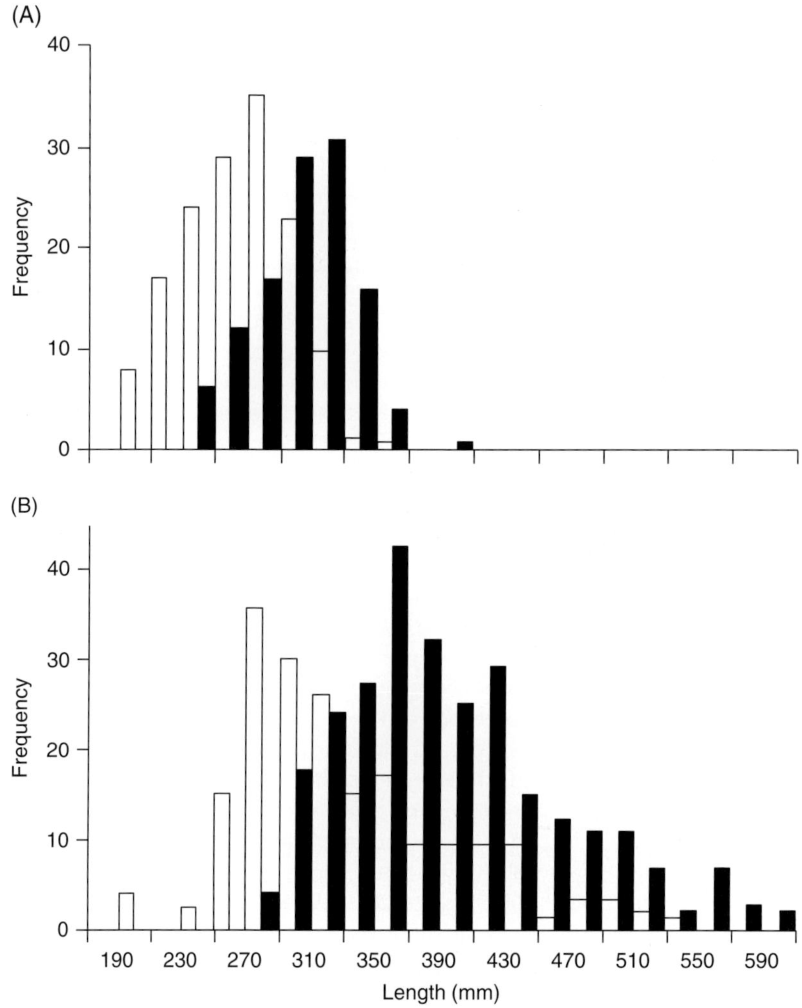

Figure 12.6 The impact of fishing on the Venus tusk fish (*Choerodon venustus*), a protogynous hermaphrodite in which sex change is socially induced, on two reefs experiencing different levels of exploitation: (A) heavily exploited; (B) lightly exploited. Despite different levels of fishing mortality, the sex ratio in the two populations was maintained at comparable ratios in both populations. (Source: Adapted from Platten *et al.* 2002. Reproduced with permission of John Wiley & Sons.)

Population size and density

Fishing reduces overall population size and fish density. A possible outcome is that where competition for resources might have limited reproduction, e.g. through food availability, a density-dependent compensatory response may occur with enhanced survival, growth and reproductive output (Trippel, 1995). Compensatory responses to harvesting are a fundamental assumption of most fishery models, although how these responses operate and are best incorporated into models are contentious issues (Rose *et al.*, 2001). There are several mechanisms by which density can influence reproductive output (reviewed by Trippel, 1995; Rose *et al.*, 2001). Predation rate, through numerical and functional responses, can vary with density, while cannibalism, which represents a special form of predation, is intimately linked to density (Smith & Reay, 1991). Changes in food availability, and intraspecific competition during feeding can drive density-dependent growth and mortality. Several reproductive parameters are responsive to density, including batch fecundity, spawning frequency, egg quality and oviposition site choice, ultimately resulting in increased levels of per capita recruitment. A number of empirical studies on marine and freshwater fishes have provided evidence of density-dependent compensatory responses (e.g. Healey, 1980; Elliott, 1994; Koslow *et al.*, 1995; Lóbon-Cerviá, 2007; Nash *et al.*, 2009). Distinguishing between the compensatory effects of reduced density on reproduction through plastic responses and fisheries-induced evolutionary changes (see the section 'Fisheries-induced evolution'), with implied genetic changes, represents a challenge (Marshall & Browman, 2007). The two processes are fundamentally distinct, although the proximate outcomes are comparable.

In contrast to compensatory responses, reduced population size and density can also have an impact on reproduction through depensatory effects (Liermann & Hilborn, 2001). Depensation describes a situation in which recruitment is suppressed at low population sizes. Depensation is sometimes termed an Allee effect (Allee, 1938) and is of interest because it can lead to unexpectedly slow rates of recovery, or even extinction, of populations reduced to low levels by overexploitation. The degree to which a fish stock is susceptible to depensation is dependent on the form of the relationship between population size and per capita growth at low population sizes (Fig. 12.7). Depensation is often attributed to reduced probability of fertilisation and can occur, for example, if adult density falls to such low levels that individuals are unable to locate mates during reproduction, or because delays in finding a suitable mate result in reproduction occurring at a time that is not optimal for egg or larval survival (Rowe & Hutchings, 2003). Delays to spawning can also result in over-ripening of gametes, which can reduce their viability (Kjørsvik *et al.*, 1990). An example of depensation is described in *G. morhua*, which have fallen to historically low stock sizes as a result of fishing mortality. In this species, low male densities resulted in reduced egg fertilisation rates

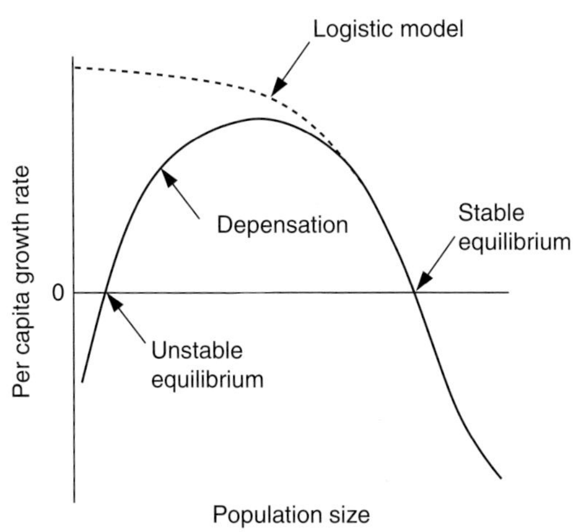

Figure 12.7 Population per capita growth rate as a function of population size with (solid line) and without depensation (dashed line). Depensation is reflected by a decline in per capita growth rate at low population size. (Source: Rowe & Hutchings 2003. Reproduced with permission of Elsevier.)

and greater individual variance in reproductive success, suggesting a decline in effective population size, such that a smaller proportion of individuals contributed genes during reproduction, with potential implications for long-term declines in population genetic variability (Rowe *et al.*, 2004). An additional depensatory mechanism can occur through reduced numbers of predatory adult cod failing to remove predators of young cod (Walters & Kitchell, 2001). Despite concerns over depensatory effects on reproduction in response to fishing, convincing case studies are largely lacking. However, the absence of substantial support for depensatory (or compensatory) effects should not be construed as evidence that these effects do not operate, as they may be difficult to demonstrate, particularly for commercial species for which key aspects of their reproductive biology are incompletely understood (Liermann & Hilborn, 2001).

Fisheries-induced evolution

Fishing, like predation, has the capacity to impose directional selection on populations experiencing fishing mortality. Given the selective nature of many fisheries, principally size selectivity, one outcome is that stocks experiencing substantial levels of fishing mortality should show evolutionary responses through changes in key life-history variables. There is good circumstantial evidence for fisheries-induced alterations to key life-history traits in many exploited fish stocks, although whether these changes represent compensatory plastic responses to reduced population size (see the section 'The impact of fisheries on reproductive traits') or evolved responses reflecting functional genetic change is a subject of ongoing debate (Law, 2000; Marshall & Browman, 2007; Palkovacs, 2011).

An outcome of fishing pressure in many stocks is for earlier ages at maturity, which are accompanied by smaller sizes at maturity (Trippel, 1995). Numerous examples of this response to fishing have been described (Kuparinen & Merilä, 2007; Palkovacs, 2011), although many fail to distinguish genetic evolutionary change from compensatory responses (Marshall & Browman, 2007). To overcome the confounding effects of compensation, a methodology based on changes to probabilistic maturation reaction norms (PMRNs) is typically used. This approach uses existing fisheries data to examine changes in the probability that individuals mature as a function of age and size, which thereby overcomes the confounding effects of variation in growth and mortality. The assumption is that any residual changes to maturation schedules can be attributed to genetic changes, and therefore represent an evolutionary response to directional selection from fishing (Olsen *et al.*, 2004; Kuparinen & Merilä, 2007). However, this assumption is flawed as shifts in maturation schedules, even after controlling for growth and mortality rate, can come from other environmental variables, such as changes in temperature or elevated exposure to pollutants (Kraak, 2007). Fishing itself may induce additional changes to maturation, not directly through size-selective mortality, but through alterations to social structure, body condition and sex ratio. An alternative approach is to use multi-dimensional PMRNs, which incorporate a greater range of confounding variables in their computation. Using this approach, Kraak (2007) was able to demonstrate that, in a stock of North Sea plaice (*Pleuronectes platessa*, Pleuronectidae), which had formerly been considered to show a significant shift in maturation to younger ages and smaller size from 1955 to 1995 (Grift *et al.*, 2003), temporal trends in sea temperature during the maturation period could account for part of the variance in observed shifts in maturation. Despite this effect of temperature, a residual trend in maturation remained unexplained, with the implication that directional selection from fishing was responsible (Kraak, 2007).

The assumptions for fisheries-induced evolution are that selection from fishing is intense, and that the heritabilities and genetic correlations of traits that respond to selection are high. However, there are few estimates of selection differentials generated by fishing combined with realistic estimates of heritabilities of life-history traits to support fisheries-induced phenotypic change (Law, 2007). Where attempts have been made to estimate these parameters, the conclusion is that evolved responses to directional selection from fishing are likely to be modest relative to compensatory responses (Law & Rowell, 1993; Law, 2007). Despite the sparse evidence for fisheries-induced evolution in commercial fish stocks, a precautionary approach to this potential problem is to ensure older age classes make a significant contribution to the spawning stock. Fishing pressure on commercial stocks is likely to persist for the foreseeable future, and its potential to alter the genetic basis of the reproductive characteristics of exploited stock needs to be considered in fisheries planning. Unlike compensatory responses to fishing, which are temporary and predicted to return to a prior state if fishing pressure is reduced, evolved changes to reproductive traits arise from alterations to the genotype, and reversing the effects of fishing is likely to be more protracted and may hinder the recovery of overfished stocks (Kuparinen & Merilä, 2007).

Another way in which fisheries can influence evolutionary change is through reducing the genetic variability of stocks, placing them at risk with regard to the deleterious effects of inbreeding and genetic drift (Hauser & Carvalho, 2008). In the orange roughy (*Hoplostethus atlanticus*, Trachichthyidae), heavy fishing that resulted in a rapid decline in population size was accompanied by a significant loss of genetic diversity, measured using polymorphic enzyme loci (Smith *et al.*, 1991). In *G. morhua*, temporal changes in microsatellite diversity between 1954 and 1998 were investigated using archived samples of otoliths (Hutchinson *et al.*, 2003). Marked genetic changes were detected over the period associated with changes to spawning stock size. These effects were attributed to genetic drift, and indicated that effective population size, the number of breeding individuals in the population, was substantially smaller than the estimated spawning stock size. Loss of genetic diversity was mitigated in this population by migration from other stocks, which may have represented an additional response to high levels of fishing and reduced population density. The implications for long-term population reproductive output of these distinct genetic changes are uncertain and further studies are warranted to examine their impact.

Control of reproduction in aquaculture

Fish cultivation is a rapidly expanding industry, with approximately 50% of total seafood supply currently coming from aquaculture (Natale *et al.*, 2012). The capacity to control and manipulate the reproductive biology of cultivated species will be key to the long-term development of the aquaculture industry, both in terms of the scale and intensity of production and in terms of the range of species under cultivation. Choice of species for cultivation is driven largely by market forces rather than ease of cultivation (Reay, 1984), and the domestication and propagation of some species represent a challenge, particularly in terms of generating adequate numbers of juveniles for ongrowing to a marketable size. Some traditional sectors of the aquaculture industry still rely on the capture of larval and juvenile stages for stocking on fish farms. Examples include milkfish (*Chanos chanos*, Chanidae) cultivation in the Philippines, and eel (*Anguilla* spp., Anguillidae) cultivation in Japan and Europe. This practice relies wholly on natural reproduction and is thus exposed to uncertainties associated with variation in natural reproduction. In other cases, sexually mature adults are collected immediately prior to spawning and are stripped of their gametes for artificial fertilisation, with subsequent rearing of eggs, larvae and juveniles for farm stocking. This practice was formerly routine in the cultivation of salmonids in Europe and North America.

Capture fisheries and aquaculture can be viewed as existing at two ends of a continuum; the distinction is simply the degree of control that is exerted over the life cycle of the species in question (Anderson, 2002). In the case of capture fisheries, control is principally in the size at which harvesting takes place. In the case of intensive aquaculture, all life stages, from fertilisation to harvesting, are under the control of the aquaculturist. Aquaculture offers substantially greater scope for intervening in reproduction and thereby for increasing productivity beyond the typical yields of capture fisheries. Extensive aquaculture practices, such as ranching, blur the edges between capture fisheries and intensive aquaculture, but still permit manipulation of reproduction at key stages.

Species used in aquaculture do not conform to any general pattern of reproductive style, with a wide range of reproductive modes represented, including egg-scattering taxa (Chinese and Indian carp species, Cyprinidae), egg-hiding species (salmon and trout, Salmonidae), migratory pelagic spawners, (black kingfish, *Rachycentron canadum*, Rachycentridae), egg guarders (channel catfish, *Ictalurus punctatus*, Ictaluridae) and mouthbrooders (tilapias, *Oreochromis* spp., Cichlidae). Semelparous and iteroparous species are both represented, and while the majority of aquaculture species are gonochorists, some are hermaphrodites (e.g. gilthead sea bream, *Sparus aurata*, Sparidae). This variety of reproductive modes suggests that no particular reproductive strategy is necessarily incompatible with aquaculture, and it is also a reflection of the wide range of aquaculture practices in operation (Reay, 1984).

The control of reproduction in aquaculture serves two different functions. Facilitating reproduction is a critical aspect of hatchery operations, in which generating further cohorts of fish for ongrowing (also termed 'grow-out') is a critical component of any aquaculture venture. By contrast, reproduction is at best an irrelevance, and more usually a hindrance, to the production of fish for market, as energy diverted to gonad development comes at a cost to somatic growth. Sexual maturation may also result in undesirable behaviours, such as territoriality and aggression, which are energy-demanding and can result in injury and death. The exception is when the gonads themselves are the chief marketable product, such as in caviar production in sturgeons (Acipenseridae). Thus, the aquaculture industry demands control of reproduction, in terms of stimulating reproductive productivity for hatcheries while simultaneously requiring its suppression for ongrowing operations. Appropriate management of broodstock in aquaculture, particularly in terms of feeding, water quality and basic husbandry, has a substantial impact on the quantity and quality of offspring produced for ongrowing operations. The impact of environmental conditions, especially diet, has been the subject of extensive research.

Facilitating reproduction

Environmental manipulation

Spontaneous reproduction is a feature of extensive aquaculture systems, where human intervention is limited and may consist simply of the removal of predators and the provision of some food supplements. Examples include the cultivation of tilapias (Cichlidae), mullets (Mugilidae) and seabream (Sparidae) in brackish lagoons. In other systems, a suitable substrate may be provided to promote natural reproduction. In Chinese carp (Cyprinidae) culture, mature fish selected for reproduction (termed 'broodstock') are placed in special spawning ponds containing aquatic vegetation that is required as a substrate on which to deposit eggs (Chapter 6). Once spawning is completed, the broodstock are removed and the eggs are harvested for rearing. Similarly, in the production of *I. punctatus* in the southern states of the USA, broodstock are provided with spawning containers, such as wooden boxes, crates or metal drums, which provide pairs of fish with a cavity in which to spawn. After spawning, the male parent may be allowed to complete care of the eggs and young stages, which are subsequently harvested for rearing, or the farmer may intervene and remove the hatchlings immediately after incubation of the eggs.

In theory, the appropriate conditions for spontaneous reproduction could be provided in any aquaculture system. However, to do so in practice might be logistically impractical, or the reproductive requirements of the species under cultivation may be too poorly known to permit provision of the appropriate conditions. One example of an ingenious aquaculture system that was specifically designed around the mating system of the species under cultivation was that developed at the Baobab Tilapia Farm in Kenya for lek-breeding mouthbrooding tilapias (Balarin & Haller, 1983). The system comprises three sections: a breeding arena in which lekking males can establish small territories for display; a brooding area, separated from the arena by narrow grills, in which females care for eggs and larval stages without interference from males; and a shallow fry release area from which young stages can be harvested after independence from the mother without disturbance to the broodstock (Fig. 12.8). The selective division of males and females is possible by stocking males with a larger body size than females, so that while females are able to swim through the grills separating the brooding area from the arena, males are solely confined to the arena. This arrangement permits females to enter the arena to spawn and, after fertilisation of her eggs, to return to the brooding area to carry out parental care. The shallow lip at the margin of the fry release area exploits the natural behaviour of newly released juvenile tilapia to shoal in shallow water to avoid fish predators, including older conspecifics. Non-return gates at the margin of the shallow lip prevent the return of juveniles to the brooding area. The fry release area can be drained independently of the breeding arena and brooding area, thereby facilitating the passive removal of juveniles from the broodstock. A typical stocking regime for this arena breeding system would be 300 female and 100 male broodstock, which could produce >7000 independent juveniles daily for ongrowing and with minimum labour requirements or disturbance to broodstock.

Endocrine manipulation

A key requirement for the establishment of a sustainable aquaculture industry is the ability to control aspects of the reproductive biology of the cultivated species, allowing control of the sex of the fish, the timing of reproduction and the quality of the gametes (Mylonas *et al.*, 2010). As knowledge of the endocrine regulation of reproduction has increased, the information has been used by the aquaculture industry to address two unrelated problems posed by the reproductive biology of species used in aquaculture systems. The first problem is that some cultivated species show a degree of reproductive dysfunction in captivity (Mylonas & Zohar, 2007). A second problem is that the aquaculture industry would prefer to rear monosex populations to a marketable size (Pandian & Sheela, 1995). A solution to the first problem exploits the activating effects of hormones in sexually mature fish, while a solution to the second problem exploits organisational effects of sex steroids during sexual differentiation.

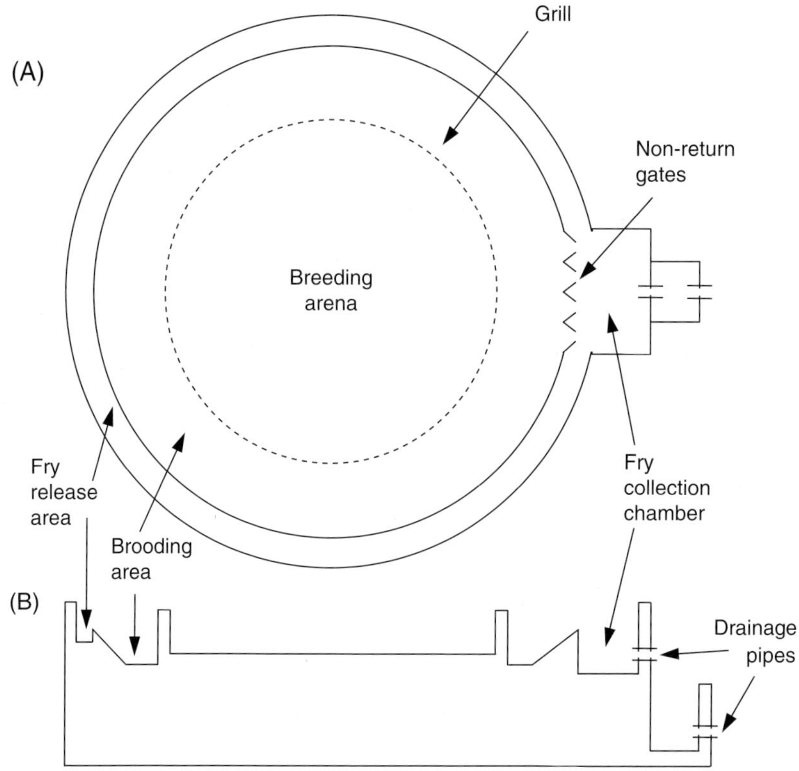

Figure 12.8 Arena system for lek-breeding tilapia. (A) Plan view showing the central breeding arena for males, brooding area for females (separated from the arena by narrow grills) and a shallow fry release area that can be drained into the fry collection chamber, independently of the rest of the arena. (B) Cross-section of the arena system.

Sexual dysfunction in cultivated fishes

Under natural conditions the endocrine–reproductive axis, linked to appropriate environmental cues, ensures that maturation and reproduction occur at a time and under conditions that maximise reproductive success (Chapter 6). In captivity, these processes are frequently disrupted, resulting in sexual dysfunction. Sexual dysfuntion can manifest itself in different ways, from partial inhibition of some reproductive processes, resulting in a reduction in the quantity or quality of gametes, to a complete failure to spawn. Female fishes seem to be more susceptible to sexual dysfunction than males.

Sexual dysfunction in females

Sexual dysfunction in cultivated females takes three forms. The rarest is the failure of vitellogenesis, such as is seen in the European eel (*Anguilla anguilla.*, Anguillidae). The second is probably the most common: the oocytes complete vitellogenesis, but do not undergo oocyte maturation and so ovulation and spawning fail. The third problem, seen in salmonids, is that ovulation occurs but spawning does not proceed naturally (Fig. 12.9). These problems are probably a result of the stress imposed by the culture conditions allied to the failure of those conditions to provide the appropriate physical and social environment required for successful spawning. Studies on striped bass

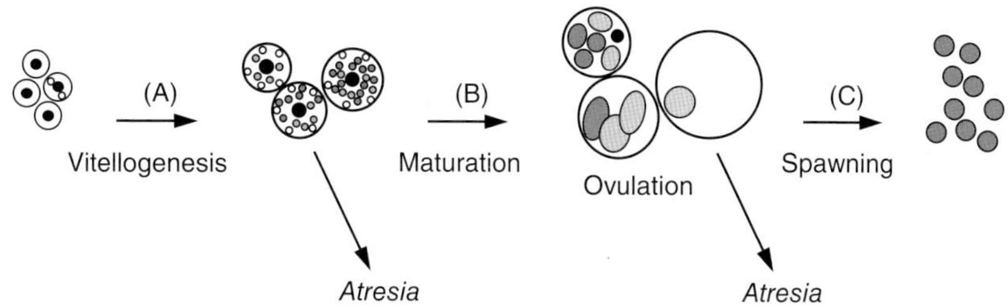

Figure 12.9 Principal sexual dysfunctions in females in aquaculture. (A) The inhibition of vitellogenesis; (B) inhibition of oocyte maturation, resulting in atresia of post-vitellogenic oocytes; (C) inhibition of spawning, resulting in retention of ovulated oocytes. (Source: Mañanós *et al.* 2009. Reproduced with permission of Taylor and Francis Group.)

(*Morone saxatilis*, Moronidae) indicate that the failure of oocyte maturation in many cultivated species is a consequence of the failure of the pituitary to release luteinising hormone (LH) in the surge that initiates the maturation, rather than effects at the level of gonadotrophin-releasing hormone (GnRH) synthesis and release or LH synthesis (Zohar & Mylonas, 2001; Mylonas & Zohar, 2007; Mylonas *et al.*, 2010). The endocrine alleviation of the female sexual dysfunction relies on the activational effects of hormones.

Most endocrine treatments focus on the problem of the failure of oocyte maturation and the succeeding ovulation and spawning (Cnaani & Levavi-Sivan, 2009; Mylonas & Zohar, 2007; Mylonas *et al.*, 2010). Two approaches are used. The first technique to be developed was hypophysation, which depends on supplying LH to the gonads. In its earliest form, hypophysation achieved this by injecting extracts of pituitaries collected from sexually mature fish. This method is still used effectively in some parts of the world because it is cheap. It is used, for example, to induce spawning in some Asian carps. There are some disadvantages. Inevitably, there is uncontrolled variation in the LH content of the pituitary extracts. Secondly, there is the danger that the extracts will transmit disease from the donor to the recipient fish, and the extracts may also induce an immunological response in the recipients.

As purified salmon and carp LH became available, these were used, but they have the disadvantage that they can only be used in closely related species because of the species specificity of teleost LHs. As an alternative, human chorionic gonadotrophin (hCG) is frequently used to induce ovarian maturation and spawning, because of its LH activity and easy availability. Both piscine LH and hCG can induce an immune response in the recipient fish, which adversely affects subsequent use in the same fish.

An advantage of hypophysation techniques is that they directly target the ovary and so can act quickly. They do not depend on the level of LH stored in the pituitary of the recipient fish or on the ability of the gonadotrophs of the recipient fish to synthesise LH.

As the understanding of the endocrine control of reproduction in teleosts increased, an alternative approach was developed. It used the hypothalamic GnRH, or more commonly, synthetic GnRH analogues (GnRHa; Zohar & Mylonas, 2001; Mylonas & Zohar, 2007; Mylonas *et al.*, 2010). The analogues are typically more potent than the native GnRH and so can be used in smaller doses. They remain in the blood circulation for longer than natural GnRH and stimulate a stronger response and a longer period of release of gonadotrophin (GTH). In application, the GnRHa may be injected or, for longer-term effect, provided as an implant, releasing the GnRHa over days or weeks.

Gonadotrophin-releasing hormone and its analogues have several advantages. As small decapeptides, they do not induce an immune response. Because of the level in the brain–pituitary–gonad (BPG) reproductive axis at which they operate, it is possible that they can overcome the physiological block at the pituitary level caused by the environmental conditions the fish experience in culture. However, in contrast to LH, they do not function directly at the level of the gonad.

The choice of the injection or the implantation of the GnRHa can relate to the dynamics of oocyte maturation. In species with the synchronous development of oocytes and total spawning, injection may be adequate. For species with group synchronous oocyte development and multiple spawning, with an interval of days between successive spawnings, injections before each spawning may be adequate, but the longer-term release of GnRHa from an implant can stimulate oocyte maturation over several cycles. Implants are also valuable in species with asynchronous oocyte development that spawn at intervals of a day or two over an extended spawning season.

Even in species such as the salmonids that do ovulate under culture conditions, the techniques of inducing ovulation may be used to synchronise the timing of spawning in broodstock. This facilitates the management of the broodstock. For species in which there is significant dopaminergic inhibition of oocyte maturation, such as the cyprinids (Chapter 5), the GnRHa treatment is combined with the use of dopamine (DA) antagonists to overcome the DA inhibition.

Sexual dysfunction in males

In males, the sexual dysfunction frequently takes the form of the production of reduced quantities of sperm (termed 'milt' in an aquaculture context) or a lower quality of sperm (Mylonas *et al.*, 2010). In some flatfishes (Pleuronectiformes), the males in culture may produce sperm that is sticky and viscous, fails to mix easily with water and does not fertilise the eggs. Males tend to spend longer on the spawning grounds than females, especially in species with total spawning. As a consequence, the use of implants releasing GnRHa may be preferred over injections. In the production of hybrid *M. saxatilis*, male white bass (*Morone chrysops*, Moronidae) are stripped of sperm at regular intervals for use in the fertilisation of eggs from female striped bass, *M. saxatilis*. GnRHa implants increase sperm production without any obvious decrease in sperm quality. In the European sea bass (*Dicentrarchus labrax*, Moronidae), a single injection of GnRHa increased sperm production for about 7 days, whereas an implant increased sperm production for about 30 days.

Preservation of gametes

The long-term preservation and storage of fish gametes has been viewed as a key development in aquaculture. Benefits come from the capacity to maintain genetic lines of fish without the economic costs and risk of disease associated with housing broodstock. Preserved gametes also provide a continuous supply of gametes throughout the year, are easier to transport, and facilitate selective breeding. Access to a bank of stored gametes also offers the aquaculturist access to a greater range of species and genetic lines than can be accommodated in a conventional hatchery (Gwo *et al.*, 2009). Preservation of gametes also has utility in fish conservation, for archiving the gametes of rare and endangered species (Maitland & Lyle, 1992), and in scientific research through providing access to inbred, transgenic, knockout and mutant lines of fish that would otherwise be difficult or expensive to generate and maintain (Yang & Tiersch, 2009).

Two techniques are used in the preservation of fish gametes, cryopreservation and supercooling. In cryopreservation, gametes are stored at subzero temperatures, thereby interrupting biological activity. Spermatozoa are readily subject to cryopreservation, and protocols for cooling and freezing the spermatozoa of numerous teleosts have been established (Cloud & Patton, 2009; Gwo *et al.*, 2009). In theory, the storage time for spermatozoa is limited only by minor external effects, such as background irradiation, making the storage time effectively unlimited (Cloud & Patton, 2009; Gwo *et al.*, 2009). In practice, the success of cryopreserved spermatozoa in fertilising eggs can deteriorate with length of storage (Kerby *et al.*, 1985; Chao *et al.*, 1987). The quality of sperm, as well as the procedures used to strip and preserve it, can determine the viability of the spermatozoa following cryopreservation. Sperm is diluted with a freezing solution containing an extender, a physiological solution that will not activate spermatozoa activity, and a protective agent to prevent cell damage during freezing and thawing (Cloud & Patton, 2009). Sperm is stored as pellets or in straws and frozen at −196 °C in liquid nitrogen. The performance of fish produced using cryopreserved spermatozoa is typically comparable to those fertilised using control spermatozoa (Tiersch *et al.*, 1994; Rana & McAndrew, 1989; Suquet *et al.*, 2000). Unexpectedly, when rainbow trout (*Oncorhynchus mykiss*, Salmonidae) eggs were inseminated with cryopreserved spermatozoa from males that were themselves produced with cryopreserved spermatozoa, the result was a significantly higher survival to the eyed egg stage than in controls. A possible explanation is that cryopreservation may inadvertently select for heterozygous zygotes (van der Bank & Steyn, 1992).

While cryopreservation techniques for spermatozoa have been successfully refined, typically using protocols that are taxon-specific, attempts to cryopreserve eggs and fertilised embryos have proven almost wholly unsuccessful. The reason is that fish eggs are relatively large and complex compared with spermatozoa, possessing membranes with different permeabilities that do not permit the rapid penetration of cryoprotectants. The result is intracellular formation of ice crystals with consequent damage to the egg or embryo.

In contrast to cryopreservation, supercooling of spermatozoa and eggs is readily feasible. Here the gametes are stored at temperatures below 0 °C in a medium lower than its theoretical freezing point, thereby avoiding ice formation. This procedure is successful for coldwater species and permits the storage of viable gametes for several weeks. The gametes of warmwater species cannot withstand supercooling (Gwo *et al.*, 2009).

Suppressing reproduction

Monosex production

One of the primary means of controlling reproduction during the ongrowing phase of aquaculture production is through the production of monosex lines. There are several possible advantages that accrue from the culture of monosex populations (Beardmore *et al.*, 2001). One sex may have faster growth rates and so reach a marketable size in a shorter time and with a greater uniformity of size. If the fish are harvested before sexual maturity, then resources are not diverted into gonadal production or reproductive behaviour. Precocial maturation of one of the sexes is a problem in some species. In monosex cultures, there may be lower frequencies of aggressive interactions. In some species, the onset of maturity may be characterised by undesirable changes in the appearance of the fish or the quality of the flesh.

There are several methods by which monosex can be produced, but hormonal manipulation is a favoured technique. Other techniques include hand sorting, which relies on the presence of morphological differences between the sexes. However, this method is laborious, time-consuming and error-prone. Some interspecies hybrids can produce single-sex progeny (see the section 'Hybridisation'). Various chromosomal manipulations such as the production of gynogenetic lines (see the section 'Gynogenesis') are also used to produce single-sex lines.

Endocrine manipulation

The lability and bipotentiality of sexual differentiation in teleosts (Chapter 2) are exploited to produce monosex lines by exposing the early developmental stages of the fish to relevant steroids, which can be either natural or synthetic. Over 30 steroids have been used and sex reversal has been achieved in over 50 species from 15 different families (Beardmore *et al.*, 2001; Piferrer, 2001). This taxonomic range indicates that lability of sexual differentiation is a widespread attribute of teleost fishes. Steroids with oestrogenic properties are used to feminise genetic males, while those with androgenic properties are used to masculinise genetic females.

Although injection of the appropriate steroid can be used to administer the hormone, the small size of the fish during the period when they are most sensitive to the hormone (the labile period) and the time consumed in injecting individual fish make this method suitable only for research rather than commercial activities. Two methods are used to treat high numbers of fish: immersion or incorporation in food (Piferrer, 2001). Immersion is suitable for the embryonic or larval developmental stages, applied either periodically or continuously. The steroid is taken up either over the gills or through the integument. In salmonids, the steroid accumulates in the yolk of the egg and is probably released as the embryo and larva develop. Food as the vehicle for the treatment with a steroid can only be used if the labile period coincides with the adoption of exogenous feeding by the treated fish. The steroid is dissolved in a solvent such as alcohol and mixed with the food. The method is easy to apply, but has several disadvantages, including a potential loss of over 99% of the hormone within 24 hours and difficulty in ensuring a uniform and consistent intake by individuals (Pandian & Sheela, 1995). Both the dietary and immersion techniques can lead to sterility or paradoxical sex reversal if the treatment is too intense. Paradoxical sex reversal occurs when a treatment designed to produce sex reversal in a given direction causes sex reversal in the preferred sex in the opposite direction.

Several variables have to be manipulated to achieve the best results, including the type and nature of steroid used, the timing and duration of the treatment, the dose applied and husbandry variables such as water temperature and rearing density. The optimal conditions have to be determined empirically, species by species (Pifferer, 2001).

Perhaps the most important variables that have to be identified are the timing and length of the labile period. This is the period when the sexually undifferentiated gonads are most sensitive to the effect of the exogenous steroids (Piferrer, 2001). Treatment in this period should ensure that the dosage and duration of exposure can be minimised. Once the gonads have differentiated as either ovaries or testes, species differ in their sensitivity to steroids, so some species do not respond to the treatment, while in others some degree of sex reversal can still be induced, again reflecting the sexual bipotentiality of teleost gonads.

Species can be divided into two groups by the timing of the labile period (Pandian & Sheela, 1995; Piferrer, 2001). In some species, the period occurs around hatching and lasts for about 10–40 days. In cichlids, such as the blue tilapia (*Oreochromis aureus*, Cichlidae), the labile period lasts from about 10 to 30 days after hatching. In the Siamese fighting fish (*Betta splendens*, Osphronemidae), the labile period extends from about 3 to 40 days after hatching. In the second group of species, the labile period is much later in development, falling in the juvenile phase of the life cycle in species such *D. labrax* and the grass carp (*Ctenopharyngodon idella*, Cyprinidae). There is no obvious phylogenetic pattern in the differences in the timing of the labile period. Generally, the labile period is earlier in development for oestrogen treatment than for androgen treatment, reflecting the earlier differentiation of the gonads in female teleosts (Chapter 3). A major question is whether the treatment with exogenous steroids mimics the natural effect of endogenous hormones on sexual differentiation or represents a pathological disturbance of the regulation of differentiation by exposing the gonad to unusual concentrations of the steroid.

Production of female monosex lines

For some species, females are preferred in aquaculture (Pandian & Sheela, 1995; Piferrer, 2001). In salmonids, females are favoured because of their better growth performance, better flesh quality and the tendency for male salmonids to undergo precocial sexual maturation. The natural sex steroid most used in feminisation is oestradiol-17β (E_2), but synthetic oestrogens, such as ethinyloestradiol (EE_2) and diethylystilbestrol (DES), are amongst the most potent feminising steroids. High dosages of aromatisable androgens can produce paradoxical feminisation, probably because of the aromatisation of the androgen to an oestrogen. The intensity of treatment with an oestrogen that is required to achieve a high rate of feminisation varies among teleost families with the following rank order: Salmonidae < Cichlidae < Anguillidae < Osphronemidae = Poeciliidae < Cyprinidae. Viviparous cyprinodonts, such as the guppy (*Poecilia reticulata*, Poeciliidae) require high doses, probably because the steroid has to penetrate the body of the mother to reach the developing embryos in the ovary (Chapter 10).

There are two protocols used to feminise fish with oestrogen, direct or indirect feminisation. In the direct protocol, the oestrogen is presented during the labile period and can be used regardless of the sex determination system or which sex is the heterogametic sex. The method has the advantages that feminisation is achieved within a generation, typically there is little or no mortality associated with the treatment, and the feminised fish cannot be distinguished phenotypically from normal females. A major disadvantage is that, at least in some countries, the public is apprehensive about buying fish that have been treated with hormones, although the evidence suggests that the hormone has been completely eliminated before the fish are harvested.

The indirect protocol ensures that the monosex fish that are harvested have not experienced exposure to the feminising steroids. However, it requires more generations of breeding to achieve the monosex culture and requires that the females are the homogametic sex (XX; see Chapter 2). The first step is to masculinise genotypic females (XX) by androgen treatment in the labile period. The sperm from these 'neomales' (female genotype, but male phenotype) is then used to fertilise eggs from normal females (genotype XX), producing progeny that all have the XX genotype. The neomales in the F_1 generation have to be identified by progeny testing. A test consists of individual crosses using the sperm from androgen-treated fish and eggs from untreated females. Only crosses using sperm from neomales will result in an all-female F_2 generation (genotypic female and phenotypic female). Crosses using normal males (genotype XY) will result in offspring of both males and females. Once the all-female stock has been generated, it can be maintained by masculinising a small proportion of the offspring each year to provide the neomales required for future crosses. The rest of the all-female stock can be ongrown to a marketable size. The indirect protocol has been used successfully to produce all female progeny in cyprinids (*Carassius auratus*), salmonids (*Salmo salar*) and cichlids (*O. niloticus*). If DNA markers are available for the identification of the genetic sex, the need for the progeny testing stage is overcome. The indirect method of feminisation has been used commercially in the culture of *O. mykiss* and Chinook salmon (*Oncorhynchus tshawytscha*, Salmonidae).

If the sex determination mechanism is polygenic or there is environmental sex determination, a range of sex ratios is generated by crosses between sex-reversed and normal fish, which makes the method difficult to establish.

Production of male monosex lines

In some species, the males have a higher growth rate and so monosex male cultures are preferred. Both natural and synthetic steroids have been used to masculinise fish, but the androgen most commonly used in the direct protocol for masculinisation is the synthetic 17α-methyltestosterone (17α-MT; Pandian & Sheela, 1995; Beardmore *et al.*, 2001). This steroid has been tested in a range of species, including Salmonidae, Cichlidae and Cyprinidae.

Oreochromis niloticus is a valuable species for aquaculture in sub-tropical and tropical regions. It can utilise a wide range of food types, is fast-growing, tolerant of poor water quality and produces flesh of good quality (Beardmore *et al.*, 2001). The main disadvantage of the species is that, under cultivation, it tends to show precocial maturation, resulting in the recruitment of small, low-quality offspring. It is also a species in which the male growth rate is higher than that of females. The male is the heterogametic sex (XY).

The indirect protocol for producing all-male *O. niloticus* starts by treating genotypic males and females with oestrogen during the labile period, so producing some progeny that are 'neofemales', with a male (XY) genotype, but a female phenotype. These neofemales are identified by progeny testing and then crossed with normal males. This cross produces progeny with the genotypes YY, XY and XX. The YY males, so-called 'supermales' are identified by progeny testing and then crossed with neofemales (XY). The progeny are treated with oestrogen during the labile period, producing XY and YY females. These YY females are again identified by progeny testing. The cross of YY females with YY males produces YY males that can be used as broodstock. The cross of YY males with normal XX females produces male monosex progeny, which can then be grown to marketable size. The success of this indirect method depends on the YY supermales being both viable and fertile. Unusually in vertebrates, the YY genotype is viable in some teleosts, exemplified by *O. mykiss*, *P. reticulata*, medaka (*Oryzias latipes*, Adrianichthyidae) and goldfish (*Carassius auratus*, Cyprinidae). By contrast, in some species, the YY supermale is inviable, as in the convict cichlid (*Amatitlania nigrofasciata*, Cichlidae).

The indirect protocol can also be used to produce male monosex lines in species in which the male is the homogametic sex (ZZ), with the female being the heterogametic sex (ZW). Genetic males and females produced by normal female (ZW) and normal male (ZZ) crosses are treated with oestrogen in the labile period, resulting in some neofemales (genotypic males, ZZ, but phenotypic females) and normal females (genotypic female, ZW, phenotypic female) in the F_1 generation. Progeny testing identifies the neofemales and these are crossed with normal males (ZZ) to produce all-male progeny in the F_2 generation. As more detail on the mechanisms of sex determination in teleosts becomes available (Chapter 2), the hormonal manipulations during the labile period should become more efficient over a greater range of sex-determining mechanisms.

Gynogenesis

In gynogenesis a haploid egg is induced to become diploid without being fertilised by a spermatozoon. If the female is the homogametic sex, the haploid egg will carry an X chromosome and all eggs induced to become diploid will develop into XX females. In the case of female heterogamety the haploid egg will carry either the Z or W sex chromosome. Gynogenetic eggs induced to be diploid will then be either WW or ZZ, with the latter developing as males. Gynogenesis occurs naturally in a small number of teleost species (Chapter 10) and is harnessed in aquaculture for producing monosexes, as well as inbred lines and clonal populations.

The production of gynogenetic individuals (gynogens) is a two-step process. Development of a normal egg must first be initiated by a spermatozoon that has had its complement of DNA inactivated in some way. Once embryo development is initiated, the haploid egg, with its exclusively maternal genes, must then be induced to achieve a diploid state for development to be completed successfully. Inactivation of spermatozoa DNA can be readily accomplished in several ways. UV radiation is an easy and effective method that is widely used. Unlike techniques using ionising radiation, which break DNA into fragments, UV radiation acts through inducing the formation of thymine dimers, thereby inhibiting DNA replication. An outcome is that UV inactivation is not associated with the expression of supernumerary paternal chromosome fragments, which is a feature of ionising radiation and chemical mutagens. The disadvantages of using UV radiation is that only small amounts of sperm can be irradiated because of its low penetration capacity. Ionising radiation, such as X and gamma radiation, has the advantage of a high penetrating capacity, but the disadvantages of being dangerous to use and of the expression of paternal traits resulting from supernumeray paternal chromosome fragments. Chemical mutagens, such as dimethylsulphate, can also be used to inactivate spermatozoa, and this technique permits the treatment of large amounts of sperm, although their use for this purpose is increasingly prohibited. Chemical mutagens are also associated with problems arising from the expression of supernumerary paternal chromosome fragments in the embryo (Urbányi *et al.*, 2009).

After activation of the egg by inactivated spermatozoa, there are two ways in which diploidisation can be achieved. After fertilisation, during normal egg development, the egg emerges from arrest in the metaphase II stage of meiosis II, which results in the formation of two haploid cells. One of these is the egg, the other is a small second polar body, which, like the first polar body, degenerates, resulting in the production of a single functional haploid gamete. If the egg is fertilised by an inactivated spermatozoon that is unable to contribute a functional haploid genome, the haploid embryo will eventually die. However, loss of the second polar body can be inhibited, thereby producing a diploid gynogen with wholly maternal chromosomes. Retention of the second polar body can be accomplished by temperature (either hot or cold) or pressure treatments, termed 'shocks'. The imposition of shocks must be applied after spermatozoon penetration, but before loss of the second polar body. The timing and nature of shocks are species-specific but usually imposed within minutes of fertilisation (Table 12.2). Gynogens produced in this way are termed 'meiotic gynogens' and are inbred. However, they are not entirely homozygous, as the chromosome sets in the egg and second polar body are not identical as a result of crossing-over and recombination during meiosis I. By contrast, 'mitotic gynogens' are entirely homozygous and are produced by blocking the first mitotic division of the egg after fertilisation, thereby doubling the chromosome number to a diploid state. Cell division can be blocked chemically or by using temperature and pressure shocks.

Table 12.2 Techniques used in the production of triploids and tetraploids. (Source: Adapted from Dunham 2011. Reproduced with permission of CABI Publishing.)

Species	Ploidy	Technique	Shock intensity	Time after fertilisation (minutes)	Duration (minutes)	Proportion of polyploids (%)
Salmo salar	Triploidy	Heat	32 °C	20	5	95
Oncorhynchus mykiss	Triploidy	Heat	26 °C	1	10	100
O. mykiss	Triploidy	Hydrostatic	7000 psi	40	4	100
Scophthalmus maximus	Triploidy	Cold	0 °C	5	20	90
Sparus aurata	Triploidy	Heat	37 °C	3	2.5	100
Ctenopharyngodon idella	Triploidy	Hydrostatic	7000 –8000 psi	4	1–2	98
C. idella	Triploidy	Heat	42 °C	4	1	67
Oreochromis aureus	Triploidy	Heat	39.5 °C	3	3.5–4	100
O. mykiss	Tetraploidy	Hydrostatic	7000 psi	350	4	100
Hypophthalmichthys nobilis	Tetraploidy	Hydrostatic	7350 psi	36	1.5	–
Labeo rohita	Tetraploidy	Heat	39 °C	–	2	70
L. rohita	Tetraploidy	Cold	10–15 °C	–	10	30–55

Meiotic gynogens can be produced more reliably than mitotic gynogens, possibly because the latter express a higher frequency of lethal recessive genotypes and die before completing embryo development. Overall the survival of gynogens is low, a result of elevated homozygosity. Thus egg mortalities typically exceed 99%, making large-scale production of monosexes in this way uneconomic. However, gynogenesis can be used in tandem with endocrine sex reversal to produce monosexes on a commercial scale. If the female is the homogametic (XX) sex, inducing gynogenesis will generate progeny that are all XX. If these gynogens are masculinised using androgens to produce neomales, and these are fertile, they can be mated with normal XX females to produce XX females that will exhibit high survival rates. This technique has been successfully applied to *C. idella* (Boney *et al.*, 1984) and silver carp (*Hypophthalmichthys molitrix*, Cyprinidae) (Mirza & Shelton, 1988). The advantage of using gynogenesis is that it makes progeny testing unnecessary. The disadvantages of this system are that production of gynogens is difficult or impossible in some species, and there is a greatly elevated risk of inbreeding (Dunham, 2011).

Androgenesis

Like gynogenesis, androgenesis can generate monosexes if males are the homogametic sex. Androgenesis involves fertilising inactivated eggs followed by the doubling of the paternal genome. Egg inactivation can be accomplished using the same techniques for inactivating sperm, i.e. UV and gamma irradiation, although lower radiation doses are required to inactivate eggs than spermatozoa. An additional consideration in the inactivation of eggs is the disabling of mitochondrial DNA (mtDNA) and damage to the egg cytoplasm. However, mtDNA appears to be more resilient to the effects of irradiation than nuclear DNA, possibly because its location in the mitochondria provides an additional protective membrane. Furthermore, its small size, circular conformation and large copy number may limit the impact of irradiation. In consequence, androgens typically have a normal complement of mtDNA of maternal origin.

The inactivated egg is fertilised using untreated sperm and a temperature shock is imposed to interrupt the first mitotic division, thereby generating a diploid state. Thus, androgens are analogous to mitotic gynogens and exhibit 100% homozygosity. Like gynogenesis, androgenesis is associated with high early mortalities as a result of a high frequency of lethal recessive genotypes and is not an economically viable technique for mass production of monosexes. However, like gynogenesis, the procedure can successfully be used to generate monosexes by combining the technique with hormonal sex reversal. Thus, in male homogametic (ZZ) species, androgens can be crossed with hormonally feminised males to produce viable male monosexes.

Hybridisation

Interspecific crossing of related species is undertaken in aquaculture to produce hybrids with more rapid growth rates, enhanced disease resistance and greater tolerance of low water quality than either of the parental species (Bartley *et al.*, 2001; Hulata, 2001). In some cases, hybridisation produces monosexes, or can at least skew the sex ratio strongly towards one sex or the other. Skews in the sex ratio arise from crosses between species with contrasting sex-determining systems. In the tilapias (*Oreochromis* spp.) both male (XX, XY) and female (ZZ, ZW) heterogamety occur. A cross between *O. niloticus*, which are male heterogametic, and *O. aureus*, which are female heterogametic, produces offspring that are predominantly (80–90%) male. In the case of this cross, the failure to produce entirely male offspring suggests some environmental and autosomal influence on sex determination. Other tilapia species used for producing monosex hybrids include the male heterogametic Mozambique tilapia (*Oreochromis mossambicus*, Cichlidae) and female heterogametic longfin tilapia (*Oreochromis macrochir*, Cichlidae) and Wami tilapia (*Oreochromis urolepis hornorum*, Cichlidae).

The attraction of using hybrids comes from its relative simplicity and because exposure to hormone treatments, which consumers may be wary about, are not required. In practice, maintaining pure lines of parental species, even in the best managed hatcheries, can be difficult and there is a tendency for the skew in the sex ratio of hybrids to erode over time as occasional inadvertent backcrossing occurs (Hulata, 2001).

Sterilisation

An alternative to sex manipulation is the inhibition of reproduction altogether through the production of sterile fish. Sterilisation can be achieved in several ways, including relatively crude techniques, such as surgery, irradiation and chemical treatment, but also through hybridisation, manipulation of chromosome number and transgenic gene manipulation.

Mechanical, immunological, chemical and irradiation techniques

Gonadectomy, surgical removal of the gonads, in teleosts is readily accomplished, but it is impractical on a commercial scale. Further, teleost fishes show remarkable powers of gonad regeneration. Complete removal of testes or ovaries, even in sexually mature individuals, does not preclude their eventual regeneration.

Another technique to achieve sterilisation has been to induce an immunological response to spermatozoa in males by injecting testicular material, usually with an adjuvant. This procedure has been attempted in salmonids (Laird *et al.*, 1981) and tilapias (Lou *et al.*, 1989), and though the procedure elicits a measureable response, including some limitations on the production of viable spermatozoa, it has never been applied in commercial aquaculture operations.

The success of using X-rays and gamma rays to sterilise fish for aquaculture purposes has been mixed. Although ionising radiation can cause sterility, it typically has only a short-term impact on inhibiting maturation and gametogenesis. Doses of irradiation that are needed to cause sterility can also cause deleterious side-effects, including mortality and an overall loss of vigour. Because irradiation is either unreliable or temporary and has negative side-effects, it has limited application in aquaculture (Stanley, 1979).

Chemical sterilisation can be accomplished through a variety of mechanisms. Hormone antagonists can inhibit the release and action of reproductive hormones, thereby inhibiting reproduction. The GTH inhibitor, methallibure, administered to stinging catfish (*Heteropneustes fossilis*, Heteropneustidae) resulted in a significant decline in levels of the GTH-II in plasma and the pituitary (Chapter 5; Senthilkumaran & Joy, 1998). Levels of E_2 and testosterone (T) were reduced as a consequence, and gonad function was disrupted. Methallibure has a comparable inhibitory effect on reproduction in other fish taxa (Leatherland, 1969; Pandey & Leatherland, 1970; Hyder, 1972; Donaldson & Hunter, 1982). Other chemical inhibitors of reproduction include diethyldithiocarbamate (DDC), cyprotene acetate, trenbolone acetate (TBA), and γ-aminobutyric acid (GABA). While these compounds are effective in halting or severely limiting reproductive function during the period of their administration, the effects are typically temporary with eventual recovery of reproductive viability. The use of these compounds in fish for human consumption is also limited on the basis of food safety issues (Dunham, 2011).

Hybridisation

Many interspecific hybrids are sterile, with increasing likelihood of sterility the greater the genetic divergence between the parental species (Bartley *et al.*, 2001). Male sterility is typically more common than female sterility, a function of stronger sexual selection on male reproductive traits (Chapter 8). Examples of sterile hybrids include crosses between the Indian major carp species (*Catla* spp., *Labeo* spp. and *Cirrhinus* spp., Cyprinidae) with the common carp (*Cyprinus carpio*, Cyprinidae). The cross between the brown trout (*Salmo trutta*, Salmonidae) and brook trout (*Salvelinus fontinalis*, Salmonidae) produces a sterile hybrid, the 'tiger trout'. Crosses between most salmonid species are sterile, but often also show low survival rates.

Sterile hybrids are characterised by chromosomal aberrations, a consequence of differences in parental chromosome number, which result in the failure of homologous chromosomes to pair during meiosis. There is also evidence for the involvement of major genes that result in hybrid sterility (Sun *et al.*, 2004). Sterile hybrids may also be polyploid (see the section 'Polyploidy'). Despite its potential as a technique for producing sterile fish in aquaculture, the commercial use of sterile hybrids is limited. The failure of sterile hybrids to make any significant contribution to aquaculture production possibly reflects the fact that they often have low viability. Low viability can be partly overcome by inducing triploidy in the hybrids. However, commercialisation of a new hybrid also faces obstacles in terms of its ease of production, growth performance, disease resistance and marketability, irrespective of the efficacy of hybridisation as a technique for generating sterility (Knibb, 2000).

Polyploidy

Sterility can be brought about through inducing polyploidy. Polyploid individuals have multiple sets of chromosomes; triploids ($3N$) have three sets, while tetraploids ($4N$) have four and hexaploids six ($6N$). Aneuploids possess the normal diploid ($2N$) set of chromosomes as well as some additional chromosomes, though not a complete additional set. While polyploidy is lethal in birds and mammals, it can produce viable offspring in other taxa, including teleosts (Le Comber & Smith, 2004). Genome duplication represents an important mechanism of evolution and speciation in some taxa.

Polyploid individuals can be created using temperature and pressure shocks. Triploids are created by inducing retention of the second polar body, the same procedures used in producing meiotic gynogens (see the section 'Gynogenesis'). Tetraploids are produced using the same techniques later in embryonic development to block the first mitotic division of the embryo's diploid nucleus. Triploids are usually sterile, a result of the imbalance in their chromosome number effectively inhibiting gamete formation, although male triploids can sometimes exhibit limited sexual behaviour. In some cases, the triploidisation process is not 100% and normal diploid individuals are produced, which can undermine some of the benefits of cultivating sterile stocks (Varadaraj & Pandian, 1990; Dunham, 2011). The advantages of triploids in aquaculture come principally from avoiding negative effects associated with sexual maturation, such as reduced growth and changes to carcass quality. However, triploids may have lower survival rates than diploids, particularly during early development. Triploids sometimes have lower tolerance of low oxygen conditions, an outcome of inferior blood characteristics (Svobodova *et al.*, 1998), and there is some evidence that their disease resistance can be compromised (Langston *et al.*, 2001). Triploids can also express physical abnormalities at unusually high rates (Benfey, 2001). Although often associated with enhanced growth performance, in some cases triploids show lower growth rates than diploids

(Dunham, 2011). Sterile triploids have been used in commercial aquaculture (Benfey, 1999), though rarely with overwhelming success (Benfey, 2001; Hulata, 2001; Dunham, 2011). Triploid *C. idella* have been produced commercially for controlling aquatic plants. The chief advantage of using triploids in this case is that they can be introduced outside their natural range without risk of naturalisation (Shelton, 1995).

In some cases, the poor performance in triploids can be attributed to the techniques used to induce triploidy, which can mechanically damage the early embryo with longer-term impacts on performance (Myers, 1986). An alternative technique for producing triploids is by crossing a tetraploid (4 *N*) with a normal diploid individual (2 *N*), thereby creating 100% triploid offspring (3 *N*). Tetraploids have a balanced set of chromosomes and are usually fertile. Stocks of tetraploid broodstock can be maintained by mating with each other. Triploids produced by tetraploid × diploid crosses are also predicted to show enhanced performance through elevated heterozygosity (Dunham, 2011).

Chapter 13
Coda

Teleost fishes present an evolutionary puzzle. They display virtually every reproductive mode found in vertebrates (Desjardins & Fernald, 2009). Why have the teleosts evolved the reproductive diversity documented in the previous pages? No convincing hypothesis has yet been proposed, but there are several factors that may be relevant.

In terms of number of species, teleosts are by far the most diverse of the vertebrates, with approximately 30 000 species representing about 50% of all vertebrate species (Helfman *et al.*, 2009). The diversity of reproductive modes in teleosts may simply reflect their species diversity. Associated with this diversity is the extremely wide range of aquatic habitats occupied by teleosts (Wootton, 1998; Helfman *et al.*, 2009). The reproductive diversity of teleosts may reflect the reproductive adaptations that are required for them to occupy this range of habitats.

There are also some unique characteristics of the teleosts that may have allowed the evolution of reproductive diversity, which is particularly striking in the Perciformes (see Chapter 10), the most evolutionarily advanced of the teleosts.

The genome-wide duplication that occurred early in the evolution of teleosts (Chapter 1) made available a reservoir of genetic variation, with the possibility of duplicated genes evolving separate functions and hence opening up the possibility of diversity in reproductive modes. Within the teleosts, extremely quick genomic evolution has occurred, with rapid duplication of genes and genomes (Desjardins & Fernald, 2009; Mank & Avise, 2009), again raising the possibility that genetic variation leads to reproductive diversity.

The mode of differentiation of the gonads in teleosts is different from the majority of jawed vertebrates, with the differentiating teleostean gonad lacking a cortical component (Chapter 3). A consequence of this is that in many teleost species, the gonads remain bipotential until later in development than in other vertebrates. The decision to become a functional male or a functional female is postponed (Adkins-Regan, 2005).

Allied to this gonadal bipotentiality is the fact that teleosts tend to retain a sexually bipotential brain. There is evidence that the sexual behavioural dimorphism seen in teleosts is a consequence of the activational effects of hormones rather than the structural organisation of the brain (Adkins-Regan, 2005). A further possible factor is that in teleosts there is adult neurogenesis, which may make flexibility in reproductive behaviour more possible (Adkins-Regan, 2005).

Reproductive Biology of Teleost Fishes, First Edition. Robert J. Wootton and Carl Smith.

The production of gonadotrophin-releasing hormone (GnRH) by the brain depends on the integration of a wide variety of neurotransmitters and other signalling molecules (Chapter 5; Desjardins & Fernald, 2009). The relationship between the brain and the anterior pituitary is unique to the teleosts. The anterior pituitary is innervated directly by neurons from the hypothalamic region of the brain, rather than a vascular link by a portal system found in other jawed vertebrates (Chapter 5). This arrangement would seem to allow for finer control of the pituitary in teleosts. Another endocrinological difference is that teleosts produce a non-aromatisable androgen, 11 ketotestosterone (11-KT), which characterises the male and is particularly important in the regulation of male reproductive behaviour (Chapters 5, 7 and 8; Adkins-Regan, 2005).

The challenge now is to use these elements, and perhaps others yet to be identified, to develop a coherent hypothesis that explains the diversity of reproductive modes in these remarkable animals.

References

Aas, Ø., Policansky, D., Einum, S. & Skurdal, J. (2010). Salmon ecological research and conservation. In: *Atlantic Salmon Ecology* (Aas, Ø., Klemetsen, A., Einum, S. & Skurdal, J., eds), pp. 445–456. Oxford: Wiley-Blackwell.

Abe, T. & Munehara, H. (2009). Adaptation and evolution of reproductive mode in copulating cottoid species. In: *Reproductive Biology and Phylogeny of Fishes (Agnathans and Bony Fishes)* (Jamieson, B.G.M., ed.), pp. 221–246. Enfield (NH): Science Publications.

Adams, E.S. (2001). Approaches to the study of territory size and shape. *Annual Review of Ecology and Systematics* **32**, 277–303.

Adams, G.L. & Johnson, J.E. (2001). Metabolic rates and natural history of Ozark cavefish. *Amblyopsis rosae*, in Logan cave. *Environmental Biology of Fishes* **62**, 97–105.

Adkins-Regan, E. (2005). *Hormones and Animal Social Behavior*. Princeton: Princeton University Press.

Agbali, M., Reichard, M., Bryjova, A., Bryja, J. & Smith, C. (2010). Mate choice for non-additive genetic benefits correlate with MHC dissimilarity in the rose bitterling (*Rhodeus ocellatus*). *Evolution* **64**, 1683–1696.

Agbali, M. & Smith, C. (2012). Long-term fitness consequences for offspring of female oviposition decisions in the rose bitterling (*Rhodeus ocellatus*). *Acta Zoologica* **93**, 367–372.

Agbali, M., Spence, R., Reichard, M. & Smith, C. (2012). Direct fitness benefits are preferred when the strength of direct and indirect sexual selection are equivalent. *Israel Journal of Ecology and Evolution* **58**, 279–288.

Ah-King, M., Kvarnemo, C. & Tullberg, B.S. (2005). The influence of territoriality and mating system on the evolution of male care: a phylogenetic study on fish. *Journal of Evolutionary Biology* **18**, 371–382.

Ahnesjö, I. (1995). Temperature affects male and female potential reproductive rates differently in the sex-role reversed pipefish, *Syngnathus typhle*. *Behavioral Ecology* **6**, 229–233.

Akagawa, I., Hara, M. & Iwamoto, T. (2008). Egg concealment in ascidians by females of the Japanese tubesnout, *Aulichthys japonicus* (Gasterosteiformes), and its subsequent copulation. *Ichthyological Research* **55**, 85–89.

Akagawa, I., Iwamoto, T., Watanabe, S. & Okiyama, M. (2004). Reproductive behaviour of Japanese tubesnout, *Aulichthys japonicus* (Gasterosteiformes), in the natural habitat compared with relatives. *Environmental Biology of Fishes* **70**, 353–361.

Akagawa, I., Tsukamoto, Y. & Okiyama, M. (1995). Sexual dimorphism and pair spawning into a sponge by the filefish, *Brachaluteres ulvarum*, with a description of the eggs and larvae. *Japanese Journal of Ichthyology* **41**, 397–407.

Ala-Honkola, O., Uddstrom, A., Diaz Pauli, B. & Lindström, K. (2009). Strong inbreeding depression in male mating behaviour in a poeciliid fish. *Journal of Evolutionary Biology* **22**, 1396–1406.

Alavi, S.M.H. & Cosson, J. (2005). Sperm motility in fishes: (I) Effects of temperature and pH: a review. *Cell Biology International* **29**, 101–110.

Alavi, S.M.H. & Cosson, J. (2006). Sperm motility in fishes: (II) Effects of ions and osmotic pressure. *Cell Biology International* **30**, 1–14.

Albrecht, U. (2012). Timing to perfection: the biology of central and peripheral circadian clocks. *Neuron* **74**, 246–266.

Ali, M. & Wootton, R.J. (1999a). Effect of variable food levels on reproductive performance of breeding female three-spined sticklebacks. *Journal of Fish Biology* **55**, 1040–1053.

Ali, M. & Wootton, R.J. (1999b). Coping with resource variation: effect of constant and variable intervals between feeding on reproductive performance at first spawning of female three-spined sticklebacks. *Journal of Fish Biology* **55**, 211–220.

Allee, W.C. (1938). *The Social Life of Animals*. London: Heinemann.

Allsop, D.J. & West, S.A. (2003). Constant relative age and size at sex change for sequentially hermaphroditic fish. *Journal of Evolutionary Biology* **16**, 921–929.

Almada, V.C., Gonalves, E.J., Oliveira, R.F. & Santos, A.J. (1995). Courting females: ecological constraints affect sex roles in a natural population of the blenniid fish *Salaria pavo*. *Animal Behaviour* **49**, 1125–1127.

Almeida, F.F.L., Kristoffersen, C., Taranger, G.L. & Schulz, R.W. (2008). Spermatogenesis in Atlantic cod (*Gadus morhua*): a novel model of cystic germ cell development. *Biology of Reproduction* **78**, 27–34.

Almeida-Toledo, L.F. & Foresti, F. (2001). Morphologically differentiated sex chromosomes in neotropical freshwater fish. *Genetica* **111**, 91–100.

Alonzo, S.H. (2008). Female mate choice copying affects sexual selection in wild populations of the ocellated wrasse. *Animal Behaviour* **75**, 1715–1723.

Alonzo, S.H. & Mangel, M. (2005). Sex change rules, stock dynamics, and the performance of spawning per recruit measures in protogynous stocks. *Fishery Bulletin* **203**, 229–245.

Alonzo, S.H., Taborsky, M. & Wirtz, P. (2000). Male alternative reproductive behaviours in a Mediterranean wrasse, *Symphodus ocellatus*: evidence from otoliths for multiple life-history pathways. *Evolutionary Ecology Research* **2**, 997–1007.

Alves, M.J., Coelho, M.M. & Collares-Pereira, M.J. (2001). Evolution in action through hybridisation and polyploidy in an Iberian freshwater fish: a genetic review. *Genetica* **111**, 375–385.

Amorim, M.C.P. (2006). Diversity of sound production in fish. In: *Fish Communication* (Ladich, F., Collin, S.P., Moller, P. & Kapoor, B.G. eds), pp. 71–105. Enfield (NH): Science Publishers.

Amorim, M.C.P., Fonseca, P.J. & Almada, V.C. (2003). Sound production during courtship and spawning of the cichlid *Oreochromis mossambicus*: male-female and male-male interactions. *Journal of Fish Biology* **62**, 658–672.

Amorim, M.C.P., Simões, J.M., Fonseca, P.J. & Turner, G.F. (2008). Species differences in courtship acoustic signals among five Lake Malawi cichlid species (*Pseudotropheus* spp.). *Journal of Fish Biology* **72**, 1355–1368.

Amundsen, T. & Forsgren, E. (2001). Male mate choice selects for female coloration in a fish. *Proceedings of the National Academy of Sciences of the USA* **98**, 13155–13160.

Amundsen, T. & Forsgren, E. (2003). Male preference for colourful females affected by male size in a marine fish. *Behavioral Ecology Sociobiology* **54**, 55–64.

Andersen, K.H., Beyer, J.E., Pedersen, M., Andersen, N.G. & Gislason, H. (2008). Life-history constraints on the success of the many small eggs reproductive strategy. *Theoretical Population Biology* **73**, 490–497.

Anderson, J.L. (2002). Aquaculture and the future: Why fisheries economists should care. *Marine Resource Economics* **17**, 133–152.

Anderson, J.L., Mari, A.R., Braasch, I., Amores, A., Hohenlohe, P., Batzel, P. & Postlethwait, J.H. (2012). Multiple sex-associated regions and a putative sex chromosome in zebrafish revealed by RAD mapping and population genomics. *PLoS One* **7**, e40701.

Andersson, M. (1994). *Sexual Selection*. Princeton: Princeton University Press.

Angers, B. & Schlosser, I.J. (2007). The origin of *Phoxinus eos-neogaeus* unisexual hybrids. *Molecular Biology* **16**, 4562–4571.

Aranzábal., M.C.U., Grier, H.J., De la Rosa Cruz, G. & Alarcón, A.G. (2009). Modifications in ovarian and testicular morphology associated with viviparity in teleosts. In: *Reproductive Biology and Phylogeny of Fishes (Agnathans and Bony Fishes)* (Jamieson, B.G.M., ed.), pp. 85–117. Enfield (NH): Science Publications.

Armstrong, R.H. & Morrow, J.E. (1980). The dolly varden. In: *Charrs: Salmonid Fishes of the Genus Salvelinus* (Balon, E.K., ed.), pp. 99–140. The Hague: Junk Publishers.

Arnold, S.J. & Duvall, D. (1994). Animal mating systems: a synthesis based on selection theory. *American Naturalist* **143**, 317–348

Arnqvist, G. & Rowe, L. (2005). *Sexual Conflict*. Princeton: Princeton University Press.

Aroua, S., Maugers, G., Jeng, S.-R., Chang, C.-F., Weltzien, F.-A., Rousseau, K. & Dufour, S. (2012). Pituitary gonadotropins FSH and LH are oppositely regulated by the activin/follistatin system in a basal teleost, the eel. *General and Comparative Endocrinology* **175**, 82–91.

van den Assem, J. (1967). Territoriality in the threespine stickleback, *Gasterosteus aculeatus* L.: an experimental study in intra-specific competition. *Behaviour* **16** (Suppl), 1–164.

Atema, J. (1996). Eddy chemotaxis and odor landscapes: Exploration of nature with animal sensors. *Biological Bulletin* **191**, 129–138.

Atz, J. (1964). Intersexuality in fishes. In: *Intersexuality in Vertebrates Including Man* (Armstrong, C.N. & Marshall, A.J., eds), pp. 145–232. London: Academic Press.

Aubin-Horth, N., Bourque, J.F., Daigle, G., Hedger, R. & Dodson, J.J. (2006). Longitudinal gradients in threshold sizes for alternative male life history tactics in a population of Atlantic salmon (*Salmo salar*). *Canadian Journal of Fisheries and Aquatic Sciences* **63**, 2067–2075.

Aubin-Horth, N. & Dodson, J.J. (2004). Influence of individual body size and variable thresholds on the incidence of a sneaker male reproductive tactic in Atlantic salmon. *Evolution* **58**, 136–144.

Aubin-Horth, N., Landry, C.R., Letcher, B.H. & Hormann, H.A. (2005). Alternative life histories shape brain gene expression profiles in males of the same population. *Proceedings of the Royal Society London, B* **272**, 1655–1662.

Aubin-Horth, N., Letcher, B.H. & Hofmann, H.A. (2009). Gene-expression signatures of Atlantic salmon's plastic life cycle. *General and Comparative Endocrinology* **163**, 278–284.

Avise, J.C. (2011). *Hermaphroditism*. New York: University of Columbia Press.

Avise, J.C. & Mank, J.E. (2009). Evolutionary perspectives on hermaphroditism in fishes. *Sexual Development* **3**, 152–163.

Axelrod, H.R. (1993). *The Most Complete Colored Lexicon of Cichlids*. New Jersey: Tropical Fish Hobbyist Publications, Inc.

Baba, R. (1994). Timing of spawning and host-nest choice for brood parasitism by the Japanese minnow, *Pungtungia herzi*, on the Japanese aucha perch, *Siniperca kawamebari*. *Ethology* **98**, 50–59.

Baba, R. & Karino, K. (1998). Countertactics of the Japanese aucha perch *Siniperca kawamebari* against brood parasitism by the Japanese minnow *Pungtungia herzi*. *Journal of Ethology* **16**, 67–72.

Babin, P.J., Carnevale, O., Lubzens, E. & Schneider, W.J. (2007). Molecular aspects of oocyte vitellogenesis in fish. In: *The Fish Oocyte: From Basic Studies in Biotechnological Applications* (Bain, P.J., Cerdà, J. & Lubzens, eds), pp. 39–76. Berlin: Springer.

Baerends, G.P. & Baerends-Van Roon, J.M. (1950). *An Introduction to the Study of the Ethology of Cichlid Fishes*. Leiden: E.J. Brill.

Bagenal, T.B. (1969). Relationship between egg size and fry survival in brown trout *Salmo trutta* L. *Journal of Fish Biology* **1**, 349–353.

Bagenal, T.B. (1971). The interrelation of the size of fish eggs, the date of spawning and the production cycle. *Journal of Fish Biology* **3**, 207–219.

Bagenal, T.B. (1973). Fish fecundity and its relations with stock and recruitment. *Rapports et Proces-verbaux des Réunions. Conseil International pour l'Éxploration de la Mer* **164**, 186–198.

Baggerman, B. (1957). An experimental study on the timing of breeding and migration in the threespined stickleback (*Gasterosteus aculeatus* Linnaeus). *Archives Néerlandaises de Zoologie* **12**, 105–317.

Baggerman, B. (1980). Photoperiodic and endogenous control of the annual reproductive cycle in teleost fishes. In: *Environmental Physiology of Fishes* (Ali, M.A., ed.), pp. 533–567. New York: Plenum.

Baggerman, B. (1990). Sticklebacks. In: *Reproductive Seasonaility in Teleosts: Environmental Influences* (Munro, A.D., Scott, A.P. & Lam, T.J., eds), pp. 79–107. Boca Raton: CRC Press.

Baker, J.A. (1994). Life history variation in female threespine stickleback. In: *The Evolutionary Biology of the Threespine Stickleback* (Bell, M.A. & Foster, S.A., eds), pp.144–187. Oxford: Oxford University Press.

Bakker, T.C.M. (1985). Two-way selection for aggression in juvenile, female and male sticklebacks (*Gasterosteus aculeatus* L.), with some notes on hormonal factors. *Behaviour* **93**, 69–81.

Bakker, T.C.M. (1993). Positive genetic correlation between female preference and preferred male ornament in sticklebacks. *Nature* **363**, 255–257.

Bakker, T.C.M. (1994a). Evolution of aggressive behaviour in the threespine stickleback. In: *The Evolutionary Biology of the Threespine Stickleback* (Bell, M.A. & Foster, S.A., eds), pp. 345–380. Oxford: Oxford University Press.

Bakker, T.C.M. (1994b). Genetic correlations and the control of behavior, exemplified by aggressiveness in sticklebacks. *Advances in the Study of Behaviour* **23**, 135–171.

Balarin, J.D. & Haller, R.D. (1983). Commercial tank culture of tilapia. In: *International Symposium on Tilapia in Aquaculture, Proceedings* (Fishelson, L. & Yaron, Z., eds), pp. 394–401. Tel Aviv: Tel Aviv University.

Ball, M.A. & Parker, G.A. (1997). Sperm competition games: inter- and intra-species results of a continuous external fertilisation model. *Journal of Theoretical Biology* **186**, 459–466.

Balon, E.K. (1975). Reproductive guilds of fishes: a proposal and definition. *Journal of the Fisheries Research Board of Canada* **32**, 821–864.

Balon, E.K. (1999). Alternative ways to become a juvenile or a definitive phenotype (and on some persisting linguistic offenses). *Environmental Biology of Fishes* **56**, 17–38.

Balsano, J.S., Rasch, E.M. & Monaco, P.J. (1989). The evolutionary ecology of *Poecilia formosa* and its triploid associate. In: *Ecology and Evolution of Livebearing Fishes (Poeciliidae)* (Meffe, G.K. & Snelson, Jr, F.F., eds), pp. 277–297. Englewood Cliffs (NJ): Prentice-Hall.

Balshine-Earn, S. (1995). The costs of parental care in Galilee St Peters fish, *Sarotherodon galilaeus*. *Animal Behaviour* **50**, 1–7.

Balshine-Earn, S. & Lotem, A. (1998). Experimental evidence for individual recognition from video playbacks in a cooperatively breeding cichlid (*Neolamprologus brichardi*). *Behaviour* **135**, 369–386.

Bănărescu, P. & Nalbant, T.T. (1973). *Pisces Teleostei Cyprinidae (Gobioninae)*. Berlin: Walter de Gruyter.

van der Bank, F.H. & Steyn, G J. (1992). The effect of cryopreservation and various cryodiluents on allozyme variation of glucose phosphate isomerase in the F_1 progeny of African catfish (*Clarias gariepinus*). *Comparative Biochemistry and Physiology, Part B* **103**, 641–643.

Barabanshchikov, E.I. (2004). Detection of eggs of the lazy gudgeon *Sarcocheilichthys sinensis* (Cyprinidae) in the mantle cavity of bivalve mollusks of the genus *Lanceolaria* (Bivalvia, Unionidae). *Voprosy Ikhtiologii* **44**, 565–566.

Baras, E., Jacobs, B. & Melard, C. (2001). Effect of water temperature on survival, growth and phenotypic sex of mixed (XX-XY) progenies of Nile tilapia *Oreochromis niloticus*. *Aquaculture* **192**, 187–199.

Baras, E., Mpo'n'tcha, A., Driouch, H. Prignon, C. & Melard, C. (2002). Ontogenetic variations of thermal optimum for growth, and its implication on thermolabile sex determination in blue tilapia. *Journal of Fish Biology* **61**, 645–660.

Baras, E., Prignon, C. Gohoungo, G. & Melard, C. (2000). Phenotypic sex differentiation of blue tilapia under constant and fluctuating thermal regimes and its adaptive and evolutionary implications. *Journal of Fish Biology* **57**, 210–223.

Barber, I., Arnott, S.A., Braithwaite, V.A., Andrew, J. & Huntingford, F.A. (2001b). Indirect fitness consequences of mate choice in sticklebacks: offspring of brighter males grow slowly but resist parasite infections. *Proceedings of the Royal Society of London, B* **268**, 71–76.

Barber, I., Nairn, D. & Huntingford, F.A. (2001a). Nests as ornaments: revealing construction by male sticklebacks. *Behavioral Ecology* **12**, 390–396.

Barlow, G.W. (1967). Social behavior of a South American leaf fish, *Polycentrus schomburgkii*, with an account of recurring pseudofemale behavior. *American Midland Naturalist* **78**, 215–34.

Barlow, G.W. (1974). Contrasts in social behavior between Central American cichlid fishes and coral-reef surgeon fishes. *American Zoologist* **14**, 9–34.

Barlow, G.W. (1991). Mating systems among cichlid fishes. In: *Cichlid Fishes: Behavior, Ecology and Evolution* (Keenleyside, M.H.A. ed.), pp. 173–190. London: Chapman and Hall.

Barlow, G.W. (1993). The puzzling paucity of feeding territories among freshwater fishes. *Marine Behavior and Physiology* **23**, 155–174.

Barlow, G.W. (2000). *The Cichlid Fishes.* Cambridge (MA): Perseus Publishing.

Baroiller, J.F. & D'Cotta, H. (2001). Environment and sex determination in farmed fish. *Comparative Biochemistry and Physiology, Part C* **130**, 399–409.

Baroiller, J.F., D'Cotta, H. & Saillant, E. (2009). Environmental effects of fish sex determination and differentiation. *Sexual Development* **3**, 118–135.

Barrett, S.C.H. & Harder, L.D. (1996). Ecology and evolution of plant mating. *Trends in Ecology and Evolution* **11**, A73–A79.

Barson, N.J., Knight, M.E. & Turner, G.F. (2007). The genetic architecture of male color differences between a sympatric Lake Malawi cichlid species pair. *Journal of Evolutionary Biology* **20**, 45–53.

Bartley, D.M., Rana, K. & Immink, A.J. (2001). The use of interspecific hybrids in aquaculture and fisheries. *Reviews in Fish Biology and Fisheries* **10**, 325–337.

Barton, N. (2001). The role of hybridization in evolution. *Molecular Ecology* **10**, 551–568.

Basolo, A.L. (1990). Female preference predates the evolution of the sword in swordtails. *Science* **250**, 808–810.

Basolo, A. & Wagner, W. (2004). Covariation between predation risk, body size and fin elaboration in the green swordtail, *Xiphophorus helleri*. *Biological Journal of the Linnean Society* **83**, 87–100.

Bass, A.H. (2008). Steroid-dependent plasticity of vocal motor systems: novel insights from teleost fish. *Brain Research Reviews* **57**, 299–308.

Bass, A.H. & McKibben, J.R. (2003). Neural mechanisms and behaviours for acoustic communication in teleost fish. *Progress in Neurobiology* **69**, 1–26.

Bateman, A.J. (1948). Intra-sexual selection in *Drosophila*. *Heredity* **2**, 349–368.

Bauer, G. & Wächtler, K. (2000). *Ecology and Evolution of the Freshwater Mussels Unionoida.* Berlin: Springer-Verlag.

Baum, D., Laughton, R., Armstrong, J.D. & Metcalfe, N.B. (2004). Altitudinal variation in the relationship between growth and maturation rate in salmon parr. *Journal of Animal Ecology* **73**, 253–260.

Baylis, J.R. (1981). The evolution of parental care in fishes with reference to Darwin's rule of male sexual selection. *Environmental Biology of Fishes* **6**, 223–251.

Beardmore, J.A., Mair, G.C. & Lewis, R.I. (2001). Monosex male production in finfish as exemplified by tilapia: applications, problems, and prospects. *Aquaculture* **197**, 283–301.

Beeching, S.C., Gross, S.H., Bretz, H.S. & Hariatis, E. (1998). Sexual dichromatism in convict cichlids: the ethological significance of female ventral coloration. *Animal Behaviour* **56**, 1021–1026

Belk, M.C. & Tuckfield, R.C. (2010). Changing costs of reproduction: age-based differences in reproductive allocation and escape performance in a livebearing fish. *Oikos* **119**, 163–169.

Bell, G. (1980). The costs of reproduction and their consequences. *American Naturalist* **116**, 45–76.

Belles-Isles, J.-C. & FitzGerald, G.J. (1991). Filial cannibalism in sticklebacks: a reproductive management strategy? *Ethology Ecology and Evolution* **3**, 49–62.

Benfey, T.J. (1999). The physiology and behavior of triploid fishes. *Reviews in Fisheries Science* **7**, 39–67.

Benfey, T.J. (2001). Use of sterile triploid Atlantic salmon (*Salmo salar* L.) for aquaculture in New Brunswick, Canada. *ICES Journal of Marine Science* **58**, 525–529.

Berg, O.K., Thronæs, E. & Bremset, G. (1998). Energetics and survival of virgin and repeat spawning brown trout (*Salmo trutta*). *Canadian Journal of Fisheries and Aquatic Sciences* **55**, 47–53.

van den Berghe, E.P. (1988). Piracy as an alternative reproductive tactic for males. *Nature* **334**, 697–698.

van den Berghe, E.P. (1992). Parental care and the cost of reproduction in a Mediterranean fish. *Behavioral Ecology and Sociobiology* **30**, 373–378.

van den Berghe, E.P. & Gross, M.R. (1989). Natural selection resulting from female breeding competition in a Pacific salmon (coho: *Oncorhynchus kisutch*). *Evolution* **43**, 125–140.

Berglund, A. (1991). Egg competition in a sex-role reversed pipefish: subdominant females trade reproduction for growth. *Evolution* **45**, 770–774.

Berglund, A. & Rosenqvist, G. (2003). Sex role reversal in pipefish. *Advances in the Study of Behavior* **32**, 131–167.

Bernier-Bourgault, I. & Magnan, P. (2002). Factors affecting redd site selection, hatching, and emergence of brook charr, *Salvelinus fontinalis*, in an artificially enhanced site. *Environmental Biology of Fishes* **64**, 333–341.

Berra, T.M. & Aday, D.D. (2004). Otolith description and age-and-growth of *Kurtus gulliveri* from northern Australia. *Journal of Fish Biology* **65**, 354–362.

Berra, T.M. & Humphrey, J.D. (2002). Gross anatomy and histology of the hook and skin of forehead brooding male nurseryfish, *Kurtus gulliveri*, from northern Australia. *Environmental Biology of Fishes* **65**, 263–270.

Berra, T.M. & Neira, F. (2003). Early life history of the nurseryfish, *Kurtus gulliveri* (Perciformes: Kurtidae), from northern Australia. *Copeia* **2003**, 384–390.

Berra, T.M., Gomelsky, B., Thompson, B.A. & Wedd, D. (2007). Reproductive anatomy, gonad development and spawning seasonality of nurseryfish, *Kurtus gulliveri* (Perciformes: Kurtidae). *Australian Journal of Zoology* **55**, 211–217.

Berra, T.M., Wedd, D., & Allsop, Q. (2004). Observation of egg carrying by male nurseryfish, *Kurtus gulliveri* (Perciformes: Kurtidae), and natural history notes from northern Australia. *The Beagle: Records of the Museums and Art Galleries of the Northern Territory* **20**, 187–193.

Beverton, R.J.H. & Holt, S.J. (1957). *On the Dynamics of Exploited Fish Populations*. London: MAFF.

Beverton, R.J.H. & Holt, S.J. (1959). A review of the lifespans and mortality rates of fish in nature and relation to growth and other physiological characteristics. *CIBA Foundation Colloquia on Ageing* **5**, 142–180.

Billard, R. (1981). The reproductive cycle in teleost fish. *Rivista Italiana di Piscicoltura e Ittiopatologia* **16**, 1–32.

Billard, R. (1986). Spermatogenesis and spermatology of some teleost species. *Reproduction Nutrition Development* **26**, 877–920.

Billard, R., Bry, C. & Gillet, C. (1981). Stress, environment and reproduction in teleost fish. In: *Stress and Fish* (Pickering, A.D., ed.), pp. 185–208. London: Academic Press.

Birkhead, T.R. & Montgomerie, R. (2009). Three centuries of sperm research. In: *Sperm Biology: An Evolutionary Perspective* (Birkhead, T.R., Hosken, D.J. & Pitnick, S., eds), pp. 1–42. Oxford: Academic Press.

Bisazza, A. & Marconato, A. (1988) Female mate choice, male–male competition and parental care in the river bullhead *Cottus gobio* L. (Pisces, Cottiidae). *Animal Behaviour* **36**, 1352–1360.

Bjørkvoll, E., Grøtan, V., Aanes, S., Saether, B.-E., Engen, S. & Aanes, R. (2012). Stochastic dynamics and life-history variation in marine fish species. *American Naturalist* **180**, 372–387.

Blaber, S.J.M. (1997). *Fish and Fisheries of Tropical Estuaries*. London: Chapman & Hall.

Blackburn, D.G. & Evans, H.E. (1986). Why are there no viviparous birds? *American Naturalist* **128**, 165–190.

Blaser, H., Eisenbeiss, S., Neumann, M., Reichman-Fried, M., Thisse, B., Thisse, C., & Raz, E. (2005). Transition from non-motile behaviour to directed migration during early PGC development in zebrafish. *Journal of Cell Science* **118**, 4027–4038.

Blaxter, J.H.S. & Hempel, G. (1963). The influence of egg size on herring larvae. *Journal du Conseil International pour l'Exploration de la Mer* **28**, 211–240.

Blažek, R., Polačik, M. & Reichard, M. (2013). Rapid growth, early maturation and short generation time in African annual fishes. *EvoDevo* **4**, 24.

Bloch, M.E. (1794). *Naturgeschichte der Auslandische Fische*. Berlin: J. Moreno & Co.

Blomenröhr, M., Goos, H. & Bogerd, J. (2005). GnRH receptors in fish: differences in structure-function relations between mammalian and non-mammalian GnRH receptors. In: *Hormones and Their Receptors in Fish Reproduction* (Melamed, P. & Sherwood, N., eds), pp. 40–75. New Jersey: World Scientific.

Blüm, V. (1986). *Vertebrate Reproduction*. Berlin: Springer-Verlag.

Blumer, L.S. (1979). Male parental care in the bony fishes. *The Quarterly Review of Biology* **54**, 149–161.

Blumer, L.S. (1982). A bibliography and categorization of bony fishes exhibiting parental care. *Zoological Journal of the Linnean Society* **76**, 1–22.

Bobe, J., Jalabert, B. & Fostier, A. (2008). Oogenesis: post vitellogenic events leading to a fertilizable oocyte. In: *Fish Reproduction* (Rocha, M.J., Arukwe, A. & Kapoor, B.G., eds), pp. 1–36. Enfied (NH): Science Publishers.

Bohlin, T., Pettersson, J. & Degerman, E. (2001). Population density of migratory and resident brown trout (*Salmo trutta*) in relation to altitude: evidence for a migration cost. *Journal of Animal Ecology* **70**, 112–121.

Bøhn, T. & Amundsen, P.-A. (2001). The competitive edge of an invading specialist. *Ecology* **82**, 2150–2163.

Bøhn, T., Terje Sandlund, O., Amundsen, P.A. & Primicerio, R. (2004). Rapidly changing life history during invasion. *Oikos* **106**, 138–150.

Bonduriansky, R. (2001). The evolution of male choice in insects: a synthesis of ideas and evidence. *Biological Reviews* **76**, 305–339.

Bone, Q. & Moore, R.H. (2008). *Biology of Fishes*, 3rd edn. Abingdon: Taylor & Francis.

Boney, S.E., Shelton, W.L., Yang, S.L. & Wilken, L.O. (1984). Sex reversal and breeding of grass carp. *Transactions of the American Fisheries Society* **113**, 348–353.

Bonnet, X., Bradshaw, D. & Shine, R. (1998). Capital versus income breeding: an ectothermic perspective. *Oikos* **83**, 333–342.

Boogert, N.J., Fawcett, T.W. & Lefebvre, L. (2011). Mate choice for cognitive traits: a review of the evidence in nonhuman vertebrates. *Behavioral Ecology* **22**, 447–459.

Borg, B. (1994). Androgens in teleost fishes. *Comparative Biochemistry and Physiology C: Comparative Pharmacology* **109**, 219–245.

Borg, B. (2007). Reproductive physiology of sticklebacks. In: *Biology of the Three-Spined Stickleback* (Östlund, S., Mayer, I. & Huntingford, F.A., eds), pp. 225–247. Boca Raton: CRC Press.

Borg, B. (2010). Photoperiodism in fishes. In: *Photoperiodism* (Nelson, R.J., Demlinger, D.L. & Somers, D.E., eds), pp. 371–398. Oxford: Oxford University Press.

Bornestaf, C. & Borg, B. (2000). Endogenous breeding cycles in male threespine sticklebacks, *Gasterosteus aculeatus*. *Behaviour* **137**, 921–932.

Boschetto, C., Gasparini, C. & Pilastro, A. (2011). Sperm number and velocity affect sperm competition success in the guppy (*Poecilia reticulata*). *Behavioral Ecology and Sociobiology* **65**, 813–821.

Both, C. & Visser, M.E. (2003). Density dependence, territoriality, and divisibility of resources: from optimality models to population processes. *American Naturalist* **161**, 326–336.

Boulcott, P. & Braithwaite, V.A. (2007). Colour perception in three-spined sticklebacks: sexes are not so different after all. *Evolutionary Ecology* **21**, 601–611.

Bowmaker, J.K. & Kunz, Y.W. (1987). Ultraviolet receptors, tetrachromatic color-vision and retinal mosaics in the brown trout *Salmo trutta*: Age-dependent changes. *Vision Research* **27**, 2101 – 2108.

Brabrand, Å., Koestler, A.G. & Borgstrøm, R. (2002). Lake spawning of brown trout related to groundwater influx. *Journal of Fish Biology* **60**, 751–763.

Bradbury, J.W. & Vehrenkamp, S.L. (1998). *Principles of Animal Communication*. Sunderland: Sinauer Associates.

Breder, C.M. & Rosen, D.E. (1966). *Modes of Reproduction in Fishes*. New York: The Natural History Press.

Bregazzi, P.R. & Kennedy, C.R. (1980). The biology of the pike, *Esox lucius* L., in a southern eutrophic lake. *Journal of Fish Biology* **17**, 91–112.

Breton, B., Horoszewicz, L., Bieniarz, K. & Epler, P. (1980). Temperature and reproduction in tench: effect of a rise in the annual temperature regime on gonadotropin level, gametogenesis and spawning. II The female. *Reproduction, Nutrition et Développment* **20**, 1011–1024.

Brett, J.R. (1986). Production energetics of a population of sockeye salmon, *Oncorhynchus nerka*. *Canadian Journal of Zoology* **64**, 555–564.

Britton, R.H. & Moser, M.E. (1982). Size specific predation by herons and its effect on the sex-ratio of natural populations of the mosquitofish *Gambusia affinis*. *Oecologia* **53**, 146–151.

Brockmann, H.J. & Taborsky, M. (2008). Alternative reproductive tactics and the evolution of alternative allocation phenotypes. In: *Alternative Reproductive Tactics* (Oliviera, R.F., Taborsky, M. & Brockmann, H.J. eds), pp. 25–51. Cambridge: Cambridge University Press.

Bromage, N.R. & Cumaranatunga, P.R.T. (1988). Egg production in rainbow trout. In: *Recent Advances in Aquaculture* (Muir, J.F. & Roberts, R.J., eds), pp. 65–138. London: Croom Helm.

Bromage, N. & Jones, J. (1991). The effects of seasonal alterations in ration on fecundity and maturation in rainbow trout (*Oncorhynchus mykiss*). In: *Reproductive Physiology of Fish* (Scott, A.P., Sumpter, J.P. Kime, D.E. & Rolfe, M.S. eds), pp. 162. Sheffield: FishSymp 91.

Bromage, N., Randall, C., Duston, J. & Thrush, M. (1993). Environmental control of reproduction in salmonids. In: *Recent Advances in Aquaculture IV* (Muir, J.F. & Roberts, R.J., eds), pp. 55–65. Oxford: Blackwell Scientific Publications.

Bromage, N., Porter, M. & Randall, C. (2001). The environmental regulation of maturation in farmed finfish with special reference to the role of photoperiod and melatonin. *Aquaculture* **197**, 63–98.

Bromley, P.J., Ravier, V. & Witthames, P.R. (2000). The influence of feeding regime on sexual maturation, fecundity and atresia in first-time spawning turbot. *Journal of Fish Biology* **56**, 264–278.

Brommer, J.E. (2000). The evolution of fitness in life-history theory. *Biological Reviews* **75**, 377–404.

Brooks, S., Tyler, C.R. & Sumpter, J.P. (1997). Egg quality in fish: what makes a good egg? *Reviews in Fish Biology and Fisheries* **7**, 387–416.

Brown, J.L. (1964). The evolution of diversity in avian territorial systems. *Wilson Bulletin* **76**, 160–169.

Brown, J.L. (1969). The buffer effect and population regulation in tit populations. *American Naturalist* **103**, 347–354.

Brown J.L. (1975). *The Evolution of Behavior*. New York: W.W. Norton & Co.

Brown J.L. & Orians, G.H. (1970). Spacing patterns in mobile animals. *Annual Review of Ecology and Systematics* **1**, 239–262.

Buckley, J., Maunder, R.J., Foey, A., Pearce, J., Val, A.L. & Sloman, K.A. (2010). Biparental mucus feeding: a unique example of parental care in an Amazonian cichlid. *Journal of Experimental Biology* **213**, 3787–3795.

Bull, J.J. (1983). *Evolution of Sex Determining Mechanisms*. London: Benjamin/Cummings Publishing Company.

Burns, J.R. (1985). The effect of low-latitude photoperiods on the reproduction of female and male *Poeciliopsis gracilis* and *Poecilia sphenops*. *Copeia* **1985**, 961–965.

Burton, M.P.M. (1994). A critical period for the nutritional control of early gametogenesis in female winter flounder, *Pseudopleuronectes americanus* (Pisces: Teleostei). *Journal of Zoology, London* **233**, 405–415.

Butts, I.A.E., Love, O.P., Farwell, M. & Pitcher, T.E. (2012). Pituitary and secondary sexual characters in alternative reproductive tactics of Chinook salmon: associations with androgens and maturation-inducing steroid. *General and Comparative Endorcinology* **175**, 449–456.

Callan, H.G. (1986). *Lampbrush Chromosomes*. Berlin: Springer-Verlag.

Calow, P. (1979). The cost of reproduction – a physiological approach. *Biological Reviews* **54**, 23–40.

Calow, P. (1985). Aspects of energy allocation. In: *Fish Energetics New Perspectives* (Tytler, P. & Calow, P. eds), pp. 13–31. London: Croom Helm.

Campbell, B., Dickey, J.T. & Swanson, P. (2003). Endocrine changes during onset of puberty in male Spring Chinook salmon, *Oncorhynchus tshawytscha*. *Biology of Reproduction* **69**, 2109–2117.

Campbell, B., Dickey, J., Beckman, B., Young, G., Pierce, A., Fukada, H. & Swanson, P. (2006). Previtellogenic oocyte growth in salmon: relationships among body growth, plasma insulin-like growth factor, estradiol-17 beta, follicle-stimulating hormones and expression of ovarian genes for insulin-like growth factors, steroidogenic acute regulatory protein and receptors for gonadotropins, growth hormone and somatolactin. *Biology of Reproduction* **75**, 34–44.

Candolin, U. (2000). Changes in expression and honesty of sexual signaling over the reproductive lifetime of sticklebacks. *Proceedings of the Royal Society of London, B* **267**, 2425–2430.

Candolin, U. & Voigt, H.-R. (2001). Correlation between male size and territory quality: consequence of male competition or predation susceptibility? *Oikos* **95**, 225–230.

Carlisle, T.R. (1982). Brood success in variable environments: implications for parental care allocation. *Animal Behaviour* **30**, 824–836.

Carlisle, T.R. (1985). Parental response to brood size in a cichlid fish. *Animal Behaviour* **33**, 234–238.

Carnevali, O., Gioacchini, G., Maradonna, F., Olivotto, I. & Migliarini, B. (2011). Melatonin induces follicle maturation in *Danio rerio*. *Plos One* **6**, e19978.

Carcupino, M., Baldacci, A., Mazzini, M. & Franzoi, P. (2002). Functional significance of the male broodpouch in the reproductive strategies of pipefish and seahorses; a morphological and ultrastructural comparative study on three anatomically different pouches. *Journal of Fish Biology* **61**, 1465–1480.

Carrasco, L.A.P., Penman, D.J. & Bromage, N. (1999). Evidence for the presence of sex chromosomes in the Nile tilapia (*Oreochromis niloticus*) from synatonemal complex analysis of XX, XY and YY genotypes. *Aquaculture* **173**, 207–218.

Carscadden, J., Nakashima, B.S. & Frank K.T. (1997). Effects of fish length and temperature on the timing of peak spawning in capelin (*Mallotus villosus*). *Canadian Journal of Fisheries and Aquatic Sciences* **54**, 781–787.

Casalini, M., Agbali, M., Reichard, M., Konečná, M., Bryjova, A. & Smith, C. (2009). Male dominance, female mate choice and intersexual conflict in the rose bitterling (*Rhodeus ocellatus*). *Evolution* **63**, 366–376.

Caselle, J.E., Hamilton, S.L., Schroeder, D.M., Love, M.S., Standish, J.D., Rosales-Casian, J.A. & Sosa-Nishizaki, O. (2011). Geographic variation in density, demography, and life history traits of a harvested, sex-changing, temperate reef fish. *Canadian Journal of Fisheries and Aquatic Sciences* **68**, 288–303.

Caswell, H. (1989). Life-history strategies. In: *Ecological Concepts* (Cherrett, J.M., ed.), pp. 285–308. Oxford: Blackwell Scientific.

Cerdà, J., Fabra, M. & Raldúa, D. (2007). Physiological and molecular basis of fish oocyte hydration. In: *The Fish Oocyte: From Basic Studies to Biotechnological Applications* (Babin, P.J., Cerdà, J. & Lubzens, E., eds), pp. 349–396. Berlin: Springer.

Cerdá-Reverter, J.M. & Canosa, L.F. (2009). Neuroendocrine systems of the fish brain. In: *Fish Neuroendocrinology. Fish Physiology Vol. 28* (Bernier, N.J., Van Der Kraak, G., Farrell, A.P. & Brauner, C.J., eds), pp. 3–74. London: Academic Press.

Chao, N.H., Chao, W.C., Liu, K.C. & Liao, I. (1987). The properties of tilapia sperm and its cryopreservation. *Journal of Fish Biology* **30**, 107–118.

Chapovetsky, V., Gattegno, T. & Admon, A. (2007). Porteomics analysis of the developing fish oocyte. In: *The Fish Oocyte: From Basic Studies to Biotechnological Applications* (Babin, P.J., Cerdà, J. & Lubzens, E., eds), pp. 99–111. Berlin: Springer.

Chapman, D.W. (1988). Critical review of variables used to define effects of fines in redds of large salmonids. *Transactions of the American Fisheries Society* **117**, 1–21.

Chapman, T., Arnqvist, G, Bangham, J. & Rowe, L. (2003). Sexual conflict. *Trends in Ecology and Evolution* **18**, 41–47.

Charlesworth, B. (1980). *Evolution in Age-Structured Populations*. Cambridge: Cambridge University Press.

Charlesworth, B. (1991). The evolution of sex chromosomes. *Science* **251**, 1030–1033.

Charlesworth, B. (2002). The evolution of chromosomal sex determination. In: *The Genetics and Biology of Sex Determination* (Short, R. ed.), pp. 207–219. Chichester: John Wiley.

Charlesworth, D., Charlesworth, B. & Marais, G. (2005). Steps in the evolution of heteromorphic sex chromosomes. *Heredity* **95**, 118–128.

Charnov, E.L. (1982). *The Theory of Sex Allocation*. Princeton: Princeton University Press.

Charnov, E.L. (1993). *Life History Invariants: Some Explorations of Symmetry in Evolutionary Ecology*. Oxford: Oxfrod University Press.

Charnov, E.L. (1997). Trade-off invariant rules for evolutionary stable life histories. *Nature* **387**, 393–394.

Charnov, E.L., Turner, T.F. & Winemiller, K.O. (2001). Reproductive constraints and the evolution of life histories with indeterminate growth. *Integrative and Comparative Biology* **44**, 494–497.

Charnov, E.L., Warne, R. & Moses, M. (2007). Lifetime reproductive effort. *American Naturalist* **170**, E129–E142.

Charnov, E.L., Gislason, H. & Pope, J.G. (2013). Evolutionary assembly rules for fish life histories. *Fish and Fisheries* **14**, 213–224.

Chauvigné, F., Tingaud-Sequeira, A., Agulleiro, M.J., Calusinka, M., Gòmex, A., Finn, R.N. & Cerdà, J. (2010). Functional and evolutionary analysis of flatfish gonadotropin receptors reveals cladal- and lineage-level divergence of the teleost glycoprotein receptor family. *Biology of Reproduction* **82**, 1088–1102.

Chellappa, S. & Huntingford, F.A. (1989). Depletion of energy reserves during reproductive aggression in male three-spined stickleback, *Gasterosteus aculeatus* L. *Journal of Fish Biology* **35**, 315–316.

Chellappa, S., Huntingford, F.A., Strang, R.H.C. & Thomson, R.Y. (1989). Annual variation in the energy reserves in male threespined stickleback, *Gasterosteus aculeatus* L. (Pisces, Gasterosteidae). *Journal of Fish Biology* **35**, 275–286.

Chellappa, S., Yamamoto, M.E., Cacho, M.S.R.F. & Huntingford, F.A. (1999). Prior residence, body size and the dynamics of territorial disputes between male freshwater angelfish. *Journal of Fish Biology* **55**, 1163–1170.

Chen, T.-R. & Reisman, H.M. (1970). A comparative chromosome study of the North American species of sticklebacks (Teleostei: Gasterosteidae). *Cytogenetic and Genome Research* **9**, 321–332.

Cheney, K.L., Grutter, A.S., Blomberg, S.P. & Marshall, N.J. (2009). Blue and yellow signal cleaning behavior in coral reef fishes. *Current Biology* **19**, 1283–1287.

Choleva, L., Janko, K., De Gelas, K., Bohlen, J., Šlechtová, V., Rábová, M. & Ráb, P. (2012). Synthesis of clonality and polyploidy in vertebrate animals by hybridization between two sexual species. *Evolution* **66**, 2191–2203.

Chong, K.L., Koh, M. & Melamed, P. (2005). Molecular regulation of gonadotropin gene expression in teleosts. In: *Hormones and their Receptors in Fish Reproduction* (Melamed, P. & Sherwood, N., eds), pp. 76–104. Singapore: World Scientific.

Chopelet, J., Waples, R.S. & Mariani, S. (2009). Sex change and the genetic structure of marine fish populations. *Fish and Fisheries* **10**, 329–343.

Christiansen, J.S., Præbel, K., Siikavuopio, S.I. & Carscadden, J.E. (2008). Facultative semelparity in capelin *Mallotus villosus* (Osmeridae) – an experimental test of a life history phenomenon in a sub-arctic fish. *Journal of Experimental Marine Biology and Ecology* **360**, 47–55.

Chung, E.-Y., Yang, Y.-W., Kang, H.-W., Choi, K.-H., Jun, J.-C. & Lee, K.-Y. (2010). Ultrastructure of germ cells and functions of Leydig cells and Sertoli cells associated with spermatogenesis in *Pampus argenteus* (Teleostei: Perciformes: Stromateidae). *Zoological Studies* **48**, 39–50.

Clelland, E. & Peng, C. (2009). Endocrine/paracrine control of zebrafish ovarian development. *Molecular and Cellular Endocrinology* **312**, 42–52.

Cloud, J. & Patton, S. (2009). Basic principles of fish spermatozoa cryopreservation. In: *Methods in Reproductive Aquaculture: Marine and Freshwater Species* (Cabrita, E., Robles, V. & Herráez, P., eds), pp. 237–250. Boca Raton: CRC Press.

Clutton-Brock, T.H. (1991). *The Evolution of Parental Care*. Princeton: Princeton University Press.

Clutton-Brock, T.H. & Parker, G.A. (1992). Potential reproductive rates and the operation of sexual selection. *The Quarterly Review of Biology* **67**, 437–456.

Cnaani, A., Lee, B.-Y., Ozouf-Costaz, C., Hulata, G., Ron, M., D'Hont, A., Baroiller, J.-F., D'Cotta, H., Penman, D.J., Tomasino, E., Coutanceau, J.-P., Pepy, E., Shirak, A. & Kocher, T.D. (2008). Genetics of sex-determination in Tilapiine species. *Sexual Development* **2**, 43–54.

Cnaani, A. & Levavi-Sivan, B. (2009). Sexual development in fish, practical applications for aquaculture. *Sexual Development* **3**, 164–175.

Cole, K.S. (2010a). Gonad morphology in hermaphroditic gobies. In: *Reproduction and Sexuality in Marine Fishes* (Cole, K.S., ed.), pp. 117–162. Berkeley: University of California Press.

Cole, K.S. (2010b). Gonad development in hermaphroditic gobies. In: *Reproduction and Sexuality in Marine Fishes* (Cole, K.S., ed.), pp. 165–202. Berkeley: University of California Press.

Cole, L.C. (1954). The population consequences of life history phenomena. *Quarterly Review of Biology* **29**, 103–137.

Coleman, R.M. (1992). Reproductive biology and female parental care in the cockscomb prickleback, *Anoplarchus purpurescens* (Pisces, Stichaenidae). *Environmental Biology of Fishes* **35**, 177–186.

Coleman, R.M. (1999). Parental care in intertidal fishes. In: *Interdial Fishes: Life in Two Worlds* (Horn, M.H., Martin, K.L.M. & Chotknowski, M.A., eds), pp. 165–180. San Diego: Academic Press.

Coleman, R.M. & Fischer, R.U. (1991). Brood size, male fanning effort and the energetics of a nonshareable parental investment in bluegill sunfish, *Lepomis macrochirus* (Teleostei: Centrarchidae). *Ethology* **87**, 177–188.

Colin, P.L. & Clavijo, I.E. (1988). Spawning activity of fishes producing pelagic eggs on a shelf-edge coral reef, southwestern Puerto Rico. *Bulletin of Marine Science* **43**, 249–279.

Collette, B.B. (1977). Epidermal breeding tubercles and bony contact organs in fishes. *Symposium of the Zoological Society of London* **39**, 225–168

Conover, D.O. (1984). Adaptive significance of temperature-dependent sex determination in a fish. *American Naturalist* **123**, 297–313.

Conover, D.O. (1992). Seasonality and the scheduling of life history at different latitudes. *Journal of Fish Biology* **41** (Suppl. B), 161–178.

Conover, D.O. & Fleischer, M. (1986). The temperature-sensitive period of sex determination in *Menidia menidia*. *Canadian Journal of Fisheries and Aquatic Sciences* **43**, 514–520.

Conover, D.O. & Heins, S.W. (1987a). The environmental and genetic components of sex ratio in *Menidia menidia* (Pisces: Atherinidae). *Copeia* 1987, 732–743.

Conover, D.O. & Heins, S.W. (1987b). Adaptive variation in environmental and genetic sex determination in a fish. *Nature* **326**, 496–498.

Conover, D.O. & Kynard, B.E. (1981). Environmental sex determination: interaction of temperature and genotype in a fish. *Science* **213**, 577–579.

Conover, D.O. & van Voorhees, D.A. (1990). Evolution of a balanced sex ratio by frequency-dependent selection in a fish. *Science* **250**, 1556–1558.

Conover, D.O., van Voorhees, D.A. & Ehtisham, A. (1992). Sex ratio selection and the evolution of environmental sex determination in laboratory populations of *Menidia menidia*. *Evolution* **46**, 1722–1730.

Constantz, G.D. (1989). Reproductive biology of Poeciliid fishes. In: *Ecology and Evolution of Livebearing Fishes (Poeciliidae)* (Meffe, G.K. & Snelson Jr, F.F., eds), pp. 33–50. New Jersey: Prentice Hall.

Consuegra, S. & de Leaniz, C.G. (2008). MHC-mediated mate choice increases parasite resistance in salmon. *Proceedings of the Royal Society of London, B* **275**, 1397–1403.

Cosson, J. (2010). Frenetic activation of fish spermatozoa flagella entails short-term motility, portending their precocious decadence. *Journal of Fish Biology* **76**, 240–279.

Cosson, J., Groison, A-L, Suquet, M., Fauvel, C., Dreanno, C. & Billard, R. (2008). Marine fish spermatozoa: racing ephemeral swimmers. *Reproduction* **136**, 277–294.

Courtenay, S.C. & Keenleyside, M.H.A. (1983). Wrigglerhanging: a response to hypoxia by brood-rearing *Herotilapia multispinosa* (Teleostei, Cichlidae). *Behaviour* **85**, 183–197.

Courtenay, S.C., Quinn, T.P., Dupuis, H.M.C., Groot, C. & Larkin, P.A. (1997). Factors affecting the recognition of population-specific odours by juvenile coho salmon. *Journal of Fish Biology* **50**, 1042–1060.

Cowen, R.K., & Sponaugle, S. (1997). Relationships between early life history traits and recruitment among coral reef fishes. In: *Early Life History and Recruitment in Fish Populations* (Chambers, R.C. & Trippel, E.A., eds), pp. 423–449. London: Chapman and Hall.

Coyne, J.A. & Orr, H.A. (2004). *Speciation*. Sunderland: Sinauer Associates.

Coyne, J.A. & Sohn, J.J. (1978). Interspecific brood care in fishes: reciprocal altruism or mistaken identity? *American Naturalist* **112**, 447–450.

Cummings, M.E., Rosenthal, G.G. & Ryan, M.J. (2003). A private ultraviolet channel in visual communication. *Proceedings of the Royal Society of London, B* **270**, 897–904.

Curry, R.A. (2005). Assessing the reproductive contribution of sympatric anadromous and freshwater-resident brook trout. *Journal of Fish Biology* **66**, 741–747.

Cury, P., Bakun, A., Crawford, R.J.M. Jarre, A., Quinones, R.A., Shannon, L.J. & Verheye, H.M. (2000). Small pelagics in upwelling systems: patterns of interaction and structural changes in "wasp-waist" ecosystems. *ICES Journal of Marine Science* **57**, 603–618.

Cushing, D.H. (1972). The production cycle and the numbers of marine fish. *Symposium of the Zoological Society of London* **29**, 213–232.

Cushing, D.H. (1975). *Marine Ecology and Fisheries*. Cambridge: Cambridge University Press.

Cushing, D.H. (1990). Plankton production and year-class strength in fish populations: an update of the match/mismatch hypothesis. *Advances in Marine Biology* **26**, 142–155.

Dahle, R., Taranger, G.L., Karlsen, Ø., Kjesbu, O.S. & Norberg, B. (2003). Gonadal development and associated changes in liver size and sexual steroids during the reproductive cycle of captive male and female Atlantic cod (*Gadus morhua* L.). *Comparative Biochemistry and Physiology, Part A* **136**, 641–653.

Darwin, C. (1871). *The Descent of Man, and Selection in Relation to Sex*. London: John Murray.

Davidson, W.S., Huang, T.-K., Fujiki, K., von Schalburg, K.R. & Koop, B.F. (2009). The sex determining loci and sex chromosomes in the family Salmonidae. *Sexual Development* **3**, 78–87.

Davie, A., Minghetti, M. & Migaud, H. (2009). Seasonal variations in clock-gene expression in Atlantic salmon (*Salmo salar*). *Chronobiology International* **26**, 379–395.

Davie, A., Porter, M.J.R., Bromage, N.R. & Migaud, H. (2007). The role of seasonally altering photoperiod in regulating physiology in Atlantic cod (*Gadus morhua*). *Part I. Sexual maturation. Canadian Journal of Fisheries and Aquatic Sciences* **64**, 84–97.

Davies, N.B. (1978). Ecological questions about territorial behaviour. In: *Behavioural Ecology: an Evolutionary Approach* (Krebs, J.R. & Davies, N.B., eds), pp. 317–350. Oxford: Blackwell Scientific Publishing.

Davis, M.R. & Fernald, R.D. (1990). Social control of neuronal soma size. *Journal of Neurobiology* **21**, 1180–1188.

Davis, R.E. & Northcutt, R.G. (1983). *Fish Neurobiology Vol. 2 Higher Brain Areas and Functions*. Ann Arbor: The University of Michigan Press.

Dawkins, R. (1982). *The Extended Phenotype*. Oxford: Freeman.

Day, J.R. & Taylor, M.H. (1984). Photoperiod and temperature interaction in the seasonal reproduction of female mummichogs. *Transactions of the American Fisheries Society* **113**, 452–457.

Deelder, C.L. (1984). *Synopsis of Biological Data on the Eel, Anguilla anguilla (Linnaeus, 1758)*. Rome: FAO.

DeFraipont, M., FitzGerald, G.J. & Guderley, H. (1992). Femme fatale – the case of the threespine stickleback. *Ethology* **91**, 147–52.

Demski, L. (1983). Behavioural effects of electrical stimulation of the brain. In: *Fish Neurobiology Vol. 2 Higher Brain Areas and Functions* (Davis, R.E. & Northcutt, R.G., eds), pp. 317–359. Ann Arbor: The University of Michigan Press.

Desjardins, J.K. & Fernald, R.D. (2009). Fish sex: why so diverse. *Current Opinion in Neurobiology* **19**, 648–653.

Desjardins, J.K., Fitzpatrick, J.L., Stiver, K.A., Van Der Kraak, G.J. & Balshine, S. (2011). Lunar and diurnal cycles in reproductive physiology and behaviour in a natural population of co-operatively breeding fish. *Journal of Zoology, London* **285**, 66–73.

Devlin, R.H. & Nagahama, Y. (2002). Sex determination and sex differentiation in fish: an overview of genetic, physiological, and environmental influences. *Aquaculture* **208**, 191–365.

DeVries, P., Goetz, F., Fresh, K.L. & Seiler, D.E. (2004). Evidence of a lunar gravitation cue on timing of estuarine entry by Pacific salmon smolts. *Transactions of the American Fisheries Society* **133**, 1379–1395.

DeWoody, J.A. & Avise, J.C. (2001). Genetic perspectives on the natural history of fish mating systems. *Journal of Heredity* **92**, 167–172.

DeWoody, J.A., Fletcher, D.E., Wilkins, S.D. & Avise, J.C. (2000). Parentage and nest guarding in the tessellated darter (*Etheostoma olmstedi*) assayed by microsatellite markers (Perciformes: Percidae). *Copeia* **2000**, 740–747.

Ding, J.L. (2005). Vitellogenesis and vitellogenin uptake in oocytes. In: *Hormones and their Receptors in Fish* (Melamed, P. & Sherwood, N., eds), pp. 254–276. New Jersey: World Scientific.

Dingle, H. (1980). Ecology and evolution of migration. In: *Animal Migration, Orientation and Navigation* (Gauthreaux, S.A. ed.), pp. 1–101. New York: Academic Press.

Dingle, H. (1996). *Migration: the Biology of Life on the Move*. Oxford: Oxford University Press.

Dittman, A.H. & Quinn, T.P. (1996). Homing in Pacific salmon: mechanisms and ecological basis. *Journal of Experimental Biology* **199**, 83–91.

Dodson, J.J., Aubin-Horth, N., Thériault, V. & Páez, D.J. (2013). The evolutionary ecology of alternative migratory tactics in salmonid fishes. *Biological Reviews* **88**, 602–625.

van Dooren, T.J.M. & Leimar, O. (2003). The evolution of environmental and genetic sex determination in fluctuating environments. *Evolution* **57**, 2667–2677.

Dominey, W.J. & Blumer, L.S. (1984). Cannibalism of early life stages in fishes. In: *Comparative and Evolutionary Perspectives: Infanticide* (Hausfaster, G. & Blaffer Hardy, S., eds), pp. 43–64. New York: Aldine.

Donaldson, E.M. & Hunter, G.A. (1982). Sex control in fish with particular reference to salmonids. *Canadian Journal of Fisheries and Aquatic Sciences* **39**, 99–110.

Dosen, L.D. & Montgomerie, R. (2004). Female size influences mate preferences of male guppies. *Ethology* **110**, 245–255.

Dubois, E.A., Zandbergen, M.A., Peute, J. & Goos, H.J.T. (2002). Evolutionary development of three gonadotropin-releasing hormone (GnRH) systems in vertebrates. *Brain Research Bulletin* **57**, 413–418.

Dufour, S., Sébert, F.M.E., Weltzien, A., Rousseau, K. & Pasqualini, C. (2010). Neuroendocrine control by dopamine of teleost reproduction. *Journal of Fish Biology* **76**, 129–160.

Dufresne, F., FitzGerald, G.J. & Lachance, S. (1990). Age- and size-related differences in reproductive success and reproductive costs in threespine stickleback (*Gasterosteus aculeatus*). *Behavioral Ecology* **1**, 140–147.

Dugatkin, L.A. (1992). Sexual selection and imitation: females copy the mate choice of others. *American Naturalist* **139**, 1384–1389.

Dugatkin, L.A. (1998). Genes, copying, and female mate choice; shifting thresholds. *Behavioral Ecology* **9**, 323–327.

Dugatkin, L.A. (2002). Animal cooperation among unrelated individuals. *Naturwissenschaften* **89**, 533–541.

Dugatkin, L.A. & Godin, J.-G.J. (1993). Female mate copying in the guppy (*Poecilia reticulata*): Age dependent effects. *Behavioral Ecology* **4**, 289–292.

Dukes, J.P., Deaville, R., Bruford, M.W., Youngson, A.F. & Jordan, W.C. (2004). Odorant receptor gene expression changes during the parr–smolt transformation in Atlantic salmon. *Molecular Ecology* **13**, 2851–2857.

Dulvy, N.K., Polunin, N.V., Mill, A.C. & Graham, N.A. (2004). Size structural change in lightly exploited coral reef fish communities: evidence for weak indirect effects. *Canadian Journal of Fisheries and Aquatic Sciences* **61**, 466–475.

Dunham, M.L., Warner, R.R. & Lawson, J.W. (1995). The dynamics of territory acquisition-a model of two coexisting strategies. *Theoretical Population Biology* **47**, 347–364.

Dunham, R.A. (2011). *Aquaculture and Fisheries Biotechnology: Genetic Approaches,* 2nd edn. Wallingford: CABI Publishing.

Duston, J. & Bromage, N. (1988). The entrainment and gating of the endogenous circannual rhythm of reproduction in female rainbow trout (*Salmo gairdneri*). *Journal of Comparative Physiology, Part A* **164**, 259–268.

van Duzer, E.M. (1939). Observations on the breeding habits of cut-lips minnow, *Exoglossum maxillingua*. *Copeia* **1939**, 65–75.

Earley, R.L., Hanninen, A.F., Fuller, A., Garcia, M.J. & Lee, E.A. (2012). Phenotypic plasticity and integration in the mangrove rivulus (*Kryptolebias marmoratus*): a prospectus. *Integrative and Comparative Biology* **52**, 814–827.

Eberhard, W.G. (1996). *Female Control: Sexual Selection by Cryptic Female Choice*. Princeton: Princeton University Press.

Einum, S. & Fleming, I.A. (1999). Maternal effects of egg size in brown trout (*Salmo trutta*): norms of reaction to environmental quality. *Proceedings of the Royal Society of London, B* **266**, 2095–2100.

Einum, S. & Fleming, I.A. (2000). Highly fecund mothers sacrifice offspring survival to maximize fitness. *Nature* **405**, 565–567.

Elder, J.F. Jr & Schlosser, I.J. (1995). Extreme clonal uniformity of *Phoxinus eos/neogaeus* gynogens (Pisces: Cyprinidae) among variable habitats in northern Minnesota beaver ponds. *Proceedings of the National Academy of Sciences of the USA* **92**, 5001–5005.

Elliott, J.M. (1990). Mechanisms responsible for population regulation in young migratory trout, *Salmo trutta*. III. The role of territorial behaviour. *Journal of Animal Ecology* **59**, 803–818.

Elliott, J.M. (1994). *Quantitative Ecology and the Brown Trout*. Oxford: Oxford University Press.

Elofsson, H., McAllister, B.G., Kime, D.E., Mayer, I. & Borg, B. (2003). Long lasting stickleback sperm; is ovarian fluid a key to success in fresh water? *Journal of Fish Biology* **63**, 240–253.

Elofsson, H., Van Look, K.J.W., Sundell, K., Sundh, H. & Borg, B. (2006). Stickleback sperm saved by salt in ovarian fluid. *Journal of Experimental Biology* **209**, 4230–4237.

Emlen, S.T. & Oring, L.W. (1977). Ecology, sexual selection, and the evolution of mating systems. *Science* **197**, 215 – 223.

Endler, J.A. (1990). On the measurement and classification of color in studies of animal color patterns. *Biological Journal of the Linnean Society* **41**, 315–352.

Enquist, M. & Leimar, O. (1987). Evolution of fighting behaviour: the effect of variation in resource value. *Journal of Theoretical Biology* **127**, 187–205.

Enquist, M., Leimar, O., Ljungberg, T., Mallner, Y. & Segerdahl, N. (1990). A test of the sequential assessment game: fighting in the cichlid fish *Nannacara anomala*. *Animal Behaviour* **40**, 1–14.

Esteve, M. (2005). Observations of spawning behaviour in Salmoninae: *Salmo*, *Oncorhynchus* and *Salvelinus*. *Reviews in Fish Biology and Fisheries* **15**, 1–21.

Evans, J.P. & Meisner, A.D. (2009). Copulatory structures: taxonomic overview and the potential for sexual selection. In: *Reproductive Biology and Phylogeny of Fishes (Agnathans and Bony Fishes)* (Jamieson, B.G.M., ed.), pp. 138–180. Enfield: Science Publishers.

Everson, A.R., Williams, H.A. & Ito, B.M. (1989). Maturation and reproduction in two Hawaiian eteline snappers, uku, *Aprion uirescens*, and onaga, *Etelis coruscans*. *Fishery Bulletin* **87**, 877–888.

Falcón, J., Besseau, L., Sauzet, S. & Boeuf, G. (2007). Melatonin effects on the hypothalamo–pituitary axis in fish. *Trends in Endocrinology and Metabolism* **18**, 81–88.

Falcón, J., Migaud, H., Muñoz-Cueto, J.A. & Carrillo, M. (2010). Current knowledge on the melatonin system in teleost fish. *General and Comparative Endocrinology* **165**, 469–482.

Falconer, D.S. & Mackay, T.F.C. (1996). *Introduction to Quantitative Genetics*, 4th edn. New York: Longman.

Fautin, D.G. (1991). The anemonefish symbiosis: what is known and what is not. *Symbiosis* **10**, 23–46.

Feldmeth, C.R. (1983). Costs of aggression in trout and pupfish. In: *Behavioural Energetics – the Cost of Survival in Vertebrates* (Asprey, W.P. & Lustick, S.I., eds), pp. 117–138. Columbus: Ohio State University Press.

Fenberg, P.B. & Roy, K. (2008). Ecological and evolutionary consequences of size-selective harvesting: how much do we know? *Molecular Ecology* **17**, 209–220.

Ferguson-Smith, M. (2006). The evolution of sex chromosomes and sex determination in vertebrates and the key role of *DMRT1*. *Sexual Development* **1**, 2–11.

Fernald, R.D. (2009). Social regulation of reproduction: what changes and why? *Hormones Brain and Behavior* **1**, 683–691.

Fernald, R.D. (2012). Social control of the brain. *Annual Review of Neuroscience* **35**, 133–351.

Fernald, R.D. & Hirata, N.R. (1977). Field study of *Haplochromis burtoni*: habitats and co-habitants. *Environmental Biology of Fishes* **2**, 299–308.

Fernando, J.I., Hattori, R.S., Shinoda, T., Kimura, H., Strobi-Mazzula, P.H., Strussmann, C.A. & Somoza, G.M. (2008). Dimorphic expression of dmrt1 and cyp19a1 (ovarian aromatase) during early gonadal development in pejerrey, *Odontesthes bonariensis*. *Sexual Development* **2**, 316–324.

Ferraro, S.P. (1980). Daily time of spawning in 12 fishes in the Peconic Bays, New York. *Fishery Bulletin* **78**, 455–464.

Feulner, P.G., Plath, M., Engelmann, J., Kirschbaum, F. & Tiedemann, R. (2009). Electrifying love: electric fish use species-specific discharge for mate recognition. *Biology Letters* **5**, 225–228.

Fishelson, L., Gon, O., Holdengraber, V. & Delarea, Y. (2006). Comparative morphology and cytology of the male sperm-transmission organs in viviparous species of clinid fishes (Clinidae: Teleostei, Perciformes). *Journal of Morphology* **267**, 1406–1414.

Finn, R.N. & Kristoffersen, B.A. (2007). Vertebrate vitellogenin gene duplication in relation to the "3R hypothesis": correlation to the pelagic egg and the oceanic radiation of teleosts. *PLoS One* **2**, e169.

Fisher, J.A.D., Frank, K.Y., Kostylev, V.E., Shackel, N.L., Horsman, T. & Hannah, C.G. (2011). Evaluating a habitat template model's predictions of marine fish diversity on the Scotian Shelf and Bay of Fundy, Northwest Atlantic. *ICES Journal of Marine System* **68**, 2096–2105.

Fisher, R.A. (1930). *The Genetical Theory of Natural Selection*. Oxford: Clarendon Press.

Fisher, R.A. (1958). *The Genetical Theory of Natural Selection*, 2nd edn. New York: Dover Publication Inc.

FitzGerald, G.J. (1983). The reproductive ecology and behaviour of three sympatric sticklebacks (Gasterosteidae) in a salt marsh. *Biology of Behaviour* **8**, 67–79.

FitzGerald, G.J., Guderley, H. & Picard, P. (1989). Hidden reproductive costs in the three-spined stickleback (*Gasterosteus aculeatus*). *Journal of Experimental Biology* **48**, 295–300.

FitzGerald, G.J. & Kedney, G.I. (1987). Aggression, fighting and territoriality in sticklebacks: three different phenomena? *Biology of Behaviour* **12**, 186–195.

FitzGerald, G.J. & van Havre, N. (1987). The adaptive significance of cannibalism in sticklebacks. *Behavioral Ecology and Sociobiology* **20**, 125–128.

Fleming, I.A. (1996). Reproductive strategies of Atlantic salmon: ecology and evolution. *Reviews in Fish Biology and Fisheries* **6**, 379–416.

Fleming, I.A. & Gross, M.R. (1990). Latitudinal clines: a trade-off between egg number and size in Pacific salmon. *Ecology* **71**, 2–11.

Fletcher, D.A. & Wootton, R.J. (1995). A hierarchical response to differences in ration size in the reproductive performance of female three-spined sticklebacks. *Journal of Fish Biology* **46**, 655–668.

Folkvord, A. (1997). Ontogeny of cannibalism in larval and juvenile fishes with special emphasis on Atlantic cod. In: *Early Life History and Recruitment in Fish Populations* (Chambers, R.C. & Trippel, E.A., eds), pp. 251–278. London: Chapman and Hall.

Fordyce, J.A. (2006). The evolutionary consequences of ecological interactions mediated through phenotypic plasticity. *Journal of Experimental Biology* **209**, 2377–2383.

Forlano, P.M. & Bass, A.H. (2011). Neural and hormonal mechanisms of reproductive-related arousal in fishes. *Hormones and Behavior* **59**, 616–629.

Forsberg, L.A., Dannewitz, J., Petersson, E. & Grahn, M. (2007). Influence of genetic dissimilarity in the reproductive success and mate choice of brown trout – females fishing for optimal MHC dissimilarity. *Journal of Evolutionary Biology* **20**, 1859–1869.

Forseth, T., Nesje, T.F., Jonsson, B. & Hårsaker, K. (1999). Juvenile migration in brown trout: a consequence of energetic state. *Journal of Animal Ecology* **68**, 783–793.

Forsgren, E. (1992). Predation risk affects mate choice in a gobiid fish. *American Naturalist* **140**, 1041–1049.

Forsgren, E. (1997). Female sand gobies prefer good fathers over dominant males. *Proceedings of the Royal Society of London, B* **264**, 1283–1286.

Forsgren, E., Karlsson, A. & Kvarnemo, C. (1996). Female sand gobies gain direct benefits by choosing males with eggs in their nests. *Behavioral Ecology and Sociobiology* **39**, 91–96.

Francis, R.C. (1984). The effects of bidirectional selection for social dominance on agonistic behaviour and sex ratios in the paradise fish (*Macropodus opercularis*). *Behaviour* **90**, 25–45.

Francis, R.C. (1992). Sexual lability in teleosts: developmental factors. *Quarterly Review of Biology* **67**, 1–18.

Fraser, S.A. (1996). The influence of predators on adoption behaviour in adult convict cichlids (*Cichlasoma nigrofasciatum*). *Canadian Journal of Zoology* **74**, 1165–1173.

Freedman, J.A. & Noakes, D.L.G. (2002). Why are there no really big bony fishes? A point-of-view on maximum body size in teleosts and elasmobranchs. *Reviews in Fish Biology and Fisheries* **12**, 403–416.

Fretwell, S.D. (1972). *Populations in a Seasonal Environment*. Princeton: Princeton University Press.

Fretwell, S.D. & Lucas, H.R. (1970). On territorial behavior and other factors influencing habitat distribution in birds. *Acta Biotheoretica* **19**, 16–36.

Frommen, J.G., Rahn, A.K., Schroth, S.H., Waltschyk, N. & Bakker, T.C.M. (2009). Mate-choice copying when both sexes face high costs of reproduction. *Evolutionary Ecology* **23**, 435–446.

Fromentin, J.M. & Fonteneau, A. (2001). Fishing effects and life history traits: a case study comparing tropical versus temperate tunas. *Fisheries Research* **53**, 133–150.

Fromentin, J.M. & Powers, J.E. (2005). Atlantic bluefin tuna: population dynamics, ecology, fisheries and management. *Fish and Fisheries* **6**, 281–306.

Fryer, G. & Iles, T.D. (1972). *The Cichlid Fishes of the Great Lakes of Africa*. Edinburgh: Oliver & Boyd.

Fu, P., Neff, B.D. & Gross, M.R. (2001). Tactic-specific success in sperm competition. *Proceedings of the Royal Society of London, B* **268**, 1105–1112.

Fujimoto, T., Nishimura, T, Goto-Kazeto, R., Kawakami, Y., Yamaha, E. & Arai, K. (2010). Sexual dimorphism of gonadal structure and gene expression in germ cell-deficient loach, a teleost fish. *Proceedings of the National Academy of Sciences of the USA* **107**, 17211–17216.

Fukushima, M., Quinn, T.J. & Smoker, W.W. (1998). Estimation of eggs lost from superimposed pink salmon (*Oncorhynchus gorbuscha*) redds. *Canadian Journal of Fisheries and Aquatic Sciences* **55**, 618–625.

Fuller, R.C., Houle, D. & Travis, J. (2005). Sensory bias as an explanation for the evolution of mate preferences. *American Naturalist* **166**, 437–446.

Fuzzen, M.L.M., Bernier, N.J. & van der Kraak, G. (2011). Stress and reproduction. In: *Hormones and Reproduction of Vertebrates Vol. 1 Fishes* (Norris, D.O. & Lopez, K.H., eds), pp. 103–117. London: Academic Press.

Gage, M.J.G., MacFarlane, C., Yeates, S., Shackelton, R. & Parker, G.A. (2002). Relationship between sperm morphometry and sperm motility in the Atlantic salmon. *Journal of Fish Biology* **61**, 1528–1539.

Gage, M.J.G., Macfarlane, C.P., Yeates, S., Ward, R.G., Searle, J.B. & Parker, G.A. (2004). Spermatozoal traits and sperm competition in Atlantic salmon: relative sperm velocity is the primary determinant of fertilization success. *Current Biology* **14**, 44–47.

Gage, M.J.G., Stockley, P. & Parker, G.A. (1995). Effects of alternative male mating strategies on characteristics of sperm production in the Atlantic salmon (*Salmo salar*): theoretical and empirical investigations. *Philosophical Transactions of the Royal Society of London, B* **350**, 391–399.

Garant, D., Dodson, J.D. & Bernatchez, L. (2005). Offspring genetic diversity increases fitness of female Atlantic salmon (*Salmo salar*). *Behavioral Ecology and Sociobiology* **57**, 240–244.

García-López, A., Fernández-Pasquier, V., Couto, E., Canario, A.V.M., Sarasquete, C. & Martínez-Rodríguez, G. (2006). Testicular development and plasma sex steroid levels in cultured male Senegalese sole *Solea senegalensis* Kaup. *General and Comparative Endocrinology* **147**, 343–351.

García-López, A., Sánchez-Amaya, M.I., Tyler, C.R. & Prat, F. (2011). Mechanisms of oocyte development in European sea bass (*Dicentrarchus labrax* L.): investigations via application of unilateral ovariectomy. *Reproduction Research* **142**, 243–253.

Garner, S.R., Bortoluzzi, R.N., Heath, D.D. & Neff, B.D. (2010). Sexual conflict inhibits female mate choice for major histocompatibility complex dissimilarity in Chinook salmon. *Proceedings of the Royal Society of London, B* **277**, 885–894.

Garrod, D.J. & Horwood, J.W. (1984). Reproductive strategies and the response to exploitation. In: *Fish Reproduction: Strategies and Tactics* (Potts, G.W. & Wootton, R.J., eds), pp. 367–384. London: Academic Press.

Gasparini, C., Pilastro, A. & Evans, J.P. (2011). Male genital morphology and its influence on female mating preferences and paternity success in guppies. *PLoS One* **6**, e22329.

Gaudreault, A., & Fitzgerald, G.J. (1985). Field observations of intraspecific and interspecific aggression among sticklebacks (Gasterosteidae). *Behaviour* **94**, 203–211.

Ge, W. (2005). Activins and its receptors in fish reproduction. In: *Hormones and their Receptors in Fish Reproduction* (Melamed, P. & Sherwood, N., eds), pp. 128–154. New Jersey: World Scientific.

Genner, M.J., Sims, D.W., Wearmouth, V.J., Southall, E.J., Southward, A.J., Henderson, P.A., & Hawkins, S.J. (2004). Regional climatic warming drives long-term community changes of British marine fish. *Proceedings of the Royal Society of London, B* **271**, 655–661.

Gerking, S.D. (1994). *The Feeding Ecology of Fish*. London: Academic Press.

Ghalambor, C.K., Reznick, D.N., & Walker, J.A. (2004). Constraints on adaptive evolution: the functional trade-off between reproduction and fast-start swimming performance in the Trinidadian guppy (*Poecilia reticulata*). *American Naturalist* **164**, 38–50.

Ghiselin, M.T. (1969). The evolution of hermaphroditism among animals. *Quarterly Review of Biology* **44**, 189–208.

Giacomello, E., Neat, F.C. & Rasotto, M.B. (2008). Mechanisms enabling sperm economy in blenniid fishes. *Behavioral Ecology & Sociobiology* **62**, 671–680.

Gilbert, D.J. (1997). Towards a new recruitment paradigm for fish stocks. *Canadian Journal of Fisheries and Aquatic Sciences* **54**, 969–977.

Gilbert, D.J. (2002). Comment: Does spawning stock affect recruitment of New England groundfish? Interpreting spawning stock and recruitment data in New England groundfish. *Canadian Journal of Fisheries and Aquatic Sciences* **59**, 191–192.

Gillet, C. & Dubois, J.P. (1995). A survey of the spawning of perch (*Perca fluviatilis*), pike (*Esox lucius*), and roach (*Rutilus rutilus*), using artificial spawning substrate in lakes. *Hydrobiologia* **300/301**, 409–414.

Ginzburg, A.S. (1968). *Fertilization of Fishes and the Problem of Polyspermy*. Moscow: Academy of Science of the USSR.

Gittleman, J.L. (1981). The phylogeny of parental care in fishes. *Animal Behaviour* **29**, 936–941.

Glebe, B.D. & Leggett, W.C. (1981). Latitudinal differences in energy allocation and use during the freshwater migrations of American shad (*Alosa sapidissima*) and their life history consequences. *Canadian Journal of Fisheries and Aquatic Sciences* **38**, 806–820.

Gluckman, P.D., Low, F.M., Buklijas, T., Hanson, M.A., & Beedle, A.S. (2011). How evolutionary principles improve the understanding of human health and disease. *Evolutionary Applications* **4**, 249–263.

Godwin, J. & Fautin, D.G. (1992). Defense of host actinians by anemonefishes. *Copeia* **1992**, 902–908.

Gonçalves, D.M. & Oliveira, R.F. (2011). Hormones and sexual behaviour of teleost fishes. In: *Hormones and Reproduction of Vertebrates Volume 1: Fishes* (Norris, D.O. & Lopez, K.H., eds), pp. 119–147. London: Academic Press.

Goodwin, J. (2010a). Neuroendocrinology of sexual plasticity in teleost fishes. *Frontiers in Neuroendocrinology* **31**, 203–216.

Goodwin, J. (2010b). Neuroendocrine regulation of sex change and alternate phenotypes in sex changing reef fishes. In: *Reproduction and Sexuality in Marine Fishes* (Cole, K.S., ed.), pp. 273–306. Berkeley: University of California Press.

Goodwin, T.W. (1980). *The Biochemistry of the Carotenoids. Volume 1 Plants*. London: Chapman & Hall.

Goodwin, T.W. (1984). *The Biochemistry of the Carotenoids*. London: Chapman & Hall.

Gönczi, A.P. (1989). A study of physical parameters at the spawning sites of the European grayling (*Thymallus thymallus* L.). *Regulated Rivers: Research and Management* **3**, 221–224.

Goodyear, C.P. (1989). Spawning stock biomass per recruit: The biological basis for a fisheries management tool. In: *ICCAT Working Document SCRS/89/82*, pp. 487–497. Miami: Southeast Fisheries Center.

Goodyear, C.P. & Ferguson, D.E. (1969). Sun-compass orientation in the mosquitofish *Gambusia affinis*. *Animal Behaviour* **17**, 636–640.

Gould, S.J. & Lewontin, R.C. (1979). The spandrels of San Marco and the Panglossian paradigm: a critique of the adaptationist programme. *Proceedings of the Royal Society of London, B* **205**, 581–598.

Graaf, M. de, Dejen, E., Osse, J.W.M. & Sibbing, F.A. (2008). Adaptive radiation of Lake Tana's (Ethiopia) *Labeobarbus* species flock (Pisces, Cyprinidae). *Marine and Freshwater Research* **59**, 391–407.

Grant, J.W.A. (1997). Territoriality. In: *Behavioural Ecology of Teleost Fishes* (Godin J.-G.J., ed.), pp. 81–103. Oxford: Oxford University Press.

Grant, J.W.A. & Green, L.D. (1996). Mate copying versus preference for actively courting males by female Japanese medaka (*Oryzias latipes*). *Behavioral Ecology* **7**, 165–167.

Graves, J.A.M. (2002). The rise and fall of *sry*. *Trends in Genetics* **18**, 259–265.

Graves, J.A.M. (2008). Weird animal genomes and the evolution of vertebrate sex and sex chromosomes. *Annual Review of Genetics* **42**, 565–586.
Graves, J.A.M. & Peichel, C.L. (2010). Are homologies in vertebrate sex determination due to shared ancestry or to limited options. *Genome Biology* **11**, 205.
Green, B.S. (2008). Maternal effects in fish populations. *Advances in Marine Biology* **54**, 1–105.
Greven, H. (2011). Gonads, genitals and reproductive biology. In: *Ecology and Evolution of Poeciliid Fishes* (Evans, J., Pilastro, A. & Schlupp, I., eds). pp. 3–17. Chicago: Chicago University Press.
Grier, H.J. (1984). Testis structure and formation of spermatophores in the atherinomorph teleost *Horaichthys setnai*. *Copeia* **4**, 833–839.
Grier, H.J. & Aranzábal, M.C.U. (2009). The testis and spermatogenesis in teleosts. In: *Reproductive Biology and Phylogeny of Fishes (Agnathans and Bony Fishes)* (Jamieson, B.G.M., ed.), pp. 121–142. Enfield (NH): Science Publications.
Grier, H.J., Aranzábal, M.C.U. & Patrio, R. (2009). The ovary, folliculogenesis, and oogenesis in teleosts. In: *Reproductive Biology and Phylogeny of Fishes (Agnathans and Bony Fishes)* (Jamieson, B.G.M., ed.), pp. 25–84. Enfield (NH): Science Publications.
Grier, H.J. & Parenti, L.R. (1994). Reproductive biology and systematics of phallostethid fishes as revealed by gonad structure. *Environmental Biology of Fishes* **41**, 287–299.
Griffiths, R., Orr, K.J., Adam, A. & Barber, I. (2000). DNA sex identification in the three-spined stickleback. *Journal of Fish Biology* **57**, 1331–1334.
Grift, R.E., Rijnsdorp, A.D., Barot, S., Heino, M. & Dieckmann, U. (2003). Fisheries-induced trends in reaction norms for maturation in North Sea plaice. *Marine Ecology Progress Series* **257**, 247–257.
Grimes, C.B., Idelberger, C.F., Able, K.W. & Turner, S.C. (1988). The reproductive biology of tilefish, *Lopholatilus chamaeleonticeps* Goode and Bean, from the United States Mid-Atlantic Bight, and the effects of fishing on the breeding system. *Fishery Bulletin* **86**, 745–762.
Grimes, C.B. & Turner, S.C. (1999). The complex life history of tilefish, *Lopholatilus chamaeleonticeps*, and vulnerability to exploitation. In: *Life in the Slow Lane: Ecology and Conservation of Long-lived Marine Animals* (Musick, J.A., ed.), pp. 17–26. Bethesda: American Fisheries Society.
Groot, C. & Margolis, L. (1991). *Pacific Salmon Life Histories*. Vancouver: University of British Columbia Press.
Gross, M.R. (1996). Alternative reproductive strategies and tactics: diversity within sexes. *Trends in Ecology and Evolution* **11**, 92–98.
Gross, M.R. (2005). The evolution of parental care. *The Quarterly Review of Biology* **80**, 37–45.
Gross, M.R. & Sargent, R. (1985). The evolution of male and female parental care in fishes. *American Zoologist* **25**, 807–22.
Guiguen, Y., Fostier, A., Piferrer, F. & Chang, C.-F. (2010). Ovarian aromatase and estrogens: a pivotal role for gonadal differentiation and sex change in fish. *General and Comparative Endocrinology* **165**, 352–366.
Gunderson, D.R. & Dygert, P.H. (1988). Reproductive effort as a predictor of natural mortality. *Rapports et Proces-verbaux des Réunions. Conseil International pour l'Éxploration de la Mer* **44**, 200–209.
Guthrie, D.M. & Muntz, W.R.A. (1993). Role of vision in fish behavior. In: *Behaviour of Teleost Fishes*, 2nd edn (Pitcher, T.P., ed.), pp. 89–121. London: Chapman and Hall.
Gwo, J.C., Jamieson, B.G.M. & Leung, L.K.P. (2009). Live preservation of fish gametes. In: *Reproductive Biology and Phylogeny of Fishes (Agnathans and Bony Fishes)* (Jamieson, B.G.M., ed.), pp. 395–484. Enfield (NH): Science Publishers.
Gwynne, D.T. & Kelly, C.D. (2004). Do the I_{mates} have it? *Evolution* **58**, 1877–1879.
Habibi, H. & Andreu-Vieyra, C.V. (2007). Hormonal regulation of follicular atresia in teleost fish. In: *The Fish Oocyte: From Basic Studies to Biotechnological Applications* (Babin, P.J., Cerdà, J. & Lubzens, E., eds), pp. 235–253. Berlin: Springer.
Haesler, M.P. & Seehausen, O. (2005). Inheritance of female mating preference in a sympatric sibling species pair of Lake Victoria cichlids: implications forspeciation. *Proceedings of the Natural Academy of Sciences of the USA* **272**, 237–245.
Hagedorn, M. & Heiligenberg, W. (1985). Court and spark: electric signals in the courtship and mating of gymnotoid fish. *Animal Behaviour* **33**, 254–265.

Haldorson, L. & Love, M. (1991). Maturity and fecundity in the rockfishes, *Sebastes* spp., a review. *Marine Fisheries Review* **53**, 25–31.

Halpern, B.S. & Warner, R.R. (2002). Marine reserves have rapid and lasting effects. *Ecology Letters* **5**, 361–366.

Hamilton, S.L., Caselle, J.E., Standish, J.D., Schroeder, D.M., Love, M.S., Rosales-Casian, J.A. & Sosa-Nishizaki, O. (2007). Size-selective harvesting alters life histories of a temperate sex-changing fish. *Ecological Applications* **17**, 2268–2280.

Hamilton, W.D. & Zuk, M. (1982). Heritable true fitness and bright birds: a role for parasites. *Science* **218**, 384–387.

Hankison, S.J. & Morris, M.R. (2002). Sexual selection and species recognition in the pygmy swordtail, *Xiphophorus pygmaeus*: conflicting preferences. *Behavioral Ecology and Sociobiology* **51**, 140–145.

Hanna, R.N. & Zhu, Y. (2011). Controls of meiotic signalling by membrane or nuclear progestin receptor in zebrafish follicle-enclosed oocytes. *Molecular and Cellular Endocrinology* **337**, 80–88.

Hansen, T., Karlsen, O., Taranger, G.L., Hemre, G.I., Holm, J.C. & Kjesbu, O.S. (2001). Growth, gonadal development and spawning time of Atlantic cod (*Gadus morhua*) reared under different photoperiods. *Aquaculture* **203**, 51–67.

Hara, T.J. (1994). The diversity of chemical stimulation in fish olfaction and gustation. *Reviews in Fish Biology and Fisheries* **4**, 1–35.

Harada, Y., Taniguchi, M., Namatame, H. & Iida, A. (2001). Magnetic materials in otoliths of bird and fish lagena and their function. *Acta Oto-laryngologica* **121**, 590–595.

Harden Jones, F.R. (1968). *Fish Migration*. London: Edward Arnold.

Harrington, R.W., Jr (1975). Sex determination and differentiation among uniparental homozygotes of the hermaphroditic fish *Rivulus marmoratus* (Cyprinodontidae: Atheriniformes). In: *Intersexuality in the Animal Kingdom* (Reinboth, R., ed.), pp. 249–262. Berlin: Springer-Verlag.

Harvey, P.H. & Pagel, M. (1991). *The Comparative Method in Evolutionary Biology*. Oxford: Oxford University Press.

Hasler, A.D. & Scholz, A.T. (1983). *Olfactory Imprinting and Homing in Salmon: Investigations into the Mechanisms of the Imprinting Processes*. Berlin: Springer-Verlag.

Hatchwell, B.J. & Komdeur, J. (2000). Ecological constrains, life history traits and the evolution of cooperative breeding. *Animal Behaviour* **59**, 1079–1086.

Hattori, R.S., Gould, R.J., Fujioka, T., Saito, T., Kurita, J., Strüssmann, C.A., Yokota, M. & Watanabe, S. (2007). Temperature-dependent sex determination in Hd-rH medaka *Oryzias latipes*: gender sensitivity, thermal threshold, critical period and *DMRTT1* expression profile. *Sexual Development* **1**, 138–146.

Hattori, R.S., Murai, Y., Oura, M., Masuda, S., Majhi, S.K., Sakamoto, T., Fernandino, J.I., Somoza, G.M., Yokota, M. & Strüssmann, C.A. (2012). A Y-linked anti-Müllerian hormone duplication takes over a critical role in sex determination. *Proceedings of the National Academy of Sciences of the USA* **109**, 2955–2959.

Hauser, L. & Carvalho, G.R. (2008). Paradigm shifts in marine fisheries genetics: ugly hypotheses slain by beautiful facts. *Fish and Fisheries* **9**, 333–362.

Hawkins, A.D., 1993. Underwater sound and fish behaviour. In: *Behaviour of Teleost Fishes*, 2nd edn (Pitcher, T.P., ed.), pp. 129–169. London: Chapman and Hall.

Hayes, J.W. (1987). Competition for spawning space between brown trout (*Salmo trutta*) and rainbow trout (*S. gairdneri*) in a lake inlet tributary, New Zealand. *Canadian Journal of Fisheries and Aquatic Sciences* **44**, 40–47.

Head, M.L., Lindholm, A. & Brooks, R. (2008). Operational sex ratio and density do not affect directional selection on male sexual ornaments and behaviour. *Evolution* **62**, 135–144.

Healey, M.C. (1980). Growth and recruitment in experimentally exploited lake whitefish (*Coregonus clupeaformis*) populations. *Canadian Journal of Fisheries and Aquatic Sciences* **37**, 255–267.

Healey, M.C. (1991). Life history of chinook salmon. In: *Pacific Salmon Life Histories* (Groot, C. & Margolis, L., eds), pp. 311–394. Vancouver: University of British Columbia Press.

Heg, D., Brouwer, L., Bachar, Z. & Taborsky, M. (2005). Large group size yields group stability in the cooperatively breeding cichlid *Neolamprologus pulcher*. *Behaviour* **142**, 1615–1641.

Heg, D., Jutzeler, E., Mitchell, J.S. & Hamilton, I.M. (2009). Helpful female subordinate cichlids are more likely to reproduce. *PLoS One* **4**, e5458.

Heibo, E., Magnhagen, C. & Vøllestad, L.A. (2005). Latitudinal variation in life-history traits in Eurasian perch. *Ecology* **86**, 3377–3386.

Helfman, G.S. (1983). Resin-coated fishes: a simple model technique for in situ studies of fish behaviour. *Copeia* **1983**, 547–549.

Helfman, G.S., Collette, B.B. & Facey, D.E. (2009). *The Diversity of Fishes*, 2nd edn. Oxford: Wiley-Blackwell.

Hellqvist, A., Schmitz, M., Mayer, I. & Borg, B. (2006). Seasonal changes in expression of LH-β and FSH-β in male and female three-spined stickleback, *Gasterosteus aculeatus*. *General and Comparative Endocrinology* **145**, 263–269.

Hendriksen, A., Lien, L., Rosseland, B.O., Traaen, T.S. & Sevaldrud, I.S. (1989). Lake acidification in Norway: present and predicted fish status. *Ambio* **18**, 314–321.

Herdman, E.J.E., Kelly, C.D. & Godin, J.-G.J. (2004). Male mate choice in the guppy (*Poecilia reticulata*): do males preferlarger females as mates? *Ethology* **110**, 97–111.

Herpin, A. & Schartl, M. (2011). *Dmrt1* genes at the crossroads: a widespread and central class of sexual development factors in fish. *FEBS Journal* **278**, 1010–1019.

Hess, H.C. (1993). Male mouthbrooding in jawfish (Opistognathidae): constraints on polygyny. *Bulletin of Marine Science* **52**, 806–818.

Hilborn, R. & Walters, C.J. (2001). *Quantitative Fisheries Stock Assessment: Choice, Dynamics and Uncertainty*. London: Chapman & Hall.

Hildemann, W.H. (1959). A cichlid fish, *Symphysodon discus*, with unique nurture habits. *American Naturalist* **93**, 27–34.

Hill, A.V.S. (1991). The genomics and genetics of human infectious disease susceptibility. *Annual Review of Genomics and Human Genetics* **2**, 373–400.

Hill, G.E. (1996). Redness as a measure of the production costs of ornamental traits. *Ethology, Ecology and Evolution* **8**, 157–175.

Hinch, S.G. & Rand, P.S. (1998). Swim speeds and energy use of upriver-migrating sockeye salmon (*Oncorhynchus nerka*): role of local environment and fish characteristics. *Canadian Journal of Fisheries and Aquatic Sciences* **55**, 1821–1831.

Hindar, K., Jonsson, B., Ryman, N. & Ståhl, G. (1991). Genetic relationships among landlocked, resident and anadromous brown trout, *Salmo trutta* L. *Heredity* **81**, 493–504.

Hirshfield, M.F. (1980). An experimental analysis of reproductive effort and cost in the Japanese medaka, *Oryzias latipes*. *Ecology* **61**, 282–292.

Hirshfield, M.F. & Tinkle, D.W. (1975). Natural selection and the evolution of reproductive effort. *Proceedings of the National Academy of Sciences of the USA* **72**, 2227–2231.

Hiyama, Y., Yoda, M. & Ohshimo, S. (2002). Stock size fluctuations in chub mackerel (*Scomber japonicus*) in the East China Sea and the Japan/East Sea. *Fisheries Oceanography* **11**, 347–353.

Hixon, M.A. (1980). Food production and competitor density as the determinants of feeding territory size. *American Naturalist* **31**, 729–735.

Hixon, M.A. (1987). Territory as a determinant of mating systems. *American Zoologist* **27**, 229–217.

Hobbs, J.-P.A., Munday, P.L. & Jones, G.P. (2004). Social induction of maturation and sex determination in a coral reef fish. *Proceedings of the Royal Society of London, B* **271**, 2109–2114.

Hoelzer, G.A. (1992). The ecology and evolution of partial-clutch cannibalism by paternal Cortez damselfish. *Oikos* **65**, 113–120.

van Hofsten, J. & Olsson, P.-E. (2005). Zebrafish sex determination and differentiation: involvement of FTZ-F1 genes. *Reproductive Biology and Endocrinology* **3**, 63.

Hogarth, P.J. (1973). Immune relations between mother and foetus in the viviparous poeciliid fish *Xiphophorus helleri* Haeckel III. Survival of embryos after ectopic transplantation. *Journal of Fish Biology* **5**, 109–113.

Holland, B. & Rice, W.R. (1998). Perspective: Chase-away sexual selection: antagonistic seduction versus resistance. *Evolution* **52**, 1–7.

Holt, G.J., Holt, S.A. & Arnold, C. (1985). Diel periodicity of spawning in sciaenids. *Marine Ecology Progress Series* **27**, 1–7.

Hontela, A. & Stacey, N.E. (1990). Cyprinidae. In: *Reproductive Seasonality in Teleosts: Environmental Influences* (Munro, A.D., Scott, A.P. & Lam, T.J., eds), pp. 53–77. Boca Raton: CRC Press.

Hoq, M.E. (2007). An analysis of fisheries exploitation and management practices in Sundarbans mangrove ecosystem, Bangladesh. *Ocean & Coastal Management* **50**, 411–427.

Hosken, D.J. & Stockley, P. (2004). Sexual selection and genital evolution. *Trends in Ecology and Evolution* **19**, 87–93.

Hoskins, L.J., Xu, Meiyu, X. & Volkoff, H. (2008). Interactions between gonadotropin-releasing hormone (GnRH) and orexin in the regulation of feeding and reproduction in goldfish (*Carassius auratus*). *Hormones and Behaviour* **54**, 379–385.

Hostache, G. & Mol, J.H. (1998). Reproductive biology of the neotropical armoured catfish, *Hoplosternum littorale* (Siluriformes, Callichthyidae): a synthesis stressing the role of the floating bubble nest. *Aquatic Living Resources* **11**, 173–185.

Houde, A.E. (1997). *Sex, Color, and Mate Choice in Guppies*. Princeton: Princeton University Press.

Hourston, A.S., Rosenthal, H. & Kerri, S. (1981). Capacity of juvenile Pacific herring (*Clupea harengus pallasi*) to feed on larvae of their own species. *Technical Report of the Fisheries Research Board of Canada* **1044**, 1–25.

Houston, A.I., Stephens, P.A., Boyd, I.L., Harding, K.C. & McNamara, J.M. (2006). Capital or income breeding? A theoretical model of female reproductive strategies? *Behavioural Ecology* **18**, 241–250.

Hsiao, S.-M., Greeley, M.S. & Wallace, R.A. (1994). Reproductive cycling in female *Fundulus heteroclitus*. *Biological Bulletin* **186**, 271–284.

Hulata, G. (2001). Genetic manipulations in aquaculture: a review of stock improvement by classical and modern technologies. *Genetica* **111**, 155–173.

Humphries, S., Evans, J.P. & Simmons, L.W. (2008). Sperm competition: linking form to function. *BMC Evolutionary Biology* **8**, 319.

Hunter, J.R. & Hasler, A.D. (1965). Spawning association of the redfin shiner, *Notropis umbratilis*, and the green sunfish, *Lepomis cyanellus*. *Copeia* **1965**, 265–281.

Hunter, J.R. & Kimbrell, C.A. (1980) Egg cannibalism in the northern anchovy, *Engraulis mordax*. *Fisheries Bulletin* **78**, 811–16.

Huntingford, F.A. (2012). The physiology of fish behaviour: a selective review of developments over the past 40 years. *Journal of Fish Biology* **81**, 2103–2126.

Hutchings, J.A. (1994). Age-and size-specific costs of reproduction within populations of brook trout, *Salvelinus fontinalis*. *Oikos* **70**, 12–20.

Hutchings, J.A. (2002). Life histories of fish. In: *Handbook of Fish Biology and Fisheries* (Hart, P.J.B. & Reynolds, J.D., eds), pp. 148–174. Oxford: Blackwell Publishing.

Hutchings, J.A. (2011). Old wine in new bottles: reaction norms in salmonid fishes. *Heredity* **106**, 421–437.

Hutchings, J.A., Bishop, T.D. & McGregor-Shaw, C.R. (1999). Spawning behaviour of Atlantic cod *Gadus morhua*: evidence of mate competition and mate choice in a broadcast spawner. *Canadian Journal of Fisheries and Aquatic Sciences* **56**, 97–104.

Hutchings, J.A. & Morris, D.W. (1985). The influence of phylogeny, size and behaviour on patterns of covariation in salmonid life histories. *Oikos* **45**, 118–124.

Hutchings, J.A. & Myers, R.A. (1994). The evolution of alternative mating strategies in variable environments. *Evolutionary Ecology* **8**, 256–268.

Hutchinson, W.F., van Oosterhout, C., Rogers, S.I. & Carvalho, G.R. (2003). Temporal analysis of archived samples indicates marked genetic changes in declining North Sea cod (*Gadus morhua*). *Proceedings of the Royal Society of London, B* **270**, 2125–2132.

Hvidsten, N.A. & Lund, R.A. (1988). Predation on hatchery-reared and wild smolts of Atlantic salmon, Salmo salar L., in the estuary of the river Orkla, Norway. *Journal of Fish Biology* **33**, 121–126.

Hvidsten, N.A. & Møkkelgjerd, P.I. (1987). Predation on salmon smolts in the estuary of the river Surna, Norway. *Journal of Fish Biology* **30**, 273–280.

Hyder, M. (1972). Endocrine regulation of reproduction in *Tilapia*. *General and Comparative Endocrinology* **3**, 729–740.

Hyoi, J. & Nagata, Y. (2000). Brood parasitism by the Japanese minnow, *Pungtungia herzi*, to the nest of the eleotrid goby, *Odontobutis obscura*, in the aquariums. *Memoirs of Osaka Kyoiku University* **48**, 127–145.

van Iersel, J.J.A. (1953). An analysis of the parental behaviour of the male three-spined stickleback (*Gasterosteus aculeatus* L.). *Behaviour* **3** (Suppl.), 1–159.

van Iersel, J.J.A. (1958). Some aspects of territorial behaviour of the male three-spined stickleback. *Archives Néerlandaises de Zoologie* **13**, 383–400.

Iles, T.D. (1984). Allocation of resources to gonad and soma in Atlantic herring *Clupea harengus* L. In: *Fish Reproduction: Strategies and Tactics* (Potts, G.W. & Wootton, R.J., eds), pp. 331–347. London: Academic Press.

Inness, C.L. & Metcalfe, N.B. (2008). The impact of dietary restriction, intermittent feeding and compensatory growth on reproductive investment and lifespan in a short-lived fish. *Proceedings of the Royal Society Series B* **275**, 1703–1708.

Itano, M., Okabayashi, N., Morishima, K., Fujimoto, T., Yoshikawa, H., Yamaha, E. & Arai, K. (2007). Cytolological mechanisms of gynogenesis and sperm incorporation in unreduced diploid eggs of the clonal loach, *Misgurnus anguillicaudatus* (Teleostei: Cobitidae). *Journal of Experimental Zoology, Part A* **307**, 35–50.

Itzkowitz, M., Itzkowitz, D.E. & Shelly, D. (1995). Territory use and disuse in the beaugregory damselfish. *Bulletin of Marine Science* **57**, 653–662.

Itzkowitz, M. & Makie, D. (1986). Habitat structure and reproductive success in the beaugregory damselfish. *Journal of Experimental Marine Biology and Ecology* **97**, 305–312.

Itzkowitz, M. & Slocum, C.J. (1995). Is the amount of algae related to success in the beaugregory damselfish? *Marine Behaviour and Physiology* **24**, 243–250.

Iwasaki, Y., Ohkawa, K., Sadakata, A., Kashiwadate, A., Takayama-Watanabe, E., Onitake, K. & Watanabe, A. (2009). Two states of active spermatogenesis switch between reproductive and non-reproductive seasons in the testes of the medaka, *Oryzias latipes*. *Development, Growth & Differentiation* **51**, 521–532.

Jackson, J.B., Kirby, M.X., Berger, W.H., Bjorndal, K.A., Botsford, L.W., Bourque, B.J., Bradbury, R.H., Cooke, R., Erlandson, J., Estes, J.A., Hughes, T.P., Kidwell, S., Lange, C.B., Lenihan, H.S., Pandolfi, J.M., Peterson, C.H., Steneck, R.S., Tegner, M.J. & Warner, R.R. (2001). Historical overfishing and the recent collapse of coastal ecosystems. *Science* **293**, 629–637.

Jackson, J.K. (2001). Coloration in female cichlids. *Journal of Aquaculture and Aquatic Sciences* **9**, S109–S129.

Jameson, E.W. Jr. (1988). *Vertebrate Reproduction*. New York: John Wiley and Sons.

Jamieson, B.G.M. (1991). *Fish Evolution and Systematics: Evidence from Spermatozoa*. Cambridge: Cambridge University Press.

Jamieson, I.G. (1995). Do female fish prefer to spawn in nests with eggs for reasons of mate choice copying or egg survival? *American Naturalist* **145**, 824–832.

Jamieson, B.G.M. (ed.) (2009). *Reproductive Biology and Phylogeny of Fishes (Agnathans and Bony Fishes)*. Enfield (NH): Science Publications.

Jamieson, I.G., Blouw, D.M. & Colgan, P.W. (1992). Field observations on the reproductive biology of a newly discovered stickleback (*Gasterosteus*). *Canadian Journal of Zoology* **70**, 1057–1063.

Jamieson, I.G., & Colgan, P.W. (1989). Eggs in the nests of males and their effect on mate choice in the three-spined stickleback. *Animal Behaviour* **38**, 859–865.

Janhunen, M., Rudolfsen, G., Kekäläinen, J., Figencshou, L., Peuhkuri, N. & Kortet, R. (2009). Spawning coloration and sperm quality in a large lake population of Arctic charr (Salmonidae: *Salvelinus alpinus* L.). *Biological Journal of the Linnean Society* **98**, 794–802.

Janko, K., Bohlen, J., Lamatsch, Flajshans, M., Epplen, J.T., Ráb, P., Kotlik, P. & Šlechtova, V. (2007). The gynogenetic reproduction of diploid and triploid hybrid spined locahes (*Cobitis*: Teleostei), and their ability to establish successful clonal lineages – on the evolution of polyploidy in asexual vertebrates. *Genetica* **131**, 185–194.

Jennions, M.D. & Kokko, H. (2010). Sexual selection. In: *Evolutionary Behavioral Ecology* (Westneat, D.F. & Fox, C.W., eds), pp. 343–364. Oxford: Oxford University Press.

Jennions, M.D. & Petrie, M. (2000). Why do females mate multiply? A review of the genetic benefits. *Biological Reviews* **75**, 21–64.

Jensen, G.L. & Shelton, W.L. (1979). Effects of estrogens on *Tilapia aurea*: Implications for production of monosex genetic male tilapia. *Aquaculture* **16**, 233–242.

Jiggins, C.D., Naisbit, R.E., Coe, R.L. & Mallet, J. (2001). Reproductive isolation caused by colour pattern mimicry. *Nature* **411**, 302–305.

Jirotkul, M. (1999). Operational sex ratio influences female preference and male-male competition in guppies. *Animal Behaviour* **58**, 287–294.

Johannes, R.E. (1978). Reproductive strategies of coastal marine fishes in the tropics. *Environmental Biology of Fishes* **3**, 65–84.

Johnson, J.B. (2001). Adaptive life-history evolution in the livebearing fish *Brachyrhaphis rhabdophora*: genetic basis for parallel divergence in age and size at maturity and a test of predator-induced plasticity. *Evolution* **55**, 1486–1491.

Johnson, J.B. & Belk, M.C. (2001). Predation environment predicts divergent life-history phenotypes among populations of the livebearing fish *Brachyrhaphis rhabdophora*. *Oecologia* **126**, 142–149.

Johnson, J.B. & Bagley, J.C. (2011). Ecological drivers of life-history divergence in Poeciliid fishes. In: *Ecology and Evolution of Poeciliid Fishes* (Evans, J., Pilastro, A. & Schlupp, I., eds), pp. 38–49. Chicago: Chicago University Press.

Johnstone, K.A., Lubieniecki, K.P., Koop, B.F. & Davidson, W.S. (2011). Expression of olfactory receptors in different life stages and life histories of wild Atlantic salmon (*Salmo salar*). *Molecular Ecology* **20**, 4059–4069.

Johnstone, R.A. (1996). Multiple displays in animal communication: 'backup signals' and 'multiple messages'. *Philosophical Transactions of the Royal Society of London, B* **352**, 329–338.

Johnstone, R.A. (1997). The tactics of mutual mate choice and competitive search. *Behavioral Ecology and Sociobiology* **40**, 51–59.

Jones, A.G. & Ratterman, N.L. (2009). Mate choice and sexual selection: what have we learned since Darwin? *Proceedings of the Natural Academy of Sciences of the USA* **106**, 10 001–10 008.

Jones, G.P. (1980). Growth and reproduction in the protogynous hermaphrodite *Pseudolabrus celidotus* (Pisces: Labridae) in New Zealand. *Copeia* **1980**, 660–675.

Jones, G.P. & Thompson, S.M. (1980). Social inhibition of maturation in females of the temperate wrasse *Pseudolabrus celidotus* and a comparison with the blennioid *Tripterygion varium*. *Marine Biology* **59**, 247–256.

Jonsson, B. (1985). Life history patterns of freshwater resident and sea-run migrant brown trout in Norway. *Transactions of the American Fisheries Society* **114**, 182–194.

Jonsson, B. & Jonsson, N. (1993). Partial migration: niche shift versus sexual maturation in fishes. *Reviews in Fish Biology and Fisheries* **3**, 348–365.

Jonsson, B. & Jonsson, N. (2006). Life-history effects of migratory costs in anadromous brown trout. *Journal of Fish Biology* **69**, 860–869.

Jonsson, B., Jonsson, N. & Finstad, A.G. (2013). Effects of temperature and food quality on age and size at maturity in ectotherms: an experimental test with Atlantic salmon. *Journal of Animal Ecology* **82**, 201–210.

Jonsson, N. (1991). Influence of water flow, water temperature and light on fish migration in rivers. *Nordic Journal of Freshwater Research* **66**, 20–35.

Jonsson, N. & Jonsson, B. (1999). Trade-off between egg mass and egg number in brown trout. *Journal of Fish Biology* **55**, 767–783.

Jonsson, N., Jonsson, B. & Hansen, L.P. (1997). Changes in proximate composition and estimates of energetic costs during upstream migration and spawning in Atlantic salmon *Salmo salar*. *Journal of Animal Ecology* **66**, 425–436.

Jørgensen, A., Morthorts, J.E., Anersen, O., Rasmussen, L.J. & Bjerregaard, P. (2008). Expression profiles for sixe zebrafish genes during gonadal sex differentiation. *Reproductive Biology and Endocrinology* **6**, 25.

Jørgensen, C., Ernande, B., Fiksen, Ø & Dieckmann, U. (2006). The logic of skipped spawning in fish. *Canadian Journal of Fisheries and Aquatic Sciences* **63**, 200–211.

Jørgensen, C. & Holt, R.E. (2013). Natural mortality: Its ecology, how it shapes fish life histories, and why it may be increased by fishing. *Journal of Sea Research* **75**, 8–18.

June, F.C. (1977). Reproductive patterns in seventeen species of warmwater fishes in a Missouri River reservoir. *Environmnetal Biology of Fishes* **2**, 285–296.

Kadri, S., Metcalfe, N.B., Huntingford, F.A. & Thorpe, J.E. (1995). What controls the onset of anorexia in maturing adult female Atlantic Salmon? *Functional Ecology* **9**, 790–797.

Kadri, S., Mitchell, D.F., Metcalfe, N.B., Huntingford, F.A. & Thorpe, J.E. (1996). Differential patterns of feeding and resource accumulation in maturing and immature Atlantic salmon, *Aquaculture* **142**, 245–257.

Kah, O. & Dufour, S. (2011). Conserved and divergent features of reproductive endocrinology of teleost fishes. In: *Hormones and Reproduction of Vertebrates, Volume 1 – Fishes* (Norris, D.O. & Lopez, K.H., eds), pp. 15–42. London: Academic Press.

Kaitala, A., Kaitala, V. & Lundberg, P. (1993). A theory of partial migration. *American Naturalist* **142**, 59–81.

Kalbe, M., Eizaguirre, C., Dankert, I., Reusch, T.B.H., Sommerfeld, R.D., Wegner, K.M. & Milinski, M. (2009). Lifetime reproductive success is maximized with optimal major histocompatibility complex diversity. *Proceedings of the Royal Society of London, B* **276**, 925–934.

Kallman, K.D. (1984). A new look at sex determination in Poeciliid fishes. In: *Evolutionary Genetics of Fishes* (Turner, B.J., ed.), pp. 95–171. New York: Plenum.

Kamiya, T., Kai, W., Tasumi, S., Oka, A., Matsunaga, T., Mizuno, N., Fujita, M., Suetake, H., Suzuki, S., Hosoya, S., Tohari, S., Brenner, S., Miyadai, T., Venkatesh, B., Suzuki, Y. & Kikuchi, K. (2012). A trans-species missense SNP in *Amhr2* is associated with sex determination in the tiger pufferfish, *Takifugu rubripes* (fugu). *PLoS Genetics* **8**, e1002798.

Kamler, E. (1992). *Early Life History of Fishes. An Energetics Approach.* London: Chapman and Hall.

Kamler, E. (2008). Resource allocation in yolk-feeding fish. *Reviews in Fish Biology and Fisheries* **18**, 143–200.

Kanoh, Y. (1996). Pre-oviposition ejaculation in externally fertilizing fish: How sneaker male rose bitterlings contrive to mate. *Ethology* **102**, 883–899.

Kanoh, Y. (2000). Reproductive success associated with territoriality, sneaking, and grouping in male rose bitterlings, *Rhodeus ocellatus* (Pisces: Cyprinidae). *Environmental Biology of Fishes* **57**, 143–154.

Karino, K., Kuwamura, T., Nakashima, Y. & Sakai, Y. (2000). Predation risk and the opportunity for female mate choice in a coral reef fish. *Journal of Ethology* **18**, 109–114.

Karlsen, Ø, Holm, J.C. & Kjesbu, O.S. (1995). Effects of periodic starvation on reproductive investment in first-time spawning Atlantic cod (*Gadus morhua* L.). *Aquaculture* **133**, 159–170.

Karplus, I. (1987). The association between gobiid fishes and burrowing alpheid shrimps. *Oceanography and Marine Biology – an Annual Review* **25**, 507–562.

Karplus, I. (1992). Obligatory and facultative goby-shrimp partnerships in the western tropical Atlantic. *Symbiosis* **12**, 275–291.

Karplus, I. (2014). *Symbiosis in Fishes. The Biology of Interspecific Partnerships.* Oxford: Wiley Blackwell.

Karplus, I. & Thompson, A.R. (2011). The partnership between gobiid fishes and burrowing alpheid shrimps. In: *Biology of Gobies* (Patzner, R.A., Tassell, J.L. van, Kovacic, M. & Kapoor, B.G., eds). Jersey: Science Publishers.

Kasahara, M. *et al.* (2007). The medaka draft genome and insights into vertebrate genome evolution. *Nature* **447**, 714–719.

Katsiadaki, I. (2007). The use of the stickleback as a sentinel and model species in ecotoxicology. In: *Biology of the Three-spined Stickleback* (Östlund-Nilsson, S., Mayer, I. & Huntingford, F.A., eds), pp. 319–351. Boca Raton: CRC Press.

Kaufmann, J.H. (1971). Is territoriality definable? In: *Behavior and Environment: The Use of Space by Animals and Men* (Esser, A.H., ed), pp. 36–40. New York: Plenum Press.

Kaufmann, J.H. (1983). On the definitions and functions of dominance and territoriality. *Biological Reviews* **58**, l–20.

Kawamura, K. & Uehara, K. (2005). Effects of temperature on free-embryonic diapause in the autumn-spawning bitterling *Acheilognathus rhombeus* (Teleostei: Cyprinidae). *Journal of Fish Biology* **67**, 684–695.

Kawase, H., Okata, Y. & Ito, K. (2013). Role of huge geometric circular structures in the reproduction of a marine pufferfish. *Scientific Reports* **3**, 2106.

Kawauchi, H., Suzuki, K., Itoh, H., Swanson, P., Naito, N., Nagahama, Y., Nozaki, M., Nakai, Y. & Itoh, S. (1989). The duality of fish gonadotropins. *Fish Physiology and Biochemistry* **7**, 29–38.

Kazancioglu, E. & Alonozo, S.H. (2010). A comparative analysis of sex change in Labridae supports the size advantage hypothesis. *Evolution* **64**, 2254–2264.

Keenleyside, M.H.A. (1972). The behaviour of *Abudefduf zonatus* (Pisces, Pomacentridae) at Heron Island, Great Barrier Reef. *Animal Behaviour* **20**, 763–774.

Keenleyside, M.H.A. (1979). *Diversity and Adaptation in Fish Behaviour*. Berlin: Springer-Verlag.

Keenleyside, M.H.A. (1983). Mate desertion in relation to adult sex-ratio in the biparental cichlid fish *Herotilapia multispinosa*. *Animal Behaviour* **31**, 683–688.

Keenleyside, M.H.A. (1985). Bigamy and mate choice in the biparental cichlid fish *Cichlasoma nigrofasciatum*. *Behavioral Ecology and Sociobiology* **17**, 285–290.

Keenleyside, M.H.A. (1991). Parental care. In: *Cichlid Fishes: Behavior, Ecology and Evolution* (Keenleyside, M.H.A., ed.), pp. 191–208. London: Chapman and Hall.

Keenleyside, M.H.A., Bailey, R.C. & Young, V.H. (1990). Variation in mating system and parental behaviour of captive and free-living *Cichlasoma nigrofasciatum* (Teleostei, Cichlidae). *Behaviour* **112**, 202–221.

Keenleyside, M.H.A. & Mackereth, R.W. (1992). Effects of loss of male parent on brood survival in a biparental cichlid fish. *Environmental Biology of Fishes* **34**, 207–212.

Kennedy, J., Witthames, P.R., Nash, R.D.M. & Fox, C.J. (2008). Is fecundity in plaice (*Pleuronectes platessa* L.) down-regulated in response to reduced food intake during autumn? *Journal of Fish Biology* **72**, 78–92.

Kennedy, J., Skjaeraasen, J.E., Nash, R.D.M., Thorsen, A., Slotte, A., Hansen, T. & Kjesbu, O.S. (2010). Do capital breeders like Atlantic herring (*Clupea harengus*) exhibit sensitive periods of nutritional control on ovary development and fecundity regulation. *Canadian Journal of Fisheries and Aquatic Sciences* **67**, 16–27.

Kerby, J.H., Bayless, J.D. & Harrell, R.M. (1985). Growth, survival, and harvest of striped bass produced with cryopreserved spermatozoa. *Transactions of the American Fisheries Society* **114**, 761–765.

Kidd, K.A., Blanchfield, P.J., Mills, K.H., Palace, V.P., Evans, R.E., Lazorchak, J.M. & Flick, R.W. (2007). Collapse of a fish population after exposure to a synthetic estrogen. *Proceedings of the Natural Academy of Sciences of the USA* **104**, 8897–8901.

Kikuchi, K. & Hamaguchi, S. (2013). Novel sex-determining genes in fish and sex chromosome evolution. *Developmental Dynamics* **242**, 339–353.

Kime, D.E. (1993). "Classical" and "non-classical" reproductive steroids in fish. *Reviews in Fish Biology and Fisheries* **3**, 160–180.

Kime, D.E. (1995). The effects of pollution on reproduction in fish. *Reviews in Fish Biology and Fisheries* **5**, 52–96,

Kingsley, D.M. *et al.* (2004). New genomic tools for molecular studies of evolutionary change in three-spine sticklebacks. *Behaviour* **141**, 1331–1344.

Kinnison, M.T., Unwin, M.J., Hendry, A.P. & Quinn, T.P. (2001). Migratory costs and the evolution of egg size and number in introduced and indigenous salmon populations. *Evolution* **55**, 1656–1667.

Kinsey, W.H., Sharma, D. & Kinsey, S.C. (2007). Fertilization and egg activation in fishes. In: *The Fish Oocyte: From Basic Studies to Biotechnological Applications* (Babin, P.J., Cerdà, J. & Lubzens, E., eds), pp. 397–409. Berlin: Springer.

Kirkpatrick, M. (1987). Sexual selection by female choice in polygynous animals. *Annual Review of Ecology and Systematics* **18**, 43–70.

Kirkpatrick, M. & Ryan, M.J. (1991). The evolution of mating preferences and the paradox of the lek. *Nature* **350**, 33–38.

Kirschbaum, F., Leyendecker, U., Nyonge, B., Schulz, C., Weitkamp, H., Didhiou, S., Thomas, M. & Schugardt, C. (2008). Environmental control of cyclical reproduction of tropical fish: evidence from comparative experimental data. *Cybium* **32**, 294–296.

Kitamura, J. (2005). Factors affecting seasonal mortality of rosy bitterling (*Rhodeus ocellatus kurumeus*) embryos on the gills of their host mussel. *Population Ecology* **47**, 41–51.

Kitamura, J., Nagata, N., Nakajima, J. & Sota, T. (2012). Divergence of ovipositor length and egg shape in a brood parasitic bitterling fish through the use of different mussel hosts. *Journal of Evolutionary Biology* **25**, 566–573.

Kitano, H., Irie, S., Ohta, K., Hirai, T., Yamaguchi, A. & Matsuyama, M. (2011). Molecular cloning of two gonadotropin receptors and their distinct mRNA profiles in daily oogenesis of the wrasse *Pseudolabrus sieboldi*. *General and Comparative Endocrinology* **172**, 268–278.

Kitano, J., Ross, J.A., Mori, S., Kume, M., Jones, F.C., Chan, Y.F., Absher, D.M., Grimwood, J., Schmutz, J., Myers, R.M., Kingsley, D.M. & Peichel, C.L. (2009). A role for a neo-sex chromosome in stickleback speciation. *Nature* **461**, 1079–1083.

Kjesbu, O.S., Klungsoyr, J., Kryvi, H., Witthames, P.R. & Greer Walker, M. (1991). Fecundity, atresia, and egg size of captive Atlantic cod (*Gadus morhua*) in relation to proximate body composition. *Canadian Journal of Fisheries and Aquatic Sciences* **48**, 2333–2343.

Kjesbu, O.S., Witthames, P.R., Solemdal, P. & Greer Walker, M. (1998). Temporal variations in the fecundity of Arcto-Norwegian cod (*Gadus morhua*) in response to natural changes in food and temperature. *Journal of Sea Research* **40**, 303–321.

Kjørsvik, E., Mangor-Jensen, A. & Holmefjord, I. (1990). Egg quality in fishes. *Advances in Marine Biology* **26**, 71–113.

Klug, H., Heuschele, J., Jennions, M.D. & Kokko, H. (2010). The mismeasurement of sexual selection. *Journal of Evolutionary Biology* **23**, 447–462.

Klug, H. & St. Mary, C.M. (2005). Reproductive fitness consequences of filial cannibalism in the flagfish, *Jordanella floridae*. *Animal Behaviour* **70**, 685–691.

Klug, H., Lindström, K. & Mary, C.M.S. (2006). Parents benefit from eating offspring: density-dependent egg survivorship compensates for filial cannibalism. *Evolution* **60**, 2087–2095.

Knapp, R. & Carlisle, S.L. (2011). Testicular function and hormonal regulation in fishes. In: *Hormones and Reproduction of Vertebrates, Vol. 1 – Fishes* (Norris, D.O. & Lopez, K.H., eds), pp. 43–63. London: Academic Press.

Knapp, R.A. & Kovach, J.T. (1991). Courtship as an honest indicator of male parental quality in the bicolor damselfish, *Stegastes partitus*. *Behavioral Ecology* **2**, 230–295.

Knibb, W. (2000). Genetic improvement of marine fish–which method for industry? *Aquaculture Research* **31**, 11–23.

Knoll-Gellida, A. & Babin, P.J. (2007). Zebrafish ovarian follicle transcriptome. In: *The Fish Oocyte: From Basic Studies to Biotechnological Applications* (Babin, P.J., Cerdà, J. & Lubzens, E., eds), pp. 77–97. Berlin: Springer.

Knouft, J.H., Page, L.M. & Plewa, M.J. (2003). Antimicrobial egg cleaning by the fringed darter (Perciformes: Percidae: Etheostoma crossopterum): implications of a novel component of parental care in fishes. *Proceedings of the Royal Society of London, B* **270**, 2405–2411.

Kobayashi, T. & Nagahama, Y. (2009). Molecular aspects of gonadal differentiation in a teleost fish, the Nile tilapia. *Sexual Development* **3**, 108–117.

Kodric-Brown, A. (1983). Determinants of male reproductive success in pupfish (*Cyprinodon peconensis*). *Animal Behaviour* **31**, 128–137.

Kohda, M., Tanimura, M., Kikue-Nakamura, M. & Yamagishi, S. (1995). Sperm drinking by female catfishes: a novel mode of insemination. *Environmental Biology of Fishes* **42**, 1–6.

Kokita, T. (2003). Potential latitudinal variation in egg size and number of a geographically widespread reef fish, revealed by common-environment experiments. *Marine Biology* **143**, 593–601.

Kokko, H., Brooks, R., Jennions, M.D. & Morley, J. (2003). The evolution of mate choices and mating biases. *Proceedings of the Royal Society of London, B* **270**, 653–664.

Kokko, H. & Johnstone, R.A. (2002). Why is mutual mate choice not the norm? Operational sex ratios, sex roles, and the evolution of sexually dimorphic and monomorphic signalling. *Philosophical Transactions of the Royal Society of London, B* **357**, 319–330

Kokko, H. & Monaghan, P. (2001). Predicting the direction of sexual selection. *Ecology Letters* **4**, 159–165

Kokko, H. & Rankin, D.J. (2006). Lonely hearts or sex in the city? Density-dependent effects in mating systems. *Philosophical Transactions of the Royal Society of London, B* **361**, 319–334.

Kolding, J. (1993). Population dynamics and life-history styles of Nile tilapia, *Oreochromis niloticus*, in Ferguson's Gulf, Lake Turkana, Kenya. *Environmental Biology of Fishes* **37**, 25–46.

Kolm, N. (2002). Male size determines reproductive output in a paternal mouthbrooding fish. *Animal Behaviour* **63**, 727–733.

Kolm, N. (2009). Parental care. In: *Reproductive Biology and Phylogeny of Fishes (Agnathans and Bony Fishes)* (Jamieson, B.G.M., ed.), pp. 351–370. Enfield (NH): Science Publishers.

Kolm, N. & Ahnesjö, I. (2005). Do egg size and parental care coevolve in fishes? *Journal of Fish Biology* **66**, 1499–1515.

Kondo, M. Nanda, I., Schmid, M. & Schartl, M. (2009). Sex determination and sex chromosome evolution: insights from medaka. *Sexual Development* **3**, 88–93.

Korta, M., Muruta, H., Kurita, Y. & Kjesbu, O.S. (2010). How are oocytes recruited in an indeterminate fish? Applications of stereological techniques along with advanced packing density theory on European hake (*Merluccius merluccius* L.). *Fisheries Research* **104**, 56–63.

Koslow, J.A., Bell, J., Virtue, P. & Smith, D.C. (1995). Fecundity and its variability in orange roughy: effects of population density, condition, egg size, and senescence. *Journal of Fish Biology* **47**, 1063–1080.

Koya, Y. (2008). Reproductive physiology in viviparous teleosts. In: *Fish Reproduction* (Rocha, M.J., Arukwe, A. & Kapoor, B.G., eds), pp. 245–276. Enfield (NH): Science Publications.

Koya, Y. & Muñoz, M. (2007). Comparative study on ovarian structures in scorpaenids: possible evolutional process of reproductive mode. *Ichthyological Research* **54**, 221–230.

van der Kraak, G., Chang, J.P. & Janz, D.M. (1998). Reproduction. In: *The Physiology of Fishes*, 2nd edn. (Evans, D.H., ed.), pp. 465–488. Boca Raton: CRC Press.

Kraak, S.B.M. (2007). Does the probabilistic maturation reaction norm approach disentangle phenotypic plasticity from genetic change? *Marine Ecology Progress Series* **335**, 295–300.

Kraak S.B.M. & Bakker, T.C.M. (1998). Mutual mate choice in sticklebacks: attractive males choose big females, which lay big eggs. *Animal Behaviour* **56**, 859–866.

Kramer, D.L. (1973). Parental behaviour in the blue gourami, *Trichogaster trichopterus* (Pisces: Belontiidae) and its induction during exposure to varying numbers of conspecific eggs. *Behaviour* **47**, 14–32.

Kramer, D.L. (1978). Reproductive seasonality in the fishes of a tropical stream. *Ecology* **59**, 976–985.

Kramer, D.L. (1987). Dissolved oxygen and fish behaviour. *Environmental Biology of Fishes* **18**, 81–92.

Kramer, D.L. & Liley, N.R. (1971). The role of spawning behaviour and stimuli from the eggs in the induction of a parental response in the blue gourami, *Trichogaster trichopterus* (Pisces, Belontiidae). *Animal Behaviour* **19**, 87–92.

Krekorian, C.O. (1976). Field observations in Guyana on the reproductive biology of the spraying characid, *Copeina arnoldi* Regan. *American Midland Naturalist* **96**, 88–97.

Krekorian, C.O. & Dunham, D.W. (1972). Preliminary observations on the reproductive and parental behavior of the spraying characid *Copeina arnoldi* Regan. *Zeitschrift für Tierpsychologie* **31**, 419–437.

Kulkarni, C.V. (1940). On the systematic position, structural modifications, bionomics and development of a remarkable new family of cyprinodont fishes from the province of Bombay. *Records of the Indian Museum, Delhi* **42**, 379–423.

Kuparinen, A. & Merilä, J. (2007). Detecting and managing fisheries-induced evolution. *Trends in Ecology & Evolution* **22**, 652–659.

Kuwamura, T. (1983). Spawning behavior and timing of fertilization in the mouthbrooding cardinal fish *Apogon notatus*. *Japanese Journal of Ichthyology* **30**, 61–71.

Kvarnemo, C. & Ahnesjö, I. (1996). The dynamics of operational sex ratios and competition for mates. *Trends in Ecology and Evolution* **11**, 404–412.

Kvarnemo, C., Mobley, K.B., Partridge, C., Jones, A.G. & Ahnesjö, I. (2011). Evidence of paternal nutrient provisioning to embryos in broad-nosed pipefish *Syngnathus typhle*. *Journal of Fish Biology* **78**, 1725–1737.

Kvarnemo, C., Svensson, O. & Forsgren, E. (1998). Parental behaviour in relation to food availability in the common goby. *Animal Behaviour* **56**, 1285–1290.

Lack, D. (1947). The significance of clutch-size. *Ibis* **89**, 302–352.

Ladich, F. & Fine, M. (2006). Sound-generating mechanisms in fishes: a unique diversity in vertebrates. In: *Fish Communication* (Ladich, F., Collin, S.P., Moller, P. & Kapoor, B.G., eds), pp. 1–43. Enfield (NH): Science Publishers.

Lahnsteiner, F. (2011). Spermatozoa of the teleost fish *Perca fluviatilis* (perch) have the ability to swim for more than 2 hours in saline solutions. *Aquaculture* **314**, 221–224.

Lahnsteiner, F., Berger, B., Horvath, A. & Urbanyi, B. (2004). Studies on the semen biology and sperm cryopreservation in the sterlet, *Acipenser ruthenus* L. *Aquaculture Research* **35**, 519–528.

Lahnsteiner, F. & Patzner, R.A. (2009). Male reproductive system: spermatic duct and accessory organs of the testis. In: *Reproductive Biology and Phylogeny of Fishes (Agnathans and Bony Fishes)* (Jamieson, B.G.M., ed.), pp. 143–187. Enfield (NH): Science Publications.

Lahnsteiner, F., Patzner, R.A. & Weismann, T. (1993). The spermatic ducts of salmonid fishes (Salmonidae, Teleostei). Morphology, histochemistry and composition of the secretion. *Journal of Fish Biology* **42**, 79–93.

Laird, L.M., Wilson, A.R. & Holliday, E.G.T. (1981). Field trials of a method of induction of auto-immune gonad rejection in Atlantic salmon (*Salmo salar* L.). *Reproduction, Nutrition, Development* **20**, 1781–1788.

Lagomarsino, I.V. & Conover, D.O. (1993). Variation in environmental and genotypic sex-determination mechanisms across a latitudinal gradient in the fish, *Menidia menidia*. *Evolution* **47**, 487–494.

Lambert, Y. & Dutil, J.D. (2000). Energetic consequences of reproduction in Atlantic cod (*Gadus morhua*) in relation to spawning level of somatic energy reserves. *Canadian Journal of Fisheries and Aquatic Sciences* **57**, 815–825.

Lampert, K.P. & Schartl, M. (2008). The origin and evolution of a unisexual hybrid: *Poecilia formosa*. *Philosophical Transactions of the Royal Society London, B* **363**, 2901–2909.

Lampert, K.P. & Schartl, M. (2010). A little bit is better than nothing: the incomplete parthenogenesis of salamanders, frogs and fish. *BMC Biology* **8**, 78.

Lande, R. (1981). Models of speciation by sexual selection on polygenic traits. *Proceedings of the National Academy of Sciences of the USA* **78**, 3721–3725.

Langston, A.L., Johnstone, R. & Ellis, A.E. (2001). The kinetics of the hypoferraemic response and changes in levels of alternative complement activity in diploid and triploid Atlantic salmon, following injection of lipopolysaccharide. *Fish & Shellfish Immunology* **11**, 333–345.

Lappalainen, J., Tarkan, A.S. & Harrod, C. (2008). A meta-analysis of latitudinal variations in life-history traits of roach, *Rutilus rutilus*, over its geographical range: linear or non-linear relationships. *Freshwater Biology* **53**, 1491–1501.

Lasker, R. (1981). Factors contributing to variable recruitment of the northern anchovy (*Engradis mordax*) in the California Current: contrasting years, 1975 through 1978. *Rapports et Proces-verbaux des Réunions. Conseil International pour l'Éxploration de la Mer* **178**, 375–388.

Lavery, R.J. & Keenleyside, M.H.A. (1990). Filial cannibalism in the biparental fish *Cichlasoma nigrofasciatum* (Pisces: Cichlidae) in response to early brood reductions. *Ethology* **86**, 326–338.

Law, R. (2007). Fisheries-induced evolution: present status and future directions. *Marine Ecology Progress Series* **335**, 271–277.

Law, R. (2000). Fishing, selection, and phenotypic evolution. *ICES Journal of Marine Science* **57**, 659–668.

Law, R. & Rowell, C.A. (1993). Cohort-structured populations, selection responses, and exploitation of the North Sea cod. In: *The Exploitation of Evolving Resources* (Stokes, T.K., McGlade, J.M. & Law, R., eds), pp. 155–173). Berlin: Springer-Verlag.

Lawson, E.T. (2011). Neuroendocrine regulation in sex-changing fishes. In: *Hormones and Reproduction of Vertebrates, Vol. 1 – Fishes* (Norris, D.O. & Lopez, K.H., eds), pp. 149–168. London: Academic Press.

Leatherland, J.F. (1969). Studies on the structure and ultrastructure of the intact and "methallibure"-treated meso-adenohypophysis of the viviparous teleost *Cymatogaster aggregata* Gibbons. *Zeitschrift für Zellforschung und Mikroskopische Anatomie* **98**, 122–134.

Leatherland, J.F., Li, M. & Barkataki, S. (2010). Stressors, glucocorticoids and ovarian function in teleosts. *Journal of Fish Biology* **76**, 86–111.

Le Comber, C., Faulkes, C.G., Formosinho, J. & Smith, C. (2003). Response of territorial males to the threat of sneaking in the three-spined stickleback: a field study. *Journal of Zoology, London* **261**, 15–20.

Le Comber, S.C., Faulkes, C.G., van Look, K.J.W., Holt, W.V. & Smith, C. (2004). Recovery of sperm activity after osmotic shock in the three-spined stickleback: implications for pre-oviposition ejaculation. *Behaviour* **141**, 1555–1569.

Le Comber, S.C. & Smith, C. (2004). Polyploidy in fishes. *Biological Journal of the Linnean Society* **82**, 25–33.

Legget, W. & Carscadden, J. (1978). Latitudinal variation in the reproductive characteristics of American shad (*Alosa sapidissima*): evidence for population specific life history strategies in fish. *Journal of the Fisheries Research Board of Canada* **35**, 1469–1478.

Leggett, W.C. & Deblois E. (1994). Recruitment in marine fishes: is it regulated by starvation and predation in the egg and larval stages? *Netherlands Journal of Sea Research* **32**, 119–134.

Leips, J., Richardson, J.M.L., Rodd, F.H. & Travis, J. (2009). Adaptive maternal adjustments of offspring size in response to conspecific density in two populations of the least killifish, *Heterandria formosa. Evolution* **63**, 1341–1347.

Le Menn, F., Cerdà, J. & Babin, P.J. (2007). Ultrastructural aspects of ontogeny and differentiation of ray-finned fish ovarian follicles. In: *The Fish Oocyte: From Basic Studies to Biotechnological Applications* (Babin, P.J., Cerdà, J. & Lubzens, E., eds), pp. 1–37. Berlin: Springer.

Leonard, J.B.K. & McCormick, S.D. (1999). Effects of migration distance on whole-body and tissue specific energy use in American shad (*Alosa sapidissima*). *Canadian Journal of Fisheries and Aquatic Sciences* **56**, 1159–1171.

Le Rouzic, P. & Prunet, P. (2005). Evidence for pleiotropic effects of prolactin in teleost fish. In: *Hormones and Their Receptors in Fish Reproduction* (Melamed, P. & Sherwood, N., eds), pp. 105–127. Singapore: World Scientific.

Leung, L.K.P. (1988). Ultrastructure of the spermatozoon of *Lepidogalaxias salamandroides* and its phylogenetic significance. *Gamete Research* **19**, 41–49.

Levavi-Sivan, B., Bogard, J., Mananos, E.L., Gomez, A. & Lareyre, J.J. (2010). Perspectives on fish gonadotropins and their receptors. *General and Comparative Endocrinology* **165**, 412–437.

Levi, L., Pekarski, I., Gutman, E., Fortina, P., Hyslop, T., Biran, J., Levavi-Sivan, B. & Lubzens, E. (2009). Revealing genes associated with vitellogenesis in the liver of the zebrafish (*Danio rerio*) by transcriptome profiling. *BMC Genomics* **10**, 141.

Levin, L.E., Belmonte, P. & Gonzalez, O. (1992). Sun-compass orientation in the characid *Cheirodon pulcher. Environmental Biology of Fishes* **35**, 321–325.

Levitan, D.R. (1998). Sperm limitation, gamete competition, and sexual selection in external fertilisers. In: *Sperm Competition and Sexual Selection* (Birkhead, T.R. & Møller, A.P., eds), pp. 173–215. London: Academic Press.

Levitan, D.R. (2000). Sperm velocity and longevity trade off each other and influence fertilization in the sea urchin *Lytechinus variegatus. Proceedings of the Royal Society of London*, B **267**, 531–534.

Lewis, Z.R., McClennan, M.C., Postlewait, J.H., Crsko, W.A. & Kaplan, R.H. (2008). Female-specific increase in primordial germ cells marks sex differentiation in threespine stickleback (*Gasterosteus aculeatus*). *Journal of Morphology* **269**, 909–921.

Liermann, M. & Hilborn, R. (2001). Depensation: evidence, models and implications. *Fish and Fisheries* **2**, 33–58.

Liew, W.C., Bartfai, R., Lim, Z., Sreenivasan, R., Siegfried, K.R. & Orban, L. (2012). Polygenic sex determination system in zebrafish. *PLoS One* **7**, e34397.

Liley, N.R., Breton, B., Fostier, A. & Tan, E.S.P. (1986). Endocrine changes associated with spawning behaviour and social stimuli in a wild population of rainbow trout (*Salmo gairdneri*). I. Males. *General and Comparative Endocrinology* **62**, 145–156.

Liley, N.R. & Stacey, N.E. (1983). Hormones, pheromones, and reproductive behaviour in fish. In: *Fish Physiology, Vol. IXB* (Hoar, W.S, Randall, D.J. & Donaldson, E.M., eds), pp. 1–63. London: Academic Press.

Lin, S.-W. & Ge, W. (2009). Differential regulation of gonadotropins (FSH and LH) and growth hormone by neuroendocrine. Endocrine and paracrine factors in the zebrafish – an *in vitro* approach. *General and Comparative Endocrinology* **160**, 183–193.

Lindholm, A. & Breden, F. (2002). Sex chromosomes and sexual selection in Poeciliid fishes. *The American Naturalist* **160**, S214-S224.

Lindström, K. & Hellström, M. (1993). Male size and parental care in the sand goby, *Pomatoschistus minutus. Ethology, Ecology and Evolution* **5**, 97–106.

Lindström, K. & Pampoulie, C. (2005). Effect of resource holding potential and resource value on tenure at nest sites in sand gobies. *Behavioral Ecology* **16**, 70–74.

Lindström, K. & Ranta, E. (1992). Predation by birds affects structure of breeding population in male sand gobies, *Pomatoschistus minutus*. *Oikos* **64**, 527–532.

Lindström, K. & Sargent, R.C. (1997). Food access, brood size and filial cannibalism in the fantail darter, *Etheostoma flabellare*. *Behavioral Ecology and Sociobiology* **40**, 107–110.

Lischka, S.A & Magnuson, J.J. (2006). Timing and site selection of spawning in a landlocked population of rainbow smelt in Wisconsin. *Environmental Biology of Fishes* **76**, 413–418.

Liu, H., Yurong, Z., Reichard, M. & Smith, C. (2006). Evidence of host specificity and congruence between phylogenies of bitterlings and freshwater mussels. *Zoological Studies* **45**, 428–434.

Lobón-Cerviá, J. (2007). Density-dependent growth in stream-living brown trout *Salmo trutta* L. *Functional Ecology* **21**, 117–124.

Locatello, L., Rasotto, M.B., Evans, J.P. & Pilastro, A. (2006). Colourful male guppies produce faster and more viable sperm. *Journal of Evolutionary Biology* **19**, 1595–1602

Longo, F.J. (1987). *Fertilization*. London: Chapman and Hall.

Lorance, P. (2011). History and dynamics of the overexploitation of the blackspot sea bream (*Pagellus bogaraveo*) in the Bay of Biscay. *ICES Journal of Marine Science* **68**, 290–301.

Losey, G.S., Cronin, T.W., Goldsmith, T.H., Hyde, D., Marshall, N.J. & McFarland, W.N. (1999). The visual world of fishes: a review. *Journal of Fish Biology* **59**, 921–943.

Lotem, A. & Nakamura, H. (1998). Evolutionary equilibria in avian brood parasitism. In: *Parasitic Birds and their Hosts: Studies in Coevolution* (Rothstein, S.I. & Robinson, S.K., eds), pp. 223–235. Oxford: Oxford University Press.

Lou, Y.H., Hara, A. & Takahashi, H. (1989). Induction of autoantibodies against spermatozoa by injection of allogeneic sperm in the Nile Tilapia, *Oreochromis niloticus*. *Comparative Biochemistry and Physiology, Part B* **94**, 829–836.

Lough R.G. & Manning J.P. (2001) Tidal-front entrainment and retention of fish larvae on the southern flank of Georges Bank. *Deep-Sea Res II* **48**:631–644

Love, D.C. & Shirley, T.C. (1993). Parasitism of the golden king crab, *Lithodes aequispinus* Benedict, 1895 (Decapoda, Anomura, Lithodidae) by a liparid fish. *Crustaceana* **65**, 97–104.

Low, R.M. (1971). Interspecific territoriality in a pomacentrid reef fish, *Pomacentrus flavicauda* Whitley. *Ecology* **52**, 649–654.

Lowe-McConnell, R.M. (1979). Ecological aspects of seasonality in fishes of tropical waters. *Symposium of the Zoological Society of London* **44**, 219–241.

Lowe-McConnell, R.M. (1987). *Ecological Studies in Tropical Fish Communities*. Cambridge: Cambridge University Press.

Lowerre-Barbieri, S.K. (2009). Reproduction in relation to conservation and exploitation of marine fishes. In: *Reproductive Biology and Phylogeny of Fishes (Agnathans and Bony Fishes)* (Jamieson, B.G.M., ed.), pp. 371–394. Enfield: Science Publishers.

Lowerre-Barbieri, S.K., Ganias, K., Saborido-Rey, F., Murua, H. & Hunter, J.R. (2011). Reproductive timing in marine fishes: variability, temporal scales, and methods. *Marine Coastal Fisheries Dynamics, Management and Ecosystem Science* **3**, 71–91.

Lubzens, E., Young, G. Bobe, J. & Cerdà, J. (2010). Oogenesis in teleosts: how eggs are formed. *General and Comparative Endocrinology* **165**, 367–389.

Lucas, M.C. & Baras, E. (2001). *Migration of Freshwater Fishes*. Oxford: Blackwell.

Luo, Y.L., Yue, P.Q. & Chen, Y.Y. (1977). Gobioninae. In: *The Cyprinid Fishes of China II* (Wu, X.W., ed.), pp. 439–549. Shanghai: Shanghai People's Publishing House.

Lyman-Gingerich, J. & Pelegri, F. (2007). Maternal factors in fish oogenesis and embryonic development. In: *The Fish Oocyte: From Basic Studies to Biotechnological Applications* (Babin, P.J., Cerdà, J. & Lubzens, E., eds), pp. 141–174. Berlin: Springer.

Maack, G. & Segner, H. (2003). Morphological development of the gonads of zebrafish. *Journal of Fish Biology* **62**, 895–906.

Maan, M.E. & Seehausen, O. (2011). Ecology, sexual selection and speciation. *Ecology Letters* **14**, 591–602.

Maatouk, D.M. & Capel, B. (2008). Sexual development of the soma in the mouse. *Current Topics in Developmental Biology* **83**, 151–183.

MacArthur, R.H. & Wilson, E.O. (1967). *The Theory of Island Biogeography*. Princeton: Princeton University Press.

Mackereth, R.W., Noakes, D.L.G. & Ridgway, M.S. (1999). Size-based variation in somatic energy reserves and parental expenditure by male smallmouth bass, *Micropterus dolomieu*. *Environmental Biology of Fishes* **56**, 263–275.

MacLaren, R.D., Rowland, W.J. & Morgan, N. (2004). Female preference for sailfin and body size in sailfin molly, *Poecilia latipinna*. *Ethology* **110**, 363–379.

Maekawa, K., Shigeru, N. & Yamamoto, S. (1994). Spawning behaviour and size-assortative mating of Japanese charr in an artificial lake-inlet stream system. *Environmental Biology of Fishes* **39**, 109–117.

Magellan, K., Pettersson, L.B. & Magurran, A.E. (2005). Quantifying male attractiveness and mating behaviour through phenotypic size manipulation in the Trinidadian guppy, *Poecilia reticulata*. *Behavioral Ecology and Sociobiology* **58**, 366–374.

Magnhagen, C. (1992). Alternative reproductive behaviour in the common goby, *Pomatoschistus microps*: an ontogenetic gradient? *Animal Behaviour* **44**, 182–184.

Magnhagen, C. & Vestergaard, K. (1993). Brood size and offspring age affect risk-taking and aggression in nest guarding common gobies. *Behaviour* **125**, 233–243.

Magurran, A.E. (2005). *Evolutionary Ecology: The Trinidadian Guppy*: Oxford: Oxford University Press.

Magurran, A.E. & Nowak, M.A. (1991). Another battle of the sexes: the consequences of sexual asymmetry in mating costs and predation risk in the guppy, *Poecilia reticulata*. *Proceedings of the Royal Society London, B* **246**, 31–38.

Magurran, A.E. & Seghers, B.H. (1990). Population differences in predator recognition and attack cone avoidance in the guppy *Poecilia reticulata*. *Animal Behaviour* **40**, 443–452.

Magurran, A.E. & Seghers, B.H. (1994). Sexual conflict as a consequence of ecology: evidence from guppy *Poecilia reticulata*, populations in Trinidad. *Proceedings of the Royal Society London, B* **255**, 31–36.

Maher C.R. & Lott, D.F. (1995). Definitions of territoriality used in the study of variation in vertebrate spacing systems. *Animal Behaviour* **49**, 1581–1597.

Maisey, J.G. (1996). *Discovering Fossil Fishes*. New York: Westview Press.

Maitland, P.S. & Campbell, R.N. (1992). *Freshwater Fishes*. London: Harper Collins.

Maitland, P.S. & Lyle, A.A. (1992). Conservation of freshwater fish in the British Isles: proposals for management. *Aquatic Conservation: Marine and Freshwater Ecosystems* **2**, 165–183.

Mañanós, E., Duncan, N. & Mylonas, C. (2009). Reproduction and control of ovulation, spermiation and spawning in cultured fish. In: *Methods in Reproductive Aquaculture: Marine and Freshwater Species* (Cabrita, E., Robles, V. & Herráez, P., eds), pp. 3–80. Boca Raton: CRC Press.

Manica, A. (2004). Parental fish change their cannibalistic behaviour in response to the cost-to-benefit ratio of parental care. *Animal Behaviour* **67**, 1015–1021.

Mank, J.E. & Avise, J.C. (2006). The evolution of reproductive and genomic diversity in ray-finned fishes: insights from phylogeny and comparative analysis. *Journal of Fish Biology* **69**, 1–27.

Mank, J.E. & Avise, J.C. (2009). Evolutionary diversity and turn-over of sex determination in teleost fishes. *Sexual Development* **3**, 60–67.

Mank, J.E., Hall, D.W., Kirkpatrick, M. & Avise, J.C. (2006b). Sex chromosomes and male ornaments: a comparative evaluation in ray-finned fishes. *Proceedings of the Royal Society of London, B* **273**, 233–236.

Mank, J.E., Promislow, D.E.L. & Avise, J.C. (2005). Phylogenetic perspectives in the evolution of parental care in ray-finned fishes. *Evolution* **59**, 1570–1578.

Mank, J.E., Promislow, D.E.L. & Avise, J.C. (2006a). Evolution of alternative sex-determining mechanisms in teleost fishes. *Biological Journal of the Linnaean Society* **87**, 83–93.

Mann, R.H.K., Mills, C.A. & Crisp, D.T. (1984). Geographical variation in the life-history tactics of some species of freshwater fish. In: *Fish Reproduction: Strategies and Tactics* (Potts, G.W. & Wootton, R.J., eds), pp. 171–186. London: Academic Press.

Manning, A.J., Burton, M.P.M. & Crim, L.W. (2004). Puberty and recrudescence in cultured male yellowtail flounder: plasma androgens and spermatogenesis. *Journal of Fish Biology* **65**, 1026–1046.

Manning, A.J., Burton, M.P.M. & Crim, L.W. (2008). The timing of puberty in cultured yellowtail flounder, *Limanda limanda* (Storer): oogenesis and sex steroid production *in vivo* and *in vitro*. *Aquaculture* **279**, 188–196.

Marconato, A. & Bisazza, A. (1988). Mate choice, egg cannibalism and reproductive success in the river bullhead, *Cottus gobio* L. *Journal of Fish Biology* **33**, 905–916.

Marconato, A., Rasotto, M.B. & Mazzoldi, C. (1996). On the mechanism of sperm release in three gobiid fishes (Teleostei: Gobiidae). *Environmental Biology of Fishes* **46**, 321–327.

Mariani, S., Sala-Bozano, M., Chopelet, J. & Benvenuto, C. (2013). Spatial and temporal patterns of size-at-sex-change in two exploited coastal fish. *Environmental Biology of Fishes* **96**, 535–541.

Marsh-Matthews, E. (2011). Matrotrophy. In: *Ecology and Evolution of Poeciliid Fishes* (Evans, J., Pilastro, A. & Schlupp, I., eds). pp. 18–27. Chicago: Chicago University Press.

Marshall, C.T. & Browman, H.I. (2007). Disentangling the causes of maturation trends in exploited fish populations. *Marine Ecology Progress Series* **335**, 249–251.

Marshall, N.B. (1967). The olfactory organs of bathypelagic fishes. *Symposium of the Zoological Society of London* **19**, 57–70.

Marshall, N.J. (2000). The visual ecology of reef fish colours. In: *Signalling and Signal Design in Animal Communication* (Espmark, Y., Amundsen, T. & Rosenqvist, G., eds), pp. 83–120. Trondheim: Tapir Academic Press.

Martin, K.L. (1999). Ready and waiting: delayed hatching and extended incubation of anamniotic vertebrate terrestrial eggs. *American Zoologist* **39**, 279–288.

Martin, K.L. & Carter, A.L. (2013). Brave new propagules: terrestrial embryos in anamniotic eggs. *Integrative and Comparative Biology* doi: 10.1093/icb/ict018

Martin, K.L., Moravek, C.L. & Walker A.J. (2011). Waiting for a sign: extended incubation postpones larval stage in the beach spawning California grunion *Leuresthes tenuis* (Ayres). *Environmental Biology of Fishes* **91**, 63–70.

Martin, K.L., Van Winkle, R.C., Drais, J.E. & Lakisic, H. (2004). Beach-spawning fishes, terrestrial eggs, and air breathing. *Physiological and Biochemical Zoology* **77**, 750–759.

Maruska, K.P. & Fernald, R.D. (2011). Social regulation of gene expression in the hypothalamic-pituitary-gonadal axis. *Physiology* **26**, 412–423.

Maruska, K.P., Boyle, K.S., Dewan, L.R. & Tricas, T.C. (2007). Sound production and spectral hearing sensitivity in the Hawaiian sergeant damselfish, *Abudefduf abdominalis*. *Journal of Experimental Biology* **210**, 3990–4004.

Maruska, K.P., Levavi-Sivan, B., Biran, J. & Fernald, R.D. (2011). Plasticity of the reproductive axis caused by social status change in an African cichlid fish: I. Pituitary gonadotropins. *Endocrinology* **152**, 281–290.

Masonjones, H.D. & Lewis, S.M. (1996). Courtship behavior in the dwarf seahorse, *Hippocampus zosterae*. *Copeia* **1996**, 634–640.

Matity, J.G., Chivers, D.P. & Smith, R.J.F. (1994). Population and sex differences in antipredator responses of breeding fathead minnows (*Pimephales promelas*) to chemical stimuli from garter snakes (*Thamnophis radix* and *T. sirtalis*). *Journal of Chemical Ecology* **20**, 2111–2121.

Matsuda, M., Nagahama, Y., Shinomiya, A., Sato, T., Matsuda, C., Kobayashi, T., Morrey, C.E., Shibata, N., Asakawa, S., Shimizu, N., Hori, H., Hamaguchi, S. & Skaizumi, M. (2002). *DMY* is a Y-specific DM-domain gene required for male development in the medaka fish. *Nature* **417**, 559–563.

Matsuda, M. & Sakaizumi, M. (2009). Sex determination. In: *Reproductive Biology and Phylogeny of Fishes (Agnathans and Bony Fishes)* (Jamieson, B.G.M., ed.), pp. 331–350. Enfied (NH): Science Publishers.

Matsuda, M., Sato, T., Toyazaki, Y., Nagahama, Y., Hamaguchi, S. & Sakaizumi, M. (2003). *Oryzias curvinotus* has DMY, a gene that is required for male development in the medaka *O. latipes*. *Zoological Science* **20**, 159–161.

Matsuyama, M. (2008). Follicular steroidogenesis in fish. In: *Fish Reproduction* (Roche, M.J., Arukwe, A. & Kapoor, B.G., eds), pp. 172–199. Enfield (NH): Science Publishers.

Matthews, I.M. (1999). Mating behaviour and reproductive biology of the guppy, *Poecilia reticulata*. PhD thesis, University of St Andrews.

Matthews, I.M., Evans, J.P. & Magurran, A.E. (1997). Male display rate reveals ejaculate characteristics in the Trinidadian guppy *Poecilia reticulata*. *Proceedings of the Royal Society London, B* **264**, 695– 700.

Maurakis, E.G., Woolcott, W.S. & Sabaj, M.H. (1991). Reproductive–behavioral phylogenetics of *Nocomis* species-groups. *American Midland Naturalist* **126**, 103–110.

Mayer, I. & Páll, M. (2007). Hormonal control of reproductive behaviour in the stickleback. In: *Biology of the Three-spined Stickleback* (Östlund-Nilsson, S., Mayer, I. & Huntingford, F.A., eds), pp. 249–269. Boca Raton: CRC Press.

Maynard Smith, J. (1974). The theory of games and the evolution of animal conflicts. *Journal of Theoretical Biology* **47**, 209–221.

Maynard Smith, J. (1977). Parental investment – a prospective analysis. *Animal Behaviour* **25**, 1-9.

Maynard Smith, J. (1978). *The Evolution of Sex*. Cambridge: Cambridge University Press.

Maynard Smith, J. & Harper, D. (2003). *Animal Signals*. Oxford: Oxford University Press.

Mazeroll, A.I. and Montgomery, W.L. (1998). Daily migrations of a coral reef fish in the Red Sea (Gulf of Aqaba, Israel): initiation and orientation. *Copeia* **1998**, 893–905.

Mazzoldi, C. (1999). Studio comparativo della dinamica di accoppiamento e fecondazione in Teleostei ad uova demerse [A comparative study of mating behaviour and fertilization dynamics in demersal spawning teleosts]. *Unpublished PhD thesis*. University of Padova.

Mazzoldi, C. (2001). Reproductive apparatus and mating system in two tropical goby species. *Journal of Fish Biology* **59**, 1686–1691.

Mboko, S.K. & Kohda, M. (1999). Piracy mating by large males in a monogamous substrate-breeding cichlid in Lake Tanganyika. *Journal of Ethology* **17**, 51–55.

McCoy, E., Syska, N., Plath, M., Schlupp, I. & Riesch, R. (2011). Mustached males in a tropical poeciliid fish: emerging female preference selects for a novel males trait. *Behavioural Ecology and Sociobiology* **65**, 1437–1445.

McGovern, J.C., Wyanski, D.M., Pashuk, O., Manooch II, C.S. & Sedberry, G.R. (1998). Changes in the sex ratio and size at maturity of gag, *Mycteroperca microlepis*, from the Atlantic coast of the southeastern United States. *Fishery Bulletin* **96**, 797–807.

McGowan, M.F. & Richards, W.J. (1989). Bluefin tuna, *Thunnus thynnus*, larvae in the Gulf Stream off the southeastern United States: satellite and shipboard observations of their environment. *Fishery Bulletin* **87**, 615–631

McKaye, K.R. (1977). Competition for breeding sites between the cichlid fishes of Lake Jiloá, Nicaragua. *Ecology* **58**, 291–302.

McKaye, K.R. (1985). Cichlid–catfish mutualistic defense of young in Lake Malawi, Africa. *Oecologia* **66**, 358–363.

McKaye, K.R. (1991). Sexual selection and the evolution of the cichlid fishes of Lake Malawi. In: *Cichlid Fishes: Behavior, Ecology and Evolution* (Keenleyside, M.H.A., ed.), pp. 241–257. London: Chapman and Hall.

McKaye, K.R. & McKaye, N.M. (1977). Communal care and kidnapping of young by parental cichlids. *Evolution* **31**, 674–681.

McKenzie, W.D., Crews, D., Kallman, K.D., Policansky, D. & Sohn, J.J. (1983). Age, weight and genetics of sexual maturation in the platyfish, *Xiphophorus maculates, Copeia* **1983**, 770–774.

McKeown, B.A. (1984) *Fish Migration*. London: Croom Helm.

McLennan, D.A. (1996). Integrating phylogenetic and experimental analyses: the evolution of male and female nuptial coloration in the stickleback fishes (Gasterosteidae). *Systematic Biology* **45**, 261–277.

McLennan, D.A. (2007). The umwelt of the three-spined stickleback. In: *Biology of the Three-Spined Stickleback* (Östlund-Nilsson, S., Mayer, I. & Huntingford, F., eds), pp. 179–224. Boca Raton: CRC Press.

McLennan, D.A. & Ryan, M.J. (1997). Responses to conspecific and heterospecific olfactory cues in the swordtail *Xiphophorus cortezi*. *Animal Behaviour* **54**, 1077–1088.

McNamara, J.M., Houston, A.I., Barta, Z., Scheuerlein, A. & Fromhage, L. (2009). Deterioration, death and the evolution of reproductive restraint in late life. *Proceedings of the Royal Society London, B* **276**, 4061–4066.

McPhail, J.D. (1993). Ecology and evolution of sympatric sticklebacks (*Gasterosteus*): origin of the species pairs. *Canadian Journal of Zoology* **71**, 515–523.

McPhee, M.V. & Quinn, T.P. (1998). Factors affecting the duration of nest defense and reproductive lifespan of female sockeye salmon. *Environmental Biology of Fishes* **51**, 369–375.

Mebs, D. (1994). Anemonefish symbiosis: vulnerability and resistance of fish to the toxin of the sea anemone. *Toxicon* **32**, 1059–1068.

Mebs, D. (2009). Chemical biology of the mutualistic relationships of sea anemones with fish and crustaceans. *Toxicon* **54**, 1071–1074.

Mehlis, M., Bakker, T.C.N., Engqvist, L. & Frommen, J.G. (2010). To eat or not to eat: egg based assessment of paternity triggers fine-tuned decisions about filial cannibalism. *Proceedings of the Royal Society London, B* **277**, 2627–2635.

Mellinger, J. (2002). *Sexualité et Reproduction des Poissons.* Paris: CNRS Editions.

Menuet, A., Adrio, F., Kah, O. & Pakdel, F. (2005). Regulation and function of estrogen receptors: comparative aspects. In: *Hormones and Their Receptors in Fish Reproduction* (Melamed, P. & Sherwood, N., eds), pp. 224–253. Singapore: World Scientific Publishing.

Merchant-Larios, H.C. (1978). Ovarian differentiation. In: *The Vertebrate Ovary* (Jones, R.E., ed.), pp. 47–81. London: Plenum Press.

Meredith, R.W., Pires, M.N., Reznick, D.N. & Springer, M.S. (2011). Molecular phylogenetic relationships and the coevolution of placentotrophy and superfetation in *Poecilia* (Poeciliidae: Cyprinodontiformes). *Molecular Phylogenetics and Evolution* **59**, 148–157.

Metcalfe, J.D., Hunter, E. & Buckley, A.A. (2006). The migratory behaviour of North Sea plaice: currents, clocks and clues. *Marine and Freshwater Behaviour and Physiology* **39**, 25–36.

Meyer, A. (1993). Phylogenetic relationships and evolutionary processes in East African cichlids. *Trends in Ecology and Evolution* **8**, 279–284.

Meyer, A. (1997). The evolution of sexually selected traits in male swordtail fishes (*Xiphophorus*: Poeciliidae). *Heredity* **79**, 329–337.

Meyers, G.S. (1960). The endemic fish fauna of Lake Lanao, and the evolution of higher taxonomic categories. *Evolution* **14**, 323–333.

Middaugh, D.P. & Hemmer, M.J. (1987). Influence of environmental temperature on sex ratios in the tidewater silverside, *Menidia peninsulae* (Pisces: Atherinidae). *Copeia* **1987**, 958–964.

Migaud, H., Davie, A. & Taylor, J.J. (2010). Current knowledge on the photoendocrine regulation in temperate fish species. *Journal of Fish Biology* **76**, 27–68.

Migaud, H., Ismail, R., Cowan, M. & Davie, A. (2012). Kisspeptin and seasonal control of reproduction in male European sea bass (*Dicentrarchus labrax*). *General and Comparative Endocrinology* **179**, 384–399.

Milinski, M. (2006) The major histocompatibility complex, sexual selection, and mate choice. *Annual Review of Ecology, Evolution, and Systematics* **37**, 159–186.

Milinski, M. & Bakker, T.C.M. (1992). Costs influence sequential mate choice in sticklebacks, *Gasterosteus aculeatus*. *Proceedings of the Royal Society London, B* **250**, 229–233.

Milinski, M., Griffiths, S., Wegner, K.M. Reusch, T.B.H., Haas-Assenbaum, A. & Boehm, T. (2005). Mate choice decisions of stickleback females predictably modified by MHC peptide ligands. *Proceedings of the National Academy of Sciences of the USA* **102**, 4414–4418.

Miller, P.J. (1984). The tokology of gobioid fishes. In: *Fish Reproduction: Strategies and Tactics* (Potts, G.W. & Wootton, R.J., eds), pp. 119–153. London: Academic Press.

Miller, R.J. (1962). Reproductive behavior of the stoneroller minnow, *Campostoma anomalum pullum*. *Copeia* **1962**, 407–417.

Miller, T.J., Crowder, L.B., Rice, J.A. & Marschall, E.A. (1988). Larval size and recruitment mechanisms in fishes: toward a conceptual framework. *Canadian Journal of Fisheries and Aquatic Sciences* **45**, 1657–1670.

M'Intosh, W.C. & Masterman, A.T. (1897). *The Life-histories of the British Marine Food-Fishes.* London: C.J. Clay & Sons.

Mirza, J.A. & Shelton, W.L. (1988). Induction of gynogenesis and sex reversal in silver carp. *Aquaculture* **68**, 1–14.

Mitani, Y., Kanda, S., Akazome, Y., Zempo, B. & Oka, Y. (2010). Hypothalamic Kiss1 but not Kiss2 neurons are involved in estrogen feedback in medaka (*Oryzias latipes*). *Neuroendocrinology* **151**, 1751–1759.

Mitchell, J.S., Jutzeler, E., Heg, D. & Taborsky, M. (2009). Dominant members of cooperatively-breeding groups adjust their behaviour in response to the sexes of subordinates. *Behaviour* **146**, 1665–1686.

Mittwoch, U. (2006). Sex is a threshold dichotomy mimicking a single gene effect. *Trends in Genetics* **22**, 96–100.

Miura, C., Higashino, T. & Miura, T. (2007). A progestin and an estrogen regulate early stages of oogenesis in fish. *Biology of Reproduction* **27**, 822–828.

Modig, C., Westerlund, L. & Olsson, P.-E. (2007). Oocyte zona pellucida proteins. . In: *The Fish Oocyte: From Basic Studies in Biotechnological Applications* (Bain, P.J., Cerdà, J. & Lubzens, eds), pp. 113–139. Berlin: Springer.

Mok, E.Y.-M. & Munro, A.D. (1997). Some anatomical and behavioural aspects of reproduction in members of an unusual teleost family: the Phallostethidae. *Journal of Natural History* **31**, 739–778.

Møller, A.P. & Pomiankowski, A. (1993). Why have birds got multiple sexual ornaments? *Behavioral Ecology and Sociobiology* **32**, 167–176.

Møller, D. (1969). The relationship between Arctic and coastal cod in their immature stages illustrated by frequencies of genetic characters. *Fiskeridirektoratets Skrifter, Serie Havundersøkelser* **15**, 220–233.

Mommsen, T.P. & Korsgard, B. (2008). Vitellogenesis. In: *Fish Reproduction* (Rocha, M.J., Arukwe, A. & Kappor, B.G., eds), pp. 115–169. Enfield (NH): Science Publishers.

Montgomerie, R. & Fitzpatrick, J.L. (2009). Testes, sperm, and sperm competition. In: *Reproductive Biology and Phylogeny of Fishes (Agnathans and Bony Fishes)* (Jamieson, B.G.M., ed.), pp. 1–53. Enfield (NH): Science Publishers.

Moore, A.J. & Moore, P.J. (1999). Balancing sexual selection through opposing mate choice and male competition. *Proceedings of the Royal Society of London, B* **266**, 711–716.

Moore, W.S. (1984). Evolutionary ecology of unisexual fishes. In: *The Evolutionary Genetics of Fishes* (Turner, B.J., ed.), pp. 329–398. New York: Plenum Press.

Moravek, C. & Martin, K. (2011). Life goes on: extended incubation and heterokairy in development of embryonic California Grunion *Leuresthes tenuis*. *Copeia* **2**, 308–314.

Moreau, D.T.R. & Fleming, I.A. (2012). Enhanced growth reduces precocial male maturation in Atlantic salmon. *Functional Ecology* **26**, 399–405.

Moreau, R.E. (1944). Clutch-size: a comparative study, with special reference to African birds. *Ibis* **86**, 286–347.

Morin, P.-P. & Doving, K.B. (1992). Changes in the olfactory function of Atlantic salmon, *Salmo salar*, in the course of smoltification. *Canadian Journal of Fisheries and Aquatic Sciences* **49**, 1704–1713.

Morisawa, M. (1985). Initation mechanism of sperm motility at spawning in teleosts. *Zoological Science* **2**, 605–615.

Morisawa, M., Suzuki, K. & Morisawa, S. (1983). Effects of potassium and osmolality on spermatozoan motility of salmonid fishes. *Journal of Experimental Biology* **107**, 105–113.

Morishima, K., Nakamura-Shiokawa, Y., Bando, E., Li, Y.-A., Boron, A., Khan, M.M. R. & Arai, K. (2008). Cryptic clonal lineages and genetic diversity in the loach *Misgurnus anguillicaudatus* (Teleostei: Cobitidae) inferred from nuclear and mitochondrial DNA analyses. *Genetica* **132**, 159–171.

Morris, D. (1955). The reproductive behavior of the river bullhead (*Cottus gobio*) with special reference to fanning activity. *Behaviour* **7**, 1–32.

Morrongiello, J.R., Bond, N.R., Crook, D.A. & Wong, B.B.M. (2012). Spatial variation in egg size and egg number reflects trade-offs and bet-hedging in a freshwater fish. *Journal of Animal Ecology* **81**, 806–817.

Moyle, P.B. & Cech, J.J. Jr (2004). *Fishes: An Introduction to Ichthyology*, 2nd edn. London: Prentice Hall.

Mrowka, W. (1987). Filial cannibalism and reproductive success in the maternal mouthbrooding cichlid fish *Pseudocrenilabrus multicolor*. *Behavioral Ecology and Sociobiology* **21**, 257–265

Mrowka, W. & Schierwater, B. (1988). Energy expenditure for mouthbrooding in a cichlid fish. *Behavioral Ecology and Sociobiology* **22**, 161–164.

Munakata, A. & Kobayashi, M. (2010). Endocrine control of sexual behaviour in teleost fish. *General and Comparative Endocrinology* **165**, 456–468.

Munday, P.L., Kuwamura, T. & Kroon, F.J. (2010). Bidirectional change in marine fishes. In: *Reproduction and Sexuality in Marine Fishes* (Cole, K.S., ed.), pp. 241–271. Berkeley: University of California Press.

Munehara, H. 1991. Utilization and ecological benefits of a sponge as a spawning bed by the little dragon sculpin *Blepsias cirrhosus*. *Japanese Journal of Ichthyology* **38**, 179–184.

Munehara, H., Takano, K. & Koya, Y. (1989). Internal gametic association and external fertilization in the Elkhorn Sculpin, *Alchichthys alcicornis*. *Copeia* **3**, 673–678.

Munger, L., Cruz, A. & Applebaum, S. (2004). Mate choice copying in female humpback Limia (*Limia nigrofasciata*, family Poeciliidae). *Ethology* **110**, 563–573.

Muñoz, R.C. & Warner, R.R. (2003). Alternative contexts of sex change with social control in the bucktooth parrotfish, *Sparisoma radians*. *Environmental Biology of Fishes* **68**, 307–319.

Munro, A.D. (1990a). Tropical freshwater fish. In: *Reproductive Seasonality in Teleosts: Environmental Influences* (Munro, A.D., Scott, A.P. & Lam, T.J., eds), pp. 145–239. Boca Raton: CRC Press.

Munro, A.D. (1990b). General introduction. In: *Reproductive Seasonality in Teleosts: Environmental Influences* (Munro, A.D., Scott, A.P. & Lam, T.J., eds), pp. 1–11. Boca Raton: CRC Press.

Murata, K. (2009). Fertilization. In: *Reproductive Biology and Phylogeny of Fishes (Agnathans and Bony Fishes)* (Jamieson, B.G.M., ed.), pp. 247–330. Enfield (NH): Science Publications.

Murua, H., Kraus, G., Saborido-Rey, Witthames, P.R., Thorsen, A. & Junquera, S. (2003). *Journal of Northwest Atlantic Fisheries Science* **33**, 33–54.

Myers, J.M. (1986). Tetraploid induction in *Oreochromis* spp. *Aquaculture* **57**, 281–287.

Myers, R.A. (1997). Comment and reanalysis: paradigms for recruitment studies. *Canadian Journal of Fisheries and Aquatic Sciences* **54**, 978–981.

Mylonas, C.C., Fostier, A. & Zanuy, S. (2010). Broodstock management and hormonal manipulation of fish reproduction. *General and Comparative Endocrinology* **165**, 516–534.

Mylonas, C.C. & Zohar, Y. (2007). Promoting oocyte maturation, ovulation and spawning in farmed fish. In: *The Fish Oocyte: From Basic Studies to Biotechnological Applications* (Babin, P.J., Cerdà, J. & Lubzens, E., eds), pp. 437–474. Berlin: Springer.

Myosho, T., Otake, H., Masuyama, H., Matsuda, M., Kuroki, Y., Fujiyama, A., Naruse, K., Hamaguchi, S. & Sakaizumi, M. (2012). Tracing the emergence of a novel sex-determining gene in medaka, *Oryzias luzonensis*. *Genetics* **191**, 163–170.

Myrberg, A.A. Jr. (1981). Sound communication and interception in fishes. In: *Communication in Fishes* (Tavolga W.N., Popper A.N. & Fay, R.R., eds), pp. 400–141. New York: Springer-Verlag.

Myrberg, A.A. Jr, Thresher, R.E. (1974). Interspecific aggression and its relevance to the concept of territoriality in reef fishes. *American Zoologist* **14**, 81–96.

Nagahama, Y. (1983). The functional morphology of teleost gonads. In: *Fish Physiology Vol. IX, Part A. Reproduction: Endocrine Tissues and Hormones* (Hoar, W.S. & Randall, D.J., eds), pp. 223–275. London: Academic Press.

Nakai, K., Yanagisawa, Y., Sato, T., Niimura, Y. & Gashagaza, M.M. (1990). Lunar synchronization in cichlid fishes of the tribe Lamprologini in Lake Tanganyika. *Journal of Fish Biology* **37**, 589–598.

Nakamura, M., Kobayashi, T., Chang, X.-T. & Nagahama, Y. (1998). Gonadal sex differentiation in teleost fish. *Journal of Experimental Zoology* **281**, 362–372.

Nakashima, B.S. & Wheeler, J.P. (2002). Capelin (*Mallotus villosus*) spawning behaviour in Newfoundland waters e the interaction between beach and demersal spawning. *ICES Journal of Marine Science* **59**, 909–916.

Nakatsuru, K. & Kramer, D.L. (1982). Is sperm cheap? Limited male-fertility and female choice in the lemon tetra (Pisces, Characidae). *Science* **216**, 753–755.

Narum, S.R., Zendt, J.S., Graves, D. & Sharp, W.R. (2008). Influence of landscape on resident and anadromous life history types of *Oncorhynchus mykiss*. *Canadian Journal of Fisheries and Aquatic Sciences* **65**, 1013–1023.

Nash, R.D., Dickey-Collas, M. & Kell, L.T. (2009). Stock and recruitment in North Sea herring (*Clupea harengus*); compensation and depensation in the population dynamics. *Fisheries Research* **95**, 88–97.

Natale, F., Fiore, G. & Hofherr, J. (2012). Mapping the research on aquaculture. A bibliometric analysis of aquaculture literature. *Scientometrics* **90**, 983–999.

Neat, F. & Lengkeek, W. (2009). Sexual selection in blennies. In: *The Biology of Blennies* (Patzner, R.A., Gonçalves, E.J., Hastings, P.A. & Kapoor, B.G., eds), pp. 249–278. Enfield (NH): Science Publishers.

Neaves, W.B. & Baumann, P. (2011). Unisexual reproduction among vertebrates. *Trends in Genetics* **27**, 81–88.

Nee, S., Colegrave, N., West, S.A. & Grafen, A. (2005). The illusion of invariant quantities in life histories. *Science* **309**, 1236–1239.

Neff, B.D. (2003). Paternity and condition affect cannibalistic behavior in nest-tending bluegill sunfish. *Behavioral Ecology and Sociobiology* **54**, 377–384.

Neff, B.D. & Pitcher, T.E. (2005). Genetic quality and sexual selection: an integrated framework for good genes and compatible genes. *Molecular Ecology* **14**, 19–38.

Nelson, J.S. (2006). *Fishes of the World*, 4th edn. London: John Wiley.

Nevitt, G.A., Dittman, A.H., Quinn, T.P. & Moody, W.J. (1994). Evidence for a peripheral olfactory memory in imprinted salmon. *Proceedings of the National Academy of Sciences of the USA* **91**, 4288–4292.

Newman, D.M., Jones, P.L. & Ingram, B.A. (2007). Temporal dynamics of oocyte development, plasma sex steroids and somatic energy reserves during seasonal ovarian maturation in captive Murray cod *Maccullochella peelii peelii*. *Comparative Biochemistry and Physiology A: Molecular & Integrative Physiology* **148**, 876–887.

Nico, L.G., Jelks, H.L. & Tuten, T. (2009). Non-native suckermouth armored catfishes in Florida: Description of nest burrows and burrow colonies with assessment of shoreline conditions. *ANSRP Bulletin, Vol-09–01*. Vicksburg: U.S. Army Engineer Research and Development Center.

Nicol, J.A.C. (1989). *The Eyes of Fishes*. Oxford: Clarendon Press.

Noakes, D.L.G. (1979). Parent-touching behaviour by young fishes: incidence, function and causation. *Environmental Biology of Fishes* **4**, 389–400.

Noakes, D.L.G. (1986). The genetic basis of fish behaviour. In: *The Behaviour of Teleost Fishes* (Pitcher, T.J., ed), pp. 3–22. London: Croom-Helm.

Noble, G.K. (1937). The rôle of dominance in the social life of birds. *Auk* **56**, 263–273.

Nocillado, J.N. & Elizur, A. (2008). Neuroendocrine regulation of puberty in fish: insights from the grey mullet (*Mugil cephalus*) model. *Molecular Reproduction and Development* **75**, 355–361.

Nordeng, H. (1977). A pheromone hypothesis for homeward migration in anadromous salmonids. *Oikos* **28**, 155–159.

Nordeng, H. (1983). Solution to the 'char problem' based on Arctic char (*Salvelinus alpinus*) in Norway. *Canadian Journal of Fisheries and Aquatic Sciences* **9**, 1372–1387.

Northcote, T.G. (1978). Migratory strategies and production in freshwater fishes. In: *Ecology of Freshwater Production* (Gerking, S.D., ed.), pp. 326–359. Oxford: Blackwell.

Northcote, T.G. (1992). Migration and residency in stream salmonids: some ecological considerations and evolutionary consequences. *Nordic Journal of Freshwater Research* **67**, 5–17.

Northcote, T.G. (1998). Migratory behaviour of fish and its significance to movement through riverine fish passage facilities. In: *Fish Migration and Fish Bypasses* (Jungwirth, M. Schmutz, S. & Weiss, S., eds), pp. 3–18. Oxford: Fishing News Books.

Northcote, T.G. (2000). Ecological interactions among an Orestiid (Pisces: Cyprinodontidae) species flock in the littoral zone of Lake Titicaca. *Advances in Ecological Research* **31**, 399–420.

Núñez, J. & Duponchelle, F. (2009). Towards a universal scale to assess sexual maturation and related life history traits in oviparous teleost fishes. *Fish Physiology and Biochemistry* **35**, 167–180.

O'Brien, C.S., Bourdo, R., Bradshaw, W.E., Holzapfel, W.E. & Cresko, W.A. (2012). Conservation of the photoperiodic neuroendocrine axis among vertebrates: evidence from the teleost fish, *Gasterosteus aculeatus*. *General and Comparative Endocrinology* **178**, 19–27.

Ocalewicz, K, Fopp-Bayat, D., Woznicki, P. & Jankun, M. (2008). Heteromorphic sex chromosomes in the ninespine stickleback *Pungitius pungitius*. *Journal of Fish Biology* **73**, 456–462.

Ochi, H. (1985). Termination of parental care due to small clutch size in the temperate damselfish, *Chromis notata*. *Environmental Biology of Fishes* **12**, 155–160.

Ochi, H. & Yanagisawa, Y. (1996). Interspecific brood-mixing in Tanganyikan cichlids. *Environmental Biology of Fishes* **45**, 141–149.

Odling-Smee, L. & Braithwaite, V.A. (2003). The role of learning in fish orientation. *Fish and Fisheries* **4**, 234–246.

Ogawa, R., Nagata, Y. & Kihira, H. (2000). Review of research on the habits of the bitterling *Acheilognathus longipinnis* in the Yodo River. *Memoirs of Osaka Kyoiku University* **49**, 33–55.

Ohno, S. (1967). *Sex Chromosomes and Sex Linkage*. Berlin: Springer.

Ohta, K., Hirano, M., Mine, T., Mizutani, H., Yamaguchi, A. & Matsuyama, M. (2008a). Body color change and serum steroid hormone levels throughout the process of sex change in the adult wrasse, *Pseudolabrus sieboldi*. *Marine Biology* **153**, 843–852.

Ohta, K., Mine, T., Yamaguchi, A. & Matsuyama, M. (2008b). Sexually dimorphic expression of pituitary glycoprotein hormones in a sex-changing fish, *Pseudolabrus sieboldi*. *Journal of Experimental Zoology, Part A* **309**, 534–541.

Oka, Y. (2009). Three-types of gonadotrophin-releasing hormone neurons and steroid-sensitive sexually dimorphic kisspeptin neurones in teleosts. *Journal of Neuroendocrinology* **21**, 334–338.

Okuda, N. (2001). The costs of reproduction to males and females of a paternal mouthbrooding cardinalfish *Apogon notatus*. *Journal of Fish Biology* **58**, 776–787.

Okuda, N., Miyazaki, M. & Yanagisawa, Y. (2002). Sexual difference in buccal morphology of the paternal mouthbrooding cardinalfish *Apogon doederleini*. *Zoological Science* **19**, 801–807

Oldfield, R.G. (2005). Genetic, abiotic and social influences on sex differentiation in cichlid fishes and the evolution of sequential hermaphroditism. *Fish and Fisheries* **6**, 93–110.

Oliveira, R.F. & Almada, V.C. (1998). Mating tactics and male–male courtship in the lek-breeding cichlid *Oreochromis mossambicus*. *Journal of Fish Biology* **52**, 1115–1129.

Oliveira, R.F., Carvalho, N., Miranda, J., Gonçalves, E.J., Grober, M. & Santos, R.S. (2002). The relationship between the presence of satellite males and nest-holder's mating success in the Azorean rock-pool blenny, *Parablennius sanguinolentus parvicornis*. *Ethology* **108**, 223–235.

Olsen, E.M., Heino, M., Lilly, G.R., Morgan, M.J., Brattey, J., Ernande, B. & Dieckmann, U. (2004). Maturation trends indicative of rapid evolution preceded the collapse of northern cod. *Nature*, **428**, 932–935.

Olson, V.A. & Owens, I.P. (1998). Costly sexual signals: are carotenoids rare, risky or required? *Trends in Ecology and Evolution* **13**, 510–514.

Olsson, I.C., Greenberg, L.A., Bergman, E. & Wysujack, K. (2006). Environmentally induced migration: the importance of food. *Ecology Letters* **9**, 645–651.

Olsson, P.-E., Karisson, J., Larsson, A., Modig, C. & Mayer, I. (2008). Molecular markers of androgen disruption. In: *Fish Reproduction* (Rocha, M.J., Arukwe, A. & Kapoor, B.G., eds), pp. 515–533. Enfield (NH): Science Publishers.

Orians, G.H. (1969). On the evolution of mating systems in birds and mammals. *American Naturalist* **103**, 589–603.

Orti, G. & Li, C. (2009). Phylogeny and classification. In: *Reproductive Biology and Phylogeny of Fishes (Agnathans and Bony Fishes)* (Jamieson, B.G.M., ed.), pp. 1–24. Enfield (NH): Science Publications.

Osório, R., Rosa, I.L. & Cabral, H. (2006). Territorial defence by the Brazilian damsel *Stegastes fuscus* (Teleostei: Pomacentridae). *Journal of Fish Biology* **69**, 233–242.

Ospina-Álvarez, N. & Piferrer, F. (2008). Temperature-dependent sex determination in fish revisited: prevalence, a single sex ratio response pattern, and possible effects of climate change. *PLoS One* **3**, e2837.

Östlund-Nilsson, S. (2001). Fifteen-spined stickleback (*Spinachia spinachia*) females prefer males with more secretional threads in their nests: an honest-condition display by males. *Behavioral Ecology and Sociobiology* **50**, 263–269.

Ottersen, G. (2008). Pronounced long-term juvenation in the spawning stock of Arcto-Norwegian cod (*Gadus morhua*) and possible consequences for recruitment. *Canadian Journal of Fisheries and Aquatic Sciences* **65**, 523–534.

Palkovacs, E.P. (2011). The overfishing debate: an eco-evolutionary perspective. *Trends in Ecology & Evolution* **26**, 616–617.

Páll, M.K., Mayer, I. & Borg, B. (2002a). Androgen and behaviour in the male three-spined stickleback, *Gasterosteus aculeatus* I. Changes in 11-ketotestosterone levels during the nesting cycle. *Hormones and Behavior* **41**, 377–383.

Páll, M.K., Mayer, I. & Borg, B. (2002b). Androgen and behaviour in the male three-spined stickleback, *Gasterosteus aculeatus* II. Castration and 11-ketoandrostenedione effects on courtship and parental care during the nesting cycle. *Hormones and Behavior* **42**, 337–344.

Pandian, T.J. (2011). *Sexuality in Fishes*. Boca Raton: CRC Press.

Pandian, T.J. & Sheela, S.G. (1995). Hormonal induction of sex reversal in fish. *Aquaculture* **138**, 1–22.

Pandey, S. & Leatherland, J.F. (1970). Comparison of the effects of methallibure and thiourea on the testis, thyroid, and adenohypophysis of the adult and juvenile guppy, *Poecilia reticulata* Peters. *Canadian Journal of Zoology* **48**, 445–450.

Pankhurst, N.W. (2008). Gonadal steroids: functions and patterns of change. In: *Fish Reproduction* (Rocha, M.J., Arukwe, A. & Kapoor, B.G., eds), pp. 67–111. Enfield (NH): Science Publishers.

Pang, Y. & Thomas, P. (2010). Role of G protein-coupled estrogens receptor 1, GPER, in inhibition of oocyte maturation by endogenous estrogens in zebrafish. *Developmental Biology* **342**, 194–206.

Parenti, L.R. & Grier, H.J. (2004). Evolution and phylogeny of gonad morphology in bony fishes. *Integrative and Comparative Biology* **44**, 333–348.

Parenti, L.R., LoNostro, F.L. & Grier, H.J. (2010). Reproductive histology of *Tomeurus gracilis* Eigenmann, 1909 (Teleostei: Atherinomorpha: Poeciliidae) with comments on evolution of viviparity in atherinomorph fishes. *Journal of Morphology* **271**, 1399–1406.

Parker, G.A. (1970). Sperm competition and its evolutionary consequences in insects. *Biological Reviews* **45**, 525–567.

Parker, G.A. (1979). Sexual selection and sexual conflict. In: *Sexual Selection and Reproductive Competition in Insects* (Blum, M.S. & Blum, N.A., eds), pp. 123–166. New York: Academic Press.

Parker, G.A. (1983). Mate quality and mating decisions. In: *Mate Choice* (Bateson, P., ed.), pp. 141–166. Cambridge: Cambridge University Press,.

Parker, G.A. (1992). The evolution of sexual size dimorphism in fish. *Journal of Fish Biology* **41** (Suppl.), 1–20.

Parker, G.A. (1998). Sperm competition and the evolution of ejaculates: towards a theory base. In: *Sperm Competition and Sexual Selection* (Birkhead, T.R. & Møller, A.P., eds), pp. 3–54. London: Academic Press.

Parker, G.A., Ball, M.A., Stockley, P. & Gage, M.J.G. (1996). Sperm competition games: individual assessment of sperm competition intensity by group spawners. *Proceedings of the Royal Society of London, B* **263**: 1291–1297.

Parker, G.A., Ball, M.A., Stockley, P. & Gage, M.J.G. (1997). Sperm competition games: a prospective analysis of risk assessment. *Proceedings of the Royal Society of London, B* **264**, 1793–1802.

Parker, G.A. & Knowlton, N. (1980). The evolution of territory size – some ESS models. *Journal of Theoretical Biology* **84**, 445–476.

Parker, G.A. & Simmons, L.W. (1996). Parental investment and the control of sexual selection: predicting the direction of sexual competition. *Proceedings of the Royal Society of London, B* **263**, 315–321.

Pascual, M.A. & Quinn, T.P. (1994). Geographical patterns of straying of fall chinook salmon (*Oncorhynchus tshawytscha*) from Columbia river (USA) hatcheries. *Aquaculture and Fisheries Management* **25**(Suppl. 2), 17–30.

Pateman-Jones, C., Rasotto, M.B., Reichard, M., Liao, C., Liu, H., Zięba G. & Smith, C. (2011). Variation in male reproductive traits among three bitterling fishes (Acheilognathinae: Cyprinidae) in relation to mating system. *Biological Journal of the Linnean Society* **103**, 622–632.

Patterson, I.J. (1985). Limitation of breeding density through territorial behaviour: experiments with convict cichlids, *Cichlasoma nigrofasciatum*. In: *Behavioural Ecology – Ecological Consequences of Adaptive Behaviour* (Sibly, R.M. & Smith, R.H., eds), pp. 393–405. Blackwell, Oxford.

Patzner, R.A. (2008). Reproductive strategies in fish. In: *Fish Reproduction* (Roche, M.J., Arukwe, A. & Kapoor, B.G., eds), pp. 311–350. Enfield (NH): Science Publishers.

Payne, A.G., Smith, C. & Campbell, A.C. (2002). Filial cannibalism improves survival and development of beaugregory damselfish embryos. *Proceedings of the Royal Society London, B* **269**, 2095–2102.

Payne, A.G., Smith, C. & Campbell, A.C. (2003). Interactions between ophiuroids and beaugregory damselfish. *Journal of the Marine Biological Association of the United Kingdom* **83**, 625–632.

Payne, A.G., Smith, C. & Campbell, A.C. (2004). A model of oxygen mediated cannibalism in fishes. *Ecological Modelling* **174**, 253–266.

Pedroso, S.S., Barber, I., Svensson, O., Fonseca, P.J., & Amorim, M.C.P. (2013). Courtship sounds advertise species identity and male quality in sympatric *Pomatoschistus* spp. gobies. *PLoS One* **8**, e64620.

Peichel, C.L., Nereng, K.S., Ohgi, K.A., Cole, B.L.E., Colosimo, P.F., Buerkle, C.A., Schluter, D. & Kingsley, D.M. (2001). The genetic architecture of divergence between threespine stickleback species. *Nature* **414**, 901–905.

Peichel, C.L., Ross, J.A., Matson, C.K., Dickson, M., Grimwood, J., Schmutz, J., Myers, R.M., Mori, S., Schluter, D. & Kingsley, D.M. (2004). The master sex-determination locus in threespine sticklebacks is on a nascent Y chromosome. *Current Biology* **14**, 1416–1424.

Penman, D.J. & Piferrer, F. (2008). Fish gonadogenesis. Part 1: genetic and environmental mechanisms of sex determination. *Reviews in Fisheries Science* **16** (Suppl.), 16–34.

Penn, D. (2002). The scent of genetic compatibility: sexual selection and the major histocompatibility complex. *Ethology* **108**, 1–21.

Penn, D. & Potts, W.K. (1999). The evolution of mating preferences and major histocompatibility complex genes. *American Naturalist* **153**, 145–164.

Pepin, P. (1991). Effect of temperature and size on development, mortality, and survival rates of the pelagic early life history stages of marine fish. *Canadian Journal of Fisheries and Aquatic Sciences* **48**, 503–518.

Pepin, P. & Myers, R.A. (1991). Significance of egg and larval size to recruitment variability of temperate marine fish. *Canadian Journal of Fisheries and Aquatic Sciences* **48**, 1820–1828.

Perrone, M. & Zaret, T.M. (1979). Parental care patterns of fishes. *American Naturalist* **113**, 351–361.

Persat, H., Olivier, J.-M. & Pont, D. (1994). Theoretical habitat templets, species traits, and species richness: fish in the Upper Rhône River and its floodplain. *Freshwater Biology* **31**, 439–454.

Peter, R.E. & Fryer, J.N. (1983). Endocrine functions of the hypothalamus of actinopterygians. In: *Fish Neurobiology Vol. 2 Higher Brain Areas and Functions* (Davis, R.E. & Northcutt, R.G., eds), pp. 165–201. Ann Arbor: The University of Michigan Press.

Petersen, C.W. (1991). Variation in fertilization rate in the tropical reef fish *Halichoeres bivittatus*: Correlates and implications. *Biological Bulletin* **181**, 232–237.

Petersen, C.W. & Marchetti, K. (1989). Filial cannibalism in the Cortez damselfish, *Stegastes rectifraenum*. *Evolution* **43**, 158–168.

Petersen, C.W., Mazzoldi, C., Zarrella, K.A. & Hale, R.B. (2005). Fertilization mode, sperm characteristics, mate choice and parental care patterns in *Artedius* spp. (Cottidae). *Journal of Fish Biology* **67**, 239–254.

Petersen, C.W. & Warner, R.R. (1998). Sperm competition in fishes. In: *Sperm Competition and Sexual Selection* (Birkhead, T.R. & Møller, A.P., eds), pp. 435–463. London: Academic Press.

Petersen, C.W., Warner, R.R., Cohen, S., Hess, H.C. & Sewell, A.T. (1992). Variable pelagic fertilization success: implications for mate choice and spatial patterns of mating. *Ecology* **73**, 391–401.

Pianka, E.R. & Parker, W.S. (1975). Age-specific reproductive tactics. *American Naturalist* **109**, 453–464.

Piché, J., Hutchings, J.A. & Blanchard, W. (2008). Genetic variation in threshold reaction norms for alternative reproductive tactics in male Atlantic salmon, *Salmo salar*. *Proceedings of the Royal Society of London, B* **275**, 1571–1575.

Pietsch, T.W. (1976). Dimorphism, parasitism and sex reproductive strategies among deep-sea ceratioid anglerfishes. *Copeia* **1976**, 781–793.

Pietsch, T.W. (2005). Dimorphism, parasitism, and sex revisited: modes of reproduction among deep-sea certioid anglerfishes (Teleostei: Lophiiformes). *Ichthyological Research* **52**, 207–236.

Pietsch, T.W. & Grobecker, D. (1980). Parental care as an alternative reproductive mode in an antennariid anglerfish. *Copeia* **1980**, 551–553.

Piferrer, F. (2001). Endocrine sex control strategies for the feminization of teleost fish. *Aquaculture* **197**, 229–281.

Piferrer, F., Blázquez, M., Navarro, L. & González, A. (2005). Genetic and endocrine, and environmental components of sex determination and differentiation in the European sea bass (*Dicentrarchus labrax* L.). *General and Compararive Endocrinology* **142**, 102–110.

Piferrer, F. & Guiguen, Y. (2008). Fish gonadogenesis. Part II. Molecular biology and genomics of sex differentiation. *Reviews in Fisheries Science* **16** (Suppl.), 35–55.

Pilastro, A., Scaggiante, M. & Rasotto, M.B. (2002). Individual adjustment of sperm expenditure accords with sperm competition theory. *Proceedings of the National Academy of Science USA* **99**, 9913–9915.

Pires, M.N., Bassar, R.D., McBride, K.E., Regus, J.U., Garland Jr, T & Reznick, D.N. (2011). Why do placentas evolve? An evaluation of the life-history facilitation hypothesis in the fish genus *Poeciliopsis*. *Functional Ecology* **25**, 757–768.

Pitcher, T.E., Rodd, F.H. & Rowe, L. (2007). Sexual colouration and sperm traits in guppies. *Journal of Fish Biology* **70**, 165–177.

Pitcher, T.J. & Hart, P.J.B. (1982). *Fisheries Ecology*. London: Croom Helm.

Pizzari, T. & Parker, G.A. (2009). Sperm competition and sperm phenotype. In: *Sperm Biology: An Evolutionary Perspective* (Birkhead, T.R., Hosken, D.J. & Pitnick, S., eds), pp. 207–245. Oxford: Academic Press.

Pizzolon, M., Giacomello, E., Marri, L., Marchini, D., Pascoli, F., Mazzoldi, C. & Rasotto, M.B. (2010). When fathers make the difference: efficacy of male sexually selected antimicrobial glands in enhancing fish hatching success. *Functional Ecology* **24**, 141–148.

Planas, J.V. & Swanson, P. (2008). Physiological function of gonadotropins in fish. In: *Fish Reproduction* (Rocha, M.J., Arukwe, A. & Kapoor, B.G., eds), pp. 37–68. Enfield (NH): Science Publishers.

Plath, M., Makowicz, A.M., Schlupp, I. & Tobler, M. (2007). Sexual harassment in live-bearing fishes (Poeciliidae): comparing courting and noncourting species. *Behavioral Ecology* **18**, 680–688.

Platten, J.R., Tibbetts, I.R. & Sheaves, M.J. (2002). The influence of increased line-fishing mortality on the sex ratio and age of sex reversal of the venus tusk fish. *Journal of Fish Biology* **60**, 301–318.

Plaut, I. (2002). Does pregnancy affect swimming performance of female mosquitofish, *Gambusia affinis*? *Functional Ecology* **16**, 290–295.

Plenderleith, M., van Oosterhout, C., Robinson, R.L. & Turner, G.F. (2005). Female preference for conspecific males based on olfactory cues in a Lake Malawi cichlid fish. *Biology Letters* **1**, 411–414.

Polačik, M., Donner, M.T. & Reichard, M. (2011). Age structure of annual *Nothobranchius* fishes in Mozambique: is there a hatching synchrony? *Journal of Fish Biology* **78**, 796–809.

Polačik, M. & Reichard, M. (2011). Asymmetric reproductive isolation between two sympatric annual killifish with extremely short lifespans. *PLoS One* **6**, e22684.

Pollux, B.J.A., Pires, M.N., Banet, A.I. & Reznick, D.N. (2009). Evolution of placentas in the fish family Poeciliidae: an empirical study of macroevolution. *Annual Review of Ecology, Evolution and Systematics* **40**, 271–280.

Poltev, Y.N. & Mukhametov, I.N. (2009). Concerning the problem of carcinophilia of *Careproctus* species (Scorpaeniformes: Liparidae) in the North Kurils. *Russian Journal of Marine Biology* **35**, 215–223.

Popesku, J.T., Martyniuk, C.J., Menningen,J., Xiong, H., Zhong, D., Xia, X., Cossins, A.R. & Trudeau, V.L. (2008). The goldfish (*Carassius auratus*) as a model for neuroendocrine signalling. *Molecular and Cellular Endocrinology* **293**, 43–56.

Porat, D. & Chadwick-Furman, N.E. (2004). Effects of anemonefish on giant sea anemones: expansion behavior, growth and survival. *Hydrobiologia* **530**/531, 513–520.

Potts, G.W. (1984). Parental behaviour in temperate marine teleosts with special reference to the development of nest structures. In: *Fish Reproduction: Strategies and Tactics* (Potts, G.W. & Wootton, R.J., eds), pp. 223–244. London: Academic Press.

Prat, F., Sumpter, J.P. & Tyler, C.R. (1996). Validation of radioimmunoassay for two salmon gonadotropins (GTHI and GTHII) and their plasma concentrations throughout the reproductive cycle in male and female rainbow trout (*Oncorhynchus mykiss*). *Biology of Reproduction* **54**, 1375–1382.

Pressley, P.H. (1981). Parental effort and the evolution of nest-guarding tactics in the threespine stickleback *Gasterosteus aculeatus* L. *Evolution* **35**, 282–295.

Preston, L. (1978). Communication systems and social interactions in a goby-shrimp symbiosis. *Animal Behaviour* **26**, 791–802.

Price, D.J. (1984). Genetics of sex determination in fishes – a brief review. In: *Fish Reproduction: Strategies and Tactics* (Potts, G.W. & Wootton, R.J., eds), pp. 77–89. London: Academic Press.

Prum, R.O. (2010). The Lande-Kirkpatrick mechanism is the null model of evolution by intersexual selection: implications for meaning, honesty and design in intersexual signals. *Evolution* **64**, 3085–3100.

Ptacek, M. & Travis, J. (1997). Mate choice in the sailfin molly. *Evolution* **51**, 1217–1231.

Purdom, C.E. (1993). *Genetics and Fish Breeding*. London: Chapman & Hall.

Quagio-Grassiotto, I. & Oliveira, C. (2008). Sperm ultrastructure and a new type of spermiogenesis in two species of *Pimelodidae*, with a comparative review of sperm ultrastructure in siluriformes (Teleostei: Ostariophysi). *Journal of Comparative Zoology* **247**, 55–66.

Qvarnström, A. & Forsgren, E. (1998). Should females prefer dominant males? *Trends in Ecology and Evolution* **13**, 498–501.

Quinn, T.P. (1984). Homing and straying in Pacific salmon. In: *Mechanisms of Migration in Fishes* (McCleave, J.D., Arnold, G.P., Dodson, J.J. & Neill, W.H., eds), pp. 357–362. New York: Plenum Press.

Quinn, T.P. & Brannon, E.L. (1982). The use of celestial and magnetic cues by orienting sockeye salmon smolts. *Journal of Comparative Physiology* **147**, 547–552.

Raine, J.C. (2011). Thyroid hormones and reproduction in fishes. In: *Hormones and Reproduction of Vertebrates* (Norris, D.O. & Lopez, K.H., eds), pp. 83–102. London: Academic Press.

Rakitin, A. Ferguson, M.M. & Trippel, E.A. (1999). Sperm competition and fertilization success in Atlantic cod (*Gadus morhua*): effect of sire size and condition factor on gamete quality. *Canadian Journal of Fisheries and Aquatic Sciences* **56**, 2315–2323.

Rana, K.J. & McAndrew, B.J. (1989). The viability of cryopreserved tilapia spermatozoa. *Aquaculture* **76**, 335–345.

Rasotto, M.B. & Shapiro, D.Y. (1998). Morphology of gonoducts and male genital papilla, in the bluehead wrasse: implications and correlates on the control of gamete release. *Journal of Fish Biology* **52**, 716–725.

Reay, P.J. (1984). Reproductive tactics: a non-event in aquaculture. In: *Fish Reproduction: Strategies and Tactics* (Potts, G.W. & Wootton, R.J., eds), pp. 291–309. London: Academic Press.

Reichard, M. (2010). *Nothobranchius kadleci* (Cyprinodontiformes: Nothobranchiidae), a new species of annual killifish from central Mozambique. *Zootaxa* **2332**, 49–60.

Reichard, M., Jurajda, P. & Ondračková, M. (2002). Interannual variability in seasonal dynamics and species composition of drifting young-of-the-year fishes in two European lowland rivers. *Journal of Fish Biology* **60**, 87–101.

Reichard, M., Jurajda, P. & Smith, C. (2004b). Male-male interference competition decreases spawning rate in the European bitterling (*Rhodeus sericeus*). *Behavioral Ecology & Sociobiology* **56**, 34–41.

Reichard, M., Le Comber, S. & Smith, C. (2007b). Sneaking from a female perspective. *Animal Behaviour* **74**, 679–688.

Reichard, M., Liu, H. & Smith, C. (2007a). The co-evolutionary relationship between bitterling fishes and freshwater mussels: insights from interspecific comparisons. *Evolutionary Ecology Research* **9**, 1–21.

Reichard, M., Ondračková, M., Bryjová, A., Smith, C. & Bryja, J. (2009a). Breeding resource distribution affects selection gradients on male phenotypic traits: experimental study on lifetime reproductive success in the bitterling fish (*Rhodeus amarus*). *Evolution* **63**, 377–390.

Reichard, M., Ondračková, M., Przybylski, M., Liu, H. & Smith, C. (2006). The costs and benefits in an unusual symbiosis: experimental evidence that bitterling fish (*Rhodeus sericeus*) are parasites of unionid mussels in Europe. *Journal of Evolutionary Biology* **19**, 788–796.

Reichard, M., Polačik, M. & Sedláček, O. (2009b). Distribution, colour polymorphism and habitat use of the African killifish, *Nothobranchius furzeri*, the vertebrate with the shortest lifespan. *Journal of Fish Biology* **74**, 198–212.

Reichard, M. Polačik, M., Tarkan, A.S., Spence, R., Gaygusuz, Ö., Ercan, E., Ondračková, M. & Smith, C. (2010). The bitterling–mussel coevolutionary relationship in areas of recent and ancient sympatry. *Evolution* **64**, 3047–3056.

Reichard, M., Smith, C. & Jordan, W.C. (2004a). Genetic evidence reveals density-dependent mediated success of alternative mating tactics in the European bitterling (*Rhodeus sericeus*). *Molecular Ecology* **13**, 1569–78.

Reichard, M., Spence, R., Bryjová, A., Bryja, J. & Smith, C. (2012). Female rose bitterling prefer MHC-dissimilar males: experimental evidence. *PLoS One* **7**, e40780.

Reimchen, T.E. (1989). Loss of nuptial color in threespine sticklebacks (*Gasterosteus aculeatus*). *Evolution* **43**, 450–460.

Reinecke, M. (2010). Insulin-like growth factors and fish reproduction. *Biology of Reproduction* **82**, 656–661.

Reusch, T.B.H., Haberli, M.A., Aeschlimann, P.B. & Milinski, M. (2001). Female sticklebacks count alleles in a strategy of sexual selection explaining Mhc polymorphism. *Nature* **414**, 300–302.

Reznick, D.N. (1983). The structure of guppy life histories: the tradeoff between growth and reproduction. *Ecology* **64**, 862–873.

Reznick, D.N. (1985). Costs of reproduction: an evaluation of the empirical evidence. *Oikos* **44**, 257–267.

Reznick, D.N. (1996). Life history evolution in guppies: a model system for the empirical study of adaptation. *Netherlands Journal of Zoology* **46**, 172–190.

Reznick, D.N., Bassar, R.D., Travis, J. & Rodd, F.H. (2012). Life-history evolution in guppies VIII: The demographics of density regulation in guppies (*Poecilia reticulata*). *Evolution* **66**, 2903–2915.

Reznick, D.N. & Braun, B. (1987). Fat cycling in the mosquitofish (*Gambusia affinis*): fat storage as a reproductive adaptation. *Oecologia* **73**, 401–413.

Reznick, D.N., Bryant, M.J. & Bashey, F. (2002a). r- and K-selection revisited: the role of population regulation in life-history evolution. *Ecology* **83**, 1509–1520.

Reznick, D.N., Bryga, H. & Endler, J.A. (1990). Experimentally induced life-history evolution in a natural population. *Nature* **346**, 357–359.

Reznick, D.N., Butler, M.J. IV, Rodd, F.H. & Ross, P. (1996b). Life-history evolution in guppies (*Poecilia reticulata*) 6. *Differential mortality as a mechanism for natural selection. Evolution* **50**, 1651–1660.

Reznick, D.N., Callahan, H. & Llauredo, R. (1996a). Maternal effects on offspring quality in poeciliid fishes. *American Zoologist* **36**, 147–156.

Reznick, D.N. & Endler, J.A. (1982). The impact of predation on life history evolution in Trinidadian guppies (*Poecilia reticulata*). *Evolution* **63**, 160–177.

Reznick, D.N., Mateos, M. & Springer, M.S. (2002b). Independent origins and rapid evolution of the placenta in the fish genus *Poeciliopsis*. *Science* **298**, 1018–1020.

Reznick, D.N., Meredith, R. & Collette, B.B. (2007). Independent evolution of complex life history adaptations in two families of fishes, live-bearing halfbeaks (Zenarchopteridae, Beloniformes) and Poeciliidae (Cyprinodontiformes). *Evolution* **61**, 2570–2583.

Reznick, D.N. & Yang, A.P. (1993). The influence of fluctuating resources on life history patterns of allocation and plasticity in female guppies. *Ecology* **74**, 2011–2019.

Ribbink, A.J. (1990). Alternative life-history styles of some African cichlid fishes. *Environmental Biology of Fishes* **28**, 87–100.

Ribbink, A.J. (1991). Distribution and ecology of the cichlids of the African great lakes. In: *Cichlid Fishes: Behavior, Ecology and Evolution* (Keenleyside, M.H.A., ed.), pp. 37–59. New York: Chapman and Hall.

Ribbink, A.J. Marsh, A.C. & Marsh, B.A. (1981). Nest-building and communal care of young by *Tilapia rendalli* Dumeril (Pisces, Cichlidae) in Lake Malawi. *Environmental Biology of Fishes* **6**, 219–222.

Ribbink, A.J. & Ribbink, A.C. (1997). Paedophagia among cichlid fishes of Lake Victoria and Lake Malawi/Nyasa. *South African Journal of Science* **93**, 509–512.

Rick, I.P. & Bakker, T.C.M. (2008). Color signalling in conspicuous red sticklebacks: Do ultraviolet signals surpass others? *BMC Evolutionary Biology* **8**, 189.

Ricker, W.E. (1954). Stock and recruitment. *Journal of the Fisheries Research Board of Canada* **11**, 559–623.

Ricker, W.E. (1975). Handbook of computations for biological statistics of fish populations. *Bulletin of the Fisheries Research Board of Canada* **119**, 1–300.

Ricklefs, R.E. & Miller, G.L. (2000). *Ecology*, 4th edn. New York: W.H. Freeman.

Ricklefs, R.E. & Starck, J.M. (1996). Applications of phylogenetically independent contrasts: a mixed progress report. *Oikos* **77**, 167–172.

Rideout, R.M. & Tomkiewicz, J. (2011). Skipped spawning in fishes: more common than you might think. *Marine and Coastal Fisheries: Dynamics, Management and Ecosystem Sciences* **3**, 176–189.

Rideout, R.M., Rose, G.A. & Burton, M.P.M. (2005). Skipped spawning in female iteroparous fishes. *Fish and Fisheries* **6**, 50–72.

Ridgway, M.S. & Shuter, B.J. (1994). The effects of supplemental food on reproduction in parental male smallmouth bass. *Environmental Biology of Fishes* **39**, 201–207.

Ridley, M. & Rechten, C. (1981). Female sticklebacks prefer to spawn with males whose nests contain eggs. *Behaviour* **76**, 152–161.

Rinchard, J., Kestemont, P. & Heine, R. (1997). Comparative study of reproductive biology in single and multiple-spaner cyprinids. II. Sex steroids and plasma protein phosphorus concentrations. *Journal of Fish Biology* **50**, 169–180.

Rios-Cardenas, O. & Webster, A.S. (2005). Paternity and paternal effort in the pumpkinseed sunfish. *Behavioral Ecology* **16**, 914–921.

Ripley, J.L. (2009). Osmoregulatory role of the paternal brood pouch for two *Syngnathus* species. *Comparative Biochemistry and Physiology, Part A* **154**, 98–104.

Ripley, J.L. & Foran, C.M. (2009). Direct evidence for embryonic uptake of paternally derived nutrients in two pipefishes (Syngnathidae: *Syngnathus* spp.). *Journal of Comparative Physiology B* **179**, 325–333.

Ritchie, M.G. (2007). Sexual selection and speciation. *Annual Review of Ecology, Evolution, and Systematics* **38**, 79–102.

Robertson, D.R. (1991). The role of adult biology in the timing of spawning of tropical reef fishes. In: *The Ecology of Fishes on Coral Reefs* (Sale, P.F., ed.), pp. 356–386. London: Academic Press.

Robertson, D.R., Petersen, C.W. & Brawn, J.D. (1990). Lunar reproductive cycles of benthic brooding reef fishes: reflections of larval or adult biology. *Ecological Monographs* **60**, 311–329.

Robertson, D.R. & Warner, R.R. (1978). Sexual patterns in the labroid fishes of the western Caribbean, II. The parrotfishes (Scaridae). *Smithsonian Contributions to Zoology* **255**, 1–26.

Robertson, O.H. (1961). Prolongation of the life span of kokanee salmon (*Oncorhynchus nerka kennerlyi*) by castration before beginning of gonad development. *Proceedings of the National Academy of Sciences of the USA* **47**, 609.

Robillard, S.R. & Marsden, J.E. (2001). Spawning substrate preference of yellow perch along a sand-cobble shoreline on southwestern Lake Michigan. *North American Journal of Fisheries Management* **21**, 208–215.

Rochet, M.J., Collie, J.S., Jennings, S. & Hall, S.J. (2011). Does selective fishing conserve community biodiversity? Predictions from a length-based multispecies model. *Canadian Journal of Fisheries and Aquatic Sciences* **68**, 469–486.

Rodd, F.H., Hughes, K.A., Grether, G.F. & Baril, C.T. (2002). A possible nonsexual origin of mate preference: are male guppies mimicking fruit? *Proceedings of the Royal Society of London, B* **269**, 475–481.

Roff, D.A. (1984). The evolution of life history parameters in teleosts. *Canadian Journal of Fisheries and Aquatic Sciences* **41**, 984–1000.

Roff, D.A. (1992). *The Evolution of Life Histories.* London: Chapman and Hall.

Roff, D.A. (2002). *Life History Evolution.* Sunderland (MA): Sinauer.

Roff, D.A. (2008). Defining fitness in evolutionary models. *Journal of Genetics* **87**, 339–348.

Roff, D.A. & Fairbairn, D.J. (2007). The evolution of trade-offs: where are we? *Journal of Evolutionary Biology* **20**, 433–447.

Roff, D.A., Heibo, E. & Vøllestad, L.A. (2006). The importance of growth and mortality costs in the evolution of optimal life history. *Journal of Evolutionary Biology* **19**, 1920–1930.

Rohmann, K.N. & Bass, A.H. (2011). Seasonal plasticity of auditory hair cell frequency correlates with plasma steroid levels in vocal fish. *Journal of Experimental Biology* **214**, 1931–1942.

Rohwer, S. (1978). Parent cannibalism of offspring and egg-raiding as a courtship strategy. *American Naturalist* **112**, 429–40.

Rollinson, N. & Hutchings, J.A. (2013). The relationship between offspring size and fitness: integrating theory and empiricism. *Ecology* **94**, 315–324.

Römer, U. & Beisenherz, W. (1996). Environmental determination of sex in *Apistogramma* (Cichlidae) and two other freshwater fishes (Teleostei). *Journal of Fish Biology* **48**, 714–725.

Romer, A.S. & Parsons, T.S. (1986). *The Vertebrate Body*, 6th edn. Philadelphia: Saunders.

Roosen-Runge, E.C. (1977). *The Process of Spermatogenesis in Animals.* Cambridge: Cambridge University Press.

Rose, G.A. (1993). Cod spawning on a migration highway in the northwest Atlantic. *Nature* **366**, 458–461.

Rose, K.A., Cowan, J.H., Winemiller, K.O., Myers, R.A. & Hilborn, R. (2001). Compensatory density dependence in fish populations: importance, controversy, understanding and prognosis. *Fish and Fisheries* **2**, 293–327.

Rose, M.R. (1984). Genetic covariation in *Drosophila* life history: untangling the data. *American Naturalist* **123**, 565–569.

Rosen, D.E. & Bailey, R.M. (1963). The poeciliid fishes (Cyprinodontiformes): their structure, zoogeography, and systematics. *Bulletin of the American Museum of Natural History* **126**, 1–176.

Rosenqvist, G. & Berglund, A. (2011). Sexual signals and mating patterns in Syngnathidae. *Journal of Fish Biology* **78**, 1647–1661.

Rosenthal, G.G. & Evans, C.S. (1998). Female preference for swords in *Xiphophorus hellerii* reflects a bias for large apparent size. *Proceedings of the National Academy of Science of the USA* **95**, 4431–4436.

Rosenthal, G.G. & Lobel, P. (2006). Communication. In: *Behaviour and Physiology of Fish* (Sloman, K., Balshine, S. & Wilson, R., eds), pp. 39–78. London: Academic Press.

Ross, J.A. & Peichel, C.L. (2008). Molecular cytogenetic evidence of rearrangements on the Y chromosome of the threespine stickleback fish. *Genetics* **179**, 2173–2182.

Ross, J.A., Urton, J.R., Boland, J., Shapiro, M.D. & Peichel, C.L. (2009). Turnover of sex chromosome in the stickleback fishes (Gasterosteidae). *PLoS Genetics* **5**, e1000391.

Rothschild, B.J. (1986). *Dynamics of Marine Fish Populations.* Cambridge (MA): Harvard University Press.

Rothstein, S.I. (1990). A model system for co-evolution: avian brood parasitism. *Annual Review of Ecology, Evolution and Systematics* **21**, 481–508.

Rousseau, K. & Dufour, S. (2007). Comparative aspects of GH and metabolic regulation in lower vertebrates. *Neuroendocrinology* **86**, 165–174.

Rowe, D.K. & Thorpe, J.E. (1990a). Differences in growth between maturing and non-maturing male Atlantic salmon, *Salmo salar* L. parr. *Journal of Fish Biology* **36**, 643–658.

Rowe, D.K. & Thorpe, J.E. (1990b). Suppression of maturation in male Atlantic salmon (*Salmo salar* L.) parr by reduction in feeding and growth during spring months. *Aquaculture* **86**, 291–313.

Rowe, D.K., Thorpe, J.E. & Shanks, A.M. (1991). The role of fat stores in the maturation in male Atlantic salmon (*Salmo salar*) parr. *Canadian Journal of Fisheries and Aquatic Sciences* **48**, 405–413.

Rowe, S. & Hutchings, J.A. (2003). Mating systems and the conservation of commercially exploited marine fish. *Trends in Ecology & Evolution* **18**, 567–572.

Rowe, S., Hutchings, J.A., Bekkevold, D. & Rakitin, A. (2004). Depensation, probability of fertilization, and the mating system of Atlantic cod (*Gadus morhua* L.). *ICES Journal of Marine Science* **61**, 1144–1150.

Rowland, W.J. (1989). Mate choice and the supernormality effect in female sticklebacks (*Gasterosteus aculeatus*). *Behavioral Ecology and Sociobiology* **24**, 433–438

Rowland, W.J. (1999). Studying visual cues in fish behavior: a review of ethological techniques. *Environmental Biology of Fishes* **56**, 285–305.

Rudolfsen, G., Figenschou, L., Folstad, I. & Kleven, O. (2008). Sperm velocity influences paternity in the Atlantic cod (*Gadus morhua* L.). *Aquaculture Research* **39**, 212–216.

Rundle, H.D. & Boughman, J.W. (2010). Behavioral ecology and speciation. In: *Evolutionary Behavioral Ecology* (Westneat, D.F. & Fox, C.W., eds), pp. 471–487. Oxford: Oxford University Press.

Rundle, H.D. & Nosil, P. (2005). Ecological speciation. *Ecology Letters* **8**, 336–352.

Rupik, W., Huszno, J. & Klag, J. (2011). Cellular organisation of the mature testes and stages of spermiogenesis in *Danio rerio* (Cyprinidae: Teleostei) – structural and ultrastructural studies. *Micron* **42**, 833–839.

Rushbrook, B.J. & Barber, I. (2008). A comparison of nest building by three-spined sticklebacks *Gasterosteus aculeatus* from still and flowing waters. *Journal of Fish Biology* **73**, 746–752.

Russ, G.R. & Alcala, A.C. (1998). Natural fishing experiments in marine reserves 1983–1993: roles of life history and fishing intensity in family responses. *Coral Reefs* **17**, 399–416.

Russell, F.S. (1976). *The Eggs and Planktonic Stages of British Marine Fishes*. London: Academic Press.

Ryan, M.J. (1990). Sexual selection, sensory systems and sensory exploitation. *Oxford Surveys in Evolutionary Biology* **7**, 157–195.

Sabat, A. (1994). Costs and benefits of parental effort in a brood-guarding fish (*Ambloplites rupestris*, Centrarchidae). *Behavioral Ecology and Sociobiology* **5**, 195–210.

Sadovy, Y. & Liu, M. (2008). Functional hermaphroditism in teleosts. *Fish and Fisheries* **9**, 1–43.

Sagebakken, G., Ahnesjö, I., Mobley, K.B., Braga Gonçalves, I. & Kvarnemo, C. (2010). Brooding fathers, not siblings, take up nutrients from embryos. *Proceedings of the Royal Society of London, B* **277**, 971–977.

Saillant, E., Chatain, B., Menu, B., Fauvel, C., Vidal, M.O. & Fostier, A. (2003). Sexual differentiation and juvenile intersexuality in the European sea bass (*Dicentrarchus labrax*). *Journal of Zoology, London* **260**, 53–63.

Saito, D. & Tanaka, M. (2009). Comparative aspects of gonadal sex differentiation in medaka: a conserved role of developing oocytes in sexual canalization. *Sexual Development* **3**, 99–107.

Saldaña, J. & Venables, B. (1983). Energy compartimentalization in a migratory fish, *Prochilodus mariae* (Prochilodontidae), of the Orinoco River. *Copeia* **1983**, 617–625.

Sale, P.F. (1975). Patterns of use of space in a guild of territorial reef fishes. *Marine Biology* **29**, 89–97.

Sandra, G.E., & Norma, M.M. (2010). Sexual determination and differentiation in teleost fish. *Reviews in Fish Biology and Fisheries* **20**, 101–121.

Santini, F., Harmon, L.J., Carnevale, G. & Alfaro, M.E. (2009). Did genome duplication drive the origin of teleosts? A comparative study of diversification in ray-finned fishes. *BMC Evolutionary Biology* **9**, 194.

Sarder, M.R.I., Penman, D.J., Myers, J.M. & McAndrew, B.J. (1999). Production and propagation of fully inbred clonal lines in the Nile tilapia (*Oreochromis niloticus* L.). *Journal of Experimental Zoology* **284**, 675–685.

Sargent, R.C. (1982). Territory quality, male quality, courtship intrusions, and female nest-choice in the threespine stickleback, *Gasterosteus aculeatus*. *Animal Behaviour* **30**, 364–374.

Sargent, R.C. (1992). Ecology of filial cannibalism in fish: theoretical perspectives. In: *Cannibalism: Ecology and Evolution Among Diverse Taxa* (Elgar, M.A. & Crespi, B.J., eds), pp. 38–62. Oxford: Oxford University Press.

Sargent, R.C. (1997). Parental care. In: *Behavioural Ecology of Teleost Fishes* (Godin J.-G., ed.), pp. 290–313. Oxford: Oxford University Press.

Sargent, R.C. & Gebler, J.B. (1980). Effects of nest site concealment on hatching success, reproductive success, and paternal behavior of the threespine stickleback, *Gasterosteus aculeatus*. *Behavioral Ecology and Sociobiology* **7**, 137–142.

Sargent, R.C. & Gross, M.R. (1993). Williams' principle: an explanation of parental care in teleost fishes. In: *Behaviour of Teleost Fishes*, 2nd edn (Pitcher, T.P., ed.), pp. 333–361. London: Chapman and Hall.

Sargent, R.C., Gross, M.R. & van den Berghe, E.P. (1986). Male mate choice in fishes. *Animal Behaviour* **34**, 545–550.

Sargent, R.C., Taylor, P.D. & Gross, M.R. (1987). Parental care and the evolution of egg size in fishes. *American Naturalist* **129**, 32–46.

Sasaki, T. (1977). The urogenital papilla of the tube-snout, *Aulichthys japonicus*. *Japanese Journal of Ichthyology* **24**, 161–166.

Sayer, M.D.J. & Davenport J. (1991). Amphibious fish: why do they leave water? *Reviews in Fish Biology and Fisheries* **1**, 159–181.

Scaggiante, M., Mazzoldi, C., Petersen, C.W. & Rasotto, M.B. (1999). Sperm competition and mode of fertilization in the grass goby *Zosterisessor ophiocephalus* (Teleostei: Gobiidae). *Journal of Experimental Zoology* **283**, 81–90.

Schaffer, W.M. (1974a). Selection for optimal life histories: the effects of age structure. *Ecology* **55**, 291–303.

Schaffer, W.M. (1974b). Optimal reproductive effort in fluctuating environments. *American Naturalist* **108**, 783–790.

Schaffer, W.M. (1979). The theory of life-history evolution and its application to Atlantic salmon. *Symposium of the Zoological Society of London* **44**, 307–326.

Schaffer, W.M. (2004). Life histories, evolution, and salmonids. In: *Evolution Illuminated. Salmon and their Relatives* (Hendry, A.P. & Stearns, S.C., eds), pp. 20–51. Oxford: Oxford University Press.

Schartl, M. (2004). A comparative view on sex determination in medaka. *Mechanisms of Development* **121**, 639–645.

Schlupp, I. (2005). The evolutionary ecology of gynogenesis. *Annual Review of Ecology, Evolution and Systematics* **36**, 399–417.

Schlupp, I. & Riesch, R. (2011). Evolution of unisexual reproduction. In: *Ecology and Evolution of Poeciliid Fishes* (Evans, J.P., Pilastro, A. & Schlupp, I., eds), pp. 50–58. Chicago: University of Chicago Press.

Schlupp, I., Riesch, R., Tobler, M., Plath, M., Parzefall, J. & Schartl, M. (2010). A novel sexually selected trait in poeciliid fishes: female preference for mustache-like, rostral filaments in male *Poecilia sphenops*. *Behavioral Ecology and Sociobiology* **64**, 1849–1855.

Schluter, D. (2000). *The Ecology of Adaptive Radiation*. Oxford: Oxford University Press.

Schluter, D. & Price, T. (1993). Honesty, perception, and population divergence in sexually selected traits. *Proceedings of the Royal Society of London, B* **253**, 117–122.

Schneider, F. & Poehland, R. (2009). Endocrinology of reproduction. In: *Reproductive Biology and Phylogeny of Fishes (Agnathans and Bony Fishes)* (Jamieson, B.G.M., ed.), pp. 54–93. Enfield (NH): Science Publications.

Schoener, T.W. (1968). Sizes of feeding territories among birds. *Ecology* **49**, 123–141.

Schoener, T.W. (1983). Simple models of optimal feeding-territory size: a reconciliation. *American Naturalist* **121**, 608–629.

Schoener, T.W. (1987). Time budgets and territory size: some simultaneous optimization models for energy maximizers. *American Zoologist* **27**, 259–291.

Scholz, S. & Klüver, N. (2009). Effects of endocrine disrupters on sexual, gonadal development in fish. *Sexual Development* **3**, 136–151.

Schradin, C. & Lamprecht, J. (2000). Female-biased immigration and male peace keeping in groups of the shell-dwelling cichlid fish *Neolamprologus multifasciatus*. *Behavioral Ecology and Sociobiology* **48**, 236–242.

Schugardt, C. & Kirschbaum, F. (2004). Control of gonadal maturation and regression by experimental variation of environmental factors in the mormyrid fish, *Mormyrus proboscirostris*. *Environmental Biology of Fishes* **70**, 227–233.

Schultheis, C., Böhne, A., Schartl, M., Volff, J.N. & Gailaina-Arnoux, D. (2009). Sex determination diversity and sex chromosome evolution in Poeciliid fish. *Sexual Development* **3**, 68–77.

Schultz, E.T. (1993a). Sexual size dimorphism at birth in *Micrometrus minimus* (Embiotocidae): a prenatal cost of reproduction. *Copeia* 1993, 456–463.

Schultz, E.T. (1993b). The effect of birth date on fitness of female dwarf perch, *Micrometrus minimus* (Perciformes: Embiotocidae). *Evolution* **47**, 520–539.

Schultz, R.J. (1989). Origins and relationships of unisexual Poeciliids. In: *Ecology and Evolution of Livebearing Fishes (Poeciliidae)* (Meffe, G.K. & Snelson, F.F. Jr, eds), pp. 69–90. Englewood Cliffs (NJ): Prentice-Hall

Schulz, R.W., de Franca, L.R., Lareyre. J.-J., Le Gac, F., Chiarini-Garcia, H., Nobrega, R.H. & Miura, T. (2010). Spermatogenesis in fish. *General and Comparative Endocrinology* **166**, 390–411.

Schulz, R.W. & Miura, T. (2002). Spermatogenesis and its endocrine regulation. *Fish Physiology and Biochemistry* **26**, 43–56.

Schütz, M. & Barlow, G.W. (1997). Young of the Midas cichlid get biologically active nonnutrients by eating mucus from the surface of their parents. *Fish Physiology and Biochemistry* **16**, 11–18.

Schwanck, E. (1986). Filial cannibalism of *Tilapia mariae*. *Journal of Applied Ichthyology* **2**, 65–74.

Scott, A.P. (1990). Salmonids. In: *Reproductive Seasonality in Teleosts: Environmental Influences* (Munro, A.D., Scott, A.P. & Lam, T.J., eds), pp. 33–51. Boca Raton: CRC Press.

Scott, A.P., Bye, V.J. & Baynes, S.M. (1980a). Seasonal variations in sex steroids of female rainbow trout (*Salmo gairdneri* Richardson). *Journal of Fish Biology* **17**, 587–592.

Scott, A.P., Bye, V.J., Baynes, S.M. & Springate, J.R.C. (1980b). Seasonal variations in plasma concentrations of 11-ketotestosterone and testosterone in male rainbow trout, *Salmo gairdneri* Richardson. *Journal of Fish Biology* **17**, 495–505.

Scott, A.P., Sumpter, J.P. & Stacey, N. (2010). The role of the maturation-inducing steroid, 17, 20β-dihydroxypregn-4-en-3-one, in male fishes: a review. *Journal of Fish Biology* **76**, 183–224.

Scott, D.B.C. (1979). Environmental timing and the control of reproduction in teleost fish. *Symposium of Zoological Society of London* **44**, 105–132.

Scrimshaw, N.S. (1944). Embyronic growth in the vivaparous Poeciliid, *Heterandria formosa*. *Biological Bulletin* **87**, 37–51.

Searcy, W.A. & Nowicki, S. (2005). *The Evolution of Animal Communication: Reliability and Deception in Signaling Systems*. Princeton: Princeton University Press.

Seehausen, O. & Schluter, D. (2004). Male-male competition and nuptial color displacement as a diversifying force in Lake Victoria cichlid fish. *Proceeding of the Royal Society of London, B* **271**, 1345–1353.

Selman, K. & Wallace, R.A. (1986). Gametogenesis in *Fundulus heteroclitus*. *American Zoologist* **26**, 173–192.

Semler, D.E. (1971). Some aspects of adaptation in a polymorphism for breeding colours in the threespine stickleback (*Gasterosteus aculeatus*). *Journal of Zoology, London* **165**, 291–302.

Senthilkumaran, B. & Joy, K.P. (1998). Methallibure-induced inhibition of hypothalamo-hypophyseal-ovarian activity in the catfish *Heteropneustes fossilis* involves changes in hypothalamic monoamine activity. *Fish Physiology and Biochemistry* **19**, 359–364.

Ser, J.R., Roberts, R.B. & Kocher, T.D. (2010). Multiple interacting loci control sex determination in Lake Malawi cichlid fish. *Evolution* **64**, 486–501.

Servili, A., Herrera-Pérez, P., Kah, O. & Muňoz-Cueto, J.A. (2012). The retina is a target for GnRH-3 system in the European sea bass, *Dicentrarchus labrax*. *General and Comparative Endocrinology* **175**, 398–406.

Sevenster, P. (1961). A causal study of displacement activity (fanning in *Gasterosteus aculeatus* L.). *Behaviour* **9** (Suppl.), 1–170.

Sevenster-Bol, A.C.A. (1962). On the causation of drive reduction after a consummatory act (in *Gasterosteus aculeatus* L.). *Archives Néerlandaises de Zoologie* **15**, 175–236.

Shapiro, D.Y. (1984). Sex reversal and sociodemographic processes in coral reef fishes. In: *Fish Reproduction: Strategies and Tactics* (Potts, G.W. & Wootton, R.J., eds), pp. 103–118. London: Academic Press.

Shapiro, D.Y., Marconato, A. & Yoshikawa, T. (1994). Sperm economy in a coral reef fish. *Ecology* **75**, 1334–1344.

Shapiro, M.D., Summers, B.R., Balabhadra, S., Aldenhoven, J.T., Miller, A.L., Cunningham, C.B., Bell, M.A. & Kingsley, D.M. (2009). The genetic architecture of skeletal convergence and sex determination in ninespine sticklebacks. *Current Biology* **19**, 1140–1145.

Shearer, K.D. & Swanson, P. (2000). The effect of whole body lipid on early sexual maturation of 1+ age male Chinook salmon (*Oncorhynchus tshawytscha*). *Aquaculture* **190**, 343–367.

Shelton, W.L. (1995). Genetic manipulation to produce sterile fishes. In: *Herbivorous Fishes: Culture and Use for Weed Management* (Opuszynski, K. & Shireman, J.V., eds), pp. 98–102. Boca Raton: CRC press.

Shepherd, J.G. (1982). A versatile new stock-recruitment relationship for fisheries, and the construction of sustainable yield curves. *Journal du Conseil International pour l'Exploration de la Mer* **40**, 67–75.

Sherbakov, D.Y. (1999). Molecular phylogenetic studies on the origin of biodiversity in Lake Baikal. *Trends in Ecology and Evolution* **14**, 92–95.

Sherwood, N.M. & Adams, B.A. (2005). Gonadotropin-releasing hormone in fish: evolution, expression and regulation of the GnRH gene. In: *Hormones and Their Receptors in Fish Reproduction* (Melamed, P. & Sherwood, N., eds), pp. 1–39. New Jersey: World Scientific.

Shibukawa, K., Tran, D.D. & Tran, L.X. (2012). *Phallostethus cuulong*, a new species of priapiumfish (Actinopterygii: Atheriniformes: Phallostethidae) from the Vietnamese Mekong. *Zootaxa* **3363**, 45–51.

Shichida, Y. & Matsuyama, T. (2009). Evolution of opsins and phototransduction. *Philosophical Transactions of the Royal Society of London, B* **364**, 2881–2895.

Shine, R. (1978). Propagule size and parental care: the 'safe harbor' hypothesis. *Journal of Theoretical Biology* **75**, 417–424.

Shine, R. (1989). Alternative models for the evolution of offspring size. *American Naturalist* **134**, 311–317.

Shohet, A.J. & Watt, P.J. (2009). Female guppies *Poecilia reticulata* prefer males that can learn fast. *Journal of Fish Biology* **75**, 1323–1330.

Shuster, S.M. (2010). Alternative mating strategies. In: *Evolutionary Behavioral Ecology* (Fox, C. & Westneat, D.F., eds), pp. 434–450. Cambridge: Cambridge University Press.

Shuster, S.M. & Wade, M.J. (2003). *Mating Systems and Strategies*. Princeton: Princeton University Press.

Siegfried, K.R. (2010). In: search of determinants: gene expression during gonadal sex differentiation. *Journal of Fish Biology* **76**, 1879–1902.

Siegfried, K.R. & Nüsslein-Volhard, C. (2008). Germ line control of female sex determination in zebrafish. *Developmental Biology* **324**, 277–287.

Sigholt, T., Staurnes, M., Jakobsen, H.J. & Asgard, T. (1995). Effects of continuous light and short-day photoperiod on smolting, seawater survival and growth in Atlantic salmon (*Salmo salar*). *Aquaculture* **130**, 373–388.

Sikkel, P. (1989). Egg presence and developmental stage influence spawning site choice by female garibaldi. *Animal Behaviour* **38**, 447–456.

Sikkel, P.C. (1993). Changes in plasma androgen levels associated with changes in male reproductive behaviour in a brood cycling marine fish. *General and Comparative Endocrinology* **89**, 229–237.

Sims, D.W., Wearmouth V.J., Genner M.J., Southward A.J. & Hawkins S.J. (2004). Low-temperature-driven early spawning migration of a temperate marine fish. *Journal of Animal Ecology* **73**, 333–341.

Sinclair, A.R.E. (1989). Population regulation in animals. In: *Ecological Concepts* (Cherrett, J.M., ed.), pp. 197–241. Oxford: Blackwell Scientific.

Slanchev, K., Stebler, J., de la Cueva-Mendez, G. & Raz, E. (2005). Development without germ cells: the role of the germ line in zebrafish sex differentiation. *Proceedings of the National Academy of Sciences of the USA* **102**, 4074 -4079.

Slotte, A. & Fiksen, Ø. (2000). State-dependent spawning migration in Norwegian spring-spawning herring. *Journal of Fish Biology* **56**, 138–162.

van der Sluijs, I., Gray, S.M., Amorim, M.C.P., Barber, I., Candolin, U., Hendry, A.P., Krahe, R., Maan, M.E., Utne-Palm, A.C., Wagner, H.-J. & Wong, B.B.M. (2011). Communication in troubled waters: responses of fish communication systems to changing environments. *Evolutionary Ecology* **25**, 623–640.

Smadja, C. & Butlin, R.K. (2011). A framework for comparing processes of speciation in the presence of gene flow. *Molecular Ecology* **20**, 5123–5140.

Smith, C. (1992). Filial cannibalism as a reproductive strategy in care-giving teleosts? *Netherlands Journal of Zoology* **42**, 607–613.

Smith, C,. Barber, I., Wootton, R.J. & Chittka, L. (2004b). A receiver bias in the origin of three-spined stickleback mate choice. *Proceedings of the Royal Society of London, B* **271**, 949–955.

Smith, C., Douglas, A. & Jurajda, P. (2001a). Oviposition site selection and embryo mortality in the perch. *Journal of Fish Biology* **58**, 880–882.

Smith, C., Douglas, A. & Jurajda, P. (2002a). Sexual conflict, sexual selection, and sperm competition in the spawning decisions of bitterling (*Rhodeus sericeus*). *Behavioral Ecology & Sociobiology* **51**, 433–439.

Smith C., Pateman-Jones, C., Zięba, G., Przybylski, M. & Reichard, M. (2009). Sperm depletion as a consequence of increased sperm competition risk in the European bitterling (*Rhodeus amarus*). *Animal Behaviour* **77**, 1227–1233.

Smith, C., Phillips, A., Polačik, M. & Reichard, M. (2013). Male nuptial coloration signals direct mate choice benefits in the European bitterling (*Rhodeus amarus*). *Environmental Biology of Fishes* DOI: 10.1007/s10641-013-0155-1

Smith, C. & Reay, P.J. (1991). Cannibalism in teleost fish. *Reviews in Fish Biology & Fisheries* **1**, 41–64.

Smith, C. & Reichard, M. (2005). Females solicit sneakers to improve fertilisation success in the bitterling (*Rhodeus sericeus*). *Proceedings of the Royal Society London, B* **272**, 1683–1688.

Smith, C. & Reichard, M. (2013). A sperm competition model for the European bitterling (*Rhodeus amarus*). *Behaviour* **150**, 1709–1730.

Smith, C., Reichard, M. & Jurajda, P. (2003). Assessment of sperm competition by bitterling (*Rhodeus sericeus*). *Behavioral Ecology and Sociobiology* **53**, 206–213.

Smith, C., Reichard, M., Jurajda, P. & Przbylski, M. (2004a). The reproductive ecology of the European bitterling (*Rhodeus sericeus*). *Journal of Zoology, London* **262**, 107–124.

Smith, C., Reynolds, J.D. & Sutherland, W.J. (2000). Population consequences of reproductive decisions. *Proceedings of the Royal Society London, B* **267**, 1327–1334.

Smith, C., Rippon, K., Douglas, A. & Jurajda, P. (2001b). A proximate cue for oviposition site choice in the bitterling (*Rhodeus sericeus*). *Freshwater Biology* **46**, 903–911.

Smith, C. & Spence, R. (2013). The potential direct and indirect benefits of mate choice in the three-spined stickleback (*Gasterosteus aculeatus*). *Evolutionary Ecology Research* **15**, 331–341.

Smith, C. & Wootton, R.J. (1994). The costs of parental care in *Haplochromis "argens"* (Cichlidae). *Environmental Biology of Fishes* **40**, 99–104.

Smith, C. & Wootton, R.J. (1995a). The costs of parental care in teleost fishes. *Reviews in Fish Biology & Fisheries* **5**, 7–22.

Smith, C. & Wootton, R.J. (1995b). Experimental analysis of some factors affecting parental expenditure in *Gasterosteus aculeatus* (Gasterosteidae). *Environmental Biology of Fishes* **43**, 63–76.

Smith, C. & Wootton, R.J. (1995c). Experimental analysis of some factors affecting parental expenditure in *Cichlasoma nigrofasciatum* (Cichlidae). *Environmental Biology of Fishes* **42**, 289–302.

Smith, C. & Wootton, R.J. (1999). Parental energy expenditure of the male three-spined stickleback. *Journal of Fish Biology* **54**, 1132–1136.

Smith, C., Yurong, Z., Liu, H. & Reichard, M. (2007). Deceptive female oviposition behaviour elicits male ejaculation in the European bitterling. *Journal of Fish Biology* **71**, 1841–1846.

Smith, C.C. & Fretwell, S.D. (1974). The optimal balance between size and number of offspring. *American Naturalist* **108**, 499–506.

Smith, C.J. & Haley, S.R. (1987). Evidence of steridogenesis in postovulatory follicles of the tilapia, *Oreochromis mossambicus. Cell Tissue Research* **247**, 675–87.

Smith, E.J., Partridge, J.C., Parsons, K.N., White, E.M., Cuthill, I.C., Bennett, A.T.D. & Church, S.C. (2002b). Ultraviolet vision and mate choice in the guppy (*Poecilia reticulata*). *Behavioral Ecology* **13**, 11–19.

Smith, P.J., Francis, R.I.C.C. & McVeagh, M. (1991). Loss of genetic diversity due to fishing pressure. *Fisheries Research* **10**, 309–316.

Smith, R.J.F. (1985). *The Control of Fish Migration*. Berling: Springer-Verlag.

Smyder, E.A. & Martin, K.L.M. (2002). Temperature effects on egg survival and hatching during the extended incubation period of the California grunion, *Leuresthes tenuis*. *Copeia* **2002**, 313–320.

Snook, R.R. (2005). Sperm in competition: not playing by the numbers. *Trends in Ecology and Evolution* **20**, 46–53.

Snyder, R.J. (1991a). Quantitative genetic analysis of life histories in two freshwater populations of the threespine stickleback. *Copeia* **2**, 526–529.

Snyder, R.J. (1991b). Migration and life histories of the threespine stickleback: evidence for adaptive variation in growth rate between populations. *Environmental Biology of Fishes* **31**, 381–388.

Sobell, S.K., MacKenzie, D.S. (2011). Reproductive endocrinology of Syngnathidae. *Journal of Fish Biology* **78**, 1662–1680.

Sogabe, A. & Ahnesjö, I. (2011). The ovarian structure and mode of egg production in two polygamous pipefishes: a link to mating pattern. *Journal of Fish Biology* **78**, 1833–1846.

Sogard, S.M. (1997). Size-selective mortality in the juvenile stage of teleost fishes: a review. *Bulletin of Marine Science* **60**, 1129–1157.

Sola, L.S., Cataudella, S. & Capanna, E. (1981). New developments in vertebrate cytotaxonomy. III. Karyology of bony fishes. A review. *Genetica* **54**, 285–328.

Somerton, D.A. & Donaldson, W. (1998). Parasitism of the golden king crab, *Lithodes aequispinus*, by two species of snailfish, genus *Careproctus*. *Fisheries Bulletin* **96**, 871–884.

Sommer, J.A., Li, C., Brozek, J., Bessert, M.L., Orti, G. & Berra, T.M. (2011). Low genetic diversity in Nurseryfish, *Kurtus gulliveri* (Perciformes: Kurtidae), and an appraisal of its breeding system using microsatellite loci. *The Beagle* **27**, 179–188.

Soucy, S. & Travis, J. (2003). Multiple paternity and population genetic structure in natural populations of the poeciliid fish, *Heterandria formosa*. *Journal of Evolutionary Biology* **16**, 1328–1336.

Southwood, T.R.E. (1977). Habitat, the templet for ecological strategies? *Journal of Animal Ecology* **46**, 337–365.

Southwood, T.R.E. (1988). Tactics, strategies and templets. *Oikos* **52**, 3–18.

Spence, R., Ashton, R. & Smith, C. (2007). Adaptive oviposition choice in the zebrafish, *Danio rerio*. *Behaviour* **144**, 953–966.

Spence, R., Gerlach, G., Lawrence, C. & Smith, C. (2008). The behaviour and ecology of the zebrafish, *Danio rerio*. *Biological Reviews* **18**, 13–34.

Spence, R. & Smith, C. (2005). Male territoriality mediates density and sex ratio effects on oviposition in the zebrafish (*Danio rerio*). *Animal Behaviour* **69**, 1317–1323.

Spence, R. & Smith, C. (2013). Rose bitterling (*Rhodeus ocellatus*) embryos parasitise freshwater mussels by competing for nutrients and oxygen. *Acta Zoologica* **94**, 113–118.

Springate, J.R.C., Bromage, N.R. & Cumarantunga, P.R.T. (1986). The effects of different ration on fecundity and egg quality in the rainbow trout (*Salmo gairdneri*). In: *Nutrition and Feeding in Fish* (Cowey, C.B. & Mackie, A.M., eds), pp. 371–393. London: Academic Press.

Springate, J.R.C., Bromage, N.R., Elliott, J.A.K. & Hudson, D.L. (1984). The timing of ovulation and stripping and their effects on the rates of fertilization and survival to eying, natch, and swim-up in rainbow trout (*Salmo gairdneri R.*). *Aquaculture* **43**, 313–322.

Stacey, N. (2011). Hormonally derived sex pheromones in fishes. In: *Hormones and Reproduction in Vertebrates Volume 1 – Fishes* (Norris, D.O. & Lopez, K.H., eds), pp. 169–192. London: Academic Press.

Stacey, N.E. & Sorensen, P.W. (2008). Hormone derived sex pheromones in fish. In: *Fish Reproduction* (Rocha, M.J., Arukwe, A. & Kapoor, B.G., eds), pp. 201–244. Enfield (NH): Science Publishers

Stadnichenko, A.P. & Stadnichenko, Y.A. (1980). On the effect of bitterling larvae on the lamellibranchid mollusc *Unio rostratus gentilis* Haas. *Gidrobiologicheskij Zhurnal* **1980**, 57–61.

Stanley, B.V. (1983). Effect of food supply on reproductive behaviour of male *Gasterosteus aculeatus*. Unpublished PhD thesis: University of Wales.

Stanley, B.V. & Wootton, R.J. (1986). Effects of ration and male density on the territoriality and nest-building of male threespined sticklebacks (*Gasteroseus aculeatus* L.). *Animal Behaviour* **34**, 527–535.

Stanley, J.G. (1979). Control of sex in fishes, with special reference to the grass carp. In: *Proceedings of the Grass Carp Conference* (Shireman, J.V., ed.), pp. 201–242. Gainesville: University of Florida.

Stearns, S.C. (1983). A natural experiment in life-history evolution: field data on the introduction of mosquitofish (*Gambusia affinis*) to Hawaii. *Evolution* **37**, 601–617.

Stearns, S.C. (1992). *The Evolution of Life Histories*. Oxford: Oxford University Press.

Stearns, S.C. (1989). The evolutionary significance of phenotypic plasticity. *BioScience* **39**, 436–445.

Stearns, S.C. & Koella, J.C. (1986). The evolution of phenotypic plasticity in life-history traits: predictions of reaction norms for age and size at maturity. *Evolution* **40**, 893–913.

Steele, C.W. (1985). Non-random, seasonal oscillations in the orientation and locomotor activity of sea catfish (*Arius felis*) in a multiple-choice situation. *Biological Bulletin* **168**, 359–376.

Steele, D.H. (1977). Correlation between egg size and development period. *American Naturalist* **111**, 371–372.

Steinhart, G.B., Sandrene, M.E., Weaver, S., Stein, R.A. & Marschall, E.A. (2005). Increased parental care cost for nest-guarding fish in a lake with hyperabundant nest predators. *Behavioral Ecology* **16**, 427–434.

Stephens, P.A., Boyd, I.L., McNamara, J.M. & Houston, A.I. (2009). Capital breeding and income breeding: their meaning, measurement and worth. *Ecology* **90**, 2057–2067.

Stewart, D.J. & Roberts, T.R. (1984). A new species of dwarf cichlid fish with reversed sexual dichromatism from Lac Mai-ndombe, Zaïre. *Copeia* **1984**, 82–86.

Stewart, K.M. (1988). Changes in condition and maturation of the *Oreochromis niloticus* L. population of Ferguson's Gulf, Lake Turkana, Kenya. *Journal of Fish Biology* **33**, 181–188.

Stock, M., Lampert, K.P., Möller, Schlupp, I. & Schartl, M. (2010). Monophyletic origin of multiple clonal lineages in an asexual fish (*Poecilia formosa*). *Molecular Ecology* **19**, 5204–5215.

Stockley, P., Gage, M.J.G., Parker, G.A. & Møller, A.P. (1997). Sperm competition in fishes: the evolution of testis size and ejaculate characteristics. *American Naturalist* **149**, 933–954.

Stöhr, S. (1998). Evolution of mate-choice copying: a dynamic model. *Animal Behaviour* **55**, 893–903.

Stölting, K.N. & Wilson, A.B. (2007). Male pregnancy in seahorses and pipefish: beyond the mammalian model. *BioEssays* **29**, 884–896.

Strüssmann, C.A., Cota, J.C.C., Phonlor, G., Higuchi, H. & Takashima, F. (1996a). Temperature effects on sex differentiation of two American atherinids, *Odontesthes argentinensis* and *Patagonina hatcheri*. *Environmental Biology of Fishes* **47**, 143–154.

Strüssmann, C.A., Moriyama, S., Hanke, E.F., Calsina Cota, J.C. & Takashima, F. (1996b). Evidence of thermolabile sex determination in pejerrey. *Journal of Fish Biology* **48**, 643–651.

Strüssmann, C.A., Takashima, F. & Toda, K. (1996c). Sex differentiation and hormonal feminization in pejerrey *Odontesthes bonariensis*. *Aquaculture* **139**, 31–45.

Strüssmann, C.A., Saito, T., Usui, M., Yamada, H. & Takashima, F. (1997). Thermal thresholds and critical period of thermolabile sex determination in two atherinid fishes, *Odontesthes bonariensis* and *Patagonina hatcheri*. *Journal of Experimental Zoology* **278**, 167–177.

Sun, S., Ting, C.T. & Wu, C.I. (2004). The normal function of a speciation gene, Odysseus, and its hybrid sterility effect. *Science* **305**, 81–83.

Sundara, R.B. (1962). The extraordinary feeding habits of the catfish *Mystus aor* (Hamilton) and *Mystus seenghala* (Sykes). *Proceedings of the National Academy of Sciences, India* **28**, 193–200.

Sundararaj, B.I. (1978). Environmental regulation of annual cycles in the catfish, *Heteropneustes fossilis*. In: *Environmental Endocrinology* (Assenmacher, I. & Farner, D.S., eds), pp. 26–27. Berlin: Springer-Verlag.

Suquet, M., Dreanno, C., Fauvel, C., Cosson, J. & Billard, R. (2000). Cryopreservation of sperm in marine fish. *Aquaculture Research* **31**, 231–243.

Suwa, K. & Yamashita, M. (2007). Regulatory mechanisms of oocyte maturation and ovulation. In: *The Fish Ooocyte: From Basic Studies to Biotechnological Applications* (Babin, P.J., Cerdà, J. & Lubzens, E., eds), pp. 323–347. Berlin: Springer.

Suzuki, K., Kawauchi, H. & Nagahama, Y. (1988). Isolation and characterization of two distinct gonadotropins from chum salmon pituitary glands. *General and Comparative Endocrinology* **71**, 292–301.

Svensson, I. (1988). Reproductive costs in two sex role pipefish species (Syngnathidae). *Journal of Animal Ecology* **57**, 929–942.

Svobodová, Z., Kolarova, J. & Flajshans, M. (1998). The first findings of the differences in complete blood count between diploid and triploid tench, *Tinca tinca*. *Acta Veterinaria* **67**, 243–248.

Swanson, P. (1991). Salmon gonadotropins: reconciling old and new ideas. In: *Reproductive Physiology of Fish* (Scott, A.P., Sumpter, J.P., Kime, D.E. & Rolfe, M.S., eds), pp. 2-7. Sheffield: FishSymp 91.

Taborsky, M. (1984). Broodcare helpers in the cichlid fish *Lamprologus brichardi*: their costs and benefits. *Animal Behaviour* **32**, 1236–1252.

Taborsky, M. (1985). Breeder-helper conflict in a cichlid fish with broodcare helpers-an experimental analysis. *Behaviour* **95**, 45–75.

Taborsky, M. (2008). Alternative reproductive tactics in fish. In: *Alternative Reproductive Tactics* (Oliviera, R.F., Taborsky, M. & Brockmann, H.J., eds), pp. 251–299. Cambridge: Cambridge University Press.

Taggart, C.T. & Leggett, W.C. (1987). Short-term mortality in post-emergent larval capelin *Mallotus villosus*. I. Analysis of multiple *in situ* estimates. *Marine Ecology Progress Series* **41**, 205–217.

Takahashi, D. & Kohda, M. (2001). Females of a stream goby choose mates that court in fast water currents. *Behaviour* **138**, 937–946.

Takahashi, D. & Kohda, M. (2004). Courtship in fast water currents by a male stream goby (*Rhinogobius brunneus*) communicates the parental quality honestly. *Behavioral Ecology and Sociobiology* **55**, 431–438.

Takemura, A., Rahman, M.S. & Park, Y. (2010). External and internal controls of lunar-related reproductive rhythms in fishes. *Journal of Biology* **76**, 7–26.

Tam, J.K.V., Lee, T.O., Cheng, C.H.K. & Chow, B.K.C. (2011). Discovery of a new reproductive hormone in teleosts: pituitary adenylate cyclase-activating polypeptide-related peptide (PRP). *General and Comparative Endocrinology* **173**, 405–410.

Tamada, K. & Iwata, K. (2005). Intra-specific variations of egg size, clutch size and larval survival related to maternal size in amphidromous *Rhinogobius goby*. *Environmental Biology of Fishes* **73**, 379–389.

Taranger, G.L., Carrillo, M., Schultz, R.W., Fontaine, P., Zanuy, S., Felip, A., Weltzien F.-A., Dufour, S., Karlsen, Ø, Norberg, B., Andersson, E. & Hansen, T. (2010). Control of puberty in farmed fish. *General and Comparative Endocrinology* **165**, 483–515.

Tatarenkov, A., Earley, R.L., Taylor, D.S. & Avise, J.C. (2012). Microevolutionary distribution of isogenicity in a self-fertilizing fish (*Kryptolebias marmoratus*) in the Florida keys. *Integrative and Comparative Biology* **52**, 743–752.

Tavolga, W.N. (1956). Visual, chemical and sound stimuli as cues in the sex discriminatory behavior of the gobiid fish, *Bathygobius soporator*. *Zoologica* **41**, 49–64.

Taylor, D.S., Turner, B.J., Davis, W.P. & Chapman, B.B. (2008). A novel terrestrial fish habitat inside emergent logs. *American Naturalist* **171**, 263–266.

Taylor, M.H. (1984). Lunar synchronization of fish reproduction. *Transactions of the American Fisheries Society* **113**, 484–493.

Taylor, M.I. & Knight, M.E. (2008). Mating systems in fishes. In: *Fish Reproduction* (Roche, M.J., Arukwe, A. & Kapoor, B.G., eds), pp. 377–309. Enfield (NH): Science Publishers.

Tedesco, P.A., Hugueny, B., Obderdorff, T., Dürr, H.H., Mérigoux, S. & de Mérona, B. (2008). River hydrological seasonality influences life history strategies of tropical riverine fishes. *Oecologia* **156**, 691–702.

Teletchea, F. & Fontaine, P. (2010). Comparison of early life-stage strategies in temperate freshwater fish species: trade-offs are directed towards first feeding of larvae in spring and early summer. *Journal of Fish Biology* **77**, 257–278.

Teletchea, F., Fostier, A., Kamler, E., Gardeur, J.-N., Le Bail, P-Y, Jalabert, B. & Fontaine, P. (2009). Comparative analysis of reproductive traits in 65 freshwater species: application to the domestication of new species. *Reviews in Fish Biology and Fisheries* **19**, 403–430.

Terzibasi Tozzini, E., Dorn, A., Ng'oma, E., Polačik, M., Blažek, R., Reichwald, K., Petzold, A., Watters, B.R., Reichard, M. & Cellerino, A. (2013). Parallel evolution of senescence in annual fishes in response to extrinsic mortality. *BMC Evolutionary Biology* **13**, 1–12.

Tesch, F.W. (1977). *The Eel – Biology and Management of Anguillid Eels*. London: Chapman & Hall.

Thorpe, J.E. (1986). Age at first maturity in Atlantic salmon, Salmo salar: freshwater period influences and conflicts with smolting. *Canadian Special Publication of Fisheries and Aquatic Sciences* **89**, 7–14.

Thorpe, J.E. (2007). Maturation responses of salmonids to changing developmental opportunities. *Marine Ecology Progress Series* **335**, 285–288.

Thorpe, J.F., Mangel, M., Metcalfe, N.B. & Huntingford, F.A. (1998). Modelling the proximate basis of salmonid life-history variation with application to Atlantic salmon, *Salmo salar* L. *Evolutionary Ecology* **12**, 581–599.

Thresher, R.E. (1978). Territoriality and aggression in the threespot damselfish (Pisces; Pomacentridae): an experimental study of causation. *Zeitschrift für Tierpsychologie* **46**, 401–434.

Thresher, R.E. (1984). *Reproduction in Reef Fishes*. Neptune City, NJ: TFH Publications.

Thresher, R.E. (1988). Latitudinal variation in egg size of tropical and sub-tropical North Atlantic shore fishes. *Environmental Biology of Fishes* **21**, 17–25.

Tiersch, T.R., Goudie, C.A. & Carmichael, G.J. (1994). Cryopreservation of channel catfish sperm: storage in cryoprotectants, fertilization trials, and growth of channel catfish produced with cryopreserved sperm. *Transactions of the American Fisheries Society* **123**, 580–586.

Timms, A.M. & Keenleyside, M.H.A. (1975). The reproductive behavior of *Aequidens paraguayensis* (Pisces, Cichlidae). *Zeitschrift für Tierpsychologie* **39**, 8–23.

Tinbergen, N. (1951). *The Study of Instinct*. Oxford: Oxford University Press.

Tinbergen, N. (1953). *The Herring Gull's World*. London: Collins.

Tobler, M. (2007). Reversed sexual dimorphism and courtship by females in the topaz cichlid *Archocentrus myrnae* (Cichlidade, Teleostei), from Costa Rica. *The Southwestern Naturalist* **52**, 371–377.

Townsend, C.R. & Hildrew, A.G. (1994). Species traits in relation to a habitat templet for river systems. *Freshwater Biology* **31**, 265–275.

Townshend, T.J. & Wootton, R.J. (1984). Effects of food supply on the reproduction of the convict cichlid, *Cichlasoma nigrofasciatum*. *Journal of Fish Biology* **24**, 91–104.

Townshend, T.J. & Wootton, R.J. (1985a). Adjusting parental investment to changing environmental conditons: the effect of food ration on parental behaviour of the convict cichlid, *Cichlasoma nigrofasciatum*. *Animal Behaviour* **33**, 494–501.

Townshend, T.J. & Wootton, R.J. (1985b). Variation in the mating system of a biparental cichlid fish, *Cichlasoma panamense*. *Behaviour* **95**, 181–197.

Tregenza, T. & Wedell, N. (2000). Genetic compatibility, mate choice and patterns of parentage. *Molecular Ecology* **9**, 1013–1027.

Trexler, J.C. (1997). Resource availability and plasticity in offspring provisioning embryo nourishment in sailfin mollies. *Ecology* **78**, 1370–1381.

Trexler, J.C. & DeAngelis, D.L. (2003). Resource allocation in offspring provisioning: an evaluation of the conditions favouring the evolution of matrotrophy. *American Naturalist* **162**, 574–585.

Trexler, J.C., DeAngelis, D.L. & Jiang, J. (2011). Community assembly and mode of reproduction: predicting the distribution of livebearing fishes. In: *Ecology and Evolution of Poeciliid Fishes* (Evans, J., Pilastro, A. & Schlupp, I., eds), pp. 95–108. Chicago: Chicago University Press.

Trippel, E.A. (1995). Age at maturity as a stress indicator in fisheries. *BioScience* **45**, 759–771.

Trippel, E.A. (1998). Egg size and viability and seasonal offspring production of young Atlantic cod. *Transactions of the American Fisheries Society* **127**, 339–359.

Trippel, E.A., Kjesbu, O.S. & Solemdal, P. (1997). Effects of adult age and size structure on reproductive output in marine fishes. In: *Early Life History and Recruitment in Fish Populations* (Chambers, R.C. & Trippel, E.A., eds), pp. 31–62. London: Chapman and Hall.

Trippel, E.A. & Neilson, J.D. (1992). Fertility and sperm quality of virgin and repeat-spawning Atlantic cod (*Gadus morhua*) and associated hatching success. *Canadian Journal of Fisheries and Aquatic Sciences* **49**, 2118–2127.

Trivers, R.L. (1972). Parental investment and sexual selection. In: *Sexual Selection and the Descent of Man: 1871–1971* (Campbell, B., ed.), pp. 136–179. Chicago: Aldine.

Tsukamoto, K. (1992). Discovery of the spawning area for Japanese eel. *Nature* **356**, 789–791.

Turelli, M. & Orr, H.A. (2000). Dominance, epistasis and the genetics of postzygotic isolation. *Genetics* **154**, 1663–1679.

Turner G.F. (1999). Explosive speciation of African cichlid fishes. In: *The Evolution of Biological Diversity* (Magurran A.E. & May, R.M., eds), pp. 217–229. Oxford: Oxford University Press.

Tyler, A.V. & Dunn, R.S. (1976). Ration, growth, and measures of somatic and organ condition in relation to meal frequency in winter flounder, *Pseudopleuronectes americanus*, with hypotheses regarding population homeostasis. *Journal of the Fisheries Research Board of Canada* **33**, 63–75.

Tyler, C.R. & Filby, A.L. (2011). Feminized fish, environmental estrogens, and wasterwater effluents in English rivers. In: *Wildlife Ecotoxicology: Forensic Approaches* (Elliott, J.E., Bishop, C.A. & Morrissey, C.A., eds), pp. 383–412. Berlin: Springer.

Tyler, C.R. & Jobling, S. (2008). Roach, sex, and gender-bending chemicals: the feminization of wild fish in English rivers. *BioScience* **58**, 1051–1059.

Tyler, C.R., Nagler, J.J., Pottinger, T.G. & Turner, M.A. (1994). Effects of unilateral ovariectomy on recruitment and growth of follicles in the rainbow trout, *Oncorhynchus mykiss. Fish Physiology and Biochemistry* **13**, 309–316.

Tyler, C.R., Pottinger, T.G., Santos, E., Sumpter, J.P., Price, S-A, Brooks, S. & Nagler, J.J. (1996). Mechanisms controlling egg size and number in the rainbow trout *Oncorhynchus mykiss. Biology of Reproduction* **54**, 8–15.

Tyler, C.R., Pottinger, T.G., Coward, K., Prat, F., Beresford, N. & Maddix, S. (1997). Salmonid follicle-stimulating hormone (GtH I) mediates vitellogenic development of oocytes in the rainbow trout, *Oncorhynchus mykiss. Biology of Reproduction* **57**, 1238–1244.

Tyler, C.R. & Sumpter, J.P. (1996). Oocyte growth an development in teleosts. *Reviews in Fish Biology and Fisheries* **6**, 287–318.

Uller, T., Pen, I. Wapstra, E. Beukeboom, L.W. & Komdeur, J. (2007). The evolution of sex ratios and sex-determining systems. *Trends in Ecology and Evolution* **22**, 292–297.

Unger, L.M. & Sargent, R.C. (1988). Alloparental care in the fathead minnow, *Pimephales promelas*: females prefer males with eggs. *Behavioral Ecology and Sociobiology* **23**, 27–32.

Urbányi, B., Horváth, Á. & Bokor, Z. (2009). Artificial fertilization in aquaculture species: from normal practice to chromosome manipulation. In: *Methods in Reproductive Aquaculture: Marine and Freshwater Species* (Cabrita, E., Robles, V. & Herráez, P., eds), pp. 183–218. Boca Raton: CRC Press.

Urbatzka, R., Rocha, M.J. & Rocha, E. (2011). Regulation of ovarian development and function in teleosts. In: *Hormones and Reproduction of Vertebrates, Volume 1 – Fishes* (Norris, D.O. & Lopez, K.H., eds), pp. 65–82. London: Academic Press.

Uribe, M.C. & Grier, H.J. (2011). Oogenesis of microlecithal oocytes in the viviparous teleost *Heterandria formosa. Journal of Morphology* **272**, 241–257.

Vajda, A.M. & Norris, D.O. (2011). Endocrine-active chemicals (EACs) in fishes. In: *Hormones and Reproduction of Vertebrates, Volume 1 – Fishes* (Norris, D.O. & Lopez, K.H., eds), pp. 245–264. London: Academic Press.

Valdes, E.S., Shelton, P.A., Armstrong, M.J. & Field, J.G. (1987). Cannibalism in South African anchovy: egg mortality and egg consumption rates. *South African Journal of Marine Science* **5**, 613–622.

Valenzuela, N. (2008). Sexual development and the evolution of sex determination. *Sexual Development* **2**, 64–72.

Valenzuela, N., Adams, D.C. & Janzen, F.J. (2003). Pattern does not equal process: exactly when is sex environmentally determined? *American Naturalist* **161**, 676–683.

Vamosi, S.M. & Schluter, D. (1999). Sexual selection against hybrids between sympatric stickleback species: evidence from a field experiment. *Evolution* **53**, 874–879.

Vandeputte, M., Dupont-Nivel, M., Chavanne, H. & Chatain, B. (2007). A polygenic hypothesis for sex determination in the European sea bass *Dicentrarchus labrax*. *Genetics* **176**, 1049–1057.

Varadaraj, K. & Pandian, T.J. (1990). Production of all-female sterile-triploid *Oreochromis mossambicus*. *Aquaculture* **84**, 117–123.

Verner, J. (1977). On the adaptive significance of territoriality. *American Naturalist* **114**, 308–312.

Videler, J.J. (1993). *Fish Swimming*. London: Chapman and Hall.

Vinyoles, D., Cöte, I.M. & De Sostoa, A. (1999). Egg cannibalism in river blennies: the role of natural prey availability. *Journal of Fish Biology* **55**, 1223–1232.

Vives, S.P. (1990). Nesting ecology and behavior of hornyhead chub *Nocomis biguttatus*, a keystone species in Allequash Creek, Wisconsin. *American Midland Naturalist* **124**, 46–56.

Vladić, T.V. & Jävri, T. (2001). Sperm quality in the alternative reproductive tactics of Atlantic salmon: the importance of the loaded raffle mechanism. *Proceedings of the Royal Society of London, B* **268**, 2375–2381.

Volkoff, H., Canosa, L.F., Unniappan, S., Cerda-Reverter, J.M., Bernier, N.J., Kelly, S.P. & Peter, R.E. (2005). Neuropeptides and control of food intake in fish. *General and Comparative Endocrinology* **142**, 3–19.

Vrijenhoek, R.C. (1984). The evolution of clonal diversity in *Poeciliopsis*. In: *Evolutionary Genetics of Fishes* (Turner, B.J., ed.), pp. 399–429. New York: Plenum Press.

Wallace, R.A. & Selman, K. (1981). Cellular and dynamic aspects of oocyte growth in teleosts. *American Zoologist* **21**, 325–343.

Walker, M.M., Diebel, C.E., Haugh, C.V., Pankhurst, P.M., Montgomery, J.C. & Green, C.R. (1997). Structure and function of the vertebrate magnetic sense. *Nature* **390**, 371–376.

Walker, M.M., Kirschvink J.L., Chang S.B.R. & Dizon, A.E. (1984). A candidate magnetic sense organ in the Yellowfin Tuna Thunnus albacares. *Science* **224**, 751–753.

Walsh, M.R., Fraser, D.F., Bassar, R.D. & Reznick, D.N. (2011). The direct and indirect effects of guppies: implications for life-history evolution in *Rivulus hartii*. *Functional Ecology* **25**, 227–237.

Walsh, M.R. & Reznick, D.N. (2010). Influence of the indirect effects of guppies on life history evolution in Rivulus hartii. *Evolution* **64**, 1583–1593.

Walters, C. & Kitchell, J.F. (2001). Cultivation/depensation effects on juvenile survival and recruitment: implications for the theory of fishing. *Canadian Journal of Fisheries and Aquatic Sciences* **58**, 39–50.

Wang, N., Gardeur, J.-N., Henrotte, E., Marie, M., Kestemont, P. & Fontaine, P. (2006). Determinism of the induction of the reproductive cycle in female Eurasian perch, *Perca fluviatilis*: identification of environmental cues and permissive factors. *Aquaculture* **261**, 706–714.

Wang, N., Teletchea, F., Kestemont, P., Milla, S. & Fontaine, P. (2010). Photothermal control of the reproductive cycle in temperate fishes. *Reviews in Aquaculture* **2**, 209–222.

Wang, S., Zhu, Y. & Melamed, P. (2009). The molecular regulation of Chinook salmon gonadotropin β-subunit gene transcription. *General and Comparative Endocrinology* **161**, 34–41.

Ward, J.A. & Wyman, R.L. (1977). Ethology and ecology of cichlid fishes of the genus *Etroplus* in Sri Lanka: preliminary findings. *Environmental Biology of Fishes* **2**, 137–145.

Warner, R.R. (1988). Sex change in fishes: hypotheses, evidence and objections. *Environmental Biology of Fishes* **22**, 81–90.

Warner, R.R. (2001). Synthesis: environment, mating systems, and life history allocations in the bluehead wrasse. In: *Model Systems in Behavioral Ecology* (Dugatkin, L.A., ed.), pp. 227–244. Princeton: Princeton University Press.

Warner, R.R. & Hoffman, S.G. (1980). Population density and the economics of territorial defense in a coral reef fish. *Ecology* **61**, 772–780.

Warner, R.R. & Robertson, D.R. (1978). Sexual patterns in the labroid fishes of the western Caribbean, I. The wrasses (Labridae). *Smithsonian Contributions to Zoology* **254**, 1–27.

Warner, R.R., Shapiro, D.Y., Marcanato, A. & Petersen, C.W. (1995). Sexual conflict: males with highest mating success convey the lowest fertilization benefits to females. *Proceedings of the Royal Society of London, B* **262**, 135–139.

Watson, P.J. (1991). Multiple paternity as genetic bet-hedging in female sierra dome spiders, *Linyphia litigiosa* (Linyphiidae). *Animal Behaviour* **41**, 343–360.
Watters, J.V. (2005). Can alternative male tactics 'fighter' and 'sneaker' be considered 'coercer' and 'cooperator'? *Animal Behaviour* **70**, 1055–1062.
Watanabe, T. (2000). The nesting site of a piscivorous cichlid *Lepidiolamprologus profundicola* as a safety zone for juveniles of a zooplanktivorous cichlid *Cyprichromis leptosoma* in Lake Tanganyika. *Environmental Biology of Fishes* **57**, 171–177.
Watanabe, S., Iida, M., Lord, C., Keith, P. & Tsukamoto, K. (2013). Tropical and temperate freshwater amphidromy: a comparison between life history characteristics of Sicydiinae, ayu, sculpins and galaxiids. *Reviews in Fish Biology and Fisheries* **1–14**.
Wedekind, C., Evanno, G., Urbach, D., Jacob, A. & Muller, R. (2008). 'Good genes' and 'compatible genes' effects in an Alpine whitefish and the information content of breeding tubercules over the course of the spawning season. *Genetica* **134**, 21–30.
Wedell, N., Gage M.J.G. & Parker, G.A. (2002). Sperm competition, male prudence and sperm-limited females. *Trends in Ecology and Evolution* **17**, 313–320.
Weihs, D. (1978). Tidal stream transport as an efficient method of migration. *Journal du Conseil International pour l'Exploration de la Mer* **38**, 92–99
Welcomme, R.L. (1979). *Fisheries Ecology of Floodplain Rivers.* London: Longman.
Welcomme, R.L. (2000) Fish biodiversity in floodplains and their associated rivers. In: *Biodiversity in Wetlands: Assessment, Function and Conservation, Vol. 1* (Gopal, B., Junk, W.J. & Davis, J.A., eds), pp. 61–87. Leiden: Backhuys Publishers.
Weltzien, F.-A., Anersson, E., Andersen, Ø., Shalchian-Tabrizi, K. & Norberg, B. (2004). The brain-pituitary-gonad axis in male teleosts, with special emphasis on flatfish (Pleuronectiformes). *Comparative Biochemistry and Physiology, Part A* **137**, 447–477.
Werner, N.Y. & Lotem, A. (2003). Choosy males in a haplochromine cichlid: first experimental evidence for male mate choice in a lekking species. *Animal Behaviour* **66**, 293–298.
Westneat, D.F., Walters, A., McCarthy, T.M., Hatch, M.I. & Hein, W.K. (2000). Alternative mechanisms of nonindependent mate choice. *Animal Behaviour* **59**, 467–476.
Wetherington, J.D., Schenck, R.A. & Vrijenhoek, R.C. (1989). The origins and ecological success of unisexual *Poeciliopsis*: the frozen-niche hypothesis. In: *Ecology and Evolution of Livebearing Fishes (Poeciliidae)* (Meffe, G.K. & Snelson, F.F. Jr, eds), pp. 259–275. Englewood Cliffs (NJ): Prentice-Hall.
Wetzel, J. & Wourms, J.P. (1995). Adaptations for reproduction and development in the skin-brooding ghost pipefishes, *Solenostomus*. *Environmental Biology of Fishes* **44**, 363–384.
Wetzel, J., Wourms, J.P. & Friel, J. (1997). Comparative morphology of cotylephores in *Platystacus* and *Solenostomus*: modifications of the integument for egg attachment in skin-brooding fishes. *Environmental Biology of Fishes* **50**, 13–25.
Wetzel, R.G. (1983). *Limnology*, 2nd edn. Philadelphia: Saunders.
White, S.A., Nguyen, T. & Fernald, R.D. (2002). Social regulation of gonadotropin-releasing hormone. *Journal of Experimental Biology* **205**, 2567–2581.
Whoriskey, F.G. & FitzGerald, G.J. (1987). Intraspecific competition in sticklebacks (Gasterosteidae: Pisces): does mother nature concur? *Journal of Animal Ecology* **56**, 939–947.
Wiebe, J.P. (1968). The reproductive cycle of the viviparous seaperch, *Cymatogaster aggregata* Gibson. *Canadian Journal of Zoology* **46**, 1221–1233.
Wiegand, M.D. (1996). Composition, accumulation and utilization of yolk lipids in teleost fish. *Reviews in Fish Biology and Fisheries* **6**, 259–286.
Wiepkema, P.R. 1961. An ethological analysis of the reproductive behaviour of the bitterling (*Rhodeus amarus* Bloch). *Archives Néerlandaises de Zoologie* **14**, 103–199.
Wilhelm, D., Palmer, S. & Koopman, P. (2007). Sex determination and gonadal development in mammals. *Physiological Reviews* **87**, 1–28.
Williams, G.C. (1966). Natural selection, the costs of reproduction, and a refinement of Lack's principle. *American Naturalist* **100**, 687–690.
Wilson, A.B., Ahnesjö, I., Vincent, A.C.J. & Meyer, A. (2003). The dynamics of male brooding, mating patterns, and sex roles in pipefishes and seahorses (Family Syngnathidae). *Evolution* **57**, 1374–1386.

Wilson, A.B. & Orr, J.W. (2011). The evolutionary origins of Syngnathidae: pipefishes and seahorses. *Journal of Fish Biology* **78**, 1603–1623.

Wiltschko, W. & Wiltschko, R. (2005). Magnetic orientation and magentoreception in birds and other animals. *Journal of Comparative Physiology, Part A* **191**, 675–693.

Winemiller, K.O. (1993). Seasonality of reproduction by livebearing fishes in tropical rainforest streams. *Oecologia* **95**, 266–276.

Winemiller, K.O. (2005). Life history strategies, population regulation, and implications for fisheries management. *Canadian Journal of Fisheries and Aquatic Sciences* **62**, 872–885.

Winemiller, K.O. & Rose, K.A. (1992). Patterns of life-history diversification in North American fishes: implications for population regulation. *Canadian Journal of Fisheries and Aquatic Sciences* **49**, 2196–2218.

Winemiller, K.O. & Rose, K.A. (1993). Why do fish produce so many tiny offspring? *American Naturalist* **142**, 585–603.

Winge, O. (1937). Succession of broods in *Lebistes*. *Nature* **140**, 467.

Wingfield, J.C. (2008). Organization of vertebrate annual cycles: implications for control mechanisms. *Philosophical Transactions of the Royal Society of London, B* **363**, 425–441.

Wisenden, B.D. (1999). Alloparental care in fishes. *Reviews in Fish Biology and Fisheries* **9**, 45–70.

Wisenden, B.D. and Keenleyside, M.H.A. (1994). The dilution effect and differential predation following brood adoption in freeranging convict cichlids (*Cichlasoma nigrofasciatum*). *Ethology* **96**, 203–212.

Wisenden, B.D., Lanfranconi-Izawa, T.L. & Keenleyside, M.H.A. (1995). Fin digging and leaf lifting by the convict cichlid, *Cichlasoma nigrofasciatum*: examples of parental food provisioning. *Animal Behaviour* **49**, 623–631.

Witte, K. & Nöbel, S. (2011). Learning and mate choice. In: *Fish Cognition and Behavior* (Brown, C., Laland, K. & Krause, J., eds), pp. 81–107. New York: Wiley-Blackwell.

Witte, K. & Ryan, M.J. (1998). Male body length influences mate-choice copying in the sailfin molly *Poecilia latipinna*. *Behavioral Ecology* **9**, 534–539.

Witte, K. & Ryan, M.J. (2002). Mate copying in the sailfin molly, *Poecilia latipinna*, in the wild. *Animal Behaviour* **63**, 943–949.

Wong, B.B.M. & Candolin, U. (2005). How is female choice affected by male competition? *Biological Reviews* **80**, 559–571.

Wong, B.B.M., Candolin, U. & Lindström, K. (2007). Environmental deterioration compromises socially enforced signals of male quality in three-spined sticklebacks. *American Naturalist* **170**, 184–189.

Wong, B.B.M. & Jennions, M.D. (2003). Costs influence male mate choice in a freshwater fish. *Proceedings of the Royal Society of London, B* **270**, S36-S38.

Wootton, R.J. (1971). A note on the nest-raiding behaviour of male sticklebacks. *Canadian Journal of Zoology* **49**, 960–962.

Wootton, R.J. (1973). The effect of size of food ration on egg production in the female three-spined stickleback, *Gasterosteus aculeatus* L. *Journal of Fish Biology* **5**, 89–96.

Wootton, R.J. (1976). *The Biology of the Sticklebacks*. London: Academic Press.

Wootton, R.J. (1977). Effect of food limitation during the breeding season on the size, body components and egg production of female sticklebacks (*Gasterosteus aculeatus*). *Journal of Animal Ecology* **46**, 823–834.

Wootton, R.J. (1979). Energy costs of egg production and environmental determinants of fecundity in teleost fishes. *Symposium of the Zoological Society of London* **44**, 133–159.

Wootton, R.J. (1984a). *A Functional Biology of Sticklebacks*. London: Croom Helm.

Wootton, R.J. (1984b). Introduction: tactics and strategies in fish reproduction. In: *Fish Reproduction: Strategies and Tactics* (Potts, G.W. & Wootton, R.J., eds), pp. 1–12. London: Academic Press.

Wootton, R.J. (1985). Energetics of reproduction. In: *Fish Energetics New Perspectives* (Tytler, P. & Calow, P., eds), pp. 231–254. Beckenham: Croom Helm.

Wootton, R.J. (1992). Constraints in the evolution of fish life histories. *Netherlands Journal of Zoology* **42**, 291–303.

Wootton, R.J. (1994a). Energy allocation in the threespine stickleback. In: *The Evolutionary Biology of the Three-spined Stickleback* (Bell, M.A. & Foster, S.A., eds), pp. 250–260. Oxford: Oxford University Press.

Wootton, R.J. (1994b). Life histories as sampling devices: optimum egg size in pelagic fishes. *Journal of Fish Biology* **45**, 1067–1077.

Wootton, R.J. (1998). *Ecology of Teleost Fishes*, 2nd edn. Dordrecht: Elsevier.

Wootton, R.J. & Fletcher, D.A. (2009). Effect of spawning number and ration on reproductive performance of the batch-spawning three-spined stickleback *Gasterosteus aculeatus*. *Journal of Fish Biology* **75**, 618–629.

Wootton, R.J. & Smith, C. (2000). Long-term study of a short-lived fish: the demography of *Gasterosteus aculeatus*. *Behaviour* **137**, 981–997.

Woram, R.A. *et al.* (2003). Comparative genome analysis of the primary sex-determining locus in salmonid fishes. *Genome Research* **13**, 272–280.

Wourms, J.P. (1972). The developmental biology of the annual fishes. *Pre-embryonic and embryonic diapause of variable duration in the eggs of annual fishes. Journal of Experimental Zoology* **182**, 389–414.

Wourms, J.P. (1981). Viviparity: the maternal foetal relationship in fishes. *American Zoologist* **21**, 473–575.

Wourms, J.P., Grove, B.D. & Lombardi, J. (1988). The maternal-embryonic relationship in viviparous fishes. In: *Fish Physiology Vol. XI, Part B. Viviparity and Posthatching Juveniles* (Hoar, W.S. & Randall, D.J., eds), pp. 1–134. London: Academic Press.

Wourms, J.P. & Lombardi, J. (1992). Reflections on the evolution of piscine viviparity. *American Zoologist* **32**, 276–293.

Wyman, J. (1859). On some unusual modes of gestation. *Amererican Journal of Science* **27**, 5–13.

Yabuta, S. (1997). Spawning migrations in the monogamous butterflyfish, *Chaetodon trifasciatus*. *Ichthyological Research* **44**, 177–182.

Yamahira, K. (2004). How do multiple environmental cycles in combination determine reproductive timing in marine organisms? A model and test. *Functional Ecology* **18**, 4–15.

Yamamoto, T. (1969). Sex differentiation. In: *Fish Physiology Vol. III* (Hoar, W.S. & Randall, D.J., eds), pp. 117–175. London: Academic Press.

Yamamoto, Y., Luckenbach, J.A., Goetz, F.W., Young, G. & Swanson, P. (2011). Disruption of the salmon reproductive endocrine axis through prolonged nutritional stress: changes in circulating hormone levels and transcripts for ovarian genes involved in steroidogenesis and apoptosis. *General and Comparative Endocrinology* **172**, 331–343.

Yamaha, E., Goto-Kazeto, R., Saito, T., Kawakami, Y., Fujimoto, T., Adachi, S. & Arai, K. (2010). Primordial germ cell in teleost fish with special reference to its specification and migration. *Journal of Applied Ichthyology* **26**, 816–822.

Yanagimachi, R., Cherr, G., Matsubara, T., Andoh, T., Harumi, T., Vines, C., Pillai, M., Griffin, F., Matsubara, H., Weatherby, T. & Kaneshiro, K. (2013). Sperm attractant in the micropyle region of fish and insect eggs. *Biology of Reproduction doi:* 10.1095/biolreprod.112.105072

Yanagisawa, Y. (1985). Parental strategy of the cichlid fish *Perissodus microlepis*, with particular reference to intraspecific brood 'farming out'. *Environmental Biology of Fishes* **12**, 241-9.

Yanagisawa, Y. & Sato, T. (1990). Active browsing by mouthbrooding females of *Tropheus duboisi* and *Tropheus moorii* (Cichlidae) to feed the young and/or themselves. *Environmental Biology of Fishes* **27**, 43–50.

Yang, H. & Tiersch, T.R. (2009). Current status of sperm cryopreservation in biomedical research fish models: Zebrafish, medaka, and *Xiphophorus*. *Comparative Biochemistry and Physiology, Part C* **149**, 224–232.

Yano, A., Guyomard, R., Nicol, B., Jouanno, E., Quillet, E., Klopp, C., Cabau, C., Bouchez, O., Fostler, A. & Gulguen, Y. (2012). An immune-related gene evolved into the master sex-determining gene in rainbow trout, *Oncorhynchus mykiss*. *Current Biology* **22**, 1423–1428.

Yaron, Z. & Levavi-Sivan, B. (2011). Endocrine regulation of fish reproduction. In: *Encyclopedia of Fish Physiology: From Genome to Environment* (Farrell, A.P., ed.), pp. 1500–1508. San Diego: Academic Press.

Yaron, Z., Gur, G., Melamed, P., Rosenfeld, H., Elizur, A. & Levavi-Sivan, B. (2003). Regulation of fish gonadotropins. *International Review of Cytology* **225**, 131–185.

Yeates, S.E., Diamond, S.E., Einum, S., Emerson, B.C., Holt, W.V. & Gage, M.J.G. (2013). Cryptic choice of conspecific sperm controlled by the impact of ovarian fluid on sperm swimming behavior. *Evolution*, **67**, 3523–3536.

Yeates, S.E., Einum, S. Fleming, I.A. Megens, H.-J. Stet, R.J.M. Hindar, K. Holt, W.V. Van Look, K.J.W. & Gage, M.J.G. (2009). Atlantic salmon eggs favour sperm in competition that have similar major histocompatibility alleles. *Proceedings of the Royal Society of London, B* **276**, 559–566.

Yom-Tov, Y. (1980). Intraspecific nest parasitism in birds. *Biological Reviews* **55**, 93–108.

Young, G., Kusakabe, M., Nakamura, I., Lokman, P.M. & Goetz, F.W. (2005). Gonadal steroidogenesis in teleost fish. In: *Hormones and Their Receptors in Fish Reproduction* (Melamed, P. & Sherwood, N., eds), pp. 155–223. Singapore: World Scientific.

Zahavi, A. (1975). Mate selection – A selection for a handicap. *Journal of Theoretical Biology* **53**, 205–214.

Zahavi, A. (1977). The cost of honesty (further remarks on the handicap principle). *Journal of Theoretical Biology* **67**, 603–605.

Zahavi, A. & Zahavi, A. (1997). *The Handicap Principle*. Oxford: Oxford University Press.

Zeh, J.A. & Zeh, D.W. (1996). The evolution of polyandry I: intragenomic conflict and genetic incompatibility. *Proceedings of the Royal Society of London, B* **263**, 1711–1717.

Zhang, L., Tang, Q.Y. & Liu, H.Z. (2008). Phylogeny and speciation of the eastern Asian cyprinid genus *Sarcocheilichthys*. *Journal of Fish Biology* **72**, 1122–1137.

Zhou, L. & Gui, J.F. (2002). Karyotypic diversity in polyploidy gibel carp, *Carassius auratus gibelio* Bloch. *Genetica* **115**, 223–232.

Zohar, Y., Muñoz-Cuerto, J.A., Elizur, A. & Kah, O. (2010). Neuroendocrinology of reproduction in teleost fish. *General and Comparative Endocrinology* **165**, 438–455.

Zohar, Y. & Mylonas, C.C. (2001). Endocrine manipulations of spawning in cultured fish: from hormones to genes. *Aquaculture* **197**, 99–136.

Systematic Index

Reproductive Biology of Teleost Fishes, First Edition. Robert J. Wootton and Carl Smith.

Subject Index

Note: Page numbers in *italics* denote tables and figures (when outside page ranges).

Reproductive Biology of Teleost Fishes, First Edition. Robert J. Wootton and Carl Smith.